中国石油勘探开发研究院出版物

全二维气相色谱及其石油地质应用

COMPREHENSIVE TWO-DIMENSIONAL GAS CHROMATOGRAPHY AND
APPLICATIONS IN PETROLEUM GEOLOGY

王汇彤　翁　娜　著

图书在版编目(CIP)数据

全二维气相色谱及其石油地质应用/王汇彤,翁娜著. —东营:中国石油大学出版社,2017.9
ISBN 978-7-5636-5769-8

Ⅰ.①全… Ⅱ.①王… ②翁… Ⅲ.①气相色谱—应用—石油天然气地质—研究 Ⅳ.①P618.130.2

中国版本图书馆CIP数据核字(2017)第244191号

书　　名:	全二维气相色谱及其石油地质应用
作　　者:	王汇彤　翁　娜
责任编辑:	高　颖　王金丽(电话　0532—86983568)
封面设计:	悟本设计
出 版 者:	中国石油大学出版社
	(地址:山东省青岛市黄岛区长江西路66号　邮编:266580)
网　　址:	http://www.uppbook.com.cn
电子邮箱:	shiyoujiaoyu@126.com
排 版 者:	青岛友一广告传媒有限公司
印 刷 者:	山东临沂新华印刷物流集团有限责任公司
发 行 者:	中国石油大学出版社(电话　0532—86981531,86983437)
开　　本:	185 mm×260 mm
印　　张:	31.75
字　　数:	765千
版 印 次:	2017年10月第1版　2017年10月第1次印刷
书　　号:	ISBN 978-7-5636-5769-8
定　　价:	238.00元

PREFACE 序言

全二维气相色谱仪作为具有全新概念的商品化分析仪器,自20世纪末问世以来,受到了学术界和工业界的广泛关注。尤其是在石油行业,更多的人希望了解和掌握可分析复杂有机化合物成分的这一利器,解决油气勘探、开发以及炼制中的一些科学难题。作为国内石油行业最早从事全二维气相色谱分析应用的科研人员,我们有责任与大家分享全二维气相色谱在石油地质样品分析、应用方面的经验和成果。希望通过本书的出版,为从事石油分析的实验人员提供基础的手册和工具,为从事石油勘探、开发、炼制等研究的科研人员拓宽解决问题的视野,为其他领域的从业人员提供应用方面的借鉴,为石油事业的发展、实验分析技术的进步尽绵薄之力。

本书内容共分为3章。第1章通过介绍全二维气相色谱的仪器特征、分离原理以及相适用的检测器,阐述全二维气相色谱的峰容量大、分辨率高、族分离特性、"瓦片效应"以及共馏峰分离等优点,旨在为从业人员提供利用全二维气相色谱解决问题的思路。第2章用大量的篇幅详细介绍石油地质样品中饱和烃、芳烃、杂原子化合物以及轻质原油的全二维气相色谱分析图谱的特征,展示全二维气相色谱-飞行时间质谱仪与常规色谱-质谱仪分析石油地质样品结果的差异,为读者介绍石油地质样品中化合物的全二维谱图识别方法。第2章所介绍的全二维谱图不仅包含油气地球化学中常见、常用的化合物,而且有大量过去未被认知的化合物,并附有对应的质谱图。该部分非常适合作为从业人员的工作手册和基础工具,是比较齐全的生物标志化合物质量色谱质谱图集。第3章主要介绍作者及同事多年来利用全二维气相色谱仪所建立的分析方法及其解决石油地质难题的相关应用,包括全二维气相色谱对常规色谱分析地球化学参数的矫正、凝析油全组分定量分析方法、生物标志化合物及金刚烷类化合物的色谱定量方法、生排烃模拟实验液态产物的定量分析方法、新化合物结构解析以及稠油中不可识别混合物的辨识与解析等。附录部分是第3章的引申,列出了作者及合作者利用全二维气相色谱仪发表的论文和授权的发明专利。

本书稿在3年前就开始动手撰写,今天才定稿,不仅是因为工作量大,水平有限、总担心出错也是一个原因。百密难免有疏,错误之处不求原谅,敬请批评指正。

本书的完稿得益于作者能够长期安心地在实验室致力于全二维气相色谱的研究工作,这要感谢中国石油股份公司科技部及中国石油集团公司油气地球化学重点实验室领导的资助和支持。书稿中介绍的成果属于中国石油,属于油气地球化学重点实验室,更属于曾经合作过及提供过帮助的同事们。

在此要特别感谢张水昌教授在仪器引进、开发和应用方面给予的指导、帮助、关心和鼓励。

感谢朱光有教授率先将全二维气相色谱技术引入地质应用，解决了一系列复杂的油气地质与地球化学难题，重建了复杂油气藏的形成演化与改造过程，推动了全二维气相色谱应用的发展；感谢陈建平教授以及魏彩云、魏小芳、王晓梅、张斌、苏劲、胡国艺等同事在工作中给予的支持和帮助。

作　者

2017年8月31日于北京

CONTENTS 目 录

第1章 全二维气相色谱基本原理 ·· 1
1.1 全二维气相色谱分离原理 ·· 2
1.2 全二维气相色谱检测器 ·· 6
1.2.1 飞行时间质谱仪 ·· 6
1.2.2 氢火焰离子化检测器 ·· 7
1.2.3 硫化学发光检测器 ·· 9
1.3 石油地质样品全二维气相色谱分析特点 ·························· 10
1.3.1 全二维气相色谱图特征 ····································· 10
1.3.2 全二维气相色谱定性方法 ··································· 12
1.3.3 全二维气相色谱定量方法 ··································· 14
1.3.4 全二维气相色谱方法的优点 ································· 14

第2章 石油地质样品的全二维气相色谱图识别 ························ 19
2.1 饱和烃组分的全二维气相色谱图特征及鉴定 ······················ 20
2.1.1 正构烷烃的全二维气相色谱图及化合物名称 ················· 23
2.1.2 异构烷烃的全二维气相色谱图及化合物名称 ················· 24
2.1.3 单环环烷烃的全二维气相色谱图及化合物名称 ··············· 24
2.1.4 双环环烷烃的全二维气相色谱图及化合物名称 ··············· 31
2.1.5 金刚烷系列的全二维气相色谱图及化合物名称 ··············· 40
2.1.6 三环萜烷和四环萜烷的全二维气相色谱图及化合物名称 ······· 82
2.1.7 甾烷类化合物的全二维气相色谱图及化合物名称 ············· 87
2.1.8 藿烷类化合物的全二维气相色谱图及化合物名称 ············· 97
2.2 芳烃组分的全二维气相色谱图特征及鉴定 ························ 127
2.1.1 苯系列化合物的全二维气相色谱图及化合物名称 ············· 130
2.2.2 萘系列化合物的全二维气相色谱图及化合物名称 ············· 135
2.2.3 联苯、二氢化苊、氧芴系列化合物的全二维气相色谱图及化合物名称
··· 153
2.2.4 芴系列化合物的全二维气相色谱图及化合物名称 ············· 155
2.2.5 菲系列化合物的全二维气相色谱图及化合物名称 ············· 155

  2.2.6 苉和莹蒽系列化合物的全二维气相色谱图及化合物名称 …………… 162
  2.2.7 䓛系列化合物的全二维气相色谱图及化合物名称 ………………… 165
  2.2.8 芘系列化合物的全二维气相色谱图及化合物名称 ………………… 166
  2.2.9 单芳甾系列化合物的全二维气相色谱图及化合物名称 …………… 168
  2.2.10 三芳甾系列化合物的全二维气相色谱图及化合物名称 ………… 175
  2.2.11 单芳断藿烷系列化合物的全二维气相色谱图及化合物名称 …… 189
  2.2.12 单芳-断-降藿烯系列化合物的全二维气相色谱图及化合物名称 … 195
  2.2.13 单芳-断-降藿烷系列化合物的全二维气相色谱图及化合物名称 … 198
  2.2.14 芳构化的五环三萜烷的全二维气相色谱图及化合物名称 ……… 201
  2.2.15 苯并藿烷的全二维气相色谱图及化合物名称 …………………… 217
 2.3 杂原子化合物的全二维气相色谱图特征及鉴定 ……………………………… 219
  2.3.1 含硫化合物的全二维气相色谱图及化合物名称 …………………… 219
  2.3.2 含氮化合物的全二维气相色谱图及化合物名称 …………………… 263

第3章 全二维气相色谱的石油地质应用 …………………………………………… 277

 3.1 石油地质样品的全二维气相色谱图识别 ……………………………………… 277
 3.2 饱和烃和芳烃同时分析方法及其全二维谱图特征 …………………………… 277
 3.3 地球化学参数校准及应用 ……………………………………………………… 279
 3.4 新的系列化合物的发现与结构鉴定 …………………………………………… 280
 3.5 凝析油的族组分定量分析方法 ………………………………………………… 281
 3.6 金刚烷类化合物的定量分析方法 ……………………………………………… 282
 3.7 生物标志化合物的定量分析方法 ……………………………………………… 283
 3.8 生排烃模拟实验液态烃全组分定量分析方法 ………………………………… 283
 3.9 稠油中"不可识别未知混合物"的成分解析 ………………………………… 284
 3.10 石油地质应用 ………………………………………………………………… 285
附 录 …………………………………………………………………………………… 287
  论文1 全二维气相色谱-飞行时间质谱对饱和烃分析的图谱识别及特征 ……
   ………………………………………………………………………………… 287
  论文2 全二维气相色谱-飞行时间质谱对原油芳烃分析的图谱识别 …… 297
  论文3 全二维气相色谱-飞行时间质谱与常规色质分析的地球化学参数对比
   ………………………………………………………………………………… 304
  论文4 Discovery and identification of a series of alkyl decalin isomers in
   petroleum geological samples ……………………………………………… 315
  论文5 Use of comprehensive two-dimensional gas chromatography for
   the characterization of ultra-deep condensate from the Bohai Bay
   Basin, China ………………………………………………………………… 325
  论文6 凝析油全二维气相色谱分析 …………………………………………… 341
  论文7 生物标志化合物甾烷、藿烷的定量分析新方法 …………………… 348
  论文8 稠油中饱和烃复杂混合物成分解析及其意义 ………………………… 359

论文 9　Insight into unresolved complex mixtures of aromatic hydrocarbons in heavy oil via two-dimensional gas chromatography coupled with time-of-flight mass spectrometry analysis ……………… 368

论文 10　Geochemistry, origin and accumulation of continental condensate in the ultra-deep-buried Cretaceous sandstone reservoir, Kuqa Depression, Tarim Basin, China ……………………… 392

论文 11　Non-cracked oil in ultra-deep high-temperature reservoirs in the Tarim Basin, China …………………………………… 411

论文 12　TSR-altered oil with high-abundance thiaadamantanes of a deep-buried Cambrian gas condensate reservoir in Tarim Basin …… 426

论文 13　Origin of diamondoid and sulphur compounds in the Tazhong Ordovician condensate, Tarim Basin, China: Implications for hydrocarbon exploration in deep-buried strata ……………… 443

专利 1　用全二维气相色谱定量石油样品中金刚烷类化合物的方法 ……… 466

专利 2　石油样品中五环三萜烷类化合物的定量分析方法 ………………… 477

专利 3　一种生排烃模拟实验液态产物的定量分析方法 …………………… 487

参考文献 …………………………………………………………………… 495

第 1 章

全二维气相色谱基本原理

全二维气相色谱法(Comprehensive Two-Dimensional Gas Chromatography,GC×GC)是 20 世纪 90 年代初出现的新方法。该方法最先基于 Jorgenson 等于 1990 年提出的全二维液相色谱-毛细管电泳联用方法,强调二维正交分离的重要性,其后 Liu 和 Phillips 利用他们以前在快速气相色谱中使用的在线热解析调制器开发出全二维气相色谱法(许国旺,2004)。随后 Phillips 和 ZOEX 公司合作,在 1999 年正式实现了全二维气相色谱仪的商品化。除 ZOEX 公司外,美国的 LECO 公司、澳大利亚的 Chromatography Concepts 公司等也有同类商品化的仪器。

全二维气相色谱法是多维色谱法的一种,但它不同于通常所说的二维色谱(GC+GC)。GC+GC 一般采用中心切割法,将第一支色谱柱预分离的部分馏分再次进样到第二支色谱柱,做进一步的分离,样品中的其他组分或被放空或被中心切割。通过增加中心切割的次数可以实现对所感兴趣组分的分离,但由于从第一支色谱柱流出进到第二支色谱柱时组分的谱带已较宽,因此第二支色谱柱的分辨率会有损失。这种方法的第二维分析速度一般较慢,不能完全利用二维气相色谱的峰容量,它只是将第一支色谱柱流出的部分馏分转移到第二支色谱柱上,进行进一步的分离。

全二维气相色谱(GC×GC)是把分离机理不同且互相独立的两支色谱柱由调制器结合而成二维气相色谱,其结构示意图如图 1-1 所示。经第一支色谱柱分离后的每个馏分都要先进入调制器聚焦,然后再以脉冲方式送到第二支色谱柱做进一步的分离。所有组分从第二支色谱柱进入检测器,信号经数据处理系统处理,得到以色谱柱 1 保留时间为第一横坐标、色谱柱 2 保留时间为第二横坐标,以信号强度为纵坐标的三维立体图(3D图,图1-2)或二维点阵图(图1-3)。经数据处理后,三维立体图上的一个峰对应二维点阵

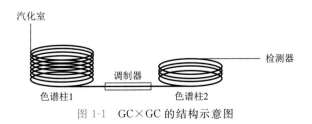

图 1-1　GC×GC 的结构示意图

图上的一个点,代表一种化合物。

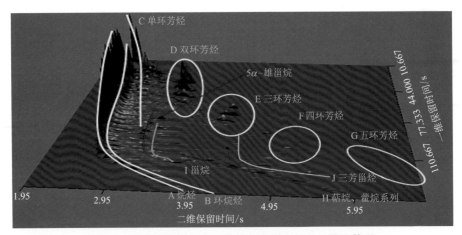

图 1-2　石油地质样品全二维气相色谱分析的三维立体图

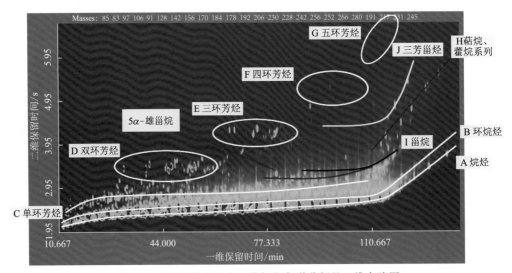

图 1-3　石油地质样品全二维气相色谱分析的二维点阵图

全二维色谱分析技术自出现以来,在复杂样品分析方面具有独特的优势,已广泛应用于烟草、中草药、酿酒、食品、石油化工和环境分析等领域。但该技术在石油地质领域中的应用起步较晚,相关方面的主要研究成果集中在 2009 年至今(高儻博等,2014)。

1.1 >> 全二维气相色谱分离原理

在全二维气相色谱方法中,如果色谱柱 1 为非极性柱,那么色谱柱 2 为极性柱,反之亦然。全二维气相色谱法通过化合物极性和沸点的差异实现气相色谱分离特性的正交化,其分离原理如图 1-4 所示:色谱柱 1 上流出的组分按保留时间大小依次进入调制器进行聚焦,然后通过快速加热的方法把聚焦后的组分快速发送到色谱柱 2 中进行再分离(许国旺,2004)。由于捕获、发送频率很高(在一个调制周期内),从外观来看,好像是从色谱柱 1

流出的峰被剁碎成一个个的碎片,聚焦后再往色谱柱 2 发送,所有组分从色谱柱 2 进入检测器,信号经数据处理系统处理后转化成全二维气相色谱图。在之后的数据处理方法中,软件会根据其二维保留时间或者质谱图将从色谱柱 2 出来的同一个化合物的一个个碎片合并成一个峰。连接色谱柱 1 和色谱柱 2,并把色谱柱 1 流出的峰切割成碎片,聚焦后再向色谱柱 2 发送的部件是全二维技术的关键,这个关键部件叫调制器。

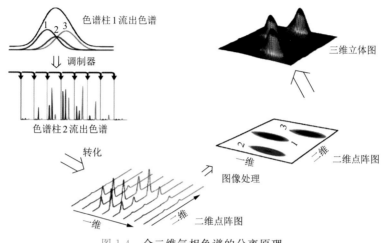

图 1-4　全二维气相色谱的分离原理

全二维气相色谱调制器需满足以下条件:
(1) 能及时浓缩从色谱柱 1 流出的分析物;
(2) 能将很窄的区带转移到色谱柱 2 的柱头,起第二维进样器的作用;
(3) 聚焦和再进样的操作应是再现的,且非歧视性的。

满足上述条件可以有很多调制方式。最早的调制方式有阀调制和热调制两类。但阀调制方式有两个缺陷:
(1) 需要很高的载气流速通过色谱柱 2;
(2) 样品中的大多数被放空,仅一小部分从色谱柱 1 流出的组分被注射进色谱柱 2,其余的被废弃。

尽管阀调制已被用于研究用化学计量学处理 GC×GC 数据,但此法并不适用于在实际中应用。目前广泛使用的调制器是热调制器。

热调制是气相色谱中最常用的调制技术。通过改变温度可以使几乎所有挥发性物质在固定相上吸附和脱附。Phillips 等设计了一个两段涂有金属的毛细管,用于对色谱柱 1 流出的溶质进行富集和快速热脱附。尽管应用该调制器获得了一些好的结果,但涂层常被烧坏,因此不得不经常替换。De Geus 等也得到了类似的结果,他们用紧密缠绕在毛细管外表面的铜线来加热调制器中的毛细管(许国旺,2004)。

为了克服金属涂层两段调制器的缺陷,Ledford 和 Phillips 设计了一种基于移动加热技术的调制器,该调制器用一个步进电机带动各加热元件("扫帚")运动,通过毛细管来达到局部加热的目的。此设计的主要优点是加热器热质足够大,可提供一个稳定且很好控制的温度(许国旺,2004)。

作为最早的商品调制器,该类热脱附调制器已应用在不同的实验室,至今大约有30%的GC×GC研究论文使用该类调制器。它的主要缺点是调制器温度必须比炉温高100 ℃(许国旺,2004)。

冷阱系统也被用作调制器。冷阱调制器由移动冷阱组成,做成径向调制冷阱系统(LMCS)。色谱柱1的谱带以很窄的区带宽度保留在冷阱调制器中,每隔几秒,调制器从T(Trap)位(捕集)到R(Release)位(释放)。在R位,冷却的毛细管在气相色谱柱温箱中加热,被捕集的馏分被立即释放,以很窄的区带在色谱柱2的柱头开始色谱分离,同时从色谱柱1流出的馏分被冷阱捕集,避免与前一周期中被释放组分在色谱柱2的重叠。几秒(调制时间)后,这个过程将重复,直到色谱柱1分析结束(许国旺,2004)。

冷阱调制器的主要好处是调制器中的毛细管加热到正常的炉温即可使其脱附,所以与"扫帚"系统相比,该系统能处理更高沸点的样品。但该方法的明显缺点是调制器中的固定相处于低达-50 ℃的物理状态。

在上述基础上,双喷液氮冷阱调制器、双喷CO_2调制器、CO_2环形调制器等被陆续推出,使得GC×GC调制器品种更加丰富,性能不断提高。

作者目前应用的是LECO公司的两级四喷口全二维调制器,具有两个独立的热喷口和两个独立的冷喷口,其中热喷口为加热的空气,冷喷口为液氮冷却的氮气;整个调制器封闭设计,独立供电;热喷最高控制温度可达400 ℃,冷喷温度低于-50 ℃;调制挥发物在$C_3 \sim C_{40}$正构烷烃范围;色谱柱2基本没有残留物;两级调制器的冷、热喷口调制时间(周期)可在1~65 s内按样品情况灵活设置,同一进样不同时间段调制时间也可灵活设置,可用于优化二维色谱分离。

图1-5(a)~(i)是两级四喷口全二维调制器在一个调制周期内的工作示意图,该图有助于进一步理解全二维气相色谱的分离原理。

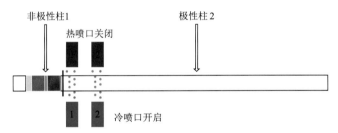

(a) 双热喷口关闭、双冷喷口开启时,色谱柱1分离的样品开始进入色谱柱2

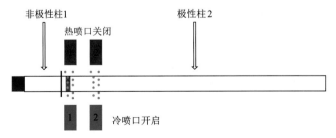

(b) 双热喷口关闭、双冷喷口开启时,色谱柱1分离的样品一个碎片被捕获在色谱柱2头端

图1-5 两级四喷口全二维调制器工作示意图

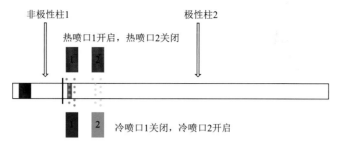

(c) 热喷口 1 开启、热喷口 2 关闭、冷喷口 1 关闭、冷喷口 2 开启时,
捕获在色谱柱 2 头端的碎片开始脱附

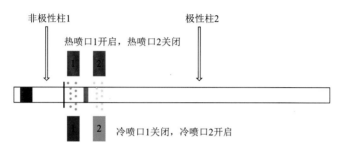

(d) 热喷口 1 开启、热喷口 2 关闭、冷喷口 1 关闭、冷喷口 2 开启时,
捕获在色谱柱 2 头端的碎片全部脱附并向前移动

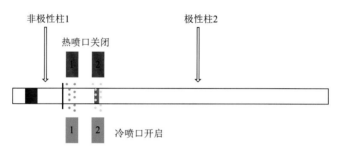

(e) 双热喷口关闭、双冷喷口开启时,捕获在色谱柱 2 头端的
碎片再次被聚焦冷冻下来

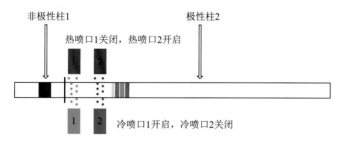

(f) 热喷口 1 关闭、热喷口 2 开启、冷喷口 1 开启、冷喷口 2 关闭时,
聚焦在色谱柱 2 头端的碎片全部进入色谱柱进行分离

图 1-5(续)　两级四喷口全二维调制器工作示意图

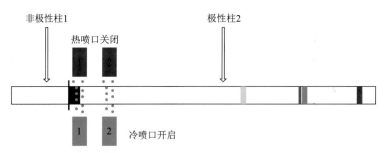

（g）双热喷口关闭、双冷喷口开启时，第二个碎片被捕获在色谱柱 2 头端

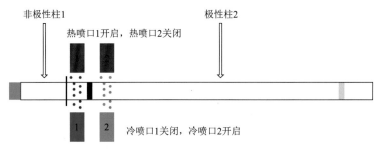

（h）热喷口 1 开启、热喷口 2 关闭、冷喷口 1 关闭、冷喷口 2 开启时，
在色谱柱 2 头端的第二个碎片开始聚焦，第一个碎片的分离将要完成

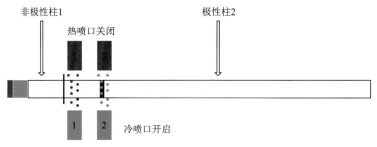

（i）双热喷口关闭、双冷喷口开启时，第二个碎片聚焦完成，在进入二维分离前
第一个碎片的分离已完成并全部流出色谱柱 2

图 1-5（续） 两级四喷口全二维调制器工作示意图

1.2 >> 全二维气相色谱检测器

全二维气相色谱中色谱柱 2 的分离非常快，在一个调制周期内要完成第二维的分离，否则前一调制周期的后流出组分可能会与后一调制周期的先流出组分交叉或重叠，造成组分识别的困难。这就要求检测器的响应时间非常快，数据处理机的采集速度至少应是 50～100 Hz。因此，所有具备死体积小、响应速度快的检测器，均可用于全二维气相色谱的检测。本节将介绍在石油地质样品分析中常用的三种检测器，即飞行时间质谱仪、氢火焰离子化检测器及硫化学发光检测器。

1.2.1 飞行时间质谱仪

飞行时间质谱仪（Time-of-Flight Mass Spectrometry，TOFMS）是全二维气相色谱检

测中最常用的质谱仪。这种质谱仪的质量分析器是一个离子漂移管(ion drift tube)。由离子源产生的离子首先被收集在收集器中,所有离子的速度变为0,然后使用一个脉冲电场加速后进入无场漂移管,并以恒定速度飞向离子接收器。离子质量越大,到达接收器所用时间越长;离子质量越小,到达接收器所用时间越短。根据这一原理,可以把不同质量的离子按质荷比(m/z)的大小进行分离。飞行时间质谱仪可检测的相对分子质量范围大,扫描速度快,仪器结构简单。飞行时间质谱仪的主要缺点是分辨率低,这是因为离子在离开离子源时初始能量不同,使得具有相同质荷比的离子到达检测器的时间在一定范围内分布,导致分辨能力下降。改进方法之一是在线性检测器前面加上一组静电场反射镜,将自由飞行中的离子反推回去,初始能量大的离子由于初始速度快,进入静电场反射镜的距离长,返回时的路程也就长,而初始能量小的离子返回时的路程短,这样就会在返回路程的一定位置聚焦,从而改善仪器的分辨能力。

飞行时间质谱仪有两种飞行模式,即平行飞行模式和垂直飞行模式。在现代质谱产品中,大都采用垂直飞行模式。质谱仪需要在真空情况下运转,用以保护检测器,同时提高测量精度。

在电子技术获得极大发展以后,TOFMS 的解析度得到了大幅度的提升。由于需要解析离子到达传感器的时间,因此要对传感器信号进行不停的扫描,以减少平均时间(averaging time)。这个过程对于数模转换器的采样率有非常高的要求。Aerodyne 和 Ionicon 的产品均采用 4 路数模转换器,每路采样率为 500 Ms/s(500 M samples/second)。通过同步协调使 4 路采样达到等效为 2 Gs/s 的采样率。仪器解析度可以达到 0.1 amu。

四级杆质谱仪在采样过程中每次只允许一个特定的 m/z 通过,因此如果要获得一个完整的质谱图,需要扫描不同的 m/z,而 TOFMS 在每次进样时可以采集样本中所有的 m/z。

1.2.2 氢火焰离子化检测器

氢火焰离子化检测器简称氢焰检测器(Flame Ionization Detector,FID),又称火焰离子化检测器,是色谱实验室中最常见的一种检测器,具有如下特点:

(1) 属于质量型检测器;
(2) 对有机化合物具有很高的灵敏度;
(3) 对无机气体、水、四氯化碳等不含氢或含氢少的物质不响应或灵敏度低;
(4) 结构简单、稳定性好、灵敏度高、响应迅速;
(5) 比热导检测器的灵敏度高出近 3 个数量级,检测下限可达 10^{-12} g。

氢火焰离子化检测器由电离室(图 1-6)和放大电路组成。FID 的电离室由金属圆筒作外罩,底座中心有喷嘴,喷嘴附近有环状金属圈(极化极,又称发射极),上端有一个金属圆筒(收集极)。环状金属圈与金属圆筒间加 100~300 V 的直流电压,形成电离电场,加速电离的

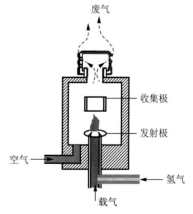

图 1-6 氢火焰离子化检测器电离室示意图

离子。收集极捕集的离子流经放大器的高阻产生信号,经放大后传送至数据采集系统;燃

烧气、辅助气和色谱柱由底座引入；燃烧气及水蒸气由外罩上方的小孔逸出。

1958年，Mewillan和Harley等分别研制成功氢火焰离子化检测器。该检测器是典型的破坏性、质量型检测器，以氢气和空气燃烧生成的火焰为能源，当有机化合物进入以氢气和氧气燃烧的火焰时，在高温下产生化学电离，产生比基流高几个数量级的离子，在高压电场的定向作用下形成离子流，微弱的离子流（$10^{-12} \sim 10^{-8}$ A）经过高阻（$10^6 \sim 10^{11}$ Ω）放大，成为与进入火焰的有机化合物量成正比的电信号，因此可以根据信号的大小对有机物进行定量分析。氢火焰离子化检测器结构简单、性能优异、稳定可靠、操作方便，虽然经过几十年的发展，但结构仍无实质性的变化。其主要特点是对几乎所有挥发性的有机化合物均有响应，对所有烃类化合物（碳数≥3）的相对响应值几乎相等，对含杂原子的烃类有机物中的同系物（碳数≥3）的相对响应值也几乎相等，这给化合物的定量带来很大的方便。

氢火焰离子化检测器需要用到3种气体：载气，用于携带试样组分；氢气，作为燃气；空气，作为助燃气。使用时需要调整三者的比例关系，使检测器的灵敏度达到最佳。根据分离及分析速度的需要选择载气（氮气）的流量，一般选择氢气流量与氮气流量比为1∶1～1∶1.4。在最佳氢、氮流量比时，检测器的灵敏度高、稳定性好。当空气流量很小时，检测器的灵敏度较低，随着空气流量的提高，检测器的灵敏度提高，但当空气流量高于某一数值后，提高空气的流量对检测器的灵敏度已没有明显影响。一般选择空气流量为氢气流量的8倍以上。

氢火焰离子化检测器对含氢的有机化合物具有较高的灵敏度，其检测机理如下：

（1）当含有机物 C_nH_m 的载气由喷嘴喷出进入火焰时，发生裂解反应，产生自由基：

$$C_nH_m \longrightarrow \cdot CH$$

（2）产生的自由基在D层火焰中与外面扩散进来的激发态原子氧或分子氧发生如下反应：

$$\cdot CH + O \longrightarrow CHO^+ + e$$

（3）生成的正离子 CHO^+ 与火焰中大量水分子碰撞而发生分子离子反应：

$$CHO^+ + H_2O \longrightarrow H_3O^+ + CO$$

（4）化学电离产生的正离子和电子在外加恒定直流电场的作用下分别向两极定向运动而产生微电流（$10^{-14} \sim 10^{-6}$ A）。

（5）在一定范围内，微电流的大小与进入离子室的被测组分的质量成正比，所以氢火焰离子化检测器是质量型检测器。

（6）组分在氢火焰中的电离效率很低，大约五十万分之一的碳原子被电离。

（7）离子电流信号输出到记录仪，得到峰面积与组分质量成正比的色谱流出曲线。

氢火焰离子化检测器具有检出限低、灵敏度高（$10^{-13} \sim 10^{-10}$ g/s）、基流小（$10^{-14} \sim 10^{-13}$ A）、线性范围宽（$10^6 \sim 10^7$）、死体积小（≤1 μL）、响应快（1 ms）、可以和毛细管柱直接联用，对气体流速、压力和温度变化不敏感等优点。FID对能在火焰中燃烧电离的有机化合物都有响应，可以直接进行定量分析，是目前应用最为广泛的气相色谱检测器之一。它的主要缺点是需要3种气源及其流速控制系统，尤其是对防爆有严格的要求，且不能检测永久性气体、水、一氧化碳、二氧化碳、氮的氧化物、硫化氢等物质。

1.2.3 硫化学发光检测器

1）检测原理和检测器构成

硫化学发光检测器（Sulfur Chemiluminescence Detector,SCD）是目前公认的检测硫最灵敏、选择性最宽的检测器。硫化学发光检测器的检测原理如下：

从柱子洗脱出的含硫化合物与载气一起流入燃烧室，在高温（>1 800 ℃）下燃烧生成 SO，然后和臭氧 O_3 发生反应形成激态 SO_2，再衰变至基态，发出特征的蓝色光谱（280～420 nm），光波 $h\nu$ 通过滤片后被光电倍增管接收进行检测，从而实现对硫的检测。

SCD 的结构示意图如图 1-7 所示，主要由燃烧室、反应室、臭氧发生器以及相关的气路组成。燃烧室为不锈钢材质，位于色谱仪的顶部，直接和色谱柱相连，以消除色谱峰的拖尾和减小系统的死体积，避免柱效降低。

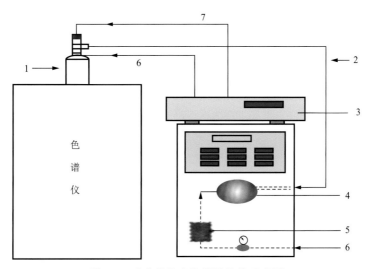

图 1-7　硫化学发光检测器结构示意图
1—燃烧室；2—传输管；3—控制器；4—反应室；5—臭氧发生器；6—空气；7—氢气

燃烧室的作用是把硫化合物裂解氧化成 SO 和其他产物。为了避免烃类物质在燃烧室内部积炭，配置有一个除焦阀，定期把积炭物除掉。反应池的作用是使 SO 和臭氧 O_3 发生反应生成 SO_2，通过一个探头把 SO 吸入反应池中进行反应。臭氧发生器的功能是为燃烧室提供反应所需要的臭氧。此外还有辅助设备真空泵，以便完成上述物质的传送。

为了同时测定烃类化合物和含硫化合物，燃烧炉可安装 FID 检测器。来自色谱柱的流出物先通过 FID，然后进入 SCD 进行检测。这种联合检测可以同时得到硫和其他烃类化合物的信息，而且省掉了分流装置，简化了操作。

2）SCD 的性能特征

（1）对硫检测的线性响应和等物质的量响应。从反应机理可以得知：SCD 对硫的响应是线性响应，其响应值随着硫浓度的增大而线性增大，并且是等物质的量响应。不管含硫化合物的结构如何，只要是物质的量相同的硫化物都产生相同的响应值。这一特征使得定量测定十分方便简单。而火焰光度检测器 FPD 的响应是非线性的，定量测定不太方便。

(2) 一流的灵敏度和选择性。SCD 的灵敏度一般小于 0.5 pg/s(S)，优于 FPD 一个量级；对烷烃、氯代烷烃的选择性高达 107 g(S)/g(C)，也优于 FPD，因而不受大量基体样品的干扰。

(3) 无猝灭作用。猝灭作用是指当非硫化物与硫化物一起或部分进入测硫检测器时，经常出现硫响应值下降，甚至完全消失的现象。

SCD 的缺点是操作比较麻烦，维护工作量比较大。

作为近年来的一个新趋势，电子俘获检测器（ECD）也在 GC×GC 中得到了应用。对 ECD 来说，主要关注如何获得较窄的峰宽，避免由池体积引起的谱带展宽。与快速色谱类似，要控制的最重要的参数是补偿气的流量，同时池体温度也应尽可能高。

1.3 >> 石油地质样品全二维气相色谱分析特点

1.3.1 全二维气相色谱图特征

为了更好地了解石油地质样品全二维气相色谱分析的特点，需要先了解石油地质样品分析谱图的一些共性特征。本节以实际样品为例，主要介绍非极性/极性（NP/P）色谱柱系统下的全二维气相色谱图特征。

图 1-8 是一个原油分离出的饱和烃样品的 GC-MS 和 GC×GC-TOFMS 检测结果谱图。图 1-8(a)和(b)分别为 GC-MS 和 GC×GC 的一维色谱图，这两张图的 Y 轴均表示色谱峰强度，只是 GC-MS 图上是相对丰度，以百分比表示，而 GC×GC 的色谱图上是绝对强度，以实际信号值表示。图 1-8(a)和(b)最大的不同之处在于 X 轴，GC-MS 的色谱图上的 X 轴只有一维保留时间（单位：min），而 GC×GC 的色谱图上的 X 轴有一维保留时间（单位：min）和二维保留时间（单位：s）两组数据。人们习惯通过色谱图上峰的强度高低判断该化合物的峰面积大小，例如图 1-8(a)中 C_{15} 的峰强度小于 C_{16} 的，则认为 C_{15} 的峰面积小于 C_{16} 的，事实也是如此。但是这种经验在 GC×GC 的色谱图上则不适用。原因是：在常规一维色谱图上，一个色谱峰代表一个化合物或者几个化合物的共馏峰，化合物的峰高和峰面积都与检测器的响应成正比；而在 GC×GC 的色谱图上，由于调制器的作用，一个化合物可能被切割成好几个碎片进入 GC×GC 的检测器被检测，因此一个色谱峰代表一个化合物峰或者一个化合物的碎片峰，要得到一个化合物的峰面积积分结果，必须把该化合物的所有碎片峰合并在一起计算出峰面积（此步骤可以通过数据处理软件完成），而不是简单地通过色谱图中峰的高低来判断。如图 1-8(b)所示，虽然 C_{15} 的峰强度大于 C_{16} 的，但经过碎片合峰发现 C_{15} 的峰面积小于 C_{16} 的。

图 1-8(b)中 83.333 min 之后抬起的黑色阴影表示的是色谱柱流失，相比于 GC-MS，GC×GC 的色谱图上柱流失明显，相对丰度较高，原因是两根色谱柱都存在柱流失现象，这是两根色谱柱柱流失在一维上叠加的结果。

图 1-8(c)是 GC×GC 的全二维点阵图，又称 GC×GC 的二维轮廓图。图中 X 轴表示一维保留时间（单位：min），Y 轴表示二维保留时间（单位：s），图中亮斑的大小和亮度对应的是化合物峰强度。化合物峰强度越高，亮斑越亮，亮斑面积越大。图中饱和烃化合物在色谱柱 2 上的出峰区域基本在 2~6 s 之间，色谱柱 1 柱流失的保留时间低于饱和烃化合物（在 1.8 s 左右），色谱柱 2 柱流失的保留时间高于饱和烃化合物（在 6.8 s 左右）。原因

是:在 GC×GC 的色谱柱 2 上化合物是按极性从小到大的顺序依次出峰,色谱柱 1 是非极性柱,它的柱流失是 100% 非极性的硅氧烷物质,极性比饱和烃物质都低,因此它的二维保留时间也低。同样地,色谱柱 2 是中等极性柱,它的柱流失多为 50% 极性的硅氧烷物质,要高于饱和烃化合物,因此它的二维保留时间也高。

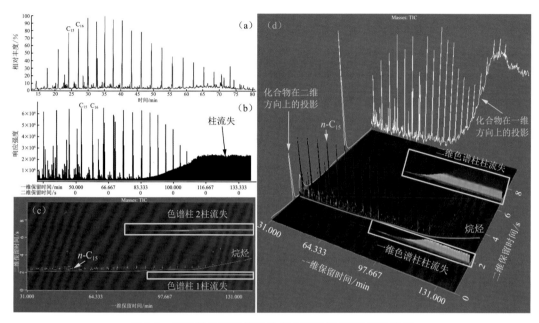

图 1-8　石油样品饱和烃组分在总离子流(TIC)下的谱图
(a) GC-MS 的色谱图;(b) GC×GC 的色谱图;(c) GC×GC 的全二维点阵图;(d) GC×GC 的三维立体图

图 1-8(d) 是 GC×GC 的三维立体图(3D 图),其中 X 轴表示一维保留时间(单位: min),Y 轴表示二维保留时间(单位: s),Z 轴表示峰强度,图中以等高线图的方式表示峰的强度,颜色从蓝到红渐变,峰强度越大,立体峰从上至下红色所占比例越多,而峰强度越小,红色所占比例越少。在三维立体图上有化合物在一维方向上的投影,该投影图是软件根据化合物在一维色谱上的出峰情况拟合出来的图像,同理化合物在二维方向上的投影是软件根据化合物在二维色谱上的出峰情况拟合出来的图像。

在 GC×GC 的二维和三维谱图上,化合物的分布情况与升温程序有关。若是一个平稳的升温程序(如以同一个温度梯度升温),则化合物在 GC×GC 的二维和三维谱图上也是平稳分布的;若升温速率发生改变(如从原有的 2 ℃/min 变成 4 ℃/min,或者从原有的 2 ℃/min 变成恒温),则化合物在 GC×GC 的二维和三维谱图上的走势会出现拐点。如图 1-8(c)所示,该样品的升温程序是从 80 ℃ 开始,以 2 ℃/min 升到 310 ℃,保持 25 min。31~115 min 对应着 2 ℃/min 的升温阶段,化合物平稳分布,图中烷烃呈直线分布;115 min 以后对应着恒温 25 min 阶段,化合物分布向上出现拐点,图中烷烃分布向上。图 1-9 是一个凝析油样品的 GC×GC 全二维点阵图。由图 1-9 可以看出,该样品有 3 个升温梯度,反映在全二维点阵图上是有 3 个明显的拐点,对应的三维立体图也是如此。

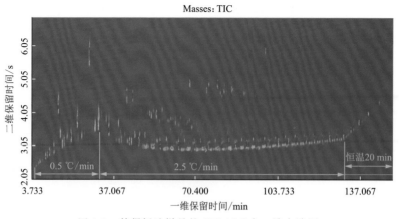

图 1-9　某凝析油样品的 GC×GC 全二维点阵图

在全二维气相色谱图特征中，族分离特征和"瓦片效应"是非常实用的定性辅助工具。在 GC×GC 的二维和三维谱图上，相同极性的化合物的二维保留时间相近，在色谱图上有规律分布，GC×GC 的二维和三维谱图被分割成不同的区带，每个区带代表一族物质，这就是 GC×GC 的族分离特征。此外，在同族带上相同碳数取代基的化合物呈线性排列，不同碳数取代基的同族化合物呈瓦片状排列，这就是 GC×GC 的"瓦片效应"。图 1-10 是原油样品饱和烃+芳烃组分在特定特征离子下的分析谱图。由图可以看出：原油中的烷烃、萘系列和菲系列化合物被划分成不同的区带，这就是族分离特性。在萘系列的区带里，按取代基碳数的不同化合物沿一维保留时间方向依次排列，按沸点由低到高依次是萘、C_1-萘（表示一个碳数取代的萘化合物）、C_2-萘、C_3-萘、C_4-萘，而相同碳数取代基的萘（如 C_2-萘）按照取代基位置的不同，沿二维保留时间方向依次排列，极性低的在下，极性高的在上，这就是"瓦片效应"。族分离特性和"瓦片效应"是 GC×GC 谱图的特点，对化合物的定性和族组分的定量有很大帮助。这两个特点在平缓的 GC×GC 谱图中比较容易被辨别，如果 GC×GC 谱图中拐点很多，化合物会随着拐点的出现破坏原有的分布规律，给未知化合物的定性和组分定量带来很大的难度。因此，在制定样品的 GC×GC 分析方案中，要尽量选择少的升温梯度。

1.3.2　全二维气相色谱定性方法

定性分析主要是对样品中未知化合物结构的鉴定。用传统一维气相色谱（1DGC）定性复杂体系中的未知化合物时，主要借助于质谱工具。首先用前处理的手法将复杂混合物按沸点、极性或化合物类型分成若干组分，再用 GC-MS 或 GC-MS-MS 对各组分进行分析，用检测到的未知化合物的质谱图与标准物质质谱图进行比对（或直接进行谱库检索），得到该化合物的定性结果。

但是在复杂的混合物（如石油样品）中，已经被鉴别出来的化合物只占少数，大多数化合物都无法被识别，且在已鉴别的化合物中被收录到标准谱库（NIST 谱库）中的化合物更少，能直接通过标准谱库检索得到定性结果的化合物仅占少数。多数化合物的识别需要依靠相关文献、行业标准等资料中提及的相对保留时间、保留指数或者文献中给出的标准质谱图定性。对完全未知的化合物定性就更加复杂，需要收集大量的资料并综合分析，利用质谱解析的方法推测未知化合物的结构。若要完全确定未知化合物的结构，还需要去

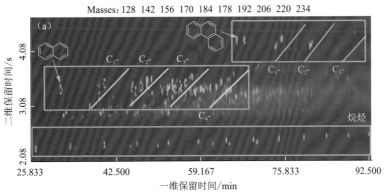

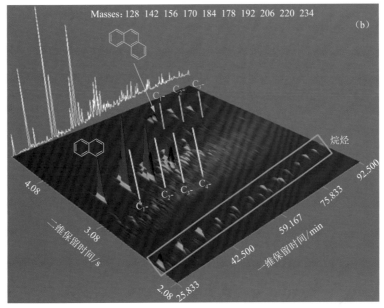

图 1-10　原油样品饱和烃+芳烃组分在特征离子 m/z 128,142,156,170,178,184,192,206,220,234 下的 GC×GC 谱图

(a) 全二维点阵图;(b) 三维立体图

合成标准物质,采用标样共注的方法确定推测的结构是否正确。

无论是标准谱库检索、相关资料质谱图比对,还是有机质谱解析,都需要化合物的质谱图够"纯净",才能有较好的匹配结果。但事实上,由于 1DGC 上共馏峰和检测基线的影响,得到的目标化合物质谱图有许多杂峰,多数化合物的分子离子不明显,进行谱库检索时与标准物质质谱图的相似度较低。为了消除共馏峰的干扰,GC-MS 常采用选择离子扫描的方式(SIM)进行分析,该方法虽然在一定程度上消除了共馏峰的干扰,但由于只采集少量离子,得不到化合物完整的质谱图,因此无法和标准物质质谱图进行比对,更不能用于质谱解析。综上,用常规气质联用的方法定性化合物,只适用于纯物质或者混合物中已有资料记载的物质。

全二维气相色谱具有高峰容量、高灵敏度、高分辨率等特点,能够最大程度消除共馏峰的干扰,族分离和"瓦片效应"等谱图特点能够帮助判定未知化合物的简单化学组成。搭配飞行时间质谱,能采集到所有化合物完整的质谱图。由于 TOFMS 和四级杆质谱都可采用电子电离(EI)源,且电离能量(electron energy)相同,因此 TOFMS 采集的质谱图可以与标准的 NIST 谱库进行比对。GC×GC-TOFMS 搭配的数据处理软件有去卷积的功能,能进一步消除原始采集的质谱图中的背景干扰,得到更为纯净的质谱图,大大提高目标化合物谱图与标准谱库比对的相似度。此外,在 GC×GC 中多数化合物都会经历调制过程,一个化合物被切割成几个碎片进入检测器,TOFMS 能捕捉到每一个碎片的质谱图并给出鉴定和确认,这相当于一个化合物被多次定性并给出相似度、反相似度等信息,可大大提高定性结果的可靠性。因此,用 GC×GC-TOFMS 对复杂混合物定性分析时,能够大大提高定性的准确率,为化合物的准确定性提供很大的帮助。

1.3.3　全二维气相色谱定量方法

全二维气相色谱的定量分析方法与常规的气相色谱定量分析方法相近,也是通过软件计算峰高或峰面积,然后采用归一化法、外标法或者内标法进行定量。不同的是,由于调制器的存在,GC×GC 的第一维色谱柱馏出的峰被切割成几个碎片峰,因此在对某个化合物定量的时候,需要把其所有碎片峰合并在一起进行积分,得到的碎片总峰面积才是常规气相色谱意义上的化合物峰面积。如果要进行族组分定量,不需要把该族中每个化合物的碎片都进行合峰,只要把该族所有化合物的碎片都划分到一个区域内(此步骤可以通过软件执行),并将该区域所有峰面积积分加和在一起就是族组分的总峰面积积分结果。值得注意的是,无论是在单体定量上还是在族组分定量上,都要保证同一个化合物或者同一族化合物的所有碎片峰面积都计算在内,否则会给定量结果带来很大的误差。用 GC×GC-TOFMS 分析样品时,其碎片峰的查找和合峰主要依据 TOFMS 提供的质谱图的相似度和保留时间,而用 GC×GC-FID 和 GC×GC-SCD 等分析样品时,其碎片峰的查找和合峰主要依据保留时间,并参考 GC×GC-TOFMS 的分析结果。

与常规的气相色谱相比,GC×GC 定量有以下几个优点:① GC×GC 的正交分离大大消除了共馏峰的干扰,一些化合物或者族组分在色谱条件下就能得到很好的分离,不用借助质谱就能得到很好的定量结果;② 能做到真正的基线分离,噪音大大降低,得到的积分结果更准确;③ 由于调制器的作用,GC×GC 的第二维色谱峰形窄而尖,灵敏度更高。

1.3.4　全二维气相色谱方法的优点

与常规一维气相色谱法相比,GC×GC 主要有以下几个优点:

(1) 分辨率高、峰容量大。GC×GC 的峰容量是两根色谱柱峰容量的乘积,而 GC+GC 的峰容量是两根色谱柱峰容量的和;GC×GC 的分辨率为两根色谱柱各自分辨率平方和的平方根。

(2) 灵敏度高,是常规的气相色谱的 20~50 倍。由于调制器的作用,化合物经一维色谱柱分离后,馏分被聚焦,再以脉冲形式进样,得到的 GC×GC 的峰型更尖锐,因此灵敏度更高。

(3) 分析时间短。由于 GC×GC 的峰容量高,复杂样品可以减少样品前处理的时间,在总体分析时间上要短于一维 GC。

(4) 定量可靠性大。GC×GC 的正交分离可以消除基线和共馏峰的干扰,得到化合物更为准确的定量结果。

(5) 定性准确性高。根据 GC×GC 提供的两个保留时间,可以推测化合物的沸点和极性信息。搭配飞行时间质谱,可以采集到化合物更为精确的质谱信息。TOFMS 提供的化合物的原始质谱图与标准谱库的质谱图对比,匹配度要远高于 GC-MS。

(6) 可实现族分离。用 GC×GC 分析时,两维的保留时间分别代表化合物不同的性质,相近性质的化合物在 GC×GC 谱图上能聚成一族,该特点就是 GC×GC 的族分离特征。在分析复杂样品时,有时不关心具体的某一个化合物,而是关心某一类化合物的定量结果,这就需要用到族分离。例如,凝析油样品的定量分析就借助 GC×GC 族分离的特点得到烷烃、环烷烃、芳烃等不同类别化合物的定量结果。

图 1-11 是某样品 GC×GC-TOFMS 的分析结果。由图可以看出:在相同的保留时间 32.667 min 时,共检测出 6 个化合物,它们在一维色谱上是共馏峰,而在 GC×GC 上由于极性的不同可在第二维色谱柱上被完全分开。与 NIST 谱库检索和专业文献比对,得到这 6 个化合物的推测定性结果(列于表 1-1 中),相似度均在 860 以上(其中 1-4 表示的 1,4-二甲基金刚烷化合物未收录在 NIST 谱库中,因此不能给出相似度数据),它们按极性由低到高的顺序依次是烷烃、单环环烷烃、双环环烷烃、多环环烷烃、单环芳烃和环烷烃取代的单环芳烃化合物。

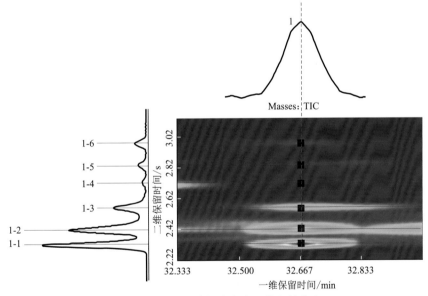

图 1-11　样品中共馏峰在全二维上的分离

表 1-1　**样品中共馏峰的定性分析结果**

峰号	一维保留时间/min	二维保留时间/s	定性质量数	化学式	相对分子质量	名　称	相似度	结构式
1-1	32.667	2.33	85	$C_{12}H_{26}$	170	4-乙基癸烷	863	
1-2	32.667	2.43	97	$C_{12}H_{24}$	168	1-甲基-2-戊基环己烷	894	

续表

峰号	一维保留时间/min	二维保留时间/s	定性质量数	化学式	相对分子质量	名　称	相似度	结构式
1-3	32.667	2.56	166	$C_{12}H_{22}$	166	2,6-二甲基十氢化萘	870	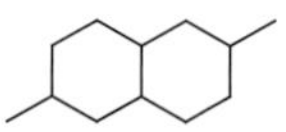
1-4	32.667	2.72	149	$C_{12}H_{20}$	164	1,4-二甲基金刚烷,顺式		
1-5	32.667	2.84	133	$C_{11}H_{16}$	148	1,3-二甲基-5-异丙基苯	930	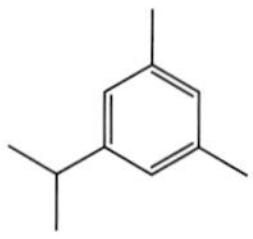
1-6	32.667	2.98	131	$C_{11}H_{14}$	146	1,2-二甲基二氢化茚	926	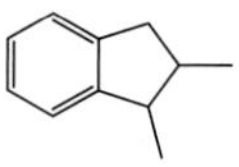

上述例子较为典型,很多化合物都可以借助全二维色谱条件得到很好的分离。常见共馏峰的区分主要有以下几种情况:

(1) 普通的共馏峰现象,如图1-11和表1-1所示,采用GC-MS的选择离子扫描方式,在各个化合物的特征离子下可以将它们区分开。

(2) 一维保留时间相同、特征离子相同的化合物,无法用GC-MS分开。如表1-2所示,甲基二苯并呋喃和二甲基联苯的一维保留时间相同,均为59.333 min,说明它们在第一维色谱上是共馏峰。这两个化合物的特征离子和分子离子都是182,在GC-MS的选择离子m/z 182下(图1-12a),这两个化合物也是共馏峰。而在GC×GC-TOFMS的选择离子m/z 182下,这两个化合物由于极性的不同在第二维色谱柱上可以被分开(图1-12b,c,d)。

表1-2　一维保留时间相同、特征离子相同的化合物定性

峰　号	一维保留时间/min	二维保留时间/s	特征离子	分子离子	名　称
1	59.333	3.72	182	182	甲基二苯并呋喃
2	59.333	3.58	182	182	二甲基联苯

(3) 特征离子相同、分子离子很弱的情况下沸点相近的两类物质。图1-13所示的C_{26}-三环萜烷和C_{24}-四环萜烷的特征离子都是m/z 191,在GC-MS的选择离子模式下(图1-13c)这3个化合物不能完全被分开。C_{26}-三环萜烷和C_{24}-四环萜烷的分子离子分别是m/z 360和m/z 330(表1-3),它们的离子丰度极低,且受柱流失的干扰,在GC-MS的选择离子m/z 360和m/z 330下无法检测到这3个化合物,因此它们不易被GC-MS分开。而在GC×GC-TOFMS中,四环萜烷和三环萜烷由于环数的差异在第二维色谱上能够被完全分开(图1-13a,b)。

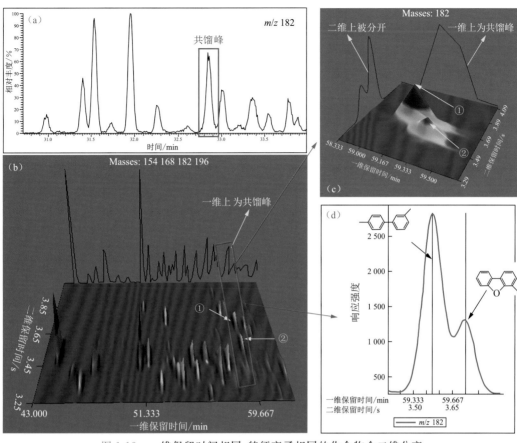

图 1-12　一维保留时间相同、特征离子相同的化合物全二维分离

表 1-3　三环萜烷和四环萜烷的全二维分离

峰 号	一维保留时间 /min	二维保留时间/s	特征离子	分子离子	名 称
1	103.17	3.15	191	360	C_{26}-三环萜烷
2	103.33	3.65	191	330	C_{24}-四环萜烷
3	103.50	3.14	191	360	C_{26}-三环萜烷

(4) 物质含量低,在特征离子下有其他强物质干扰而不易鉴别的峰。如原油中的含氮化合物,其含量很低,用普通的 GC-MS 分析饱和烃和芳烃的方法一般无法检测到。咔唑类物质会被相对含量很高的菲类物质所"掩盖"(图 1-14),苯并咔唑会被同在 m/z 217 上出峰的甾烷等物质所"掩盖"。因此用一维气相色谱分析原油中含氮化合物时,首先要用特殊的前处理方法把含氮化合物分离出来,这一过程对实验条件及流程的要求很严格,一般需要花费大量时间。但用 GC×GC-TOFMS 分析时,含氮化合物的极性要高于同沸点的碳氢化合物,在二维色谱上可以很好地被分开(图 1-14),节省了前处理的时间及过程中化合物的损耗。

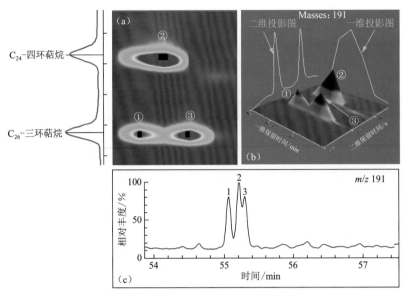

图 1-13 三环萜烷和四环萜烷的全二维分离与常规色质分析对比

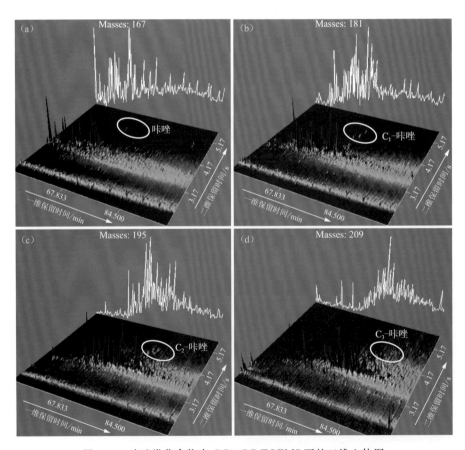

图 1-14 咔唑类化合物在 GC×GC-TOFMS 下的三维立体图

(a) m/z 167 下的三维立体图,图中标记出咔唑的出峰位置;(b) m/z 181 下的三维立体图,图中标记出 C_1 取代的咔唑的出峰位置;(c) m/z 195 下的三维立体图,图中标记出 C_2 取代的咔唑的出峰位置;(d) m/z 209 下的三维立体图,图中标记出 C_3 取代的咔唑的出峰位置

第2章

石油地质样品的全二维气相色谱图识别

认识石油地质样品的全二维气相色谱图是利用全二维气相色谱开展工作的基础。本章将介绍石油地质样品中饱和烃、芳烃以及部分杂原子化合物的全二维气相色谱图的识别方法和规律,为读者提供一些图谱识别的技巧。

本章的全二维气相色谱图及常规的色谱-质谱图来自不同地区的不同样品,这些样品包括四川盆地的凝析油、塔里木盆地的轻质油、玉门油田的正常原油、辽河油田的降解原油以及渤海湾盆地南堡凹陷地区的泥岩抽提物等。

使用的仪器包括:7890A气相色谱仪和双喷口冷热调制器组成的全二维气相色谱仪(GC×GC,美国LECO公司产品)、氢火焰离子化检测器(FID,美国Agilent公司产品)、Pegasus 4D飞行时间质谱仪(TOFMS,美国LECO公司产品)、Trace气相色谱/DSQⅡ质谱仪(GC-MS,美国Thermo Fisher Scientific公司产品)。工作站软件为Chroma TOF(美国LECO公司产品,支持GC×GC与TOFMS和FID的联用)、Xcalibur(美国Thermo Fisher Scientific公司产品,GC-MS仪器使用)。

样品前处理使用的化学试剂为正己烷(分析纯)和二氯甲烷(分析纯),使用前需要进一步提纯。分离柱填料为细硅胶(100~200目),使用前要在200℃下活化4 h。

凝析油采用直接进样法分析。全二维气相色谱分析条件如下:一维色谱柱为Petro柱(50 m×0.2 mm×0.5 μm),升温程序为35 ℃保持0.2 min,以1.5 ℃/min的速率升到230 ℃保持0.2 min,再以4 ℃/min的速率升到300 ℃保持15 min;二维色谱柱是DB-17HT柱(3 m×0.1 mm×0.1 μm),采用与一维色谱柱相同的升温速率,温度比一维色谱柱温箱高10 ℃。调制器温度比一维色谱柱温箱高35 ℃。进样口温度为280 ℃,采用分流进样模式,分流比为700∶1,进样量为0.5 μL。以氦气为载气,流速设定为1.5 mL/min。调制周期为10 s,其中2.5 s为热吹时间。飞行时间质谱的传输线和离子源温度分别为280 ℃和240 ℃,检测器电压为1 475 V,质量扫描范围为40~520 amu,采集速率为100谱图/s,溶剂延迟时间为0。

其他样品分别制备出饱和烃和芳烃后再进行仪器分析。饱和烃和芳烃的制备步骤如下:

(1) 取原油或者岩石抽提物样品20 mg左右,用适量正己烷溶解,加入配制好的氘代二十四烷烃、5α-雄甾烷和氘代蒽标准溶液作为内标化合物。

(2) 取细硅胶 3 g 转入玻璃柱中,震荡压实,用少许正己烷淋洗后,将样品转入玻璃柱中。

(3) 当柱子下端溶液流出时,分批加 10 mL 正己烷,收集饱和烃馏分;加 20 mL 二氯甲烷淋洗,收集芳烃馏分。各馏分用氮吹仪浓缩至约 1.5 mL 后转至进样瓶。

饱和烃和芳烃的仪器分析条件如下:全二维色谱的一维色谱柱为 50 m×0.2 mm×0.5 μm 的 Petro 柱,升温程序为 80 ℃保持 0.2 min,以 2 ℃/min 的速率升到 290 ℃保持 0.2 min,再以 0.2 ℃/min 的速率升到 300 ℃保持 20 min;二维色谱柱是 3 m×0.1 mm×0.1 μm 的 DB-17HT 柱,采用与一维色谱柱相同的升温速率,温度比一维色谱柱温箱高 5 ℃。调制器温度比一维色谱柱温箱高 35 ℃。进样口温度为 300 ℃,采用分流进样模式,分流比为 15∶1,进样量为 1 μL。以氦气为载气,流速设定为 1.8 mL/min。调制周期为 10 s,其中 2.5 s 为热吹时间。飞行时间质谱的传输线和离子源温度分别为 300 ℃和 240 ℃,检测器电压为 1 500 V,质量扫描范围为 40~520 amu,采集速率为 100 谱图/s,溶剂延迟时间为 10 min。

常规色谱-质谱柱为 60 m×0.25 mm×0.25 μm 的 HP-5MS 柱,载气为氦气,流速为 1 mL/min。升温程序为 100 ℃保持 5 min,以 4 ℃/min 的速率升到 220 ℃,再以 2 ℃/min 的速率升到 320 ℃保持 15 min。进样口温度为 280 ℃,采用不分流进样模式,进样量为 1.0 μL。质谱检测器电压为 1 700 V,选择离子方式扫描,溶剂延迟时间为 8 min。

2.1 >> 饱和烃组分的全二维气相色谱图特征及鉴定

利用全二维气相色谱-飞行时间质谱分析石油地质样品中的饱和烃组分或者凝析油样品,可以检测到饱和烃组分中所有类型的化合物,不同类型化合物的名称、同系物分子式、基本化学结构以及质谱碎片的主要特征离子见表 2-1。

表 2-1 饱和烃组分中的化合物类型及化学结构

序号	化合物类型	同系物分子式	化合物名称	基本化学结构	主要特征离子 (m/z)
1	正构烷烃	C_nH_{2n+2}	正构烷烃		57,85
2	异构烷烃	C_nH_{2n+2}	异构烷烃		57,183
3	单环环烷烃	C_nH_{2n}	烷基环戊烷		69
			烷基环己烷		83,97,111

续表

序号	化合物类型	同系物分子式	化合物名称	基本化学结构	主要特征离子（m/z）
4	双环环烷烃	C_nH_{2n-2}	双环戊烷		110
			十氢化萘		138,137,152,166
			二环倍半萜		123,179,193
5	三环环烷烃	C_nH_{2n-4}	三环萜烷		191
			三环己烷		178,192
6	四环环烷烃	C_nH_{2n-6}	四环萜烷		191
			甾烷		217,218
			断藿烷		123

续表

序号	化合物类型	同系物分子式	化合物名称	基本化学结构	主要特征离子（m/z）
7	五环环烷烃	C_nH_{2n-8}	藿烷/莫烷		191
		C_nH_{2n-8}	降藿烷		177
			伽马蜡烷		191
		C_nH_{2n-10}	藿烯		191
8	金刚烷类化合物	C_nH_{2n-4}	单金刚烷		136,135,149,163,177,191
		C_nH_{2n-8}	双金刚烷		188,187,201,215,229
		C_nH_{2n-12}	三金刚烷		240,239,253,267
		C_nH_{2n-16}	四金刚烷		292,291,305,319
		$C_nH_{2n-4}S$	硫代金刚烷		154,168,182,196,210

2.1.1 正构烷烃的全二维气相色谱图及化合物名称

大多数石油地质样品饱和烃中的正构烷烃含量较高,其在全二维气相色谱三维立体图中的特征同常规的色谱分析图几乎一致,如图 2-1 所示。利用姥鲛烷和植烷的位置特征与正构烷烃的分布规律可以对所有的正构烷烃定性,结果见表 2-2。

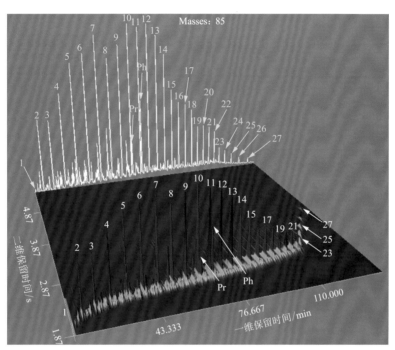

图 2-1　饱和烃中正构烷烃的全二维气相色谱图

表 2-2　饱和烃中正构烷烃的定性结果表

峰　号	一维保留时间 /min	二维保留时间 /s	特征离子 (m/z)	分子式	相对分子质量	名　　称
1	10.50	2.05	85	C_8H_{18}	114	n-Octane 正辛烷
2	14.67	2.17	85	C_9H_{20}	128	n-Nonane 正壬烷
3	20.17	2.26	85	$C_{10}H_{22}$	142	n-Decane 正癸烷
4	26.67	2.32	85	$C_{11}H_{24}$	156	n-Undecane 正十一烷
5	33.33	2.36	85	$C_{12}H_{26}$	170	n-Dodecane 正十二烷
6	40.00	2.39	85	$C_{13}H_{28}$	184	n-Tridecane 正十三烷
7	46.50	2.41	85	$C_{14}H_{30}$	198	n-Tetradecane 正十四烷
8	52.67	2.43	85	$C_{15}H_{32}$	212	n-Pentadecane 正十五烷
9	58.67	2.44	85	$C_{16}H_{34}$	226	n-Hexadecane 正十六烷
10	64.17	2.46	85	$C_{17}H_{36}$	240	n-Heptadecane 正十七烷
11	69.50	2.48	85	$C_{18}H_{38}$	254	n-Octadecane 正十八烷

续表

峰 号	一维保留时间 /min	二维保留时间 /s	特征离子 （m/z）	分子式	相对分子质量	名 称
12	74.50	2.49	85	$C_{19}H_{40}$	268	n-Nonadecane 正十九烷
13	79.33	2.51	85	$C_{20}H_{42}$	282	n-Eicosane 正二十烷
14	83.83	2.53	85	$C_{21}H_{44}$	296	n-Heneicosane 正二十一烷
15	88.17	2.54	85	$C_{22}H_{46}$	310	n-Docosane 正二十二烷
16	92.33	2.56	85	$C_{23}H_{48}$	324	n-Tricosane 正二十三烷
17	96.33	2.58	85	$C_{24}H_{50}$	338	n-Tetracosane 正二十四烷
18	100.17	2.60	85	$C_{25}H_{52}$	352	n-Pentacosane 正二十五烷
19	103.83	2.62	85	$C_{26}H_{54}$	366	n-Hexacosane 正二十六烷
20	107.50	2.65	85	$C_{27}H_{56}$	380	n-Heptacosane 正二十七烷
21	110.83	2.68	85	$C_{28}H_{58}$	394	n-Octacosane 正二十八烷
22	114.17	2.71	85	$C_{29}H_{60}$	408	n-Nonacosane 正二十九烷
23	117.50	2.81	85	$C_{30}H_{62}$	422	n-Triacontane 正三十烷
24	121.00	2.99	85	$C_{31}H_{64}$	436	n-Hentriacontane 正三十一烷
25	125.17	3.20	85	$C_{32}H_{66}$	450	n-Dotriacontane 正三十二烷
26	130.00	3.43	85	$C_{33}H_{68}$	464	n-Tritriacontane 正三十三烷
27	135.83	3.72	85	$C_{34}H_{70}$	478	n-Tetratriacontane 正三十四烷

2.1.2 异构烷烃的全二维气相色谱图及化合物名称

在非极性+极性色谱柱组合的全二维气相色谱的图谱中，异构烷烃的二维保留时间是相对最小的，异构烷烃分布在全二维谱图的最下方，即正构烷烃的下方，如图2-1和表2-2所示，姥鲛烷、植烷的二维保留时间要小于相邻的正十七烷和正十八烷。异构烷烃的同分异构体比较复杂，相关的文献和研究较少，本书未介绍其定性结果。

2.1.3 单环环烷烃的全二维气相色谱图及化合物名称

在石油地质样品的全二维气相色谱分析图谱中，存在两类二维保留时间大于正构烷烃且小于双环环烷烃的、有明显分布规律的化合物，它们与正构烷烃如影相随，碳数分布从C_7开始一直到C_{30}以上。依据二维保留时间、一维分布规律和质谱解析，判断这两类化合物分别是烷基环戊烷和烷基环己烷。

烷基环戊烷的特征离子是m/z 69，由于在m/z 69下干扰物质较多，因此一般选择在特征离子m/z 68下对其进行鉴定。图2-2是GC-MS下的色谱-质谱图，图中除了烷基环戊烷，还能明显看到正构烷烃，这是因为正构烷烃也有碎片离子m/z 68。图中用符号"＋"标记正构烷烃，用符号"＊"标记烷基环戊烷。烷基环戊烷与正构烷烃一样，也具有随碳数增加化合物呈规律分布的特征。从图2-2上可以看出，相差一个碳数的烷基环戊烷和正构烷烃在色谱图上的相对出峰位置,随着升温程序的增加会发生规律改变。C_{31}烷烃晚于二十五烷基环戊烷($C_{30}H_{60}$)出峰,C_{32}烷烃和二十六烷基环戊烷($C_{31}H_{62}$)重叠出峰,C_{33}烷烃和二十七烷基环戊烷($C_{32}H_{64}$)重叠出峰,C_{34}烷烃和二十八烷基环戊烷($C_{33}H_{66}$)重叠出峰,C_{35}烷烃早于二十九烷基环戊烷($C_{34}H_{68}$)出峰。烷基环戊烷与正构烷烃的这种分布特点使得在常规的色谱-质谱分析时总有部分烷基环戊烷与正构烷烃形成共馏峰,无法被完全分开。

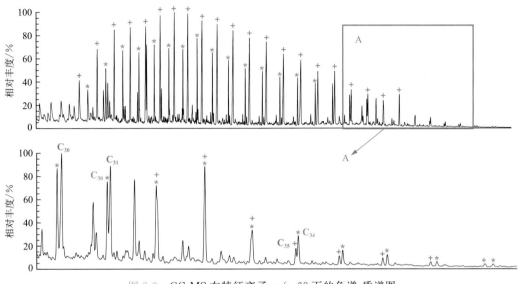

图 2-2 GC-MS 在特征离子 m/z 68 下的色谱-质谱图

符号"+"标记正构烷烃,符号"*"标记烷基环戊烷

在全二维气相色谱中,由于烷基环戊烷的极性大于正构烷烃,因此两者的二维保留时间有明显差距。图 2-3 是正构烷烃和烷基环戊烷的全二维点阵图,从图中可以看出正构烷烃和烷基环戊烷互不干扰,可以被完全分离。相比于常规色质分析,用全二维气相色谱分析烷基环戊烷可以消除正构烷烃的干扰,得到该类化合物的完整信息。

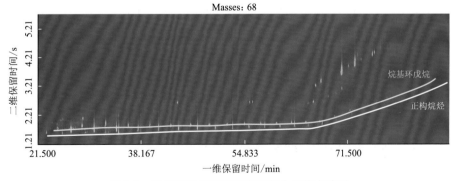

图 2-3 正构烷烃与烷基环戊烷的全二维点阵图

图 2-4 是用全二维气相色谱-飞行时间质谱分析得到的烷基环戊烷的全二维点阵图和三维立体图,其中共鉴定出 29 个化合物,其定性结果见表 2-3。

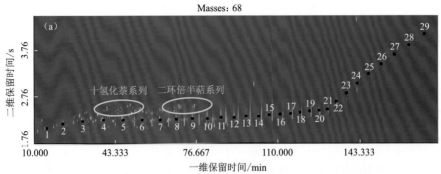

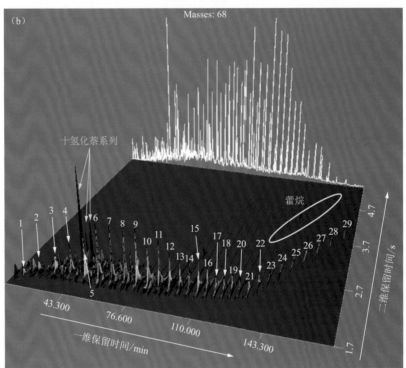

图 2-4 烷基环戊烷的全二维点阵图(a)和三维立体图(b)

表 2-3 饱和烃中烷基环戊烷的定性结果表

峰号	一维保留时间 /min	二维保留时间 /s	特征离子 (m/z)	分子式	相对分子质量	名 称
1	14.93	2.10	68	C_7H_{14}	98	Cyclopentane,ethyl- 乙基环戊烷
2	21.87	2.20	68	C_8H_{16}	112	Cyclopentane,propyl- 丙基环戊烷
3	30.00	2.26	68	C_9H_{18}	126	Cyclopentane,butyl- 丁基环戊烷
4	38.53	2.28	68	$C_{10}H_{20}$	140	Cyclopentane,pentyl- 戊基环戊烷
5	46.67	2.28	68	$C_{11}H_{22}$	154	Cyclopentane,hexyl- 己基环戊烷

续表

峰 号	一维保留时间 /min	二维保留时间 /s	特征离子 (m/z)	分子式	相对分子质量	名 称
6	54.40	2.28	68	$C_{12}H_{24}$	168	Cyclopentane, heptyl- 庚基环戊烷
7	61.73	2.28	68	$C_{13}H_{26}$	182	Cyclopentane, octyl- 辛基环戊烷
8	68.53	2.30	68	$C_{14}H_{28}$	196	Cyclopentane, nonyl- 壬基环戊烷
9	74.93	2.31	68	$C_{15}H_{30}$	210	Cyclopentane, decyl- 癸基环戊烷
10	80.93	2.31	68	$C_{16}H_{32}$	224	Cyclopentane, undecyl- 十一烷基环戊烷
11	86.67	2.33	68	$C_{17}H_{34}$	238	Cyclopentane, dodecyl- 十二烷基环戊烷
12	92.00	2.34	68	$C_{18}H_{36}$	252	Cyclopentane, tridecyl- 十三烷基环戊烷
13	97.07	2.36	68	$C_{19}H_{38}$	266	Cyclopentane, tetradecyl- 十四烷基环戊烷
14	102.00	2.36	68	$C_{20}H_{40}$	280	Cyclopentane, pentadecyl- 十五烷基环戊烷
15	106.53	2.39	68	$C_{21}H_{42}$	294	Cyclopentane, hexadecyl- 十六烷基环戊烷
16	110.93	2.40	68	$C_{22}H_{44}$	308	Cyclopentane, heptadecyl- 十七烷基环戊烷
17	115.07	2.42	68	$C_{23}H_{46}$	322	Cyclopentane, octadecyl- 十八烷基环戊烷
18	119.07	2.44	68	$C_{24}H_{48}$	336	Cyclopentane, nonadecyl- 十九烷基环戊烷
19	122.93	2.47	68	$C_{25}H_{50}$	350	Cyclopentane, eicosyl- 二十烷基环戊烷
20	126.67	2.49	68	$C_{26}H_{52}$	364	Cyclopentane, heneicosyl- 二十一烷基环戊烷
21	130.13	2.52	68	$C_{27}H_{54}$	378	Cyclopentane, docosyl- 二十二烷基环戊烷
22	133.73	2.68	68	$C_{28}H_{56}$	392	Cyclopentane, tricosyl- 二十三烷基环戊烷
23	137.73	2.85	68	$C_{29}H_{58}$	406	Cyclopentane, tetracosyl- 二十四烷基环戊烷
24	142.00	3.06	68	$C_{30}H_{60}$	420	Cyclopentane, pentacosyl- 二十五烷基环戊烷
25	146.67	3.27	68	$C_{31}H_{62}$	434	Cyclopentane, hexacosyl- 二十六烷基环戊烷
26	151.73	3.47	68	$C_{32}H_{64}$	448	Cyclopentane, heptacosyl- 二十七烷基环戊烷

续表

峰号	一维保留时间 /min	二维保留时间 /s	特征离子 (m/z)	分子式	相对分子质量	名　称
27	157.20	3.68	68	$C_{33}H_{66}$	462	Cyclopentane, octacosyl- 二十八烷基环戊烷
28	163.20	3.90	68	$C_{34}H_{68}$	476	Cyclopentane, nonacosyl- 二十九烷基环戊烷
29	169.60	4.12	68	$C_{35}H_{70}$	490	Cyclopentane, triacontyl- 三十烷基环戊烷

由于六元环比五元环的构型更加稳定，在石油地质样品中烷基环己烷的含量一般会高于烷基环戊烷的含量，在全二维气相色谱的谱图中更容易被识别。对应同碳数的烷基环戊烷，其二维保留时间更长一些，因此烷基环己烷的出峰位置在烷基环戊烷的上方。

烷基环己烷的特征离子是 m/z 82 或 m/z 83，在色谱图上和正构烷烃一样具有随碳数增加化合物呈规律分布的特征。图 2-5(a) 是常规色谱-质谱的谱图，从图中可以看出，烷基环己烷和正构烷烃的相对分布与烷基环戊烷具有几乎一致的规律，只是取代基的碳数不同。C_{27} 烷烃晚于二十烷基环戊烷 ($C_{26}H_{52}$) 出峰，C_{28} 烷烃和二十一烷基环己烷 ($C_{27}H_{54}$) 重叠出峰，C_{29} 烷烃和二十二烷基环戊烷 ($C_{28}H_{56}$) 重叠出峰，C_{30} 烷烃早于二十三烷基环戊烷 ($C_{29}H_{58}$) 出峰。烷基环戊烷与正构烷烃的这种分布特点使得常规的色谱-质谱分析时，总有部分烷基环戊烷与正构烷烃形成共馏峰，无法完全被分开。

图 2-5(b) 和 (c) 是同一样品的全二维分析谱图，烷基环己烷和正构烷烃极性的不同使它们在二维色谱柱上被完全分开，消除常规色质中共馏峰的干扰。所有烷基环己烷定性结果见表 2-4。

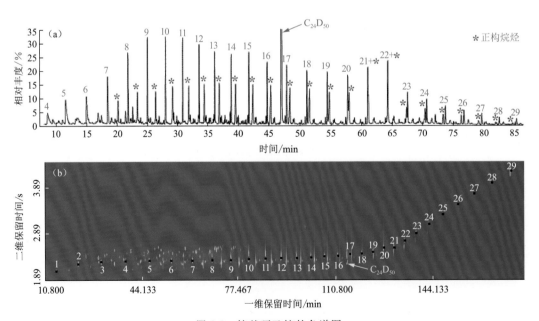

图 2-5　烷基环己烷的色谱图
(a) GC-MS 色谱图（符号"＊"标记正构烷烃，$C_{24}D_{50}$ 是内标化合物）；
(b) 全二维点阵图；(c) 三维立体图

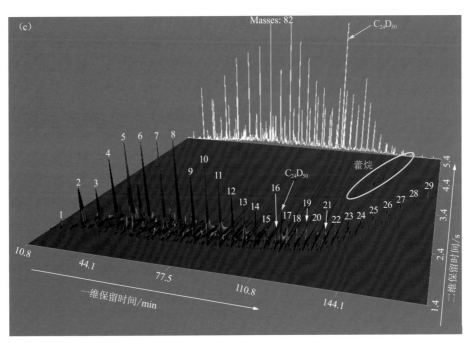

图 2-5(续) 烷基环己烷的色谱图

(a) GC-MS 色谱图(符号"*"标记正构烷烃,$C_{24}D_{50}$ 是内标化合物);
(b) 全二维点阵图;(c) 三维立体图

表 2-4 饱和烃中烷基环己烷的定性结果表

峰 号	一维保留时间 /min	二维保留时间 /s	特征离子 (m/z)	分子式	相对分子质量	名 称
1	14.40	2.10	82	C_7H_{14}	98	Cyclohexane, methyl- 甲基环己烷
2	22.00	2.25	82	C_8H_{16}	112	Cyclohexane, ethyl- 乙基环己烷
3	29.73	2.29	82	C_9H_{18}	126	Cyclohexane, propyl- 丙基环己烷
4	38.27	2.31	82	$C_{10}H_{20}$	140	Cyclohexane, butyl- 丁基环己烷
5	46.40	2.32	82	$C_{11}H_{22}$	154	Cyclohexane, pentyl- 戊基环己烷
6	54.27	2.32	82	$C_{12}H_{24}$	168	Cyclohexane, hexyl- 己基环己烷
7	61.60	2.32	82	$C_{13}H_{26}$	182	Cyclohexane, heptyl- 庚基环己烷
8	68.53	2.33	82	$C_{14}H_{28}$	196	Cyclohexane, octyl- 辛基环己烷
9	75.07	2.33	82	$C_{15}H_{30}$	210	Cyclohexane, nonyl- 壬基环己烷
10	81.07	2.35	82	$C_{16}H_{32}$	224	Cyclohexane, decyl- 癸基环己烷

续表

峰 号	一维保留时间 /min	二维保留时间 /s	特征离子 (m/z)	分子式	相对分子质量	名 称
11	86.80	2.36	82	$C_{17}H_{34}$	238	Cyclohexane,undecyl- 十一烷基环己烷
12	92.27	2.37	82	$C_{18}H_{36}$	252	Cyclohexane,dodecyl- 十二烷基环己烷
13	97.33	2.38	82	$C_{19}H_{38}$	266	Cyclohexane,tridecyl- 十三烷基环己烷
14	102.27	2.39	82	$C_{20}H_{40}$	280	Cyclohexane,tetradecyl- 十四烷基环己烷
15	106.80	2.41	82	$C_{21}H_{42}$	294	Cyclohexane,pentadecyl- 十五烷基环己烷
16	111.33	2.42	82	$C_{22}H_{44}$	308	Cyclohexane,hexadecyl- 十六烷基环己烷
17	115.47	2.45	82	$C_{23}H_{46}$	322	Cyclohexane,heptadecyl- 十七烷基环己烷
18	119.47	2.47	82	$C_{24}H_{48}$	336	Cyclohexane,octadecyl- 十八烷基环己烷
19	123.33	2.5	82	$C_{25}H_{50}$	350	Cyclohexane,nonadecyl- 十九烷基环己烷
20	127.07	2.59	82	$C_{26}H_{52}$	364	Cyclohexane,eicosyl- 二十烷基环己烷
21	130.67	2.59	82	$C_{27}H_{54}$	378	Cyclohexane,heneicosyl- 二十一烷基环己烷
22	134.27	2.74	82	$C_{28}H_{56}$	392	Cyclohexane,docosyl- 二十二烷基环己烷
23	138.27	2.91	82	$C_{29}H_{58}$	406	Cyclohexane,tricosyl- 二十三烷基环己烷
24	142.67	3.10	82	$C_{30}H_{60}$	420	Cyclohexane,tetracosyl- 二十四烷基环己烷
25	147.33	3.31	82	$C_{31}H_{62}$	434	Cyclohexane,pentacosyl- 二十五烷基环己烷
26	152.53	3.53	82	$C_{32}H_{64}$	448	Cyclohexane,hexacosyl- 二十六烷基环己烷
27	158.13	3.76	82	$C_{33}H_{66}$	462	Cyclohexane,heptacosyl- 二十七烷基环己烷
28	164.27	3.99	82	$C_{34}H_{68}$	476	Cyclohexane,octacosyl- 二十八烷基环己烷
29	170.67	4.25	82	$C_{35}H_{70}$	490	Cyclohexane,nonacosyl- 二十九烷基环己烷

2.1.4 双环环烷烃的全二维气相色谱图及化合物名称

本节介绍的双环环烷烃包括十氢化萘系列和二环倍半萜类化合物。

1) 十氢化萘的全二维气相色谱图及化合物名称

十氢化萘系列化合物包含十氢化萘及其不同取代基的化合物，取代基的碳数从 C_1 到 C_6，甚至更多。十氢化萘的特征离子是 m/z 138，其他系列化合物的特征离子依碳数增加呈规律变化。图 2-6 是十氢化萘系列化合物的全二维点阵图。从图中可以看出，随着取代基碳数的增加，化合物的数量增多，不同碳数取代基的化合物在全二维点阵图上呈"瓦片状"排列。

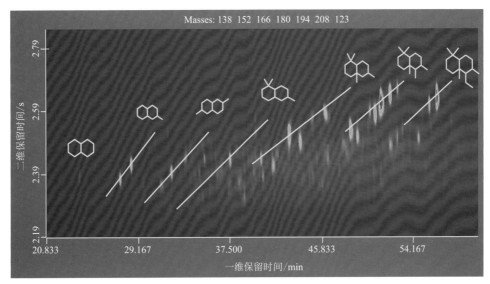

图 2-6 十氢化萘系列化合物的全二维点阵图

将正构烷烃的全二维点阵图（图 2-7a）与十氢化萘系列化合物的全二维点阵图（图 2-7b,c,d,e,f,g）叠放在一起，以方便鉴定，如图 2-7 所示。

作者曾在石油地质样品中发现了系列的单取代十氢化萘化合物，其结构解析见附录论文 4。

2) 二环倍半萜的全二维气相色谱图及化合物名称

二环倍半萜类化合物是地球化学研究中常用的生物标志化合物，其主要特征离子为 m/z 123（图 2-8b），其中一些化合物在特征离子 m/z 179（图 2-8c）和 m/z 193（图 2-8d）下丰度更高。在常规色质分析中，一般仅鉴定图 2-8(a)中红色标注的 1,2,3,7,8,9+10,11+12,15,16,17 共 10 个峰，而在全二维气相色谱分析中，不仅原来的共馏峰完全分开（图 2-8b），而且依据化合物在正构烷烃（图 2-8e）上方的相对位置和对 TOFMS 提供的质谱信息（图 2-9～图 2-29）解析，鉴定出更多的二环倍半萜，共鉴定化合物 21 个，见表 2-5。

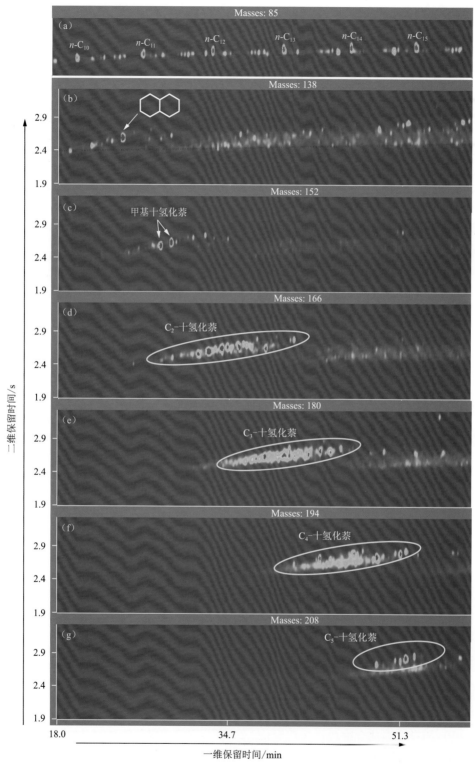

图 2-7 正构烷烃与十氢化萘系列化合物的全二维点阵图

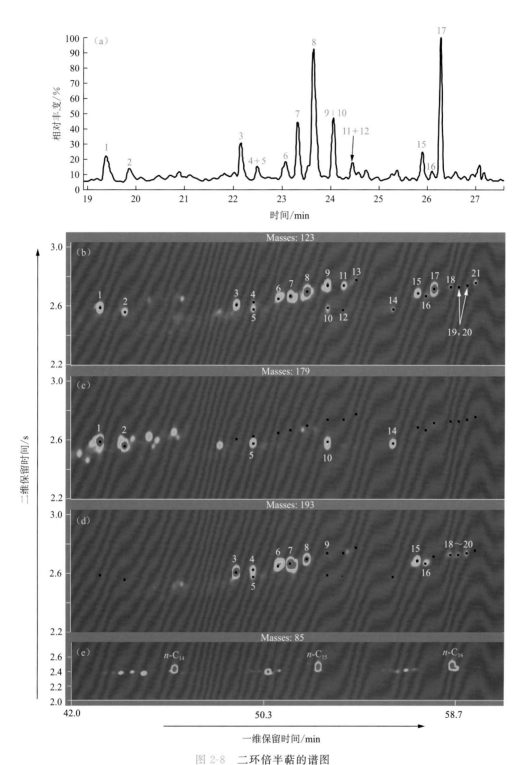

图 2-8　二环倍半萜的谱图

(a) GC-MS 在 m/z 123 下的色谱图；(b)~(d) GC×GC-TOFMS 在 m/z 123,179 和 193 下的全二维点阵图；
(e) 正构烷烃的全二维点阵图

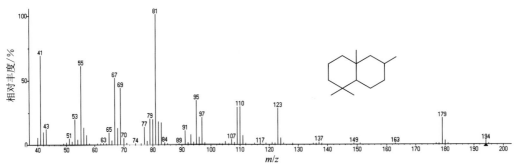

图 2-9　1 号峰　C_{14}，二环倍半萜质谱图

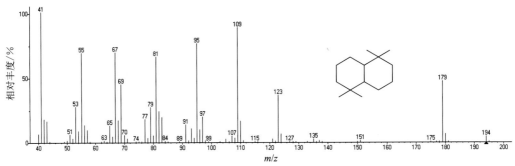

图 2-10　2 号峰　C_{14}，二环倍半萜质谱图

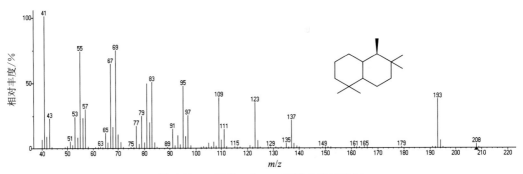

图 2-11　3 号峰　C_{15}，二环倍半萜质谱图

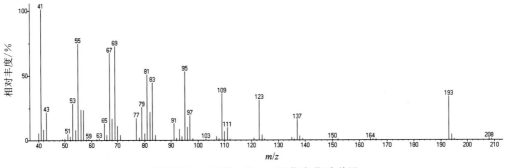

图 2-12　4 号峰　C_{15}，二环倍半萜质谱图

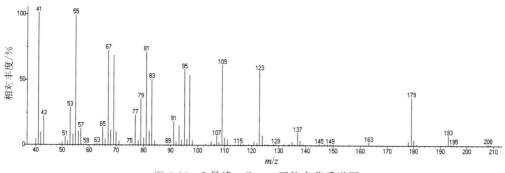

图 2-13　5 号峰　C_{15}，二环倍半萜质谱图

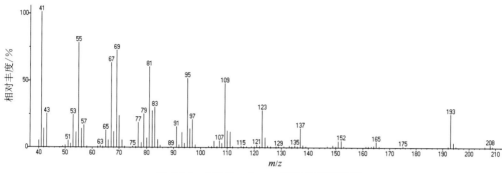

图 2-14　6 号峰　C_{15}，二环倍半萜质谱图

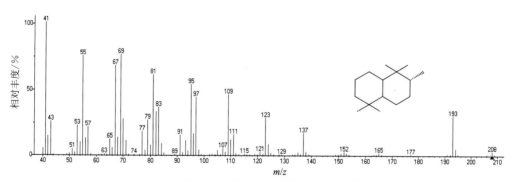

图 2-15　7 号峰　C_{15}，二环倍半萜质谱图

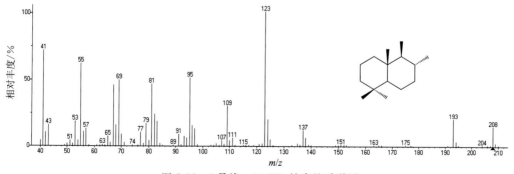

图 2-16　8 号峰　$8\beta(H)$-补身烷质谱图

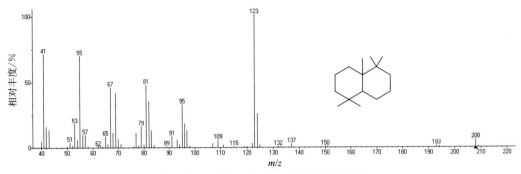

图 2-17　9 号峰　C_{15}，二环倍半萜质谱图

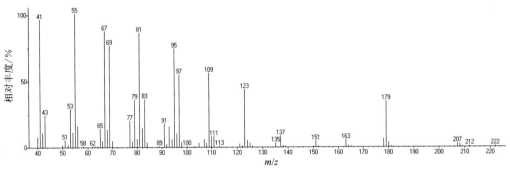

图 2-18　10 号峰　C_{16}，二环倍半萜质谱图

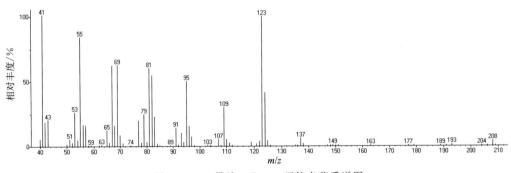

图 2-19　11 号峰　C_{15}，二环倍半萜质谱图

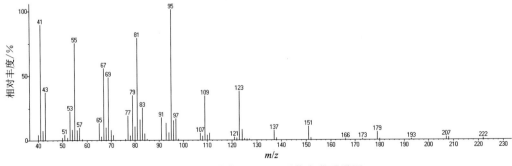

图 2-20　12 号峰　C_{16}，二环倍半萜质谱图

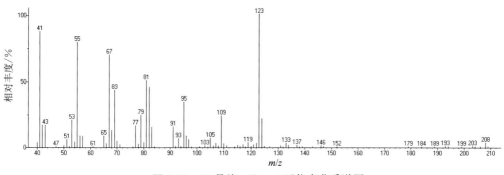

图 2-21 13 号峰 C_{15}，二环倍半萜质谱图

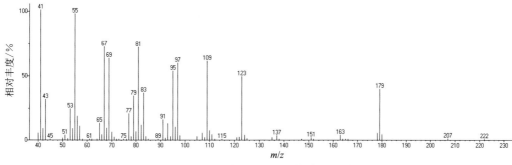

图 2-22 14 号峰 C_{16}，二环倍半萜质谱图

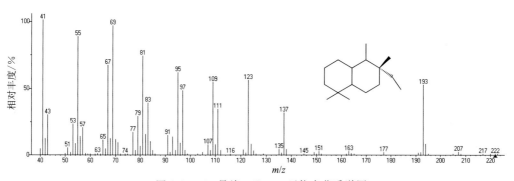

图 2-23 15 号峰 C_{16}，二环倍半萜质谱图

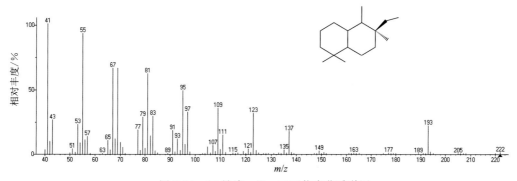

图 2-24 16 号峰 C_{16}，二环倍半萜质谱图

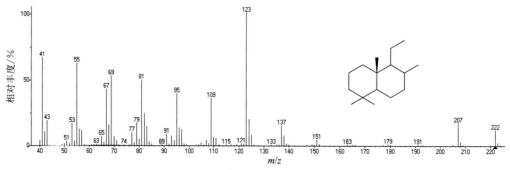

图 2-25　17 号峰　8β(H)-升补身烷质谱图

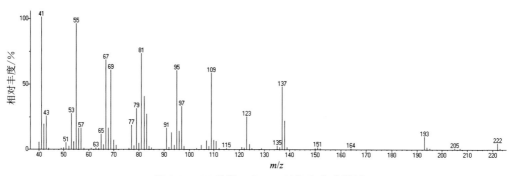

图 2-26　18 号峰　C_{16}，二环倍半萜质谱图

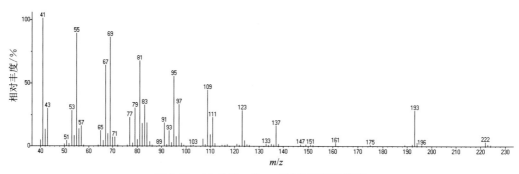

图 2-27　19 号峰　C_{16}，二环倍半萜质谱图

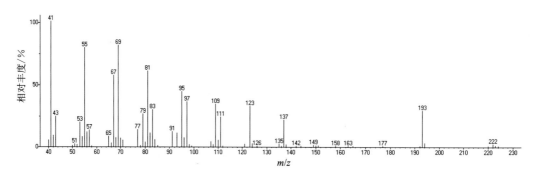

图 2-28　20 号峰　C_{16}，二环倍半萜质谱图

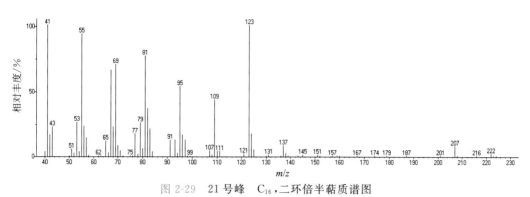

图 2-29　21 号峰　C_{16}，二环倍半萜质谱图

表 2-5　二环倍半萜化合物鉴定结果表

峰号	一维保留时间 /min	二维保留时间 /s	特征离子 (m/z)	分子式	相对分子质量	名　称
1	43.50	2.66	81	$C_{14}H_{26}$	194	C_{14}，Bicyclic sesquiterpane C_{14}，二环倍半萜
2	44.50	2.63	179	$C_{14}H_{26}$	194	C_{14}，Bicyclic sesquiterpane C_{14}，二环倍半萜
3	49.00	2.68	193	$C_{15}H_{28}$	208	C_{15}，Bicyclic sesquiterpane C_{15}，二环倍半萜
4	49.67	2.70	193	$C_{15}H_{28}$	208	C_{15}，Bicyclic sesquiterpane C_{15}，二环倍半萜
5	49.67	2.65	179	$C_{15}H_{28}$	208	C_{15}，Bicyclic sesquiterpane C_{15}，二环倍半萜
6	50.67	2.72	193	$C_{15}H_{28}$	208	C_{15}，Bicyclic sesquiterpane C_{15}，二环倍半萜
7	51.17	2.74	193	$C_{15}H_{28}$	208	C_{15}，Bicyclic sesquiterpane C_{15}，二环倍半萜
8	51.83	2.77	123	$C_{15}H_{28}$	208	$8\beta(H)$-Drimane $8\beta(H)$-补身烷
9	52.67	2.81	123	$C_{15}H_{28}$	208	C_{15}，Bicyclic sesquiterpane C_{15}，二环倍半萜
10	52.67	2.66	179	$C_{16}H_{30}$	222	C_{16}，Bicyclic sesquiterpane C_{16}，二环倍半萜
11	53.33	2.81	123	$C_{15}H_{28}$	208	C_{15}，Bicyclic sesquiterpane C_{15}，二环倍半萜
12	53.33	2.64	123	$C_{16}H_{30}$	222	C_{16}，Bicyclic sesquiterpane C_{16}，二环倍半萜
13	53.83	2.85	123	$C_{15}H_{28}$	208	C_{15}，Bicyclic sesquiterpane C_{15}，二环倍半萜

续表

峰 号	一维保留时间 /min	二维保留时间 /s	特征离子 (m/z)	分子式	相对分子质量	名 称
14	55.33	2.65	179	$C_{16}H_{30}$	222	C_{16}, Bicyclic sesquiterpane C_{16}, 二环倍半萜
15	56.33	2.76	193	$C_{16}H_{30}$	222	C_{16}, Bicyclic sesquiterpane C_{16}, 二环倍半萜
16	56.67	2.74	193	$C_{16}H_{30}$	222	C_{16}, Bicyclic sesquiterpane C_{16}, 二环倍半萜
17	57.00	2.79	123	$C_{16}H_{30}$	222	$8\beta(H)$-Homodrimane $8\beta(H)$-升补身烷
18	57.67	2.80	193	$C_{16}H_{30}$	222	C_{16}, Bicyclic sesquiterpane C_{16}, 二环倍半萜
19	58.00	2.80	193	$C_{16}H_{30}$	222	C_{16}, Bicyclic sesquiterpane C_{16}, 二环倍半萜
20	58.33	2.81	193	$C_{16}H_{30}$	222	C_{16}, Bicyclic sesquiterpane C_{16}, 二环倍半萜
21	58.67	2.83	123	$C_{16}H_{30}$	222	C_{16}, Bicyclic sesquiterpane C_{16}, 二环倍半萜

2.1.5 金刚烷系列的全二维气相色谱图及化合物名称

金刚烷系列化合物包括单金刚烷类化合物、双金刚烷类化合物、三金刚烷类化合物、四金刚烷类化合物等。每类金刚烷在全二维气相色谱图上都具有"瓦片效应"和族分离特征，且随着金刚烷环数的增加，在二维上的保留时间不断增加。它们与其他种类的化合物在全二维气相色谱图上位置的相对关系如图2-30所示，除了单金刚烷区域内有烷基取代的苯类化合物混入外，其他金刚烷类化合物的出峰区域相对独立，因此通过前处理的手段除去芳烃类化合物，再利用全二维气相色谱定性、定量分析金刚烷类化合物，得到的结果准确性会更高。

1) 单金刚烷化合物的全二维气相色谱图及化合物名称

单金刚烷类化合物的特征离子有 m/z 136, m/z 135, m/z 149, m/z 163, m/z 177 及 m/z 191, 同一相对分子质量的化合物在图谱上呈瓦片状规则排列。图2-31是单金刚烷系列化合物的全二维气相色谱-飞行时间质谱下的选择离子图谱，依据化合物在全二维谱图上的相对位置及质谱图(图2-32～图2-67)，除了内标峰IS，共鉴定了36个单金刚烷类化合物，鉴定结果见表2-6。图2-31(b)中的14和15号峰在Wei等(2006)看来是一个峰，鉴定为2,6-+2,4-二甲基金刚烷，在全二维点阵图中它们可以被分开，但是无法确定它们哪个是2,6-二甲基金刚烷，哪个是2,4-二甲基金刚烷。

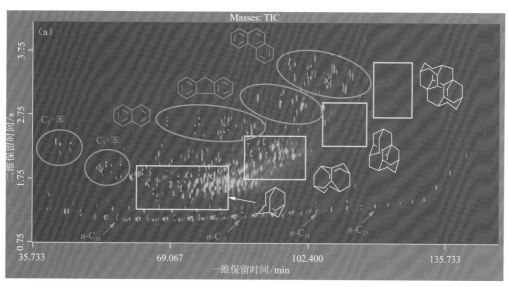

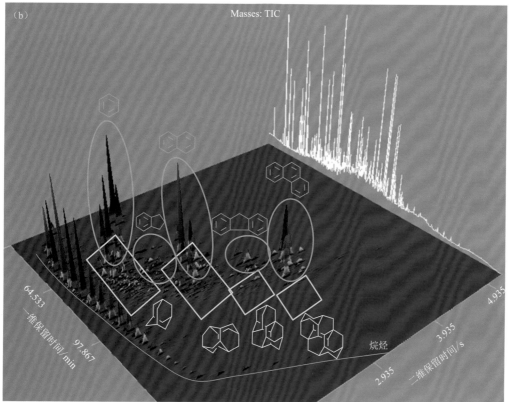

图 2-30　原油样品中金刚烷类化合物的谱图
（a）全二维点阵图；（b）三维立体图

全二维气相色谱及其石油地质应用

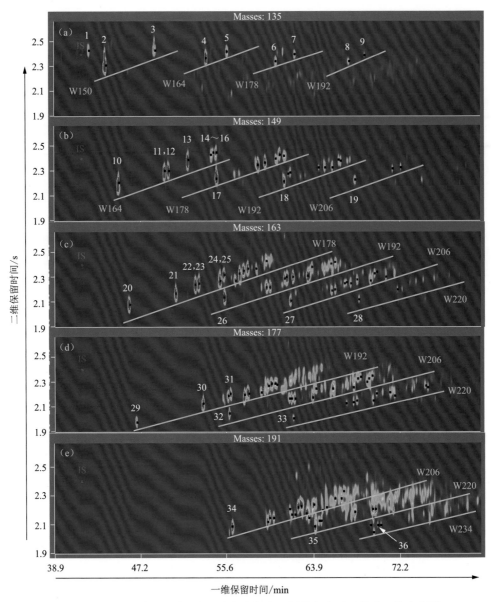

图 2-31 不同取代基的单金刚烷类化合物在不同特征离子下的全二维点阵图
IS 表示内标化合物，W 表示相对分子质量，如 W150 表示此类化合物的相对分子质量是 150

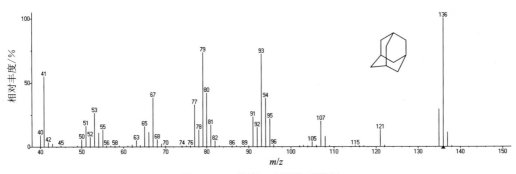

图 2-32 1 号峰 金刚烷质谱图

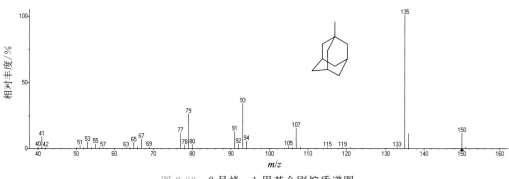

图 2-33　2 号峰　1-甲基金刚烷质谱图

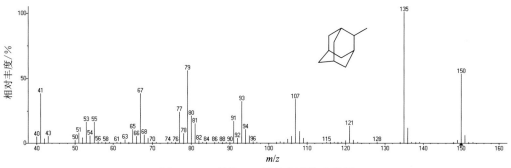

图 2-34　3 号峰　2-甲基金刚烷质谱图

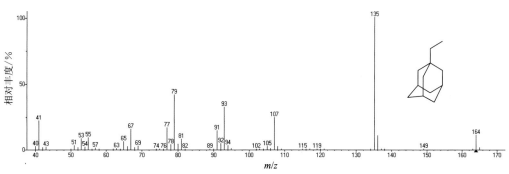

图 2-35　4 号峰　1-乙基金刚烷质谱图

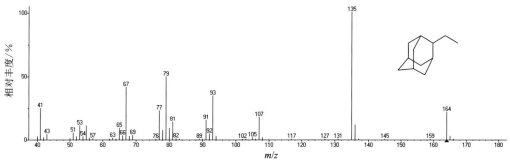

图 2-36　5 号峰　2-乙基金刚烷质谱图

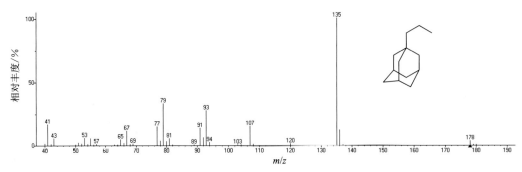

图 2-37　6 号峰　1-正丙基金刚烷质谱图

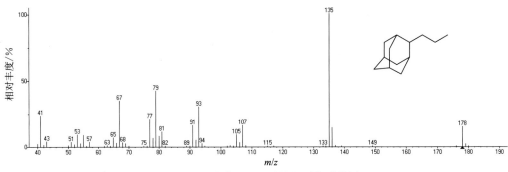

图 2-38　7 号峰　2-正丙基金刚烷质谱图

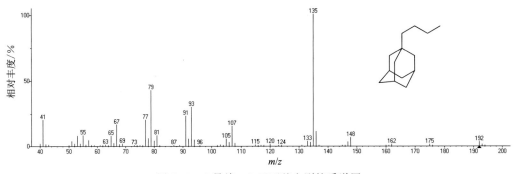

图 2-39　8 号峰　1-正丁基金刚烷质谱图

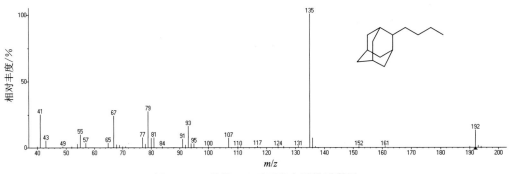

图 2-40　9 号峰　2-正丁基金刚烷质谱图

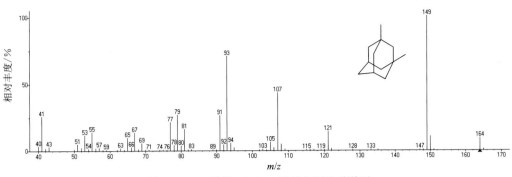

图 2-41 10 号峰 1,3-二甲基金刚烷质谱图

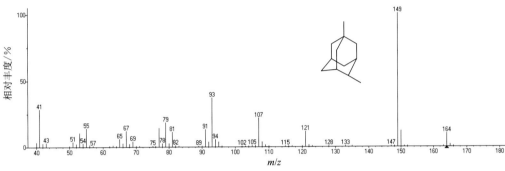

图 2-42 11 号峰 1,4-二甲基金刚烷(顺式)质谱图

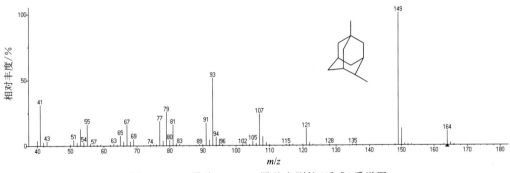

图 2-43 12 号峰 1,4-二甲基金刚烷(反式)质谱图

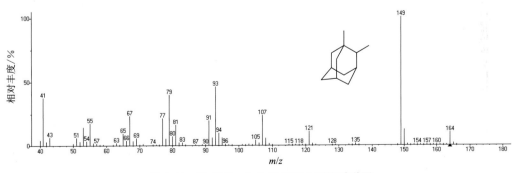

图 2-44 13 号峰 1,2-二甲基金刚烷质谱图

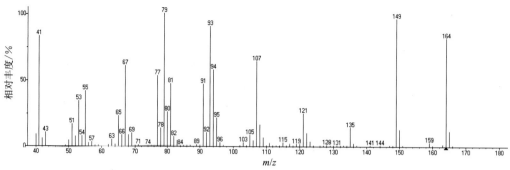

图 2-45　14 号峰　2,6-或 2,4-二甲基金刚烷质谱图

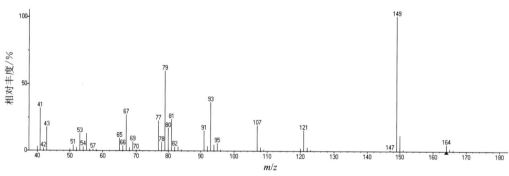

图 2-46　15 号峰　2,6-或 2,4-二甲基金刚烷质谱图

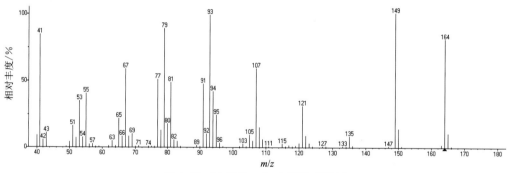

图 2-47　16 号峰　C_2-金刚烷质谱图

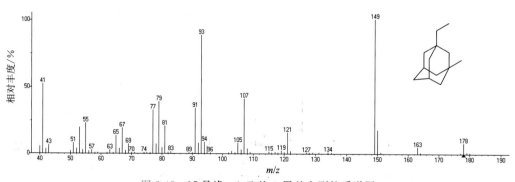

图 2-48　17 号峰　1-乙基-3-甲基金刚烷质谱图

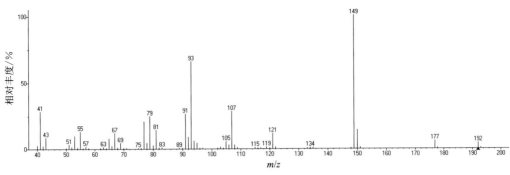

图 2-49　18 号峰　C_4-金刚烷质谱图

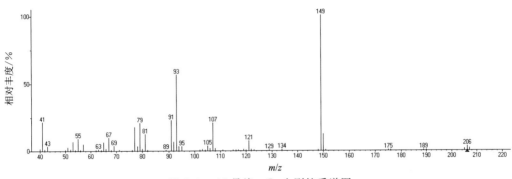

图 2-50　19 号峰　C_5-金刚烷质谱图

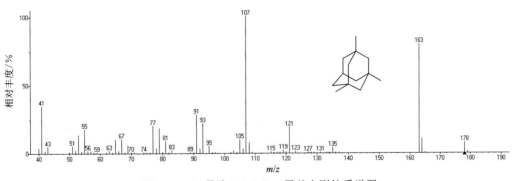

图 2-51　20 号峰　1,3,5-三甲基金刚烷质谱图

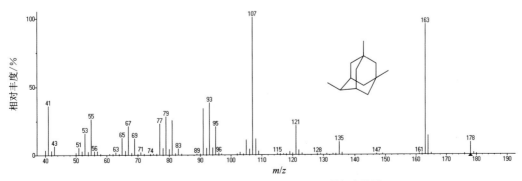

图 2-52　21 号峰　1,3,6-三甲基金刚烷质谱图

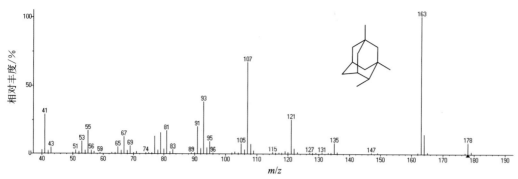

图 2-53　22 号峰　1,3,4-三甲基金刚烷(顺式)质谱图

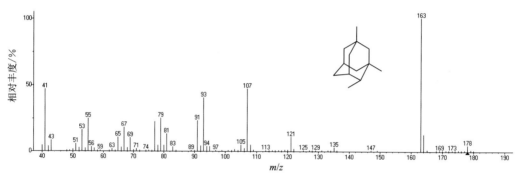

图 2-54　23 号峰　1,3,4-三甲基金刚烷(反式)质谱图

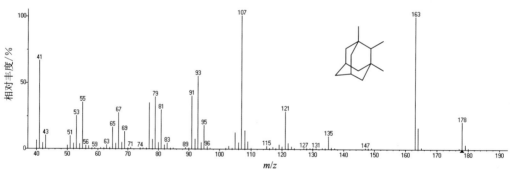

图 2-55　24 号峰　1,2,3-三甲基金刚烷质谱图

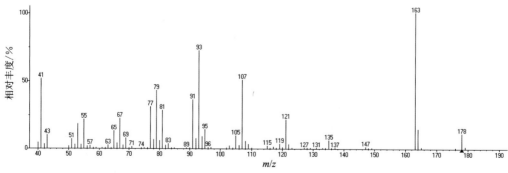

图 2-56　25 号峰　C_3-金刚烷质谱图

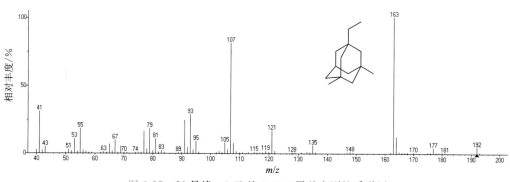

图 2-57　26 号峰　1-乙基-3,5-二甲基金刚烷质谱图

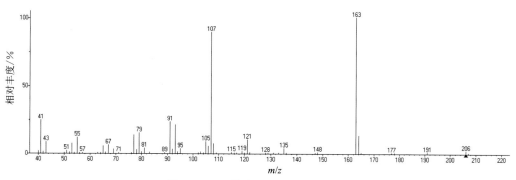

图 2-58　27 号峰　C_5-金刚烷质谱图

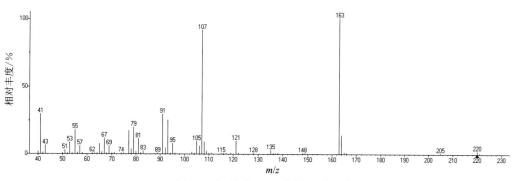

图 2-59　28 号峰　C_6-金刚烷质谱图

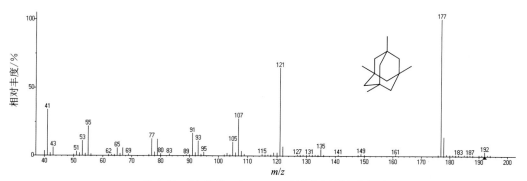

图 2-60　29 号峰　1,3,5,7-四甲基金刚烷质谱图

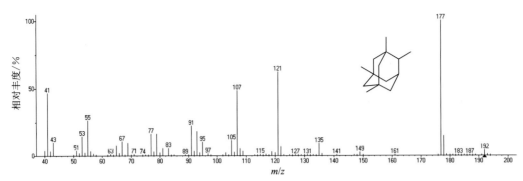

图 2-61　30 号峰　1,2,5,7-四甲基金刚烷质谱图

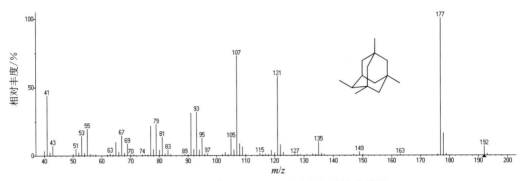

图 2-62　31 号峰　1,3,5,6-四甲基金刚烷质谱图

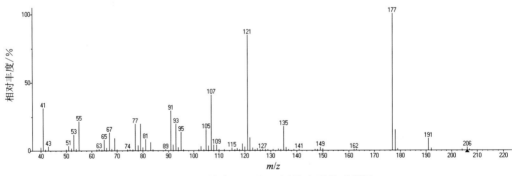

图 2-63　32 号峰　乙基-三甲基金刚烷质谱图

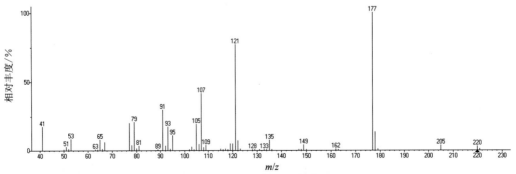

图 2-64　33 号峰　C_6-金刚烷质谱图

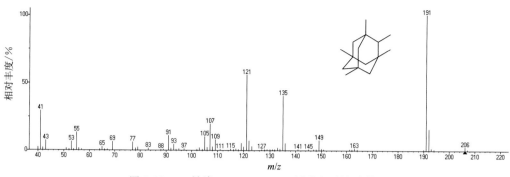

图 2-65 34 号峰 1,2,3,5,7-五甲基金刚烷质谱图

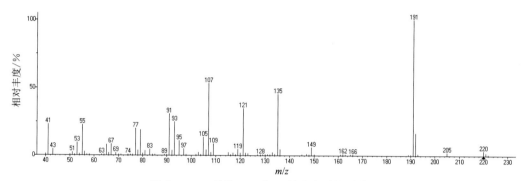

图 2-66 35 号峰 乙基-四甲基金刚烷质谱图

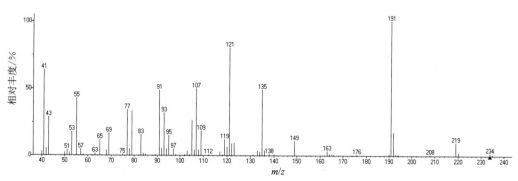

图 2-67 36 号峰 C_7-金刚烷质谱图

表 2-6 单金刚烷系列化合物鉴定结果表

峰号	一维保留时间 /min	二维保留时间 /s	特征离子 (m/z)	分子式	相对分子质量	名 称
IS	41.50	2.39	152	$C_{10}D_{16}$	152	D_{16}-Adamantane D_{16}-金刚烷
1	42.30	2.42	136	$C_{10}H_{16}$	136	Adamantane 金刚烷
2	43.90	2.30	135	$C_{11}H_{18}$	150	1-Methyladamantane 1-甲基金刚烷

续表

峰 号	一维保留时间 /min	二维保留时间 /s	特征离子 (m/z)	分子式	相对分子质量	名 称
3	48.70	2.42	135	$C_{11}H_{18}$	150	2-Methyladamantane 2-甲基金刚烷
4	53.60	2.36	135	$C_{12}H_{20}$	164	1-Ethyladamantane 1-乙基金刚烷
5	55.60	2.41	135	$C_{12}H_{20}$	164	2-Ethyladamantane 2-乙基金刚烷
6	60.30	2.33	135	$C_{13}H_{22}$	178	1-n-Propyladamantane 1-正丙基金刚烷
7	62.10	2.39	135	$C_{13}H_{22}$	178	2-n-Propyladamantane 2-正丙基金刚烷
8	67.30	2.325	135	$C_{14}H_{24}$	192	1-n-Butyladamantane 1-正丁基金刚烷
9	68.80	2.38	135	$C_{14}H_{24}$	192	2-n-Butyladamantane 2-正丁基金刚烷
10	45.00	2.17	149	$C_{12}H_{20}$	164	1,3-Dimethyladamantane 1,3-二甲基金刚烷
11	49.60	2.29	149	$C_{12}H_{20}$	164	1,4-Dimethyladamantane(cis) 1,4-二甲基金刚烷(顺式)
12	50.00	2.29	149	$C_{12}H_{20}$	164	1,4-Dimethyladamantane(trans) 1,4-二甲基金刚烷(反式)
13	51.90	2.38	149	$C_{12}H_{20}$	164	1,2-Dimethyladamantane 1,2-二甲基金刚烷
14	54.10	2.42	149	$C_{12}H_{20}$	164	2,6- or 2,4-Dimethyladamantane 2,6-或 2,4-二甲基金刚烷
15	54.30	2.44	149	$C_{12}H_{20}$	164	2,6- or 2,4-Dimethyladamantane 2,6-或 2,4-二甲基金刚烷
16	54.60	2.44	149	$C_{12}H_{20}$	164	C_2-Adamantane C_2-金刚烷
17	54.60	2.23	149	$C_{13}H_{22}$	178	1-Ethyl-3-methyladamantane 1-乙基-3-甲基金刚烷
18	61.10	2.21	149	$C_{14}H_{24}$	192	C_4-Adamantane C_4-金刚烷
19	67.90	2.22	149	$C_{15}H_{26}$	206	C_5-Adamantane C_5-金刚烷
20	46.10	2.07	163	$C_{13}H_{22}$	178	1,3,5-Trimethyladamantane 1,3,5-三甲基金刚烷
21	50.60	2.16	163	$C_{13}H_{22}$	178	1,3,6-Trimethyladamantane 1,3,6-三甲基金刚烷
22	52.50	2.23	163	$C_{13}H_{22}$	178	1,3,4-Trimethyladamantane(cis) 1,3,4-三甲基金刚烷(顺式)

续表

峰 号	一维保留时间 /min	二维保留时间 /s	特征离子 (m/z)	分子式	相对分子质量	名 称
23	52.90	2.24	163	$C_{13}H_{22}$	178	1,3,4-Trimethyladamantane(trans) 1,3,4-三甲基金刚烷(反式)
24	54.90	2.26	163	$C_{13}H_{22}$	178	1,2,3-Trimethyladamantane 1,2,3-三甲基金刚烷
25	55.30	2.31	163	$C_{13}H_{22}$	178	C_3-Adamantane C_3-金刚烷
26	55.30	2.13	163	$C_{14}H_{24}$	192	1-Ethyl-3,5-dimethyladamantane 1-乙基-3,5-二甲基金刚烷
27	61.70	2.1	163	$C_{15}H_{26}$	206	C_5-Adamantane C_5-金刚烷
28	68.20	2.11	163	$C_{16}H_{28}$	220	C_6-Adamantane C_6-金刚烷
29	46.80	1.97	177	$C_{14}H_{24}$	192	1,3,5,7-Tetramethyladamantane 1,3,5,7-四甲基金刚烷
30	53.30	2.12	177	$C_{14}H_{24}$	192	1,2,5,7-Tetramethyladamantane 1,2,5,7-四甲基金刚烷
31	55.90	2.20	177	$C_{14}H_{24}$	192	1,3,5,6-Tetramethyladamantane 1,3,5,6-四甲基金刚烷
32	55.80	2.04	177	$C_{15}H_{26}$	206	Ethyl-trimethyladamantane 乙基-三甲基金刚烷
33	62.00	2.00	177	$C_{16}H_{28}$	220	C_6-Adamantane C_6-金刚烷
34	56.10	2.08	191	$C_{15}H_{26}$	206	1,2,3,5,7-Pentamethyladamantane 1,2,3,5,7-五甲基金刚烷
35	64.00	2.08	191	$C_{16}H_{28}$	220	Ethyl-tetramethyladamantane 乙基-四甲基金刚烷
36	69.60	2.04	191	$C_{17}H_{30}$	234	C_7-Adamantane C_7-金刚烷

2) 双金刚烷系列化合物的全二维气相色谱图及化合物名称

双金刚烷系列化合物在全二维气相色谱图上位于正构烷烃 n-C_{15}～n-C_{19} 的上方,单金刚烷的右上方,如图 2-30 和图 2-68 所示。双金刚烷的特征离子有 m/z 188,m/z 187,m/z 201,m/z 215 及 m/z 229。依据双金刚烷化合物的全二维分布特征、相关参考文献(Wei Zhibin 等,2007)及质谱图(图 2-69～图 2-86)解析,共鉴定出双金刚烷化合物 18 个,鉴定结果见表 2-7。

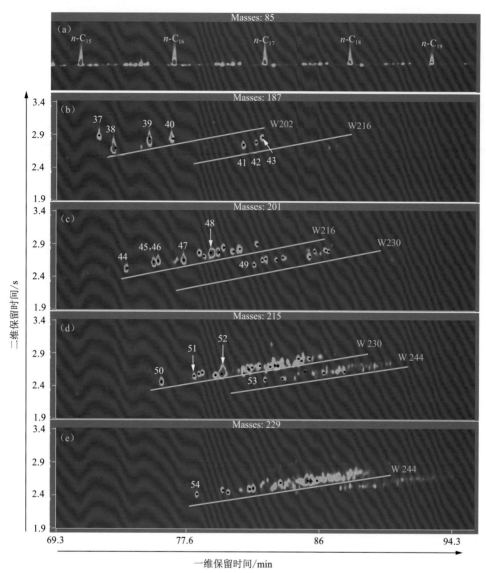

图 2-68　正构烷烃（a）和双金刚烷类化合物（b～e）在不同特征离子下的全二维点阵图
W 表示相对分子质量

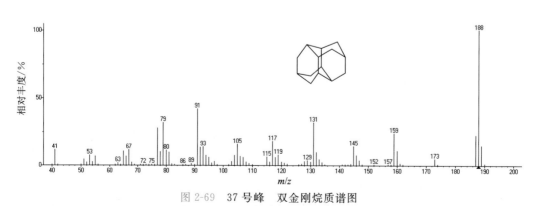

图 2-69　37 号峰　双金刚烷质谱图

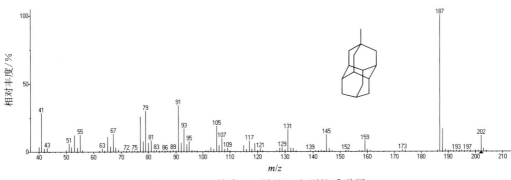

图 2-70　38 号峰　4-甲基双金刚烷质谱图

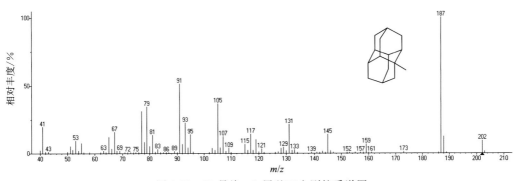

图 2-71　39 号峰　1-甲基双金刚烷质谱图

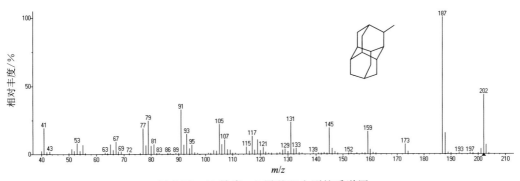

图 2-72　40 号峰　3-甲基双金刚烷质谱图

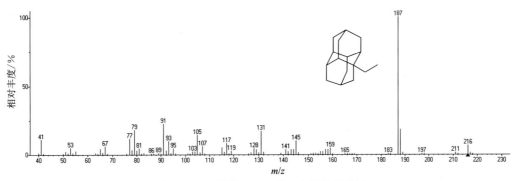

图 2-73　41 号峰　1-乙基双金刚烷质谱图

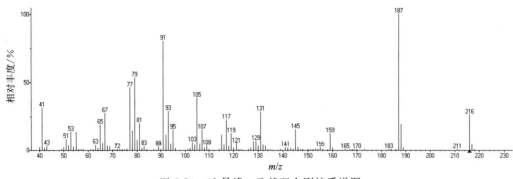

图 2-74　42 号峰　乙基双金刚烷质谱图

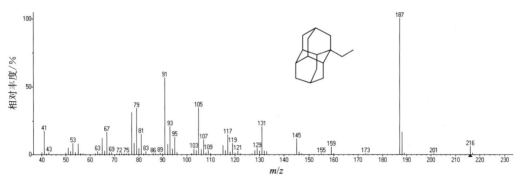

图 2-75　43 号峰　2-乙基双金刚烷质谱图

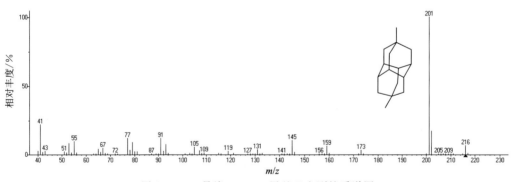

图 2-76　44 号峰　4,9-二甲基双金刚烷质谱图

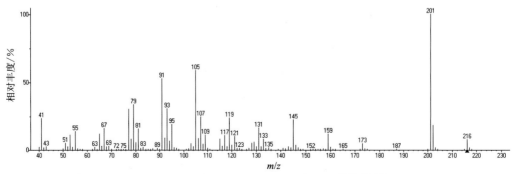

图 2-77　45 号峰　1,2-＋2,4-二甲基双金刚烷质谱图

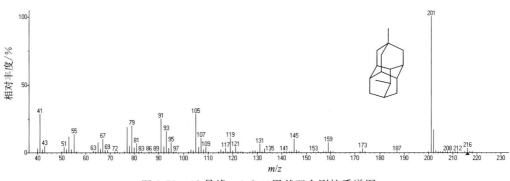

图 2-78　46 号峰　4,8-二甲基双金刚烷质谱图

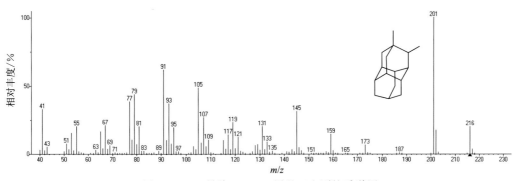

图 2-79　47 号峰　3,4-二甲基双金刚烷质谱图

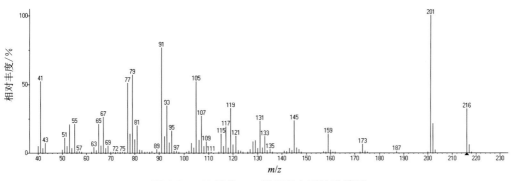

图 2-80　48 号峰　二甲基双金刚烷质谱图

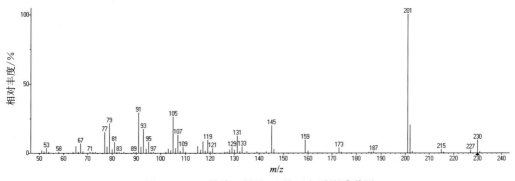

图 2-81　49 号峰　甲基-乙基双金刚烷质谱图

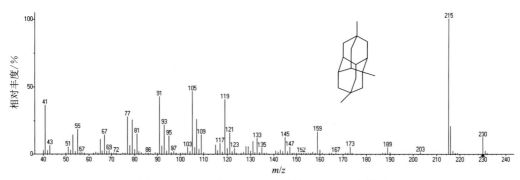

图 2-82　50 号峰　1,4,9-三甲基双金刚烷质谱图

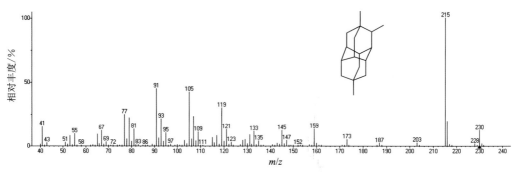

图 2-83　51 号峰　3,4,9-三甲基双金刚烷质谱图

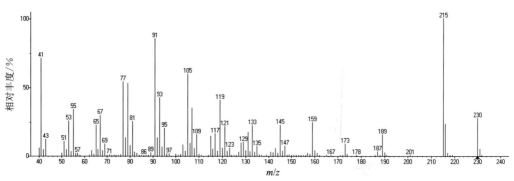

图 2-84　52 号峰　三甲基双金刚烷质谱图

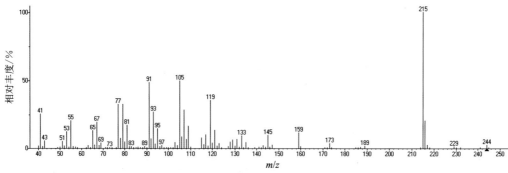

图 2-85　53 号峰　乙基-二甲基双金刚烷质谱图

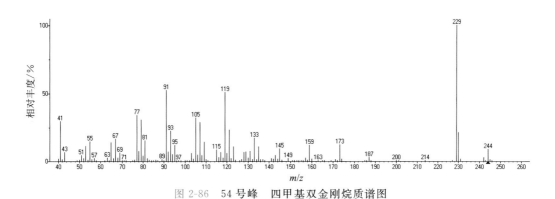

图 2-86　54 号峰　四甲基双金刚烷质谱图

表 2-7　双金刚烷系列化合物鉴定结果表

峰　号	一维保留时间 /min	二维保留时间 /s	特征离子 （m/z）	分子式	相对分子质量	名　　称
37	72.00	2.93	188	$C_{14}H_{20}$	188	Diamantane 双金刚烷
38	73.00	2.74	187	$C_{15}H_{22}$	202	4-Methyldiamantane 4-甲基双金刚烷
39	75.20	2.88	187	$C_{15}H_{22}$	202	1-Methyldiamantane 1-甲基双金刚烷
40	76.70	2.88	187	$C_{15}H_{22}$	202	3-Methyldiamantane 3-甲基双金刚烷
41	81.10	2.79	187	$C_{16}H_{24}$	216	1-Ethyldiamantane 1-乙基双金刚烷
42	81.90	2.83	187	$C_{16}H_{24}$	216	Ethyldiamantane 乙基双金刚烷
43	82.30	2.9	187	$C_{16}H_{24}$	216	2-Ethyldiamantane 2-乙基双金刚烷
44	73.80	2.57	201	$C_{16}H_{24}$	216	4,9-Dimethyldiamantane 4,9-二甲基双金刚烷
45	75.50	2.67	201	$C_{16}H_{24}$	216	1,2-＋2,4-Dimethyldiamantane 1,2-＋2,4-二甲基双金刚烷*
46	75.80	2.69	201	$C_{16}H_{24}$	216	4,8-Dimethyldiamantane 4,8-二甲基双金刚烷
47	77.40	2.70	201	$C_{16}H_{24}$	216	3,4-Dimethyldiamantane 3,4-二甲基双金刚烷
48	79.10	2.81	201	$C_{16}H_{24}$	216	Dimethyldiamantane 二甲基双金刚烷
49	81.80	2.63	201	$C_{17}H_{26}$	230	Methyl-ethyldiamantane 甲基-乙基双金刚烷
50	76.00	2.51	215	$C_{17}H_{26}$	230	1,4,9-Trimethyldiamantane 1,4,9-三甲基双金刚烷
51	78.10	2.60	215	$C_{17}H_{26}$	230	3,4,9-Trimethyldiamantane 3,4,9-三甲基双金刚烷

续表

峰号	一维保留时间/min	二维保留时间/s	特征离子(m/z)	分子式	相对分子质量	名称
52	79.80	2.63	215	$C_{17}H_{26}$	230	Trimethyldiamantane 三甲基双金刚烷
53	82.50	2.54	215	$C_{18}H_{28}$	244	Ethyl-dimethyldiamantane 乙基-二甲基双金刚烷
54	78.30	2.46	229	$C_{18}H_{28}$	244	Tetramethyldiamantane 四甲基双金刚烷

注：表中 45 号化合物的定性结果引自 Wei 等(2006)文章，但王培荣(1993)的书中鉴定的化合物为 1,4-＋2,4-二甲基双金刚烷。

3) 三金刚烷系列化合物的全二维气相色谱图及化合物名称

三金刚烷系列化合物在全二维气相色谱图上位于正构烷烃 n-C_{19}～n-C_{22} 的上方，双金刚烷的右上方，如图 2-30 和图 2-87 所示。三金刚烷的特征离子有 m/z 240，m/z 239，m/z 253 及 m/z 267。依据三金刚烷化合物的全二维分布特征、相关参考文献及质谱图(图 2-88～图 2-99)解析，共鉴定出双金刚烷化合物 13 个，峰号从 55 开始至 67，鉴定结果见表 2-8。

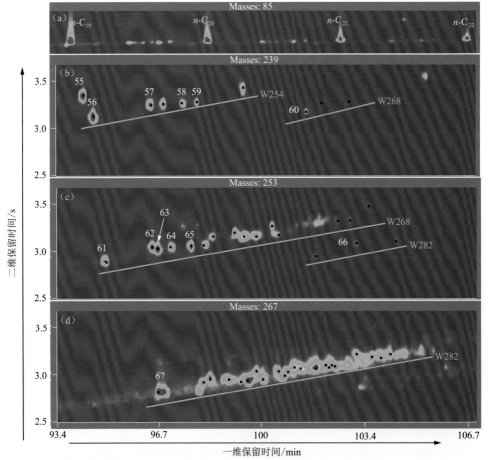

图 2-87　正构烷烃(a)和三金刚烷类化合物(b～d)在不同特征离子下的全二维点阵图

W 表示相对分子质量

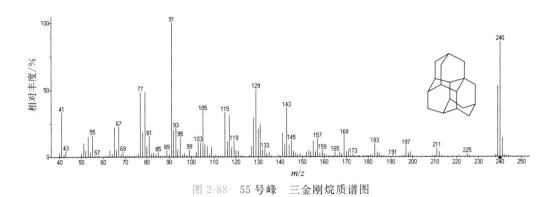

图 2-88　55 号峰　三金刚烷质谱图

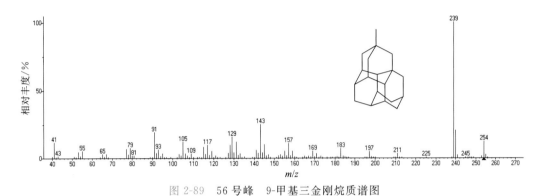

图 2-89　56 号峰　9-甲基三金刚烷质谱图

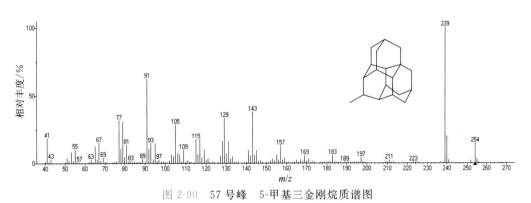

图 2-90　57 号峰　5-甲基三金刚烷质谱图

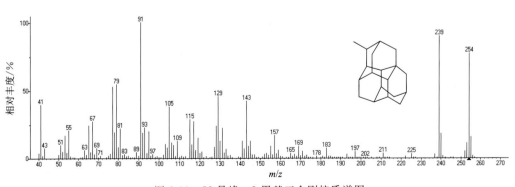

图 2-91　58 号峰　8-甲基三金刚烷质谱图

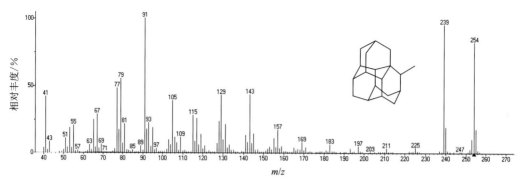

图 2-92 59 号峰 16-甲基三金刚烷质谱图

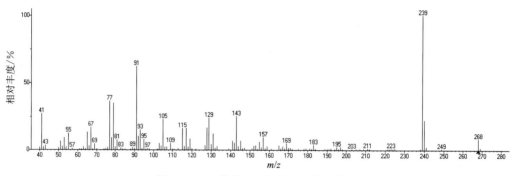

图 2-93 60 号峰 乙基三金刚烷质谱图

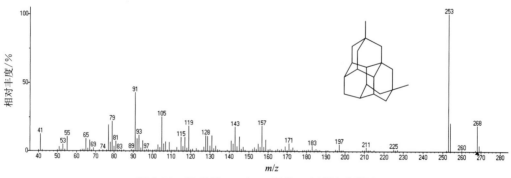

图 2-94 61 号峰 9,15-二甲基三金刚烷质谱图

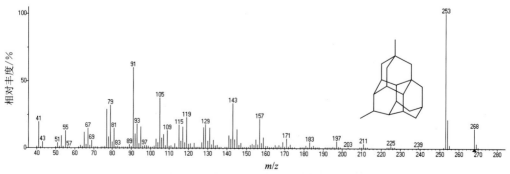

图 2-95 62 号峰 5,9-二甲基三金刚烷质谱图

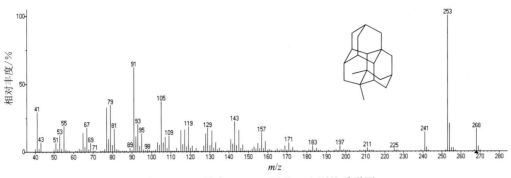

图 2-96　63 号峰　3,4-二甲基三金刚烷质谱图

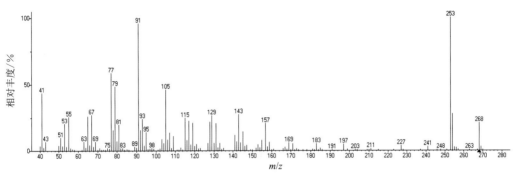

图 2-97　64 号峰　二甲基三金刚烷质谱图

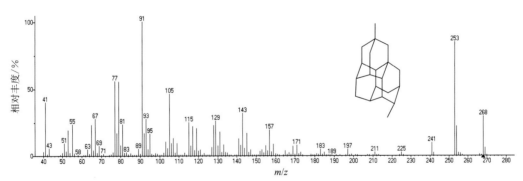

图 2-98　65 号峰　9,14-二甲基三金刚烷质谱图

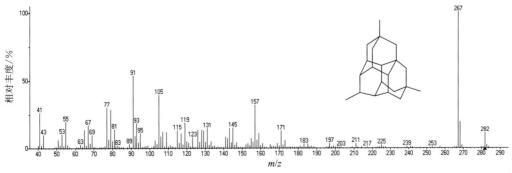

图 2-99　67 号峰　5,9,15-三甲基三金刚烷质谱图

表 2-8　三金刚烷系列化合物鉴定结果表

峰 号	一维保留时间 /min	二维保留时间 /s	特征离子 (m/z)	分子式	相对分子质量	名　　称
55	94.30	3.36	240	$C_{18}H_{24}$	240	Triamantane 三金刚烷
56	94.60	3.14	239	$C_{19}H_{26}$	254	9-Methyltriamantane 9-甲基三金刚烷
57	96.50	3.26	239	$C_{19}H_{26}$	254	5-Methyltriamantane 5-甲基三金刚烷
58	97.50	3.28	239	$C_{19}H_{26}$	254	8-Methyltriamantane 8-甲基三金刚烷
59	98.00	3.29	239	$C_{19}H_{26}$	254	16-Methyltriamantane 16-甲基三金刚烷
60	101.50	3.19	239	$C_{20}H_{28}$	268	Ethyltriamantane 乙基三金刚烷
61	95.00	2.90	253	$C_{20}H_{28}$	268	9,15-Dimethyltriamantane 9,15-二甲基三金刚烷
62	96.50	3.06	253	$C_{20}H_{28}$	268	5,9-Dimethyltriamantane 5,9-二甲基三金刚烷
63	96.70	3.04	253	$C_{20}H_{28}$	268	3,4-Dimethyltriamantane 3,4-二甲基三金刚烷
64	97.10	3.06	253	$C_{20}H_{28}$	268	Dimethyltriamantane 二甲基三金刚烷
65	97.80	3.07	253	$C_{20}H_{28}$	268	9,14-Dimethyltriamantane 9,14-二甲基三金刚烷
66	103.10	3.10	253	$C_{21}H_{30}$	282	Ethyl-methyltriamantane 乙基-甲基三金刚烷
67	96.70	2.83	267	$C_{21}H_{30}$	282	5,9,15-Trimethyltriamantane 5,9,15-三甲基三金刚烷

4）四金刚烷系列化合物的全二维气相色谱图及化合物名称

　　四金刚烷系列化合物在含气页岩抽提物及部分凝析油中常被检测到。四金刚烷系列化合物在全二维气相色谱图上位于正构烷烃 $n\text{-}C_{23} \sim n\text{-}C_{26}$ 的上方，如图 2-30 和图 2-100 所示。四金刚烷的特征离子有 $m/z\ 292$，$m/z\ 291$，$m/z\ 305$ 及 $m/z\ 319$。依据四金刚烷化合物的全二维分布特征、相关参考文献及质谱图（图 2-101～图 2-120）解析，共鉴定出四金刚烷化合物 20 个，峰号从 68 开始至 87，鉴定结果见表 2-9。其中一些组成相同、取代基位置不同、构型不同的峰，由于无法判定是何种构型或者准确的取代基位置，无法给出具体的命名。

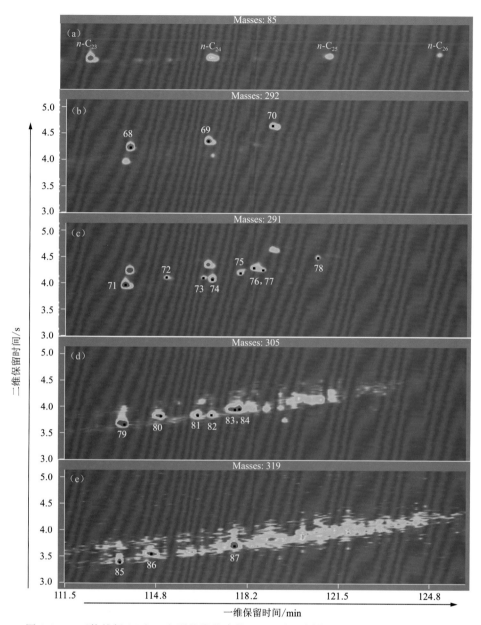

图 2-100　正构烷烃(a)和四金刚烷类化合物(b~e)在不同特征离子下的全二维点阵图

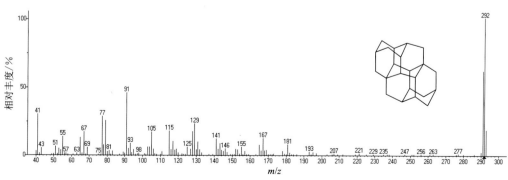

图 2-101　68号峰　四金刚烷质谱图

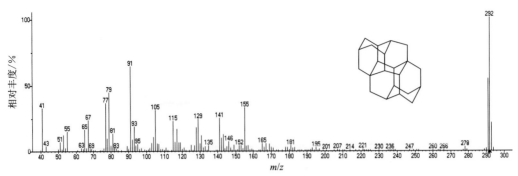

图 2-102　69 号峰　四金刚烷质谱图

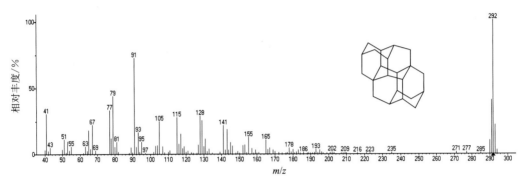

图 2-103　70 号峰　四金刚烷质谱图

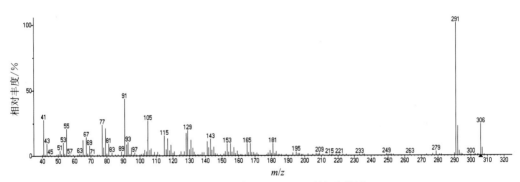

图 2-104　71 号峰　甲基四金刚烷质谱图

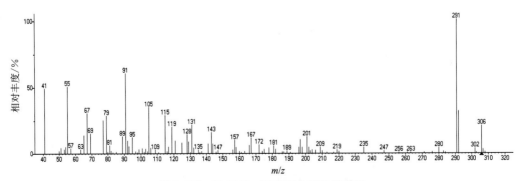

图 2-105　72 号峰　甲基四金刚烷质谱图

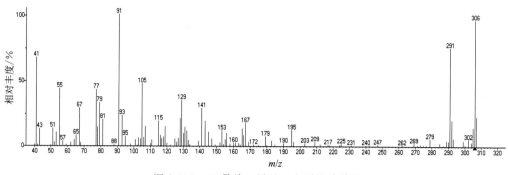

图 2-106　73 号峰　甲基四金刚烷质谱图

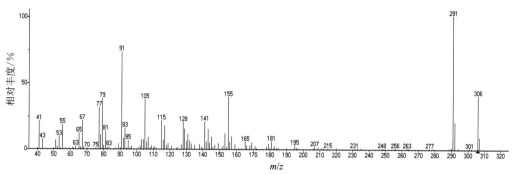

图 2-107　74 号峰　甲基四金刚烷质谱图

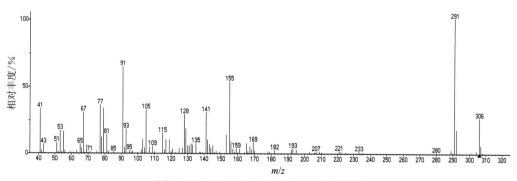

图 2-108　75 号峰　甲基四金刚烷质谱图

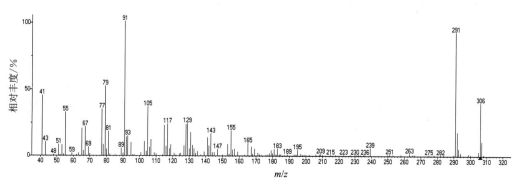

图 2-109　76 号峰　甲基四金刚烷质谱图

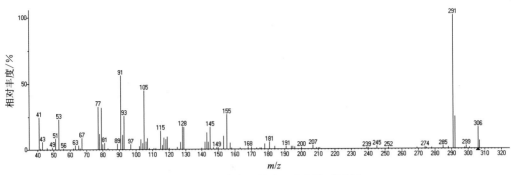

图 2-110　77 号峰　甲基四金刚烷质谱图

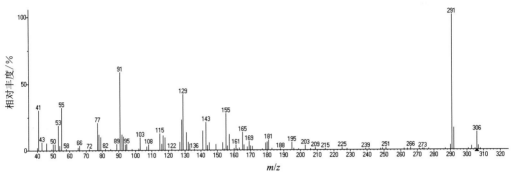

图 2-111　78 号峰　甲基四金刚烷质谱图

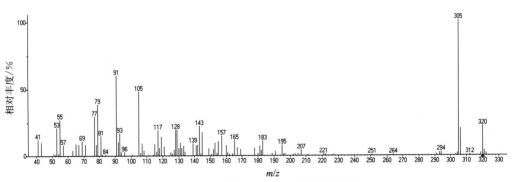

图 2-112　79 号峰　C_2-四金刚烷质谱图

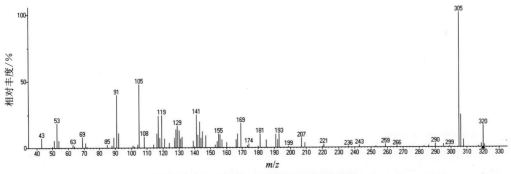

图 2-113　80 号峰　C_2-四金刚烷质谱图

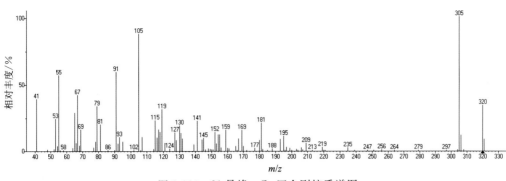

图 2-114　81 号峰　C_2-四金刚烷质谱图

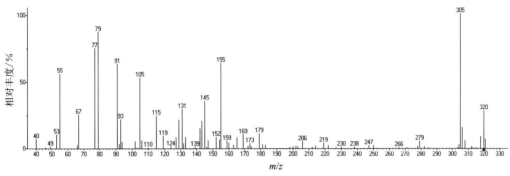

图 2-115　82 号峰　C_2-四金刚烷质谱图

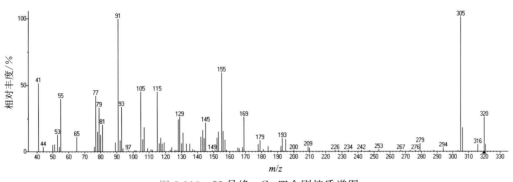

图 2-116　83 号峰　C_2-四金刚烷质谱图

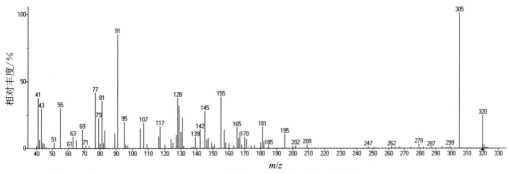

图 2-117　84 号峰　C_2-四金刚烷质谱图

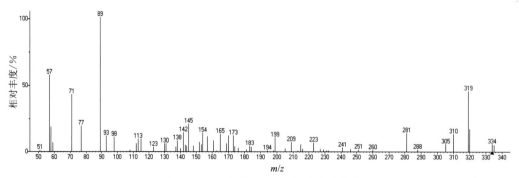

图 2-118　85 号峰　C_3-四金刚烷质谱图

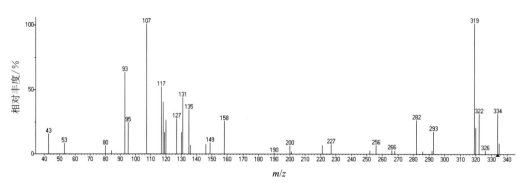

图 2-119　86 号峰　C_3-四金刚烷质谱图

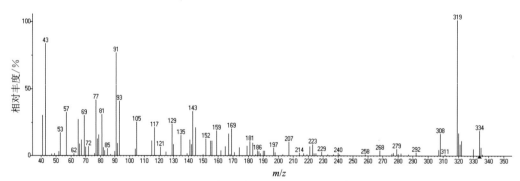

图 2-120　87 号峰　C_3-四金刚烷质谱图

表 2-9　四金刚烷鉴定结果表

峰号	一维保留时间 /min	二维保留时间 /s	特征离子 (m/z)	分子式	相对分子质量	名　称
68	113.83	4.30	292	$C_{22}H_{28}$	292	Tetramantane 四金刚烷
69	116.67	4.42	292	$C_{22}H_{28}$	292	Tetramantane 四金刚烷
70	119.00	4.71	292	$C_{22}H_{28}$	292	Tetramantane 四金刚烷
71	113.67	4.03	291	$C_{23}H_{30}$	306	Methyl-tetramantane 甲基四金刚烷

续表

峰 号	一维保留时间 /min	二维保留时间 /s	特征离子 (m/z)	分子式	相对分子质量	名 称
72	115.17	4.18	291	$C_{23}H_{30}$	306	Methyl-tetramantane 甲基四金刚烷
73	116.50	4.17	306	$C_{23}H_{30}$	306	Methyl-tetramantane 甲基四金刚烷
74	116.83	4.14	291	$C_{23}H_{30}$	306	Methyl-tetramantane 甲基四金刚烷
75	117.83	4.26	291	$C_{23}H_{30}$	306	Methyl-tetramantane 甲基四金刚烷
76	118.33	4.36	291	$C_{23}H_{30}$	306	Methyl-tetramantane 甲基四金刚烷
77	118.67	4.32	291	$C_{23}H_{30}$	306	Methyl-tetramantane 甲基四金刚烷
78	120.67	4.55	291	$C_{23}H_{30}$	306	Methyl-tetramantane 甲基四金刚烷
79	113.67	3.73	305	$C_{24}H_{32}$	320	C_2-Tetramantane C_2-四金刚烷
80	115.00	3.89	305	$C_{24}H_{32}$	320	C_2-Tetramantane C_2-四金刚烷
81	116.33	3.90	305	$C_{24}H_{32}$	320	C_2-Tetramantane C_2-四金刚烷
82	116.83	3.91	305	$C_{24}H_{32}$	320	C_2-Tetramantane C_2-四金刚烷
83	117.67	4.01	305	$C_{24}H_{32}$	320	C_2-Tetramantane C_2-四金刚烷
84	117.83	4.02	305	$C_{24}H_{32}$	320	C_2-Tetramantane C_2-四金刚烷
85	113.50	3.47	319	$C_{25}H_{34}$	334	C_3-Tetramantane C_3-四金刚烷
86	114.67	3.62	319	$C_{25}H_{34}$	334	C_3-Tetramantane C_3-四金刚烷
87	117.67	3.77	319	$C_{25}H_{34}$	334	C_3-Tetramantane C_3-四金刚烷

5) 硫代单金刚烷系列化合物的全二维气相色谱图及化合物名称

硫代单金刚烷系列化合物在发生了特定次生蚀变(如硫酸盐的热还原反应等)的原油中可以检测到,它是研究原油次生蚀变的有效指标。硫代单金刚烷系列化合物极性相对较高,在全二维气相色谱图上位于单金刚烷系列化合物的上方、萘系列化合物的下方这一中间区域,具体位置关系如图2-121所示。由于硫原子的存在,硫代单金刚烷的特征离子都是偶数,分别是 m/z 168, m/z 182, m/z 196 及 m/z 210,如图2-122所示。依据硫代单金刚烷化合物的全二维分布特征、相关参考文献及质谱图(图2-123~图2-153)解析,共鉴定出硫代单金刚烷化合物31个,鉴定结果见表2-10。由于精确的结构构型、取代基的取代位置等无法确定,又没有相关参考文献,很多同分异构体无法给出具体名字。

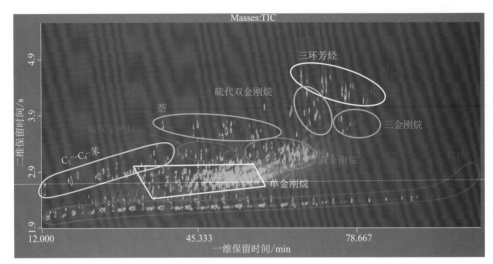

图 2-121　硫代单金刚烷化合物在不同特征离子下的全二维点阵图

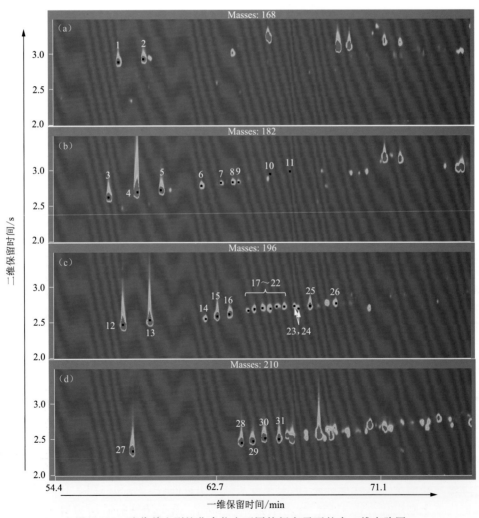

图 2-122　硫代单金刚烷化合物在不同特征离子下的全二维点阵图

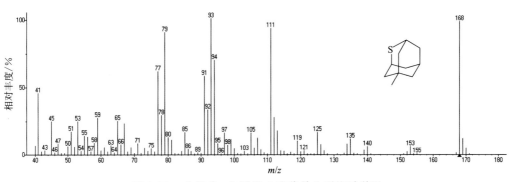

图 2-123 1 号峰 5-甲基-2-硫代单金刚烷质谱图

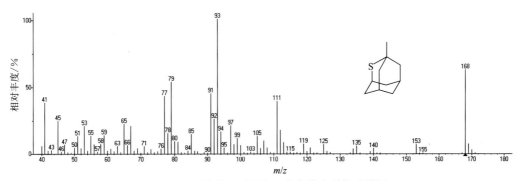

图 2-124 2 号峰 1-甲基-2-硫代单金刚烷质谱图

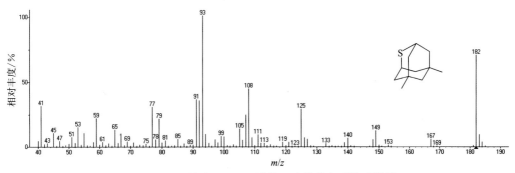

图 2-125 3 号峰 5,7-二甲基-2-硫代单金刚烷质谱图

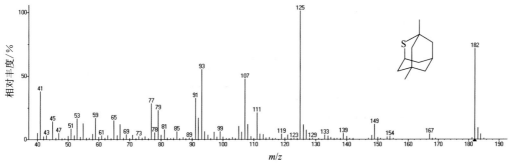

图 2-126 4 号峰 1,5-二甲基-2-硫代单金刚烷质谱图

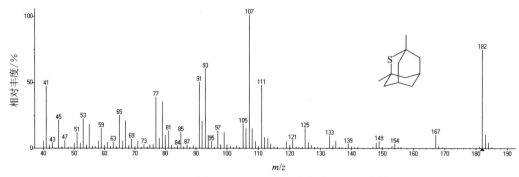

图 2-127　5 号峰　1,3-二甲基-2-硫代单金刚烷质谱图

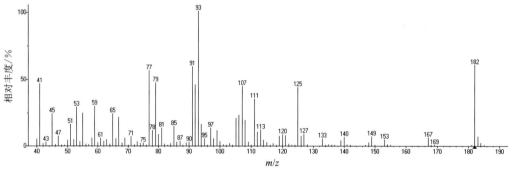

图 2-128　6 号峰　C_2-2-硫代单金刚烷质谱图

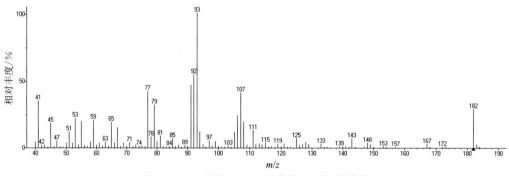

图 2-129　7 号峰　C_2-2-硫代单金刚烷质谱图

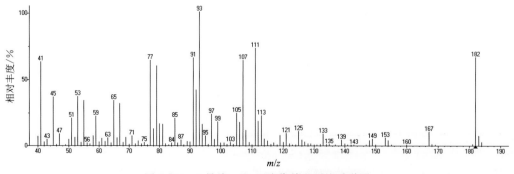

图 2-130　8 号峰　C_2-2-硫代单金刚烷质谱图

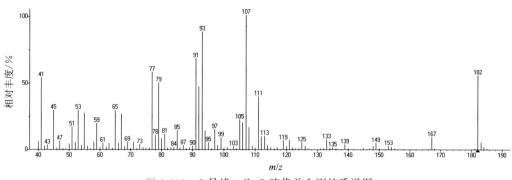

图 2-131　9 号峰　C_2-2-硫代单金刚烷质谱图

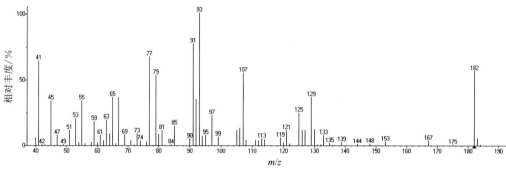

图 2-132　10 号峰　C_2-2-硫代单金刚烷质谱图

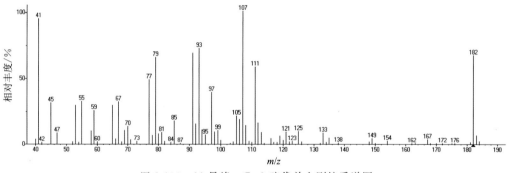

图 2-133　11 号峰　C_2-2-硫代单金刚烷质谱图

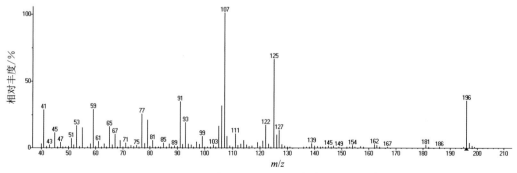

图 2-134　12 号峰　三甲基-2-硫代单金刚烷质谱图

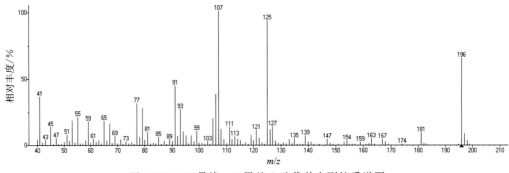

图 2-135　13 号峰　三甲基-2-硫代单金刚烷质谱图

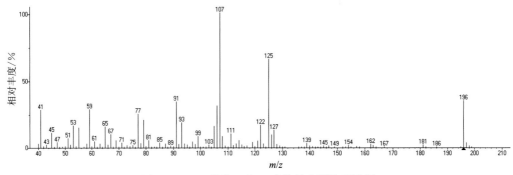

图 2-136　14 号峰　C_3-2-硫代单金刚烷质谱图

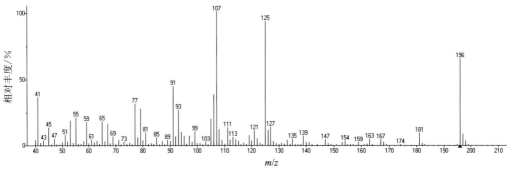

图 2-137　15 号峰　C_3-2-硫代单金刚烷质谱图

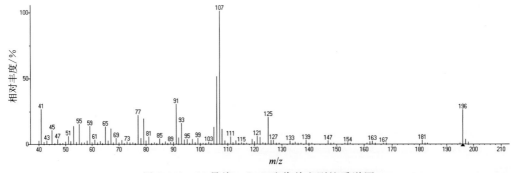

图 2-138　16 号峰　C_3-2-硫代单金刚烷质谱图

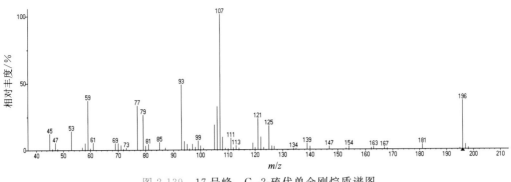

图 2-139　17 号峰　C_3-2-硫代单金刚烷质谱图

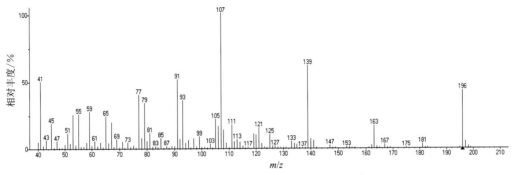

图 2-140　18 号峰　C_3-2-硫代单金刚烷质谱图

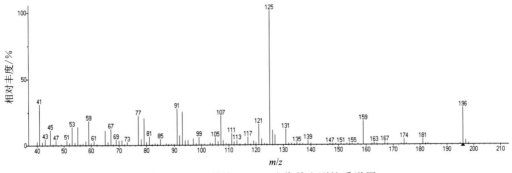

图 2-141　19 号峰　C_3-2-硫代单金刚烷质谱图

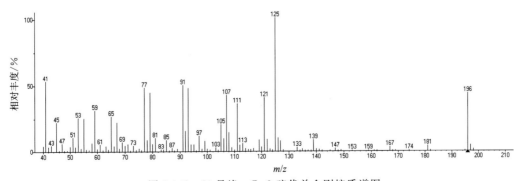

图 2-142　20 号峰　C_3-2-硫代单金刚烷质谱图

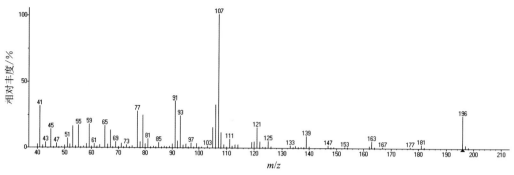

图 2-143　21 号峰　C_3-2-硫代单金刚烷质谱图

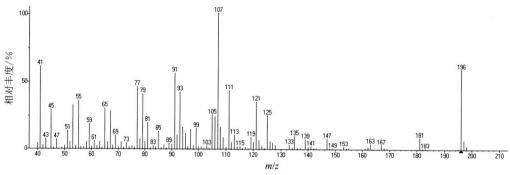

图 2-144　22 号峰　C_3-2-硫代单金刚烷质谱图

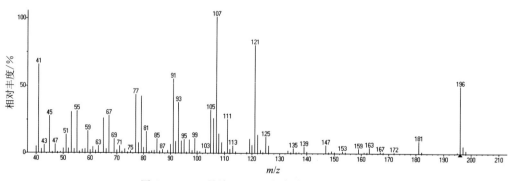

图 2-145　23 号峰　C_3-2-硫代单金刚烷质谱图

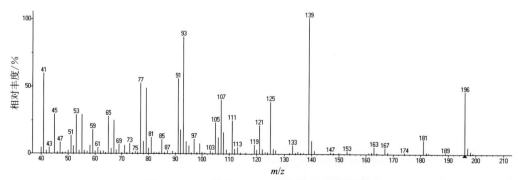

图 2-146　24 号峰　C_3-2-硫代单金刚烷质谱图

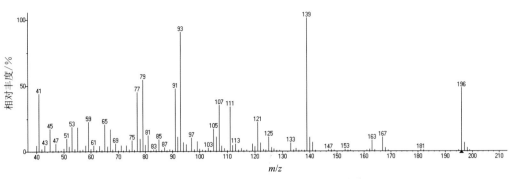

图 2-147 25 号峰 C_3-2-硫代单金刚烷质谱图

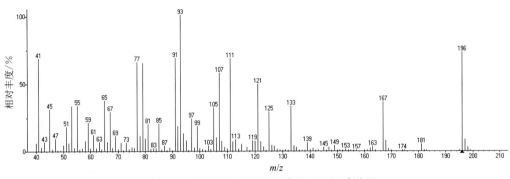

图 2-148 26 号峰 C_3-2-硫代单金刚烷质谱图

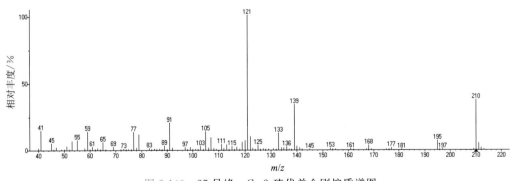

图 2-149 27 号峰 C_4-2-硫代单金刚烷质谱图

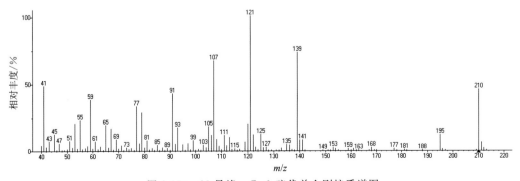

图 2-150 28 号峰 C_4-2-硫代单金刚烷质谱图

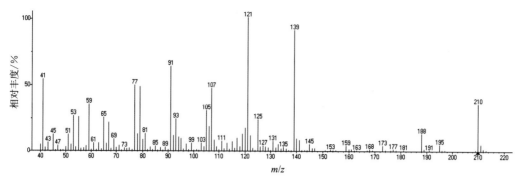

图 2-151　29 号峰　C_4-2-硫代单金刚烷质谱图

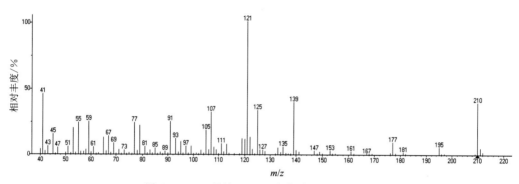

图 2-152　30 号峰　C_4-2-硫代单金刚烷质谱图

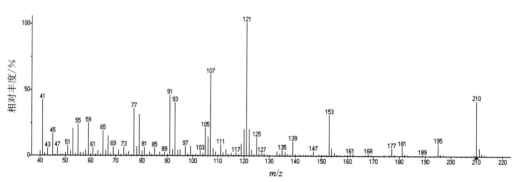

图 2-153　31 号峰　C_4-2-硫代单金刚烷质谱图

表 2-10　硫代单金刚烷鉴定结果表

峰　号	一维保留时间 /min	二维保留时间 /s	特征离子 (m/z)	分子式	相对分子质量	名　称
1	57.70	2.95	168	$C_{10}H_{16}S$	168	5-Methyl-2-thiaadamantane 5-甲基-2-硫代单金刚烷
2	59.00	2.99	168	$C_{10}H_{16}S$	168	1-Methyl-2-thiaadamantane 1-甲基-2-硫代单金刚烷
3	57.20	2.69	182	$C_{11}H_{18}S$	182	5,7-Dimethyl-2-thiaadamantane 5,7-二甲基-2-硫代单金刚烷
4	58.70	2.76	182	$C_{11}H_{18}S$	182	1,5-Dimethyl-2-thiaadamantane 1,5-二甲基-2-硫代单金刚烷

续表

峰 号	一维保留时间 /min	二维保留时间 /s	特征离子 (m/z)	分子式	相对分子质量	名 称
5	59.90	2.79	182	$C_{11}H_{18}S$	182	1,3-Dimethyl-2-thiaadamantane 1,3-二甲基-2-硫代单金刚烷
6	62.00	2.85	182	$C_{11}H_{18}S$	182	C_2-2-Thiaadamantane C_2-2-硫代单金刚烷
7	63.00	2.89	182	$C_{11}H_{18}S$	182	C_2-2-Thiaadamantane C_2-2-硫代单金刚烷
8	63.60	2.90	182	$C_{11}H_{18}S$	182	C_2-2-Thiaadamantane C_2-2-硫代单金刚烷
9	63.90	2.90	182	$C_{11}H_{18}S$	182	C_2-2-Thiaadamantane C_2-2-硫代单金刚烷
10	65.50	3.01	182	$C_{11}H_{18}S$	182	C_2-2-Thiaadamantane C_2-2-硫代单金刚烷
11	66.50	3.05	182	$C_{11}H_{18}S$	182	C_2-2-Thiaadamantane C_2-2-硫代单金刚烷
12	57.90	2.54	196	$C_{12}H_{20}S$	196	Trimethyl-2-thiaadamantane 三甲基-2-硫代单金刚烷
13	59.30	2.60	196	$C_{12}H_{20}S$	196	Trimethyl-2-thiaadamantane 三甲基-2-硫代单金刚烷
14	62.20	2.62	196	$C_{12}H_{20}S$	196	C_3-2-Thiaadamantane C_3-2-硫代单金刚烷
15	62.80	2.65	196	$C_{12}H_{20}S$	196	C_3-2-Thiaadamantane C_3-2-硫代单金刚烷
16	63.40	2.68	196	$C_{12}H_{20}S$	196	C_3-2-Thiaadamantane C_3-2-硫代单金刚烷
17	64.40	2.73	196	$C_{12}H_{20}S$	196	C_3-2-Thiaadamantane C_3-2-硫代单金刚烷
18	64.70	2.75	196	$C_{12}H_{20}S$	196	C_3-2-Thiaadamantane C_3-2-硫代单金刚烷
19	65.10	2.77	196	$C_{12}H_{20}S$	196	C_3-2-Thiaadamantane C_3-2-硫代单金刚烷
20	65.50	2.76	196	$C_{12}H_{20}S$	196	C_3-2-Thiaadamantane C_3-2-硫代单金刚烷
21	65.80	2.79	196	$C_{12}H_{20}S$	196	C_3-2-Thiaadamantane C_3-2-硫代单金刚烷
22	66.20	2.79	196	$C_{12}H_{20}S$	196	C_3-2-Thiaadamantane C_3-2-硫代单金刚烷
23	66.70	2.80	196	$C_{12}H_{20}S$	196	C_3-2-Thiaadamantane C_3-2-硫代单金刚烷
24	66.90	2.76	196	$C_{12}H_{20}S$	196	C_3-2-Thiaadamantane C_3-2-硫代单金刚烷

续表

峰 号	一维保留时间 /min	二维保留时间 /s	特征离子 (m/z)	分子式	相对分子质量	名 称
25	67.50	2.80	196	$C_{12}H_{20}S$	196	C_3-2-Thiaadamantane C_3-2-硫代单金刚烷
26	68.80	2.83	196	$C_{12}H_{20}S$	196	C_3-2-Thiaadamantane C_3-2-硫代单金刚烷
27	58.40	2.40	210	$C_{13}H_{22}S$	210	C_4-2-Thiaadamantane C_4-2-硫代单金刚烷
28	64.00	2.52	210	$C_{13}H_{22}S$	210	C_4-2-Thiaadamantane C_4-2-硫代单金刚烷
29	64.60	2.54	210	$C_{13}H_{22}S$	210	C_4-2-Thiaadamantane C_4-2-硫代单金刚烷
30	65.20	2.58	210	$C_{13}H_{22}S$	210	C_4-2-Thiaadamantane C_4-2-硫代单金刚烷
31	65.90	2.57	210	$C_{13}H_{22}S$	210	C_4-2-Thiaadamantane C_4-2-硫代单金刚烷

2.1.6 三环萜烷和四环萜烷的全二维气相色谱图及化合物名称

三环萜烷是石油地质样品中普遍存在的一类化合物,常用于油源对比。在质谱分析中,三环萜烷的特征离子为 m/z 191,与四环萜烷一致,因此常放在一起讨论。在常规色质分析中,一般可检测 15 个(13 个三环萜烷,2 个四环萜烷)化合物(图 2-154a),其中的 8,9,10 号峰共馏。在全二维气相色谱-飞行时间质谱分析图谱中,由于不同环数的萜烷二维保留时间不同,不仅 8,9,10 号峰无共馏,而且相对分子质量较高的三环萜烷不再受五环三萜烷的干扰(图 2-154b 和 c)。全二维气相色谱上共检测三环萜烷和四环萜烷 25 个化合物,新增的 10 个三环萜烷化合物质谱图与前面的三环萜烷化合物质谱图的规律性非常好,如图 2-155~图 2-164 所示。三环萜烷和四环萜烷的鉴定结果见表 2-11。

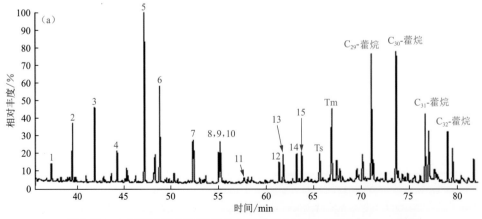

图 2-154 三环萜烷和四环萜烷的色质分析(a)和全二维分析(b,c)谱图

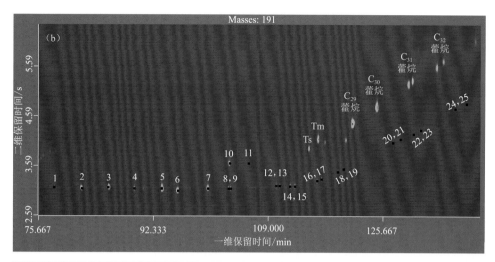

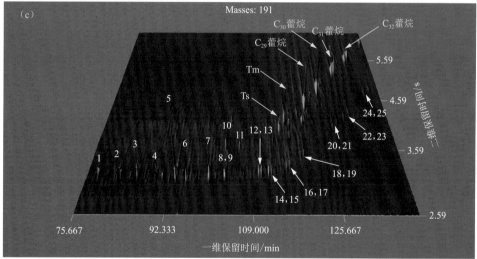

图 2-154（续） 三环萜烷和四环萜烷的色质分析(a)和全二维分析(b,c)谱图

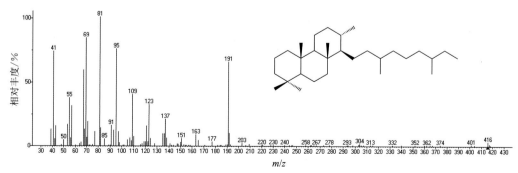

图 2-155　16 号峰　C_{30},13β(H),14α(H)-三环萜烷(20S)质谱图

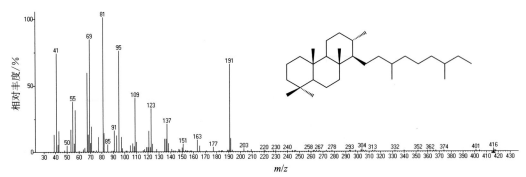

图 2-156　17 号峰　C_{30},13β(H),14α(H)-三环萜烷(20R)质谱图

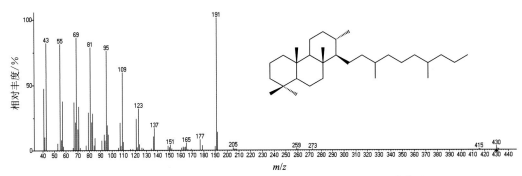

图 2-157　18 号峰　C_{31},13β(H),14α(H)-三环萜烷(20S)质谱图

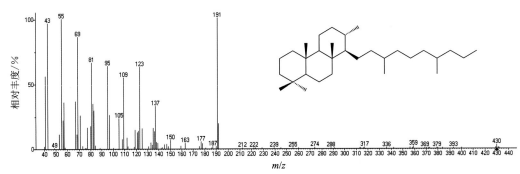

图 2-158　19 号峰　C_{31},13β(H),14α(H)-三环萜烷(20R)质谱图

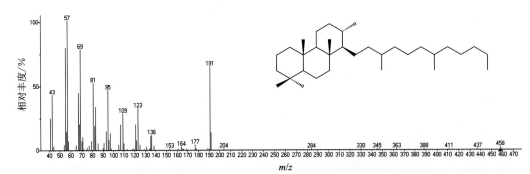

图 2-159　20 号峰　C_{33},13β(H),14α(H)-三环萜烷(20S)质谱图

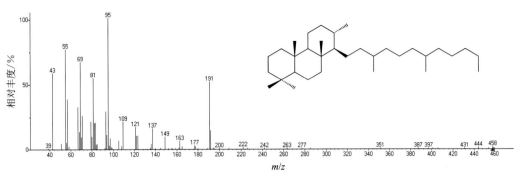

图 2-160　21 号峰　C_{33},13β(H),14α(H)-三环萜烷(20R)质谱图

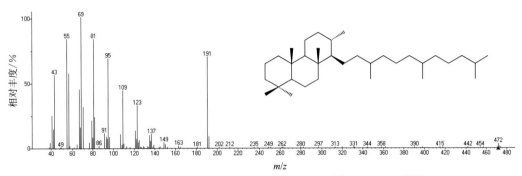

图 2-161　22 号峰　C_{34},13β(H),14α(H)-三环萜烷(20S)质谱图

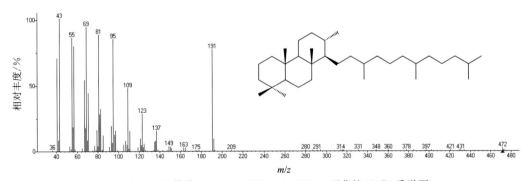

图 2-162　23 号峰　C_{34},13β(H),14α(H)-三环萜烷(20R)质谱图

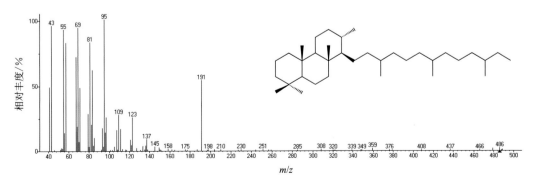

图 2-163　24 号峰　C_{35},13β(H),14α(H)-三环萜烷(20S)质谱图

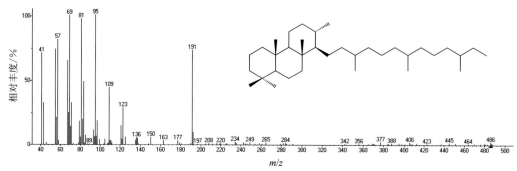

图 2-164 25 号峰 C_{35}, $13\beta(H)$, $14\alpha(H)$-三环萜烷(20R)质谱图

表 2-11 三环萜烷和四环萜烷鉴定结果表

峰 号	一维保留时间/min	二维保留时间/s	特征离子(m/z)	分子式	相对分子质量	名 称
1	78.00	3.17	191	$C_{19}H_{34}$	262	C_{19}, $13\beta(H)$, $14\alpha(H)$-Tricyclic terpane C_{19}, $13\beta(H)$, $14\alpha(H)$-三环萜烷
2	82.00	3.14	191	$C_{20}H_{36}$	276	C_{20}, $13\beta(H)$, $14\alpha(H)$-Tricyclic terpane C_{20}, $13\beta(H)$, $14\alpha(H)$-三环萜烷
3	85.83	3.17	191	$C_{21}H_{38}$	290	C_{21}, $13\beta(H)$, $14\alpha(H)$-Tricyclic terpane C_{21}, $13\beta(H)$, $14\alpha(H)$-三环萜烷
4	89.50	3.14	191	$C_{22}H_{40}$	304	C_{22}, $13\beta(H)$, $14\alpha(H)$-Tricyclic terpane C_{22}, $13\beta(H)$, $14\alpha(H)$-三环萜烷
5	93.50	3.14	191	$C_{23}H_{42}$	318	C_{23}, $13\beta(H)$, $14\alpha(H)$-Tricyclic terpane C_{23}, $13\beta(H)$, $14\alpha(H)$-三环萜烷
6	95.83	3.10	191	$C_{24}H_{44}$	332	C_{24}, $13\beta(H)$, $14\alpha(H)$-Tricyclic terpane C_{24}, $13\beta(H)$, $14\alpha(H)$-三环萜烷
7	100.17	3.15	191	$C_{25}H_{46}$	346	C_{25}, $13\beta(H)$, $14\alpha(H)$-Tricyclic terpane C_{25}, $13\beta(H)$, $14\alpha(H)$-三环萜烷
8	103.17	3.15	191	$C_{26}H_{48}$	360	C_{26}, $13\beta(H)$, $14\alpha(H)$-Tricyclic terpane(20S) C_{26}, $13\beta(H)$, $14\alpha(H)$-三环萜烷(20S)
9	103.50	3.14	191	$C_{26}H_{48}$	360	C_{26}, $13\beta(H)$, $14\alpha(H)$-Tricyclic terpane(20R) C_{26}, $13\beta(H)$, $14\alpha(H)$-三环萜烷(20R)
10	103.33	3.65	191	$C_{24}H_{42}$	330	C_{24}, Tetracyclic terpane C_{24}, 四环萜烷
11	106.17	3.65	191	$C_{25}H_{44}$	344	C_{25}, Tetracyclic terpane C_{25}, 四环萜烷
12	110.17	3.20	191	$C_{28}H_{52}$	388	C_{28}, $13\beta(H)$, $14\alpha(H)$-Tricyclic terpane(20S) C_{28}, $13\beta(H)$, $14\alpha(H)$-三环萜烷(20S)
13	110.67	3.20	191	$C_{28}H_{52}$	388	C_{28}, $13\beta(H)$, $14\alpha(H)$-Tricyclic terpane(20R) C_{28}, $13\beta(H)$, $14\alpha(H)$-三环萜烷(20R)
14	112.17	3.19	191	$C_{29}H_{54}$	402	C_{29}, $13\beta(H)$, $14\alpha(H)$-Tricyclic terpane(20S) C_{29}, $13\beta(H)$, $14\alpha(H)$-三环萜(20S)
15	112.83	3.19	191	$C_{29}H_{54}$	402	C_{29}, $13\beta(H)$, $14\alpha(H)$-Tricyclic terpane(20R) C_{29}, $13\beta(H)$, $14\alpha(H)$-三环萜烷(20R)
16	116.00	3.29	191	$C_{30}H_{56}$	416	C_{30}, $13\beta(H)$, $14\alpha(H)$-Tricyclic terpane(20S) C_{30}, $13\beta(H)$, $14\alpha(H)$-三环萜烷(20S)

续表

峰 号	一维保留时间/min	二维保留时间/s	特征离子(m/z)	分子式	相对分子质量	名 称
17	116.67	3.34	191	$C_{30}H_{56}$	416	C_{30},13β(H),14α(H)-Tricyclic terpane(20R) C_{30},13β(H),14α(H)-三环萜烷(20R)
18	119.00	3.49	191	$C_{31}H_{58}$	430	C_{31},13β(H),14α(H)-Tricyclic terpane(20S) C_{31},13β(H),14α(H)-三环萜烷(20S)
19	119.83	3.54	191	$C_{31}H_{58}$	430	C_{31},13β(H),14α(H)-Tricyclic terpane(20R) C_{31},13β(H),14α(H)-三环萜烷(20R)
20	127.00	4.07	191	$C_{33}H_{62}$	458	C_{33},13β(H),14α(H)-Tricyclic terpane(20S) C_{33},13β(H),14α(H)-三环萜烷(20S)
21	128.17	4.14	191	$C_{33}H_{62}$	458	C_{33},13β(H),14α(H)-Tricyclic terpane(20R) C_{33},13β(H),14α(H)-三环萜烷(20R)
22	130.00	4.24	191	$C_{34}H_{64}$	472	C_{34},13β(H),14α(H)-Tricyclic terpane(20S) C_{34},13β(H),14α(H)-三环萜烷(20S)
23	131.17	4.32	191	$C_{34}H_{64}$	472	C_{34},13β(H),14α(H)-Tricyclic terpane(20R) C_{34},13β(H),14α(H)-三环萜烷(20R)
24	136.17	4.75	191	$C_{35}H_{66}$	486	C_{35},13β(H),14α(H)-Tricyclic terpane(20S) C_{35},13β(H),14α(H)-三环萜烷(20S)
25	137.83	4.85	191	$C_{35}H_{66}$	486	C_{35},13β(H),14α(H)-Tricyclic terpane(20R) C_{35},13β(H),14α(H)-三环萜烷(20R)

2.1.7 甾烷类化合物的全二维气相色谱图及化合物名称

甾烷是来源于真核生物的生物标志化合物,其在质谱分析中的特征离子为 m/z 217,m/z 218也常作为特征离子使用。在正常原油中,无论是常规的色质分析还是全二维气相色谱-飞行时间质谱分析都可以检测到24个化合物,但全二维分析谱图去除了链烷烃和萜烷的碎片离子干扰,各化合物之间的分离度远好于常规色质分析的分离度,如图 2-165 所示。在经历生物降解的原油中,可能检测到更多数量的甾烷类化合物。将全二维分析的高碳数区域内甾烷图谱放大,如图 2-166 所示,全二维气相色谱对甾烷类化合物的分析优势会更加明显。参照常规色质分析的结果和前人的成果(王培荣,1993;王培荣,周光甲,1995),结合甾烷类各化合物的质谱图(图 2-167~图 2-190),甾烷类化合物的定性结果见表 2-12。

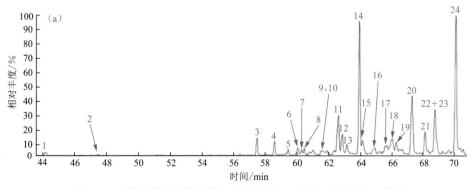

图 2-165 甾烷类化合物的常规色质分析(a)和全二维分析(b)谱图

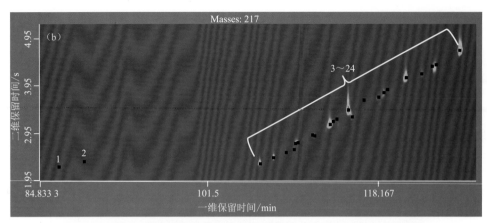

图 2-165(续)　甾烷类化合物的常规色质分析(a)和全二维分析(b)谱图

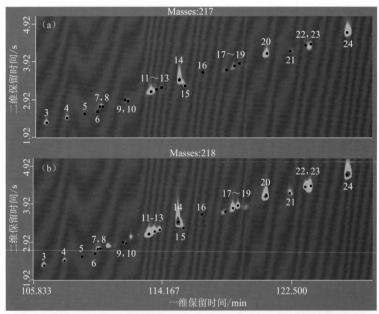

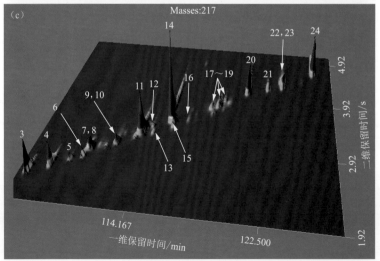

图 2-166　高碳数区域内甾烷类化合物全二维点阵图(a,b)和三维立体图(c)

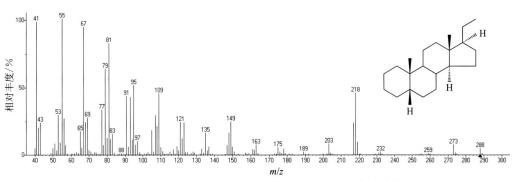

图 2-167 1号峰 $5\beta(H),14\alpha(H),17\alpha(H)$-$C_{21}$孕甾烷质谱图

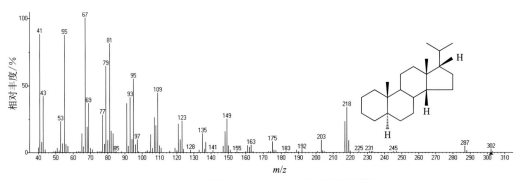

图 2-168 2号峰 $5\alpha(H)$-C_{22}升孕甾烷质谱图

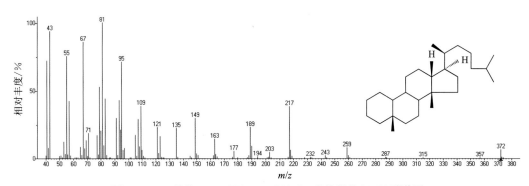

图 2-169 3号峰 $13\beta(H),17\alpha(H)$-C_{27}重排甾烷(20S)质谱图

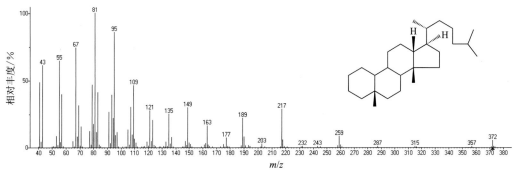

图 2-170 4号峰 $13\beta(H),17\alpha(H)$-C_{27}重排甾烷(20R)质谱图

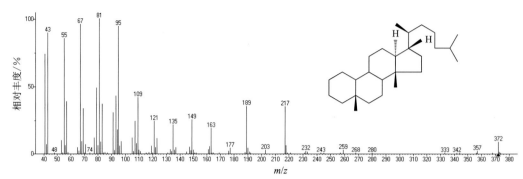

图 2-171　5 号峰　$13\alpha(H),17\beta(H)$-C_{27} 重排甾烷 (20S) 质谱图

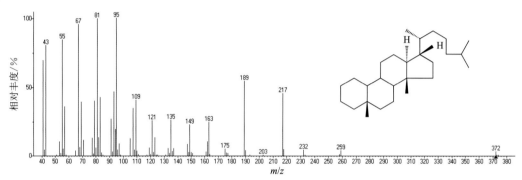

图 2-172　6 号峰　$13\alpha(H),17\beta(H)$-C_{27} 重排甾烷 (20R) 质谱图

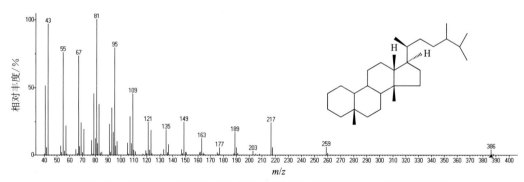

图 2-173　7 号峰　$13\beta(H),17\alpha(H)$-C_{28} 重排甾烷 [20S(24S+24R)] 质谱图

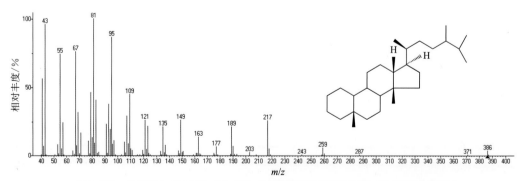

图 2-174　8 号峰　$13\beta(H),17\alpha(H)$-C_{28} 重排甾烷 [20S(24S+24R)] 质谱图

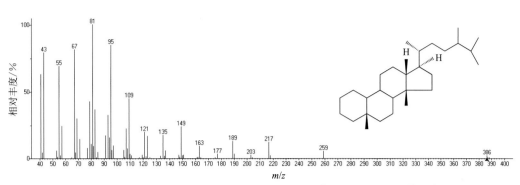

图 2-175　9 号峰　$13\beta(H),17\alpha(H)$-C_{28} 重排甾烷[$20R(24S+24R)$]质谱图

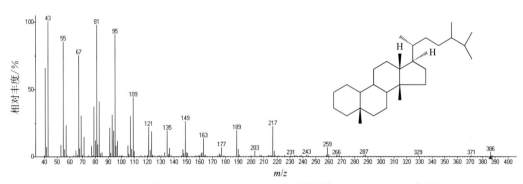

图 2-176　10 号峰　$13\beta(H),17\alpha(H)$-C_{28} 重排甾烷[$20R(24S+24R)$]质谱图

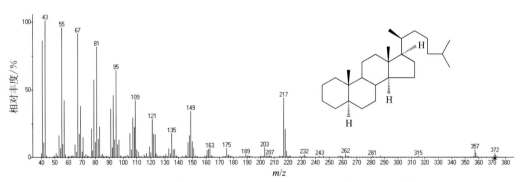

图 2-177　11 号峰　$5\alpha(H),14\alpha(H),17\alpha(H)$-$C_{27}$ 甾烷($20S$)质谱图

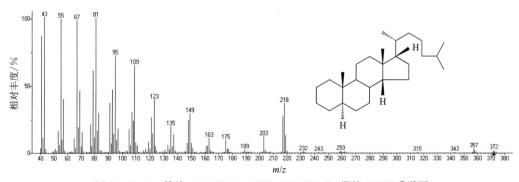

图 2-178　12 号峰　$5\alpha(H),14\beta(H),17\beta(H)$-$C_{27}$ 甾烷($20R$)质谱图

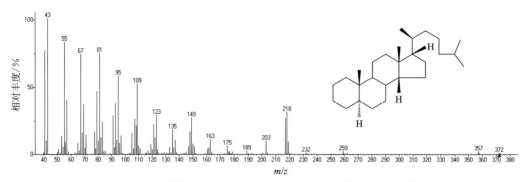

图 2-179　13 号峰　$5\alpha(H),14\beta(H),17\beta(H)$-$C_{27}$ 甾烷(20S)质谱图

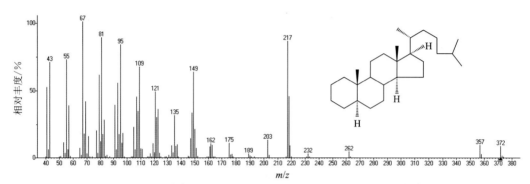

图 2-180　14 号峰　$5\alpha(H),14\alpha(H),17\alpha(H)$-$C_{27}$ 甾烷(20R)质谱图

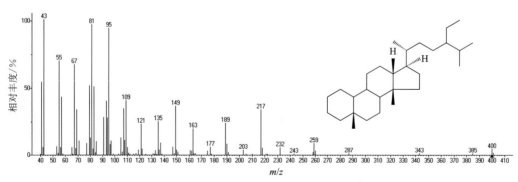

图 2-181　15 号峰　24-乙基-$13\beta(H),17\alpha(H)$-C_{29} 重排甾烷(20R)质谱图

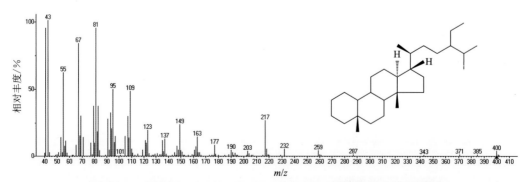

图 2-182　16 号峰　24-乙基-$13\alpha(H),17\beta(H)$-C_{29} 重排甾烷(20S)质谱图

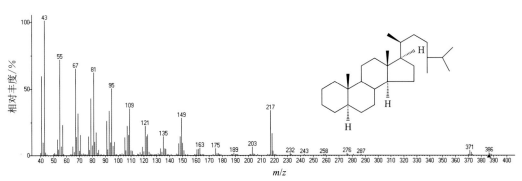

图 2-183　17 号峰　24-甲基-5α(H),14α(H),17α(H)-C$_{28}$甾烷(20S)质谱图

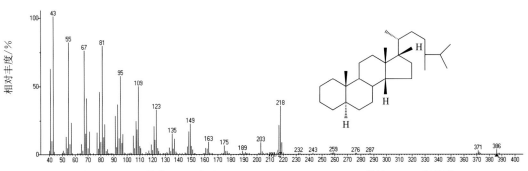

图 2-184　18 号峰　24-甲基-5α(H),14β(H),17β(H)-C$_{28}$甾烷(20R)质谱图

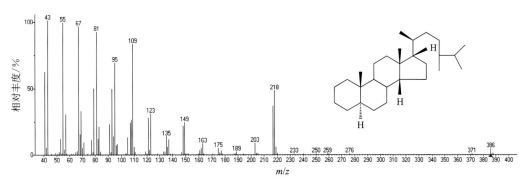

图 2-185　19 号峰　24-甲基-5α(H),14β(H),17β(H)-C$_{28}$甾烷(20S)质谱图

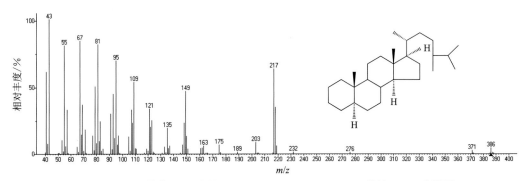

图 2-186　20 号峰　24-甲基-5α(H),14α(H),17α(H)-C$_{28}$甾烷(20R)质谱图

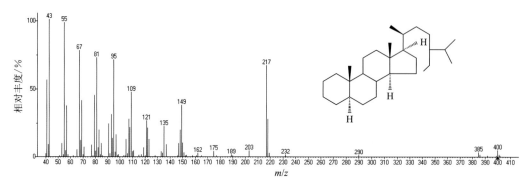

图 2-187　21 号峰　24-乙基-5α(H),14α(H),17α(H)-C$_{29}$ 甾烷(20S)质谱图

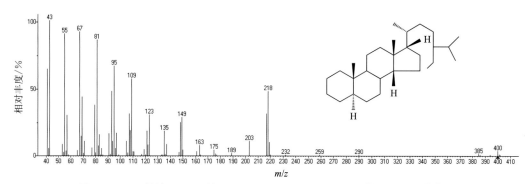

图 2-188　22 号峰　24-乙基-5α(H),14β(H),17β(H)-C$_{29}$ 甾烷(20R)质谱图

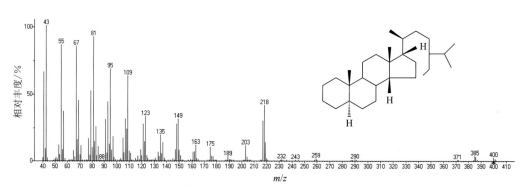

图 2-189　23 号峰　24-乙基-5α(H),14β(H),17β(H)-C$_{29}$ 甾烷(20S)质谱图

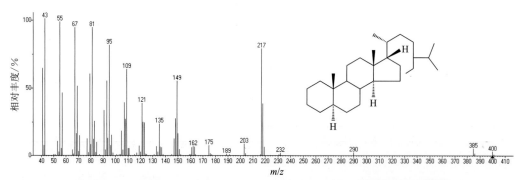

图 2-190　24 号峰　24-乙基-5α(H),14α(H),17α(H)-C$_{29}$ 甾烷(20R)质谱图

表 2-12 甾烷类化合物鉴定结果表

峰 号	一维保留时间 /min	二维保留时间 /s	特征离子 (m/z)	分子式	相对分子质量	名　称
1	86.67	2.26	218	$C_{21}H_{36}$	288	$5\beta(H),14\alpha(H),17\alpha(H)$-$C_{21}$ Pregnane $5\beta(H),14\alpha(H),17\alpha(H)$-$C_{21}$ 孕甾烷
2	89.17	2.36	218	$C_{22}H_{38}$	302	$5\alpha(H)$-C_{22} Homopregnane $5\alpha(H)$-C_{22} 升孕甾烷
3	106.50	2.32	217	$C_{27}H_{48}$	372	$13\beta(H),17\alpha(H)$-C_{27} Diasterane(20S) $13\beta(H),17\alpha(H)$-C_{27} 重排甾烷(20S)
4	107.83	2.44	217	$C_{27}H_{48}$	372	$13\beta(H),17\alpha(H)$-C_{27} Diasterane(20R) $13\beta(H),17\alpha(H)$-C_{27} 重排甾烷(20R)
5	109.00	2.55	217	$C_{27}H_{48}$	372	$13\alpha(H),17\beta(H)$-C_{27} Diasterane(20S) $13\alpha(H),17\beta(H)$-C_{27} 重排甾烷(20S)
6	109.83	2.62	217	$C_{27}H_{48}$	372	$13\alpha(H),17\beta(H)$-C_{27} Diasterane(20R) $13\alpha(H),17\beta(H)$-C_{27} 重排甾烷(20R)
7	110.00	2.75	217	$C_{28}H_{50}$	386	$13\beta(H),17\alpha(H)$-C_{28} Diasterane [20S(24S+24R)] $13\beta(H),17\alpha(H)$-C_{28} 重排甾烷 [20S(24S+24R)]
8	110.17	2.76	217	$C_{28}H_{50}$	386	$13\beta(H),17\alpha(H)$-C_{28} Diasterane [20S(24S+24R)] $13\beta(H),17\alpha(H)$-C_{28} 重排甾烷 [20S(24S+24R)]
9	111.67	2.92	217	$C_{28}H_{50}$	386	$13\beta(H),17\alpha(H)$-C_{28} Diasterane [20R(24S+24R)] $13\beta(H),17\alpha(H)$-C_{28} 重排甾烷 [20R(24S+24R)]
10	111.83	2.90	217	$C_{28}H_{50}$	386	$13\beta(H),17\alpha(H)$-C_{28} Diasterane [20R(24S+24R)] $13\beta(H),17\alpha(H)$-C_{28} 重排甾烷 [20R(24S+24R)]
11	113.33	3.14	217	$C_{27}H_{48}$	372	$5\alpha(H),14\alpha(H),$ $17\alpha(H)$-C_{27} Sterane(20S) $5\alpha(H),14\alpha(H),$ $17\alpha(H)$-C_{27} 甾烷(20S)
12	113.67	3.21	218	$C_{27}H_{48}$	372	$5\alpha(H),14\beta(H),$ $17\beta(H)$-C_{27} Sterane(20R) $5\alpha(H),14\beta(H),$ $17\beta(H)$-C_{27} 甾烷(20R)
13	114.00	3.25	218	$C_{27}H_{48}$	372	$5\alpha(H),14\beta(H),$ $17\beta(H)$-C_{27} Sterane(20S) $5\alpha(H),14\beta(H),$ $17\beta(H)$-C_{27} 甾烷(20S)

续表

峰 号	一维保留时间 /min	二维保留时间 /s	特征离子 (m/z)	分子式	相对分子质量	名 称
14	115.17	3.45	217	$C_{27}H_{48}$	372	$5\alpha(H),14\alpha(H),17\alpha(H)$-$C_{27}$ sterane(20R) $5\alpha(H),14\alpha(H),17\alpha(H)$-$C_{27}$甾烷(20R)
15	115.50	3.30	217	$C_{29}H_{52}$	400	24-Ethyl-$13\beta(H),17\alpha(H)$-C_{29} Diasterane(20R) 24-乙基-$13\beta(H),17\alpha(H)$-C_{29}重排甾烷(20R)
16	116.67	3.648	217	$C_{29}H_{52}$	400	24-Ethyl-$13\alpha(H),17\beta(H)$-C_{29} Diasterane(20S) 24-乙基-$13\alpha(H),17\beta(H)$-C_{29}重排甾烷(20S)
17	118.17	3.719	217	$C_{28}H_{50}$	386	24-Methyl-$5\alpha(H),14\alpha(H),17\alpha(H)$-$C_{28}$ sterane(20S) 24-甲基-$5\alpha(H),14\alpha(H),17\alpha(H)$-$C_{28}$甾烷(20S)
18	118.67	3.818	218	$C_{28}H_{50}$	386	24-Methyl-$5\alpha(H),14\beta(H),17\beta(H)$-$C_{28}$ sterane(20R) 24-甲基-$5\alpha(H),14\beta(H),17\beta(H)$-$C_{28}$甾烷(20R)
19	119.00	3.88	218	$C_{28}H_{50}$	386	24-Methyl-$5\alpha(H),14\beta(H),17\beta(H)$-$C_{28}$ sterane(20S) 24-甲基-$5\alpha(H),14\beta(H),17\beta(H)$-$C_{28}$甾烷(20S)
20	120.83	4.14	217	$C_{28}H_{50}$	386	24-Methyl-$5\alpha(H),14\alpha(H),17\alpha(H)$-$C_{28}$ sterane(20R) 24-甲基-$5\alpha(H),14\alpha(H),17\alpha(H)$-$C_{28}$甾烷(20R)
21	122.33	4.21	217	$C_{29}H_{52}$	400	24-Ethyl-$5\alpha(H),14\alpha(H),17\alpha(H)$-$C_{29}$ sterane(20S) 24-乙基-$5\alpha(H),14\alpha(H),17\alpha(H)$-$C_{29}$甾烷(20S)
22	123.33	4.37	218	$C_{29}H_{52}$	400	24-Ethyl-$5\alpha(H),14\beta(H),17\beta(H)$-$C_{29}$ sterane(20R) 24-乙基-$5\alpha(H),14\beta(H),17\beta(H)$-$C_{29}$甾烷(20R)

续表

峰 号	一维保留时间 /min	二维保留时间 /s	特征离子 (m/z)	分子式	相对分子质量	名 称
23	123.67	4.40	218	$C_{29}H_{52}$	400	24-Ethyl-5α(H),14β(H), 17β(H)-C_{29} sterane(20S) 24-乙基-5α(H),14β(H), 17β(H)-C_{29}甾烷(20S)
24	126.00	4.70	217	$C_{29}H_{52}$	400	24-Ethyl-5α(H),14α(H), 17α(H)-C_{29} sterane(20R) 24-乙基-5α(H),14α(H), 17α(H)-C_{29}甾烷(20R)

2.1.8 藿烷类化合物的全二维气相色谱图及化合物名称

藿烷类化合物是来自原核生物细胞膜的一类化合物,是油气地球化学研究中广泛使用的一类生物标志化合物。藿烷原指含有30个碳原子的五环三萜烷,所有含有五环三萜烷分子骨架的化合物统称为藿烷类化合物。

本节涉及的藿烷类化合物包括常规藿烷、早流出物、重排藿烷、断藿烷以及降藿烷等。

1) 常规藿烷类化合物的全二维气相色谱图及化合物名称

这里的常规藿烷是指常规地球化学分析中常常见到的一些藿烷类化合物,其特征离子为 m/z 191,正常的色质分析可以检测到21个化合物,包括伽马蜡烷、奥利烷、莫烷等,如图2-191(a)所示;全二维气相色谱-飞行时间质谱分析可以检测到23个化合物(图2-191b和c);由于全二维分析谱图去除了链烷烃和甾烷的干扰、二维上保留时间的差异,所以常规藿烷利用全二维分析时各化合物得到完全分离。将常规色质分析结果和各化合物的质谱图(图2-192~图2-214)与标准图谱比对,得到全二维分析的常规藿烷定性结果,见表2-13。由于藿烷类化合物的学名相对较长,表中部分名称添加了化合物的俗称。

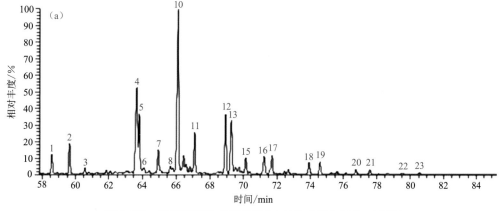

图2-191 常规藿烷的色质分析(a)与全二维分析(b,c)谱图

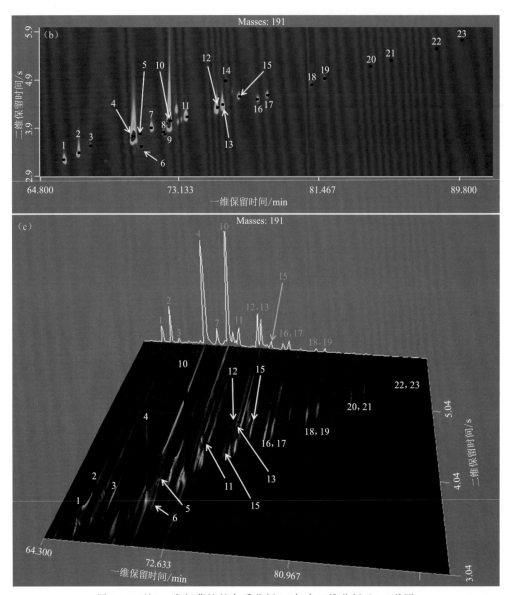

图 2-191(续)　常规藿烷的色质分析(a)与全二维分析(b,c)谱图

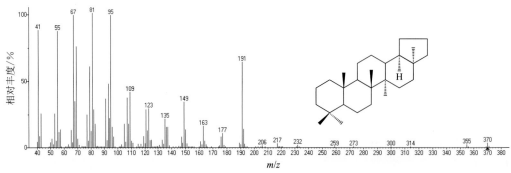

图 2-192　1 号峰　$18\alpha(H)$-22,29,30-三降藿烷(Ts)质谱图

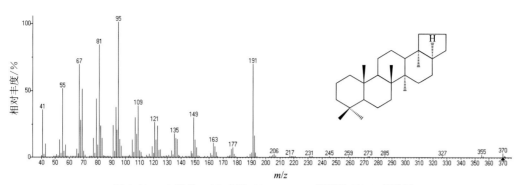

图 2-193　2 号峰　17α(H)-22,29,30-三降藿烷(Tm)质谱图

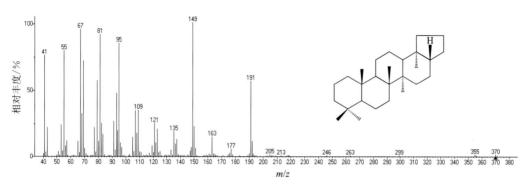

图 2-194　3 号峰　17β(H)-22,29,30-三降莫烷质谱图

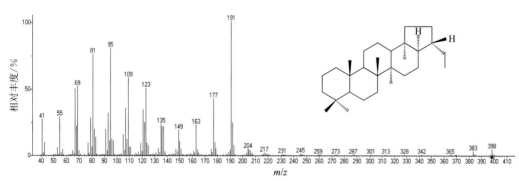

图 2-195　4 号峰　17α(H),21β(H)-30-降藿烷质谱图

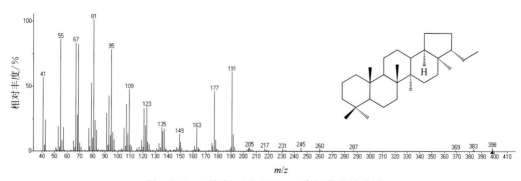

图 2-196　5 号峰　18α(H)-30-降新藿烷质谱图

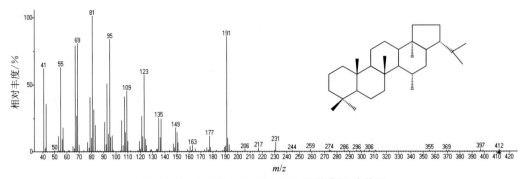

图 2-197 6 号峰 C_{30},$17\alpha(H)$-重排藿烷质谱图

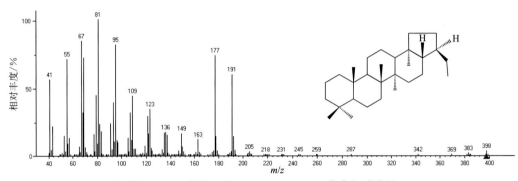

图 2-198 7 号峰 $17\beta(H)$,$21\alpha(H)$-30-降莫烷质谱图

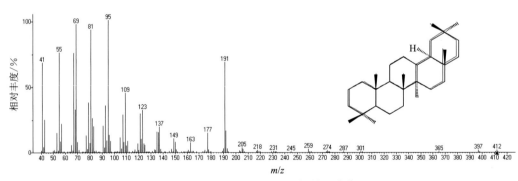

图 2-199 8 号峰 $18\alpha(H)$-奥利烷质谱图

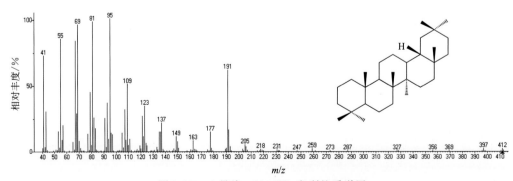

图 2-200 9 号峰 $18\beta(H)$-奥利烷质谱图

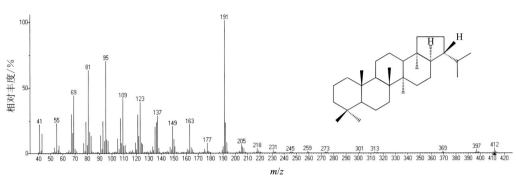

图 2-201　10 号峰　$17\alpha(H),21\beta(H)$-藿烷质谱图

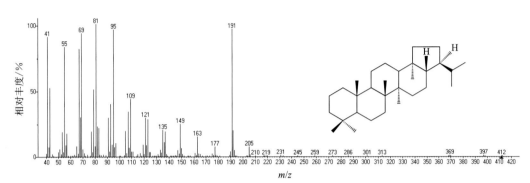

图 2-202　11 号峰　$17\beta(H),21\alpha(H)$-莫烷质谱图

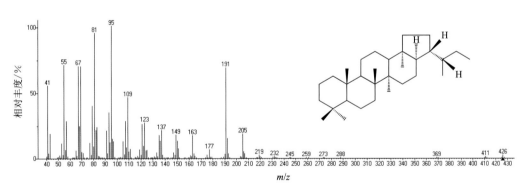

图 2-203　12 号峰　$17\alpha(H),21\beta(H)$-升藿烷(22S)质谱图

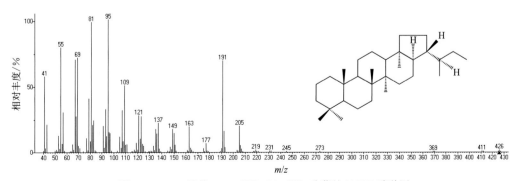

图 2-204　13 号峰　$17\alpha(H),21\beta(H)$-升藿烷(22R)质谱图

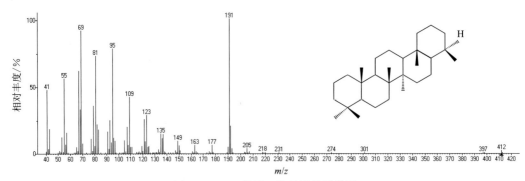

图 2-205　14 号峰　伽马蜡烷质谱图

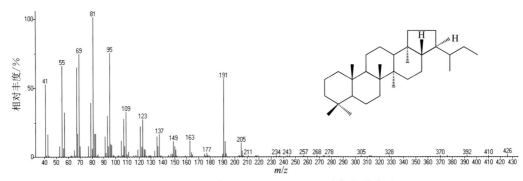

图 2-206　15 号峰　17β(H),21α(H)-升莫烷质谱图

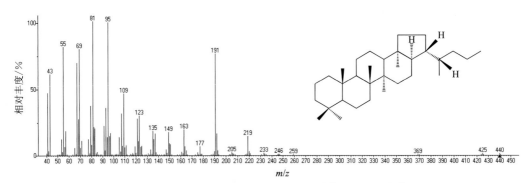

图 2-207　16 号峰　17α(H),21β(H)-双升藿烷(22S)质谱图

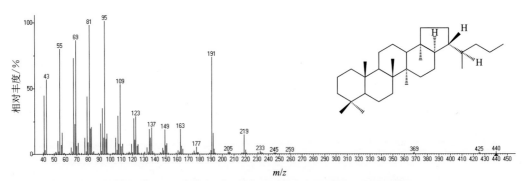

图 2-208　17 号峰　17α(H),21β(H)-双升藿烷(22R)质谱图

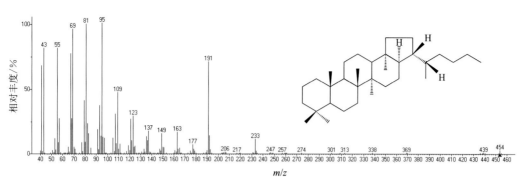

图 2-209　18 号峰　17α(H),21β(H)-三升藿烷(22S)质谱图

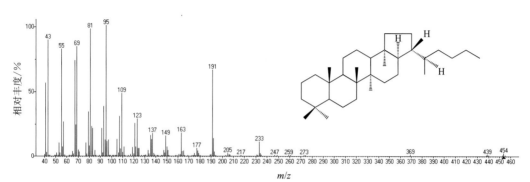

图 2-210　19 号峰　17α(H),21β(H)-三升藿烷(22R)质谱图

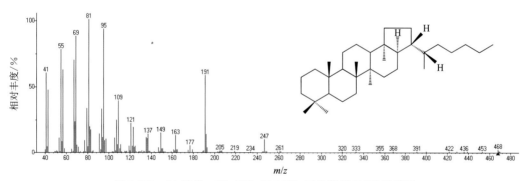

图 2-211　20 号峰　17α(H),21β(H)-四升藿烷(22S)质谱图

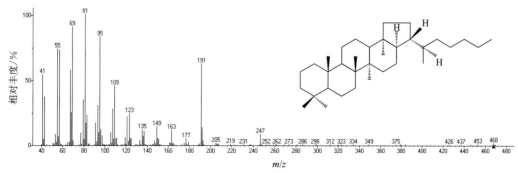

图 2-212　21 号峰　17α(H),21β(H)-四升藿烷(22R)质谱图

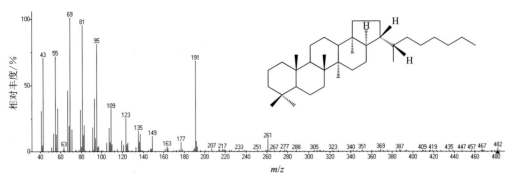

图 2-213　22 号峰　17α(H),21β(H)-五升藿烷(22S)质谱图

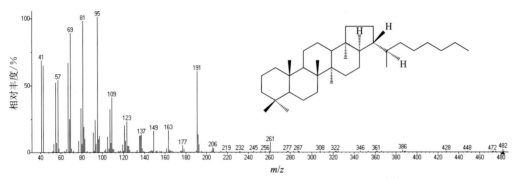

图 2-214　23 号峰　17α(H),21β(H)-五升藿烷(22R)质谱图

表 2-13　常规藿烷类化合物鉴定结果表

峰号	一维保留时间 /min	二维保留时间 /s	特征离子 (m/z)	分子式	相对分子质量	名　　称
1	66.20	3.26	191	$C_{27}H_{46}$	370	18α(H)-22,29,30-Trisnorhopane 18α(H)-22,29,30-三降藿烷 俗称 Ts
2	67.10	3.40	191	$C_{27}H_{46}$	370	17α(H)-22,29,30-Trisnorhopane 17α(H)-22,29,30-三降藿烷 俗称 Tm
3	67.80	3.55	149	$C_{27}H_{46}$	370	17β(H)-22,29,30-Trisnormoretane 17β(H)-22,29,30-三降莫烷 俗称 C_{27} 莫烷
4	70.40	3.75	191	$C_{29}H_{50}$	398	17α(H),21β(H)-30-Norhopane 17α(H),21β(H)-30-降藿烷 俗称 C_{29} 藿烷
5	70.80	3.80	191	$C_{29}H_{50}$	398	18α(H)-30-Norneohopane 18α(H)-30-降新藿烷 俗称 C_{29} Ts
6	70.90	3.54	191	$C_{30}H_{52}$	412	C_{30},17α(H)-Diahopane C_{30},17α(H)-重排藿烷
7	71.50	3.92	177	$C_{29}H_{50}$	398	17β(H),21α(H)-30-Normoretane 17β(H),21α(H)-30-降莫烷 俗称 C_{29} 莫烷

续表

峰 号	一维保留时间 /min	二维保留时间 /s	特征离子 (m/z)	分子式	相对分子质量	名　称
8	72.20	3.83	191	$C_{30}H_{52}$	412	18α(H)-Oleanane 18α(H)-奥利烷
9	72.30	3.79	191	$C_{30}H_{52}$	412	18β(H)-Oleanane 18β(H)-奥利烷
10	72.60	4.06	191	$C_{30}H_{52}$	412	17α(H),21β(H)-Hopane 17α(H),21β(H)-藿烷 俗称 C_{30}藿烷
11	73.60	4.15	191	$C_{30}H_{52}$	412	17β(H),21α(H)-Moretane 17β(H),21α(H)-莫烷 俗称 C_{30}莫烷
12	75.40	4.34	191	$C_{31}H_{54}$	426	17α(H),21β(H)-Homohopane,22S 17α(H),21β(H)-升藿烷(22S) 俗称 C_{31}藿烷(22S)
13	75.70	4.40	191	$C_{31}H_{54}$	426	17α(H),21β(H)-Homohopane,22R 17α(H),21β(H)-升藿烷(22R) 俗称 C_{31}藿烷(22R)
14	75.90	4.89	191	$C_{30}H_{52}$	412	Gammacerane 伽马蜡烷
15	76.70	4.57	191	$C_{31}H_{54}$	426	17β(H),21α(H)-Homomoretane 17β(H),21α(H)-升莫烷 俗称 C_{31}莫烷
16	77.80	4.51	191	$C_{32}H_{56}$	440	17α(H),21β(H)-Bishomohopane,22S 17α(H),21β(H)-双升藿烷(22S) 俗称 C_{32}藿烷(22S)
17	78.40	4.59	191	$C_{32}H_{56}$	440	17α(H),21β(H)-Bishomohopane,22R 17α(H),21β(H)-双升藿烷(22R) 俗称 C_{32}藿烷(22R)
18	81.00	4.81	191	$C_{33}H_{58}$	454	17α(H),21β(H)-Trishomohopane,22S 17α(H),21β(H)-三升藿烷(22S) 俗称 C_{33}藿烷(22S)
19	81.80	4.94	191	$C_{33}H_{58}$	454	17α(H),21β(H)-Trishomohopane,22R 17α(H),21β(H)-三升藿烷(22R) 俗称 C_{33}藿烷(22R)
20	84.50	5.19	191	$C_{34}H_{60}$	468	17α(H),21β(H)-Tetrashomohopane,22S 17α(H),21β(H)-四升藿烷(22S) 俗称 C_{34}藿烷(22S)
21	85.70	5.31	191	$C_{34}H_{60}$	468	17α(H),21β(H)-Tetrashomohopane,22R 17α(H),21β(H)-四升藿烷(22R) 俗称 C_{34}藿烷(22R)

续表

峰号	一维保留时间/min	二维保留时间/s	特征离子(m/z)	分子式	相对分子质量	名称
22	88.40	5.55	191	$C_{35}H_{62}$	482	17α(H),21β(H)-Pentashomohopane,22S 17α(H),21β(H)-五升藿烷(22S) 俗称 C_{35} 藿烷(22S)
23	89.90	5.72	191	$C_{35}H_{62}$	482	17α(H),21β(H)-Pentashomohopane,22R 17α(H),21β(H)-五升藿烷(22R) 俗称 C_{35} 藿烷(22R)

2）重排藿烷类化合物的全二维气相色谱图及化合物名称

重排藿烷一般指与黏土矿物伴生的、藿烷分子骨架上14位的甲基重排至15位的一类化合物。事实上，新藿烷系列化合物，也就是藿烷分子骨架上18位的甲基重排至17位的一类化合物(Ts系列)也属于重排藿烷类，文献报道的早流出物亦是重排藿烷的一种。为了区分及方便使用，我们把它们分别称为重排藿烷、新藿烷及早流出物。重排藿烷的特征离子与常规藿烷的特征离子一样都是 m/z 191。当用常规色质分析含有所有重排藿烷的石油地质样品时，会出现部分峰的重叠，如图2-215所示。但在全二维气相色谱分析时，由于不同种类的重排藿烷极性存在差异，二维保留时间不同，各类化合物很容易彻底分开。图2-216是多类藿烷的全二维分析谱图，重排藿烷在二维空间上位于正常藿烷的下方，同碳数的重排藿烷先于对应的正常藿烷出峰，易于定性，本节不再介绍。早流出物重排藿烷在二维空间位于重排藿烷的下方，说明其极性最低；相同碳数的早流出物比对应的正常藿烷早3个碳数间隔，这也是称其为早流出物的原因。早流出物的具体结构尚不明确，但根据其质谱图谱(图2-217～图2-222)及相对分子质量的信息可以推断它属于重排藿烷，具体结构确认正在尝试中，目前名称可依据碳数多少称为 C_x 早流出物。

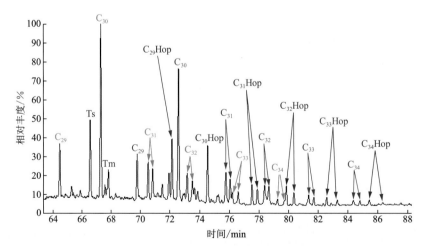

图2-215　常规色质分析多种重排藿烷谱图

绿色字标记早流出物，红色字标记重排藿烷，蓝色字标记常规藿烷

第 2 章　石油地质样品的全二维气相色谱图识别

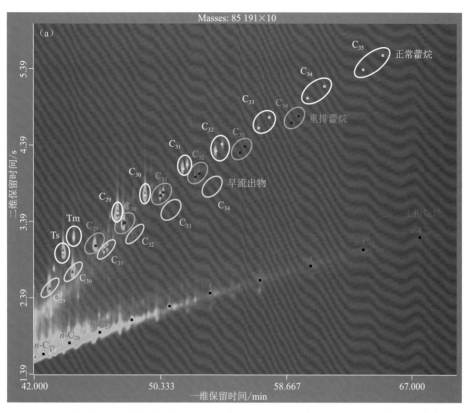

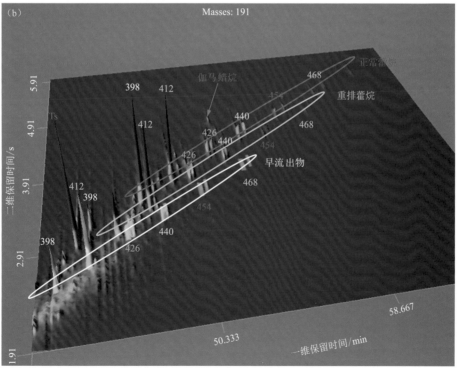

图 2-216　多种重排藿烷类化合物的全二维分析谱图

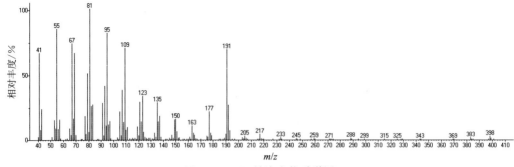

图 2-217　C_{29} 早流出物质谱图

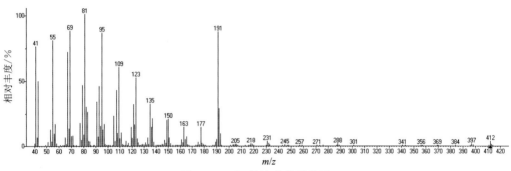

图 2-218　C_{30} 早流出物质谱图

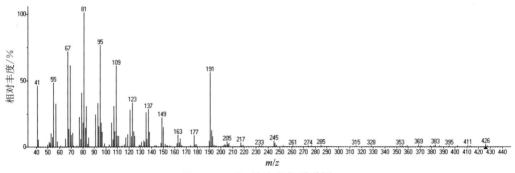

图 2-219　C_{31} 早流出物质谱图

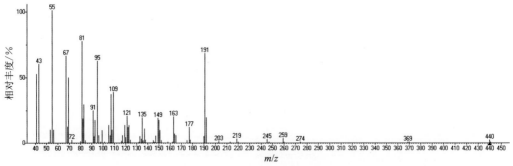

图 2-220　C_{32} 早流出物质谱图

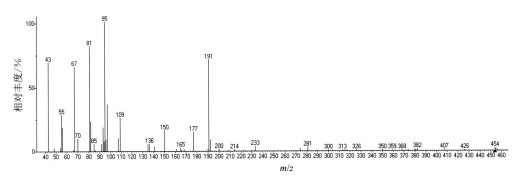

图 2-221　C_{33} 早流出物质谱图

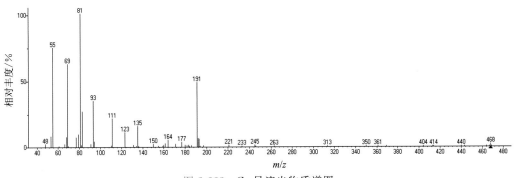

图 2-222　C_{34} 早流出物质谱图

3）断藿烷类化合物的全二维气相色谱图及化合物名称

藿烷在微生物或者酸性介质催化下发生环上的 C—C 键断裂，这种开环加氢作用形成的化合物即断藿烷类化合物。常见的断藿烷有两种，一种是 E 环上 17,21 位 C—C 键断裂，另一种是 C 环上 8,14 位 C—C 断裂。由于作者尚未遇到含有 E 环断裂的断藿烷样品，本节仅介绍 C 环上 8,14 位 C—C 断裂形成的断藿烷。

断藿烷的特征离子是 m/z 123，在常规色质分析中仅能检测到少量的几个峰，如图 2-223(a)所示；在全二维分析谱图中，断藿烷与其他四个环的化合物在二维保留时间上比较接近，处于三环化合物和五环化合物之间的区域内，如图 2-223(b)和(c)所示，此区域可以检测到 31 个断藿烷化合物。在图 2-223(b)中，与断藿烷二维保留时间非常接近的甲基甾烷和甲藻甾烷化合物（特征离子 m/z 231，m/z 232）处于断藿烷峰的区域内，但不会影响断藿烷的检测。

检测到的 31 个断藿烷化合物中有较多的同分异构体，但同分异构体的具体构型目前尚无法确认（王培荣，1993），只能根据不同同分异构体质谱图（图 2-224~图 2-246）的差异将其分成 6 个类别。各化合物的鉴定结果见表 2-14。

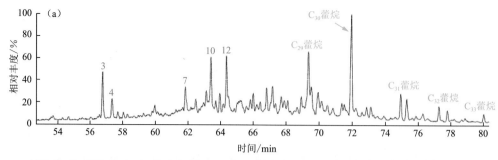
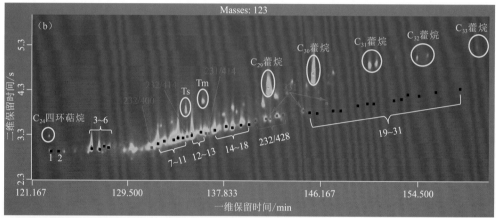
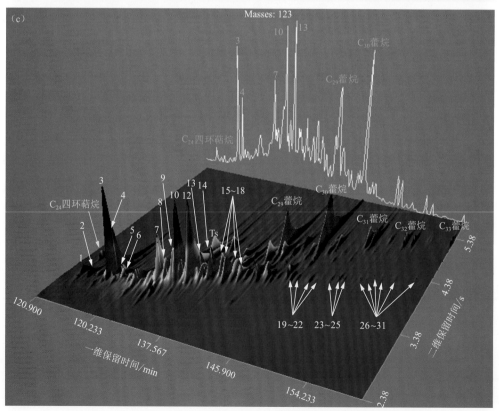

图 2-223　断藿烷的色质分析谱图(a)、全二维点阵图(b)和全二维立体图(c)

图中标记 232/400 表示此化合物的特征离子是 m/z 232，分子离子是 m/z 400

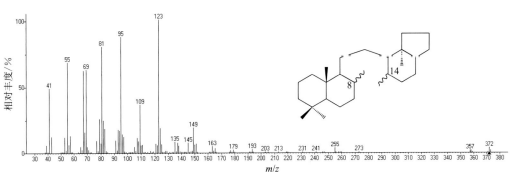

图 2-224　1 号峰　C_{27},8,14-断藿烷质谱图

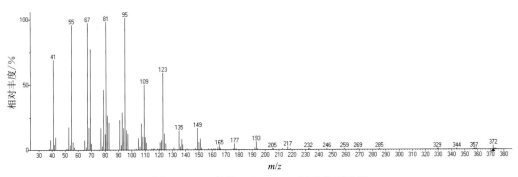

图 2-225　2 号峰　C_{27},8,14-断藿烷质谱图

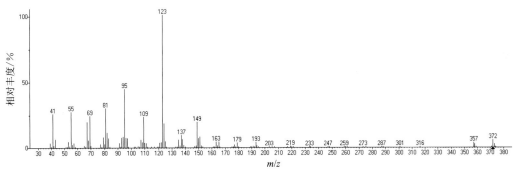

图 2-226　3 号峰　C_{27},8,14-断藿烷质谱图

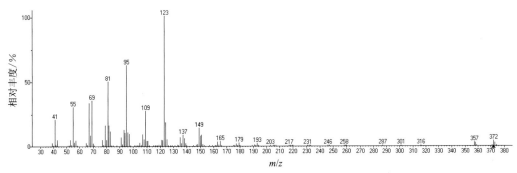

图 2-227　4 号峰　C_{27},8,14-断藿烷质谱图

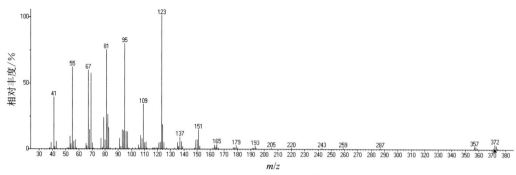

图 2-228　5 号峰　C_{27},8,14-断藿烷质谱图

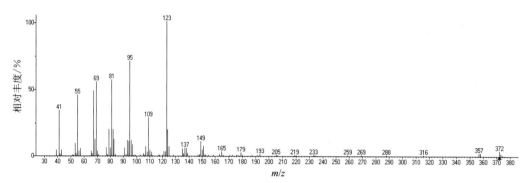

图 2-229　6 号峰　C_{27},8,14-断藿烷质谱图

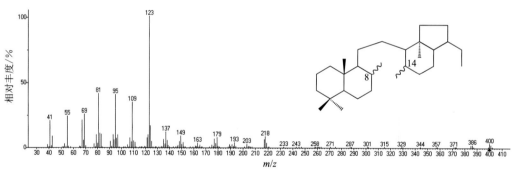

图 2-230　7 号峰　C_{29},8,14-断藿烷质谱图

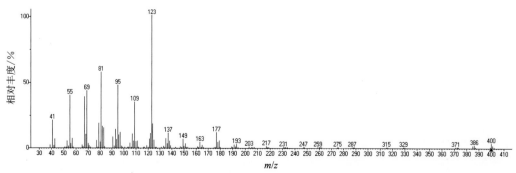

图 2-231　8 号峰　C_{29},8,14-断藿烷质谱图

第 2 章 石油地质样品的全二维气相色谱图识别

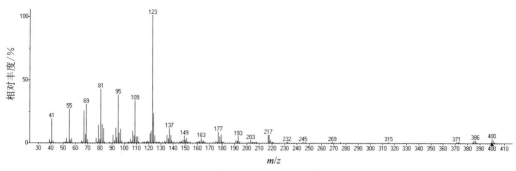

图 2-232　9 号峰　C_{29},8,14-断藿烷质谱图

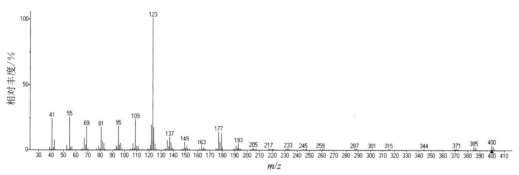

图 2-233　10 号峰　C_{29},8,14-断藿烷质谱图

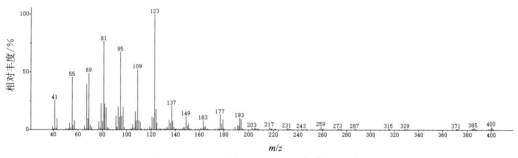

图 2-234　11 号峰　C_{29},8,14-断藿烷质谱图

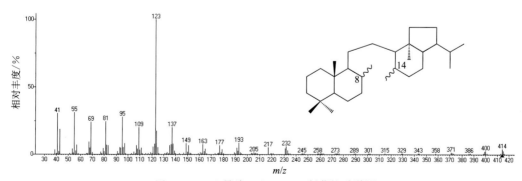

图 2-235　12 号峰　C_{30},8,14-断藿烷质谱图

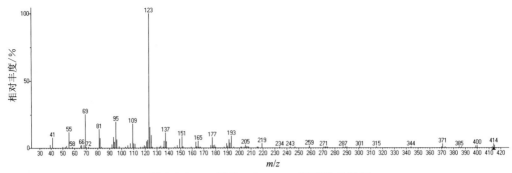

图 2-236　13 号峰　C_{30}, 8,14-断藿烷质谱图

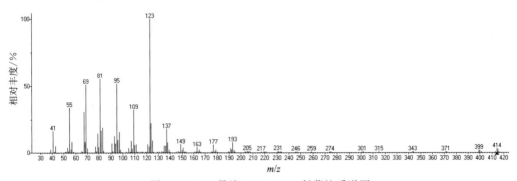

图 2-237　14 号峰　C_{30}, 8,14-断藿烷质谱图

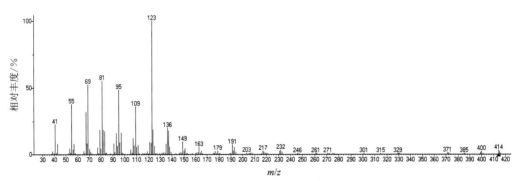

图 2-238　15 号峰　C_{30}, 8,14-断藿烷质谱图

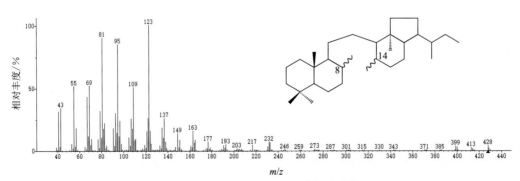

图 2-239　16 号峰　C_{31}, 8,14-断藿烷质谱图

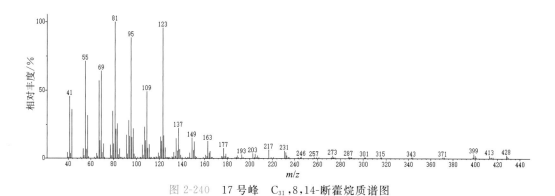

图 2-240　17 号峰　C_{31}, 8,14-断藿烷质谱图

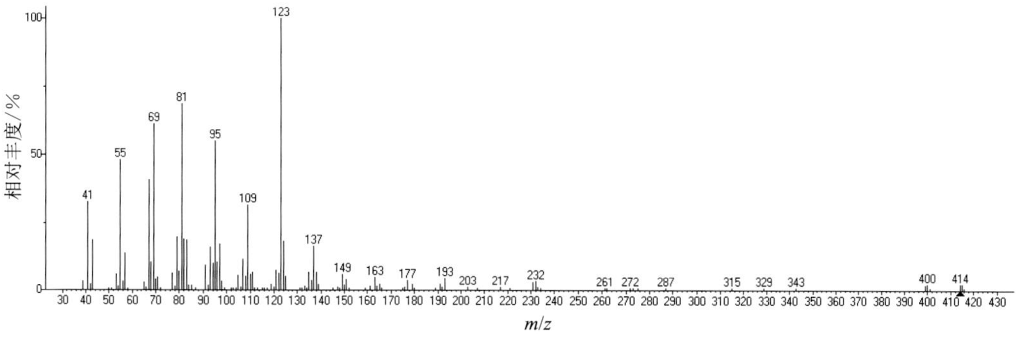

图 2-241　18 号峰　C_{30}, 8,14-断藿烷质谱图

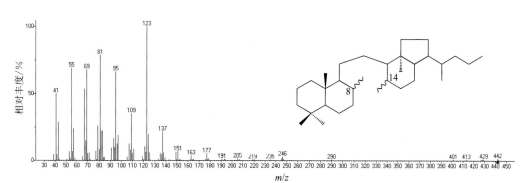

图 2-242　19 号峰　C_{32}, 8,14-断藿烷质谱图

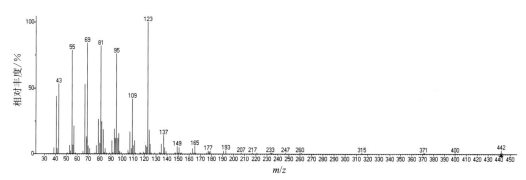

图 2-243　20 号峰　C_{32}, 8,14-断藿烷质谱图

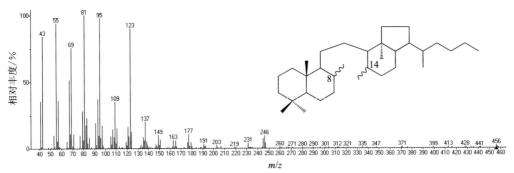

图 2-244　21 号峰　C_{33},8,14-断藿烷质谱图

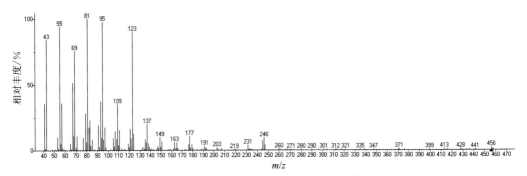

图 2-245　22 号峰　C_{33},8,14-断藿烷质谱图

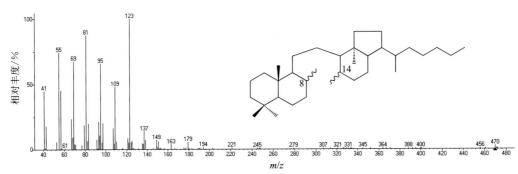

图 2-246　26 号峰　C_{34},8,14-断藿烷质谱图

表 2-14　断藿烷系列化合物鉴定结果表

峰号	一维保留时间 /min	二维保留时间 /s	特征离子 (m/z)	分子式	相对分子质量	名　称
1	122.77	2.95	123	$C_{27}H_{48}$	372	C_{27},8,14-Secohopane- C_{27},8,14-断藿烷
2	123.43	2.95	123	$C_{27}H_{48}$	372	C_{27},8,14-Secohopane- C_{27},8,14-断藿烷
3	126.37	3.01	123	$C_{27}H_{48}$	372	C_{27},8,14-Secohopane- C_{27},8,14-断藿烷
4	127.03	3.00	123	$C_{27}H_{48}$	372	C_{27},8,14-Secohopane- C_{27},8,14-断藿烷

续表

峰 号	一维保留时间 /min	二维保留时间 /s	特征离子 (m/z)	分子式	相对分子质量	名 称
5	127.43	3.06	123	$C_{27}H_{48}$	372	C_{27},8,14-Secohopane- C_{27},8,14-断藿烷
6	127.83	3.03	123	$C_{27}H_{48}$	372	C_{27},8,14-Secohopane- C_{27},8,14-断藿烷
7	132.10	3.13	123	$C_{29}H_{52}$	400	C_{29},8,14-Secohopane C_{29},8,14-断藿烷
8	132.63	3.18	123	$C_{29}H_{52}$	400	C_{29},8,14-Secohopane C_{29},8,14-断藿烷
9	133.43	3.21	123	$C_{29}H_{52}$	400	C_{29},8,14-Secohopane C_{29},8,14-断藿烷
10	133.83	3.25	123	$C_{29}H_{52}$	400	C_{29},8,14-Secohopane C_{29},8,14-断藿烷
11	134.23	3.25	123	$C_{29}H_{52}$	400	C_{29},8,14-Secohopane C_{29},8,14-断藿烷
12	135.03	3.31	123	$C_{30}H_{54}$	414	C_{30},8,14-Secohopane C_{30},8,14-断藿烷
13	135.83	3.39	123	$C_{30}H_{54}$	414	C_{30},8,14-Secohopane C_{30},8,14-断藿烷
14	137.03	3.43	123	$C_{30}H_{54}$	414	C_{30},8,14-Secohopane C_{30},8,14-断藿烷
15	137.97	3.51	123	$C_{30}H_{54}$	414	C_{30},8,14-Secohopane C_{30},8,14-断藿烷
16	138.50	3.49	123	$C_{31}H_{56}$	428	C_{31},8,14-Secohopane C_{31},8,14-断藿烷
17	139.30	3.54	123	$C_{31}H_{56}$	428	C_{31},8,14-Secohopane C_{31},8,14-断藿烷
18	139.97	3.55	123	$C_{30}H_{54}$	414	C_{30},8,14-Secohopane C_{30},8,14-断藿烷
19	145.17	3.83	123	$C_{32}H_{58}$	442	C_{32},8,14-Secohopane C_{32},8,14-断藿烷
20	145.83	3.81	123	$C_{32}H_{58}$	442	C_{32},8,14-Secohopane C_{32},8,14-断藿烷
21	147.17	3.88	123	$C_{33}H_{60}$	456	C_{33},8,14-Secohopane C_{33},8,14-断藿烷
22	147.83	3.88	123	$C_{33}H_{60}$	456	C_{33},8,14-Secohopane C_{33},8,14-断藿烷

续表

峰 号	一维保留时间/min	二维保留时间/s	特征离子（m/z）	分子式	相对分子质量	名 称
23	149.30	3.99	123	$C_{33}H_{60}$	456	C_{33}，8,14-Secohopane C_{33}，8,14-断藿烷
24	150.10	4.03	123	$C_{33}H_{60}$	456	C_{33}，8,14-Secohopane C_{33}，8,14-断藿烷
25	150.63	4.03	123	$C_{33}H_{60}$	456	C_{33}，8,14-Secohopane C_{33}，8,14-断藿烷
26	152.37	4.11	123	$C_{34}H_{62}$	470	C_{34}，8,14-Secohopane C_{34}，8,14-断藿烷
27	153.03	4.13	123	$C_{34}H_{62}$	470	C_{34}，8,14-Secohopane C_{34}，8,14-断藿烷
28	153.57	4.24	123	$C_{35}H_{64}$	484	C_{35}，8,14-Secohopane C_{35}，8,14-断藿烷
29	154.50	4.21	123	$C_{34}H_{62}$	470	C_{34}，8,14-Secohopane C_{34}，8,14-断藿烷
30	155.97	4.29	123	$C_{35}H_{64}$	484	C_{35}，8,14-Secohopane C_{35}，8,14-断藿烷
31	158.10	4.37	123	$C_{35}H_{64}$	484	C_{35}，8,14-Secohopane C_{35}，8,14-断藿烷

4）降藿烷类化合物的全二维气相色谱图及化合物名称

藿烷分子骨架上脱掉含碳基团形成的化合物都叫降藿烷。本节介绍的是处于C-10的编号为25的甲基被溶解，形成的25-降藿烷系列。

25-降藿烷系列化合物的特征离子为m/z 177，常规色质分析时正常藿烷类化合物的峰会影响部分25-降藿烷化合物的检测，如图2-247（a）所示。在全二维分析谱图上（图2-247b,c），断藿烷系列化合物位于正常藿烷的下方，不受其干扰。检测到的18个25-降藿烷化合物的质谱图如图2-248～图2-265所示，鉴定结果见表2-15。

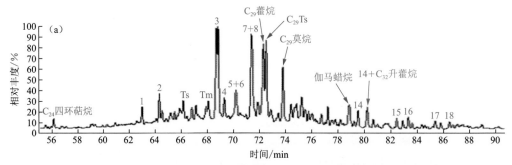

图2-247　25-断藿烷化合物的色质分析谱图（a）、全二维点阵图（b）和全二维立体图（c）

第 2 章　石油地质样品的全二维气相色谱图识别

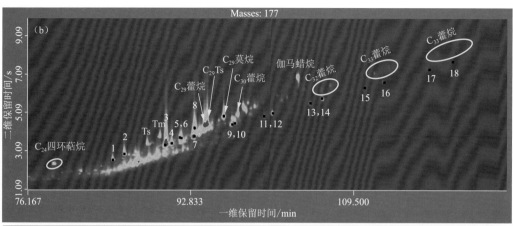

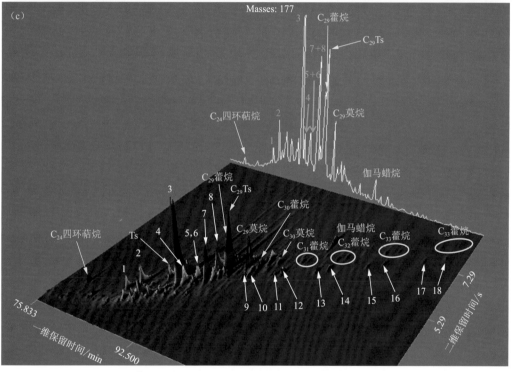

图 2-247(续)　25-断藿烷化合物的色质分析谱图(a)、全二维点阵图(b)和全二维立体图(c)

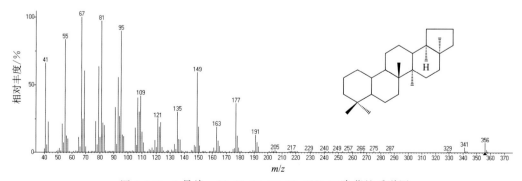

图 2-248　1 号峰　22,25,29,30-18α(H)-四降藿烷质谱图

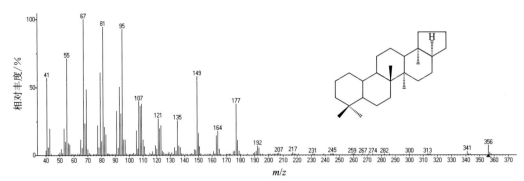

图 2-249 2号峰 22,25,29,30-17α(H)-四降藿烷质谱图

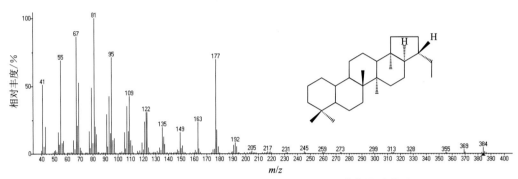

图 2-250 3号峰 17α(H),21β(H)-25,30-二降藿烷质谱图

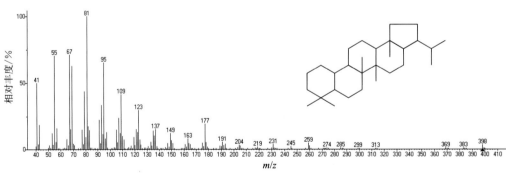

图 2-251 4号峰 C_{29},降藿烷质谱图

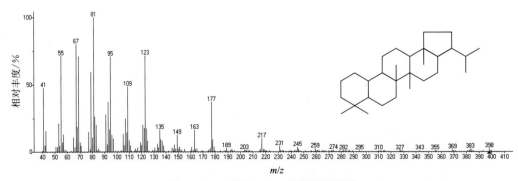

图 2-252 5号峰 C_{29},降藿烷质谱图

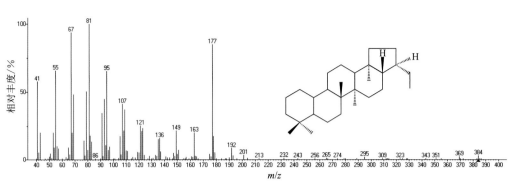

图 2-253　6 号峰　$17\beta(H),21\alpha(H)$-25,30-二降莫烷质谱图

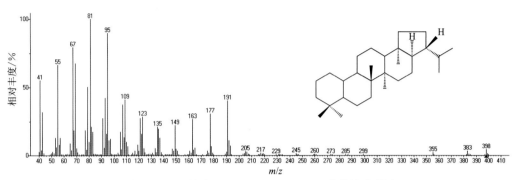

图 2-254　7 号峰　$17\alpha(H),21\beta(H)$-25-降藿烷质谱图

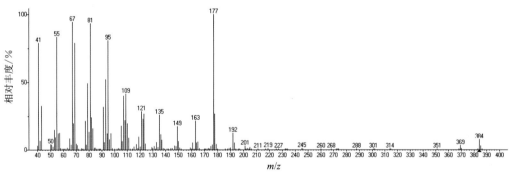

图 2-255　8 号峰　C_{28} 二降藿烷质谱图

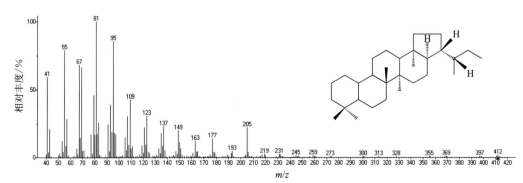

图 2-256　9 号峰　$17\alpha(H),21\beta(H)$-25-降升藿烷(22S)质谱图

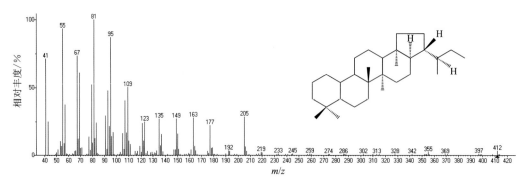

图 2-257 10 号峰 17α(H),21β(H)-25-降升藿烷(22R)质谱图

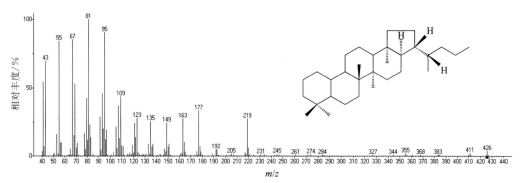

图 2-258 11 号峰 17α(H),21β(H) 25 降二升藿烷(22S)质谱图

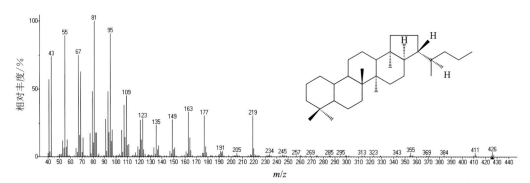

图 2-259 12 号峰 17α(H),21β(H)-25-降二升藿烷(22R)质谱图

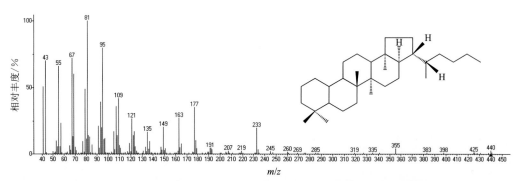

图 2-260 13 号峰 17α(H),21β(H)-25-降三升藿烷(22S)质谱图

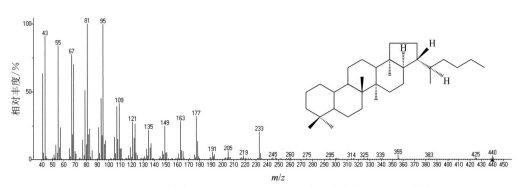

图 2-261　14 号峰　17α(H),21β(H)-25-降三升藿烷(22R)质谱图

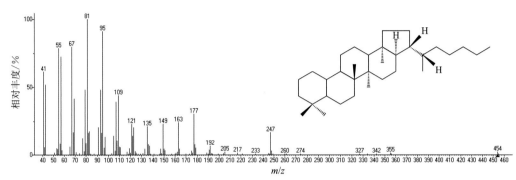

图 2-262　15 号峰　17α(H),21β(H)-25-降四升藿烷(22S)质谱图

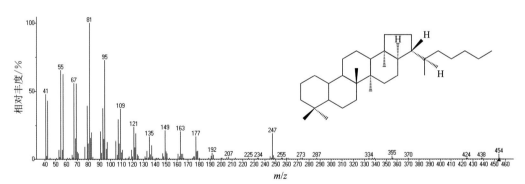

图 2-263　16 号峰　17α(H),21β(H)-25-降四升藿烷(22R)质谱图

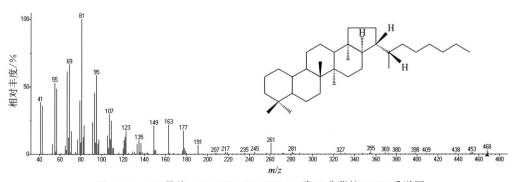

图 2-264　17 号峰　17α(H),21β(H)-25-降五升藿烷(22S)质谱图

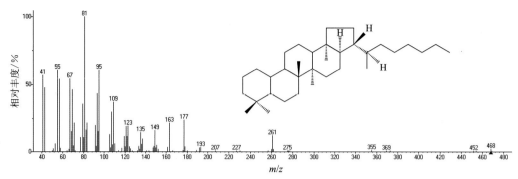

图 2-265　18 号峰　$17\alpha(H),21\beta(H)$-25-降五升藿烷(22R)质谱图

表 2-15　25-降藿烷系列化合物鉴定结果表

峰 号	一维保留时间 /min	二维保留时间 /s	特征离子 (m/z)	分子式	相对分子质量	名　称
1	84.83	2.67	177	$C_{26}H_{44}$	356	22,25,29,30-$18\alpha(H)$-Tetrasnorhopane 22,25,29,30-$18\alpha(H)$-四降藿烷
2	86.00	2.97	177	$C_{26}H_{44}$	356	22,25,29,30-$17\alpha(H)$-Tetrasnorhopane 22,25,29,30-$17\alpha(H)$-四降藿烷
3	90.17	3.42	177	$C_{28}H_{48}$	384	$17\alpha(H),21\beta(H)$-25,30-Bisnorhopane $17\alpha(H),21\beta(H)$-25,30-二降藿烷
4	90.83	3.51	177	$C_{29}H_{50}$	398	C_{29},Norhopane C_{29},降藿烷
5	91.67	3.82	177	$C_{29}H_{50}$	398	C_{29},Norhopane C_{29},降藿烷
6	91.83	3.81	177	$C_{28}H_{48}$	384	$17\beta(H),21\alpha(H)$-25,30-Bisnormoretane $17\beta(H),21\alpha(H)$-25,30-二降莫烷
7	93.17	3.88	177	$C_{29}H_{50}$	398	$17\alpha(H),21\beta(H)$-25-Norhopane $17\alpha(H),21\beta(H)$-25-降藿烷
8	93.17	4.32	177	$C_{28}H_{48}$	384	C_{28},Bisnorhopane C_{28},二降藿烷
9	97.00	4.49	177	$C_{30}H_{52}$	412	$17\alpha(H),21\beta(H)$- 25-Norhomohopane(22S) $17\alpha(H),21\beta(H)$-25-降升藿烷(22S)
10	97.33	4.56	177	$C_{30}H_{52}$	412	$17\alpha(H),21\beta(H)$- 25-Norhomohopane(22R) $17\alpha(H),21\beta(H)$-25-降升藿烷(22R)
11	100.33	4.92	177	$C_{31}H_{54}$	426	$17\alpha(H),21\beta(H)$- 25-Norbishomohopane(22S) $17\alpha(H),21\beta(H)$-25-降二升藿烷(22S)
12	101.17	5.09	177	$C_{31}H_{54}$	426	$17\alpha(H),21\beta(H)$- 25-Norbishomohopane(22R) $17\alpha(H),21\beta(H)$-25-降二升藿烷(22R)
13	105.00	5.60	177	$C_{32}H_{56}$	440	$17\alpha(H),21\beta(H)$- 25-Nortrishomohopane(22S) $17\alpha(H),21\beta(H)$-25-降三升藿烷(22S)

续表

峰号	一维保留时间/min	二维保留时间/s	特征离子(m/z)	分子式	相对分子质量	名称
14	106.17	5.86	177	$C_{32}H_{56}$	440	$17\alpha(H),21\beta(H)$-25-Nortrishomohopane(22R) $17\alpha(H),21\beta(H)$-25-降三升藿烷(22R)
15	110.50	6.40	177	$C_{33}H_{58}$	454	$17\alpha(H),21\beta(H)$-25-Nortrashomohopane(22S) $17\alpha(H),21\beta(H)$-25-降四升藿烷(22S)
16	112.50	6.70	177	$C_{33}H_{58}$	454	$17\alpha(H),21\beta(H)$-25-Nortrashomohopane(22R) $17\alpha(H),21\beta(H)$-25-降四升藿烷(22R)
17	117.17	7.36	177	$C_{34}H_{60}$	468	$17\alpha(H),21\beta(H)$-25-Norpentashomohopane(22S) $17\alpha(H),21\beta(H)$-25-降五升藿烷(22S)
18	119.67	7.77	177	$C_{34}H_{60}$	468	$17\alpha(H),21\beta(H)$-25-Norpentashomohopane(22R) $17\alpha(H),21\beta(H)$-25-降五升藿烷(22R)

5)$\beta\beta$ 藿烷及藿烯的全二维气相色谱图及化合物名称

在一些特殊的低熟样品中,利用全二维气相色谱-飞行时间质谱的 m/z 191 离子可以检测到 $\beta\beta$ 藿烷及藿烯(特征离子 m/z 189)类化合物,如图 2-266 所示。$\beta\beta$ 藿烷及藿烯的质谱图及鉴定结果如图 2-267~图 2-271 所示。

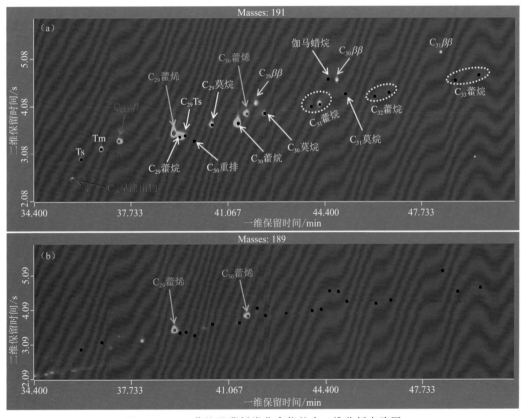

图 2-266 $\beta\beta$ 藿烷及藿烯类化合物的全二维分析点阵图

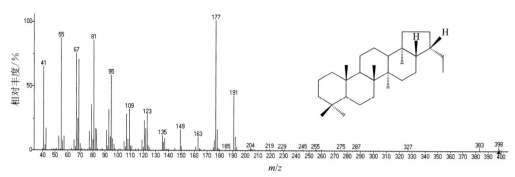

图 2-267　17β(H),21β(H)-30-降藿烷质谱图

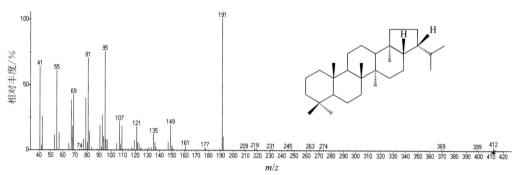

图 2-268　17β(H),21β(H)-藿烷质谱图

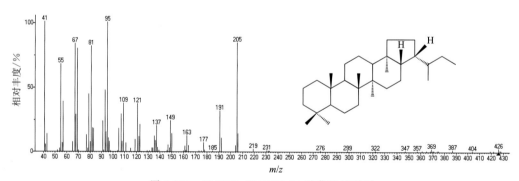

图 2-269　17β(H),21β(H)-30-升藿烷质谱图

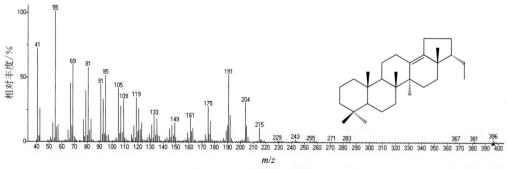

图 2-270　C_{29}藿烯质谱图

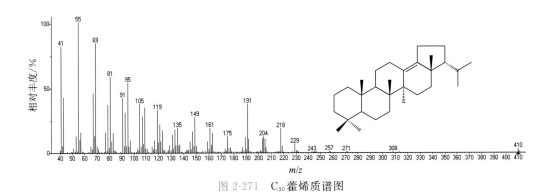

图 2-271　C_{30}藿烯质谱图

2.2 >> 芳烃组分的全二维气相色谱图特征及鉴定

采用全二维气相色谱-飞行时间质谱分析石油地质样品中的芳烃组分或者凝析油样品，可以检测到芳烃组分中从单环到多环的各种类型化合物。不同类型化合物的名称、同系物分子式、基本化学结构以及质谱碎片的主要特征离子见表 2-16。

表 2-16　芳烃组分中的化合物类型及化学结构

序号	化合物类型	同系物分子式	典型化合物	基本化学结构	主要特征离子（m/z）
1	单环芳烃	C_nH_{2n-6}	苯		91,105,106,119,120,133,134,…
		C_nH_{2n-8}	环烷苯：环己基取代苯、二氢化茚		160,174,117,132,131
		C_nH_{2n-10}	茚、二氢萘、双环烷苯：如八氢菲、六氢芴、四氢-二氢苊等		116,186,130,172,158
2	双环芳烃	C_nH_{2n-12}	萘		128,142,156,170,184,198
		C_nH_{2n-14}	二氢化苊、联苯、四氢菲		154,168,182
		C_nH_{2n-16}	芴、二氢菲、六氢芘		166,180,194,208

续表

序号	化合物类型	同系物分子式	典型化合物	基本化学结构	主要特征离子（m/z）
3	三环芳烃	C_nH_{2n-18}	菲、蒽		178,192,206,220,234,248
		C_nH_{2n-20}	苯基萘		204,218,232
4	四环芳烃	C_nH_{2n-22}	芘、荧蒽		202,216,230,244
		C_nH_{2n-24}	䓛		228,242,256
5	五环芳烃	C_nH_{2n-28}	苯并荧蒽,苯并芘,苝		252,266,280
		C_nH_{2n-30}	苯并䓛,芘		278,292,306
6	六环芳烃	C_nH_{2n-32}	苯并苝		276,290,304
7	芳构化甾萜烷烯	C_nH_{2n-12}	单芳甾	C环-单芳甾	253,267
		C_nH_{2n-12}	单芳断藿烷		159,365

续表

序号	化合物类型	同系物分子式	典型化合物	基本化学结构	主要特征离子 (m/z)
7	芳构化甾萜烷烯	C_nH_{2n-12}	单芳断降藿烷		145,351
		C_nH_{2n-14}	单芳断降藿烯		143,349 157,363
		C_nH_{2n-18}	双芳五环三萜烷		181,195,209
		C_nH_{2n-20}	三芳甾		231,245
		C_nH_{2n-22}	三芳五环三萜烷		271,285,299
		C_nH_{2n-26}	四芳五环三萜烷		281
		C_nH_{2n-30}	五芳五环三萜烷（二甲基苉）		306

续表

序号	化合物类型	同系物分子式	典型化合物	基本化学结构	主要特征离子（m/z）
7	芳构化甾萜烷烯	C_nH_{2n-16}	苯并藿烷		191

图 2-272 是芳烃组分的全二维分析谱图，从图中可以看出，不同种类的芳烃化合物分布在谱图的不同区域，族分离的特性和"瓦片效应"明显。最下方一族是相对极性最弱的苯系列化合物（长侧链和支链取代苯），其在一维方向随保留时间增加，取代基碳数增加。在它的上方随着二维保留时间的增加依次排列的是双环芳烃系列，芴、氧芴等系列，三环芳烃系列，四环芳烃系列、五环芳烃系列，单芳甾和三芳甾等。每一族化合物都有不同的结构特征和特征离子，这些化合物中少部分可用谱库软件直接识别，但大部分需要用质谱图与相关文献的标准谱图比对进行定性识别。

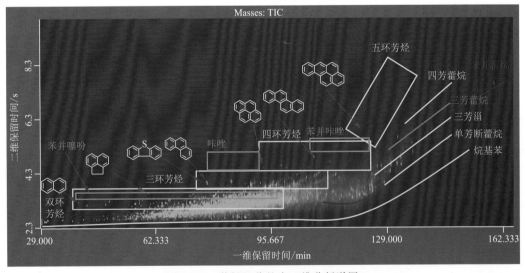

图 2-272　芳烃组分的全二维分析谱图

2.1.1　苯系列化合物的全二维气相色谱图及化合物名称

苯系列化合物中主要包括苯及其苯环上"1 个位"~"4 个位"被烷基取代的系列化合物。取代位数不同的苯系列化合物质谱分析的特征离子分别为 $m/z\ 91, m/z\ 105, m/z\ 119$ 等，但当取代基长度较长时，因发生 γ 氢重排使得偶数碎片离子强度增加，特征离子由

奇数 m/z 91，m/z 105，m/z 119 等变为偶数 m/z 92，m/z 106，m/z 120 等。

在"1 个位"取代的苯系列化合物中，正烷基苯系列化合物极性最小，在二维保留时间上位于芳烃全二维气相色谱分析谱图的最下方，如图 2-273 所示。正烷基苯系列化合物质谱分析的特征离子为 m/z 92，由于一般芳烃样品中正烷基苯含量较高，用常规色质分析（图 2-273a）和全二维分析（图 2-273b，c）效果都比较好，23 个化合物的鉴定结果见表 2-17。

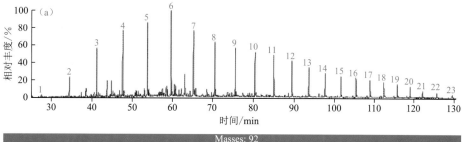

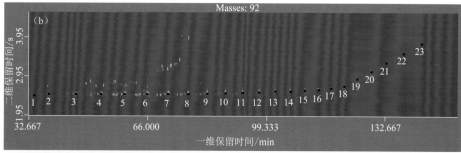

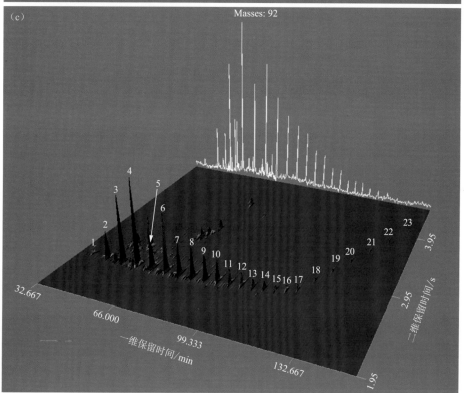

图 2-273　正烷基苯的色质分析谱图(a)、全二维点阵图(b)和三维立体图(c)

表 2-17 正烷基苯化合物鉴定结果表

峰号	一维保留时间 /min	二维保留时间 /s	特征离子 (m/z)	分子式	相对分子质量	名称
1	38.13	2.52	92	$C_{11}H_{16}$	148	Benzene, pentyl- 戊基苯
2	45.60	2.51	92	$C_{12}H_{18}$	162	Benzene, hexyl- 己基苯
3	52.67	2.51	92	$C_{13}H_{20}$	176	Benzene, heptyl- 庚基苯
4	59.33	2.51	92	$C_{14}H_{22}$	190	Benzene, octyl- 辛基苯
5	65.60	2.51	92	$C_{15}H_{24}$	204	Benzene, nonyl- 壬基苯
6	71.60	2.52	92	$C_{16}H_{26}$	218	Benzene, decyl- 癸基苯
7	77.33	2.52	92	$C_{17}H_{28}$	232	Benzene, undecyl- 十一烷基苯
8	82.67	2.53	92	$C_{18}H_{30}$	246	Benzene, dodecyl- 十二烷基苯
9	87.73	2.54	92	$C_{19}H_{32}$	260	Benzene, tridecyl- 十三烷基苯
10	92.53	2.56	92	$C_{20}H_{34}$	274	Benzene, tetradecyl- 十四烷基苯
11	97.07	2.57	92	$C_{21}H_{36}$	288	Benzene, pentadecyl- 十五烷基苯
12	101.47	2.59	92	$C_{22}H_{38}$	302	Benzene, hexadecyl- 十六烷基苯
13	105.73	2.60	92	$C_{23}H_{40}$	316	Benzene, heptadecyl- 十七烷基苯
14	109.73	2.62	92	$C_{24}H_{42}$	330	Benzene, octadecyl- 十八烷基苯
15	113.60	2.64	92	$C_{25}H_{44}$	344	Benzene, nonadecyl- 十九烷基苯
16	117.20	2.67	92	$C_{26}H_{46}$	358	Benzene, eicosyl- 二十烷基苯
17	120.80	2.72	92	$C_{27}H_{48}$	372	Benzene, heneicosyl- 二十一烷基苯
18	124.53	2.91	92	$C_{28}H_{50}$	386	Benzene, docosyl- 二十二烷基苯
19	128.53	3.10	92	$C_{29}H_{52}$	400	Benzene, tricosyl- 二十三烷基苯
20	132.80	3.32	92	$C_{30}H_{54}$	414	Benzene, tetracosyl- 二十四烷基苯
21	137.73	3.55	92	$C_{31}H_{56}$	428	Benzene, pentacosyl- 二十五烷基苯

续表

峰 号	一维保留时间 /min	二维保留时间 /s	特征离子 (m/z)	分子式	相对分子质量	名 称
22	142.93	3.80	92	$C_{32}H_{58}$	442	Benzene, hexacosyl- 二十六烷基苯
23	148.53	4.05	92	$C_{33}H_{60}$	456	Benzene, heptacosyl- 二十七烷基苯

苯系的"2个位"被取代和"3个位"被取代的化合物种类更多,如果"2个位"中有1个位的取代基是甲基,那么其质谱分析的特征离子为 m/z 105 或者 m/z 106;如果"3个位"中有2个位的取代基是甲基,那么其质谱分析的特征离子为 m/z 119 或者 m/z 120。3 种不同取代形式的苯系化合物在全二维分析谱图上依据相对分子质量的大小规则排列(图 2-274)。化合物的鉴定可以根据标准质谱图比对确定,各化合物在全二维上的分布规律也可为其结构确定提供帮助。

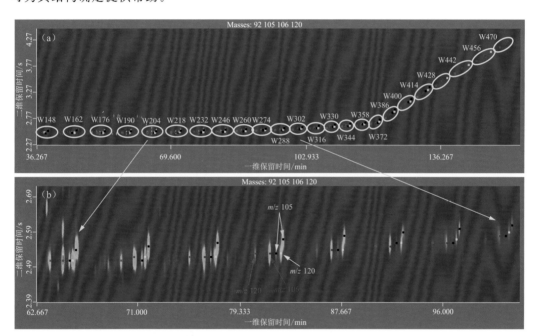

图 2-274 不同取代的苯系化合物全二维分析谱图

图中特征离子是 m/z 105 的化合物用黑色点标记,特征离子是 m/z 106 的化合物用红色点标记,
特征离子是 m/z 120 的化合物分别用玫红色和黄色点标记。图(a)中"W148"表示该区域的
化合物的相对分子质量是 148,其他同理;图(b)是图(a)中相对分子质量 204~302 之间化合物的放大图

苯环上"4个位"被取代的化合物(其中有3个位的取代基是甲基)的质谱分析特征离子为 m/z 133 或者 m/z 134,此类化合物最典型的代表是三甲基苯基类异戊二烯化合物。事实上,m/z 133 和 m/z 134 对化合物同分异构体的响应灵敏度有差异,无论是在常规色质分析谱图(图 2-275b)上还是在全二维分析谱图(图 2-275a)上,都可以看到不同极性的两种同分异构体响应的差异,不同同分异构体对应的质谱图(图 2-276 和图 2-277)特征离子差异明显。

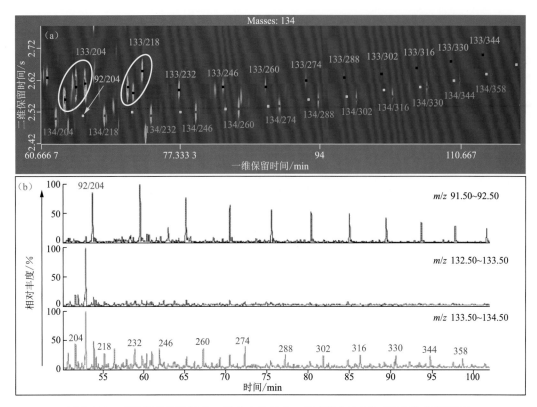

图 2-275　三甲基苯基类异戊二烯化合物的全二维点阵图(a)和 GC-MS 色谱图(b)

图(a)中特征离子是 m/z 92 的化合物用黄色点标记,特征离子是 m/z 133 的化合物用黑色点标记,特征离子是 m/z 134 的化合物用红色点标记,其中"133/204"表示该化合物的特征离子是 m/z 133,相对分子质量是 204,其他同理;图(b)中"204"表示该化合物的相对分子质量,其他同理

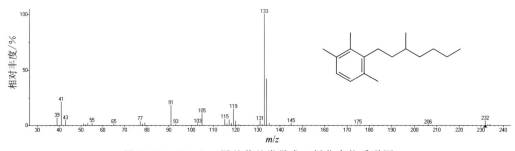

图 2-276　2,3,6-三甲基苯基类异戊二烯化合物质谱图

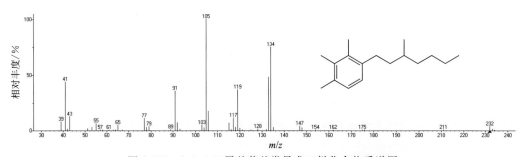

图 2-277　2,3,4-三甲基苯基类异戊二烯化合物质谱图

其他取代基的苯系化合物检测多用其分子离子。图 2-278 是分子离子检测到的单环苯系化合物。

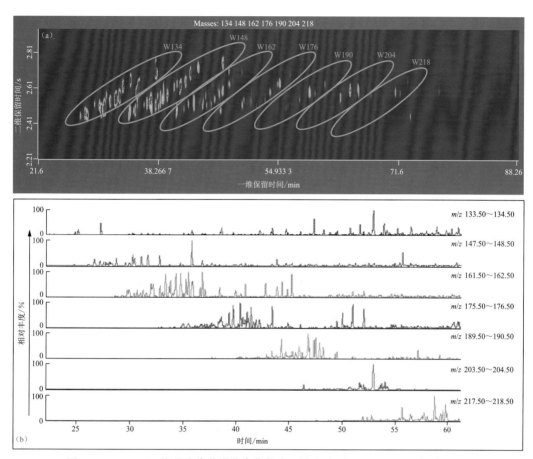

图 2-278　$C_4 \sim C_{10}$ 烷基取代苯类化合物的全二维点阵图(a)和 GC-MS 色谱图(b)

图(a)中 W134 表示该区域的化合物的相对分子质量是 134，其他同理

2.2.2　萘系列化合物的全二维气相色谱图及化合物名称

萘系列化合物在全二维气相色谱图中处于苯系列化合物的上方，一般包括萘、甲基萘以及取代基碳数从 2 到 5 的系列化合物，如图 2-279 所示。与其他系列化合物一样，萘系列化合物依据取代基的碳数在全二维谱图上有规律地分布，其分子离子可实现对应化合物的鉴定，即 m/z 128，m/z 142，m/z 156，m/z 170，m/z 184 以及 m/z 198 可分别用于萘、甲基萘及取代基碳数 2～5 的萘系列化合物的检测。

萘及 2-甲基萘、1-甲基萘在全二维谱图上很容易被识别，如图 2-279(a)所示。

取代基碳数为 2 的萘系列化合物检测到 11 个，峰号自 4 至 14，如图 2-280 所示。从图中可以看出，传统色质分析时的共馏峰(图 2-280a)在全二维气相色谱-飞行时间质谱分析谱图(图 2-280b)中被完全分开。各化合物的鉴定结果见表 2-18。

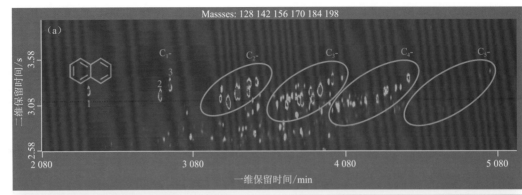

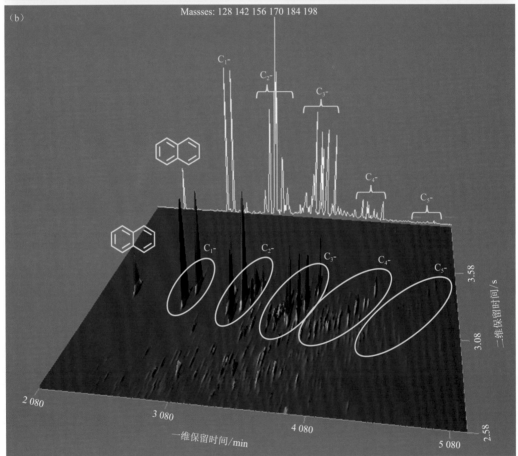

图 2-279 萘系列化合物的全二维点阵图(a)和三维立体图(b)

取代基碳数为 3 的萘系列化合物在图 2-281 上共标记出 23 个(如图 2-281b 上黑点标记所示),通过与专业书籍比对,给出其中 12 个化合物的定性结果,峰号自 15 至 26,其他化合物的质谱图与这 12 个化合物的相似度极高,根据全二维谱图的族分离和"瓦片效应"特点推测它们也是取代基碳数为 3 的萘系列化合物。从图中可以看出,传统色质检测时分离度不高的化合物(图 2-281a)在全二维气相色谱-飞行时间质谱图(图 2-281b)中实现了较好的分离。15,16 号峰是丙基萘的不同异构体,其质谱图如图 2-282 和图 2-283 所示。其他各化合物的鉴定结果见表 2-18。

第 2 章 石油地质样品的全二维气相色谱图识别

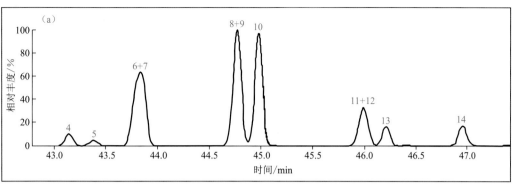

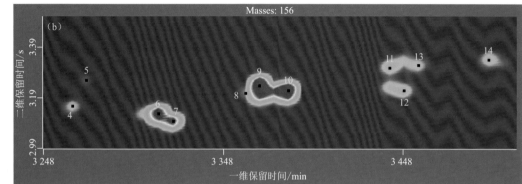

图 2-280 取代基碳数为 2 的萘系列化合物的 GC-MS 谱图(a)和全二维点阵图(b)

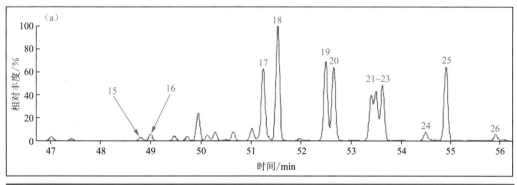

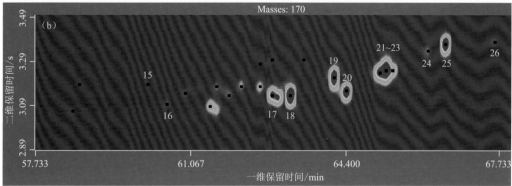

图 2-281 取代基碳数为 3 的萘系列化合物的 GC-MS 谱图(a)和全二维点阵图(b)

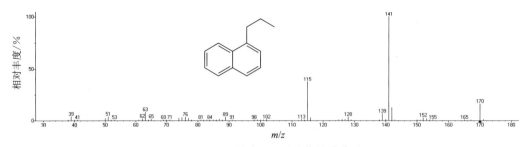

图 2-282　15 号峰　1-丙基萘的质谱图

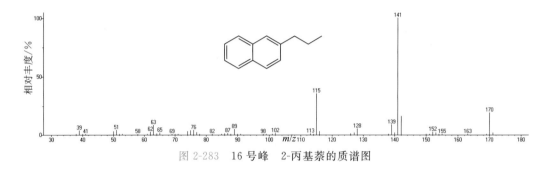

图 2-283　16 号峰　2-丙基萘的质谱图

取代基碳数为 4 的萘系列化合物数量很多(如图 2-284b 上黑点标记所示),通过与专业书籍比对,给出其中 12 个化合物的定性结果,峰号自 27 至 38,其他化合物的结构未知,如图 2-284 所示。12 个化合物的质谱图如图 2-285～图 2-296 所示,鉴定结果见表 2-18。

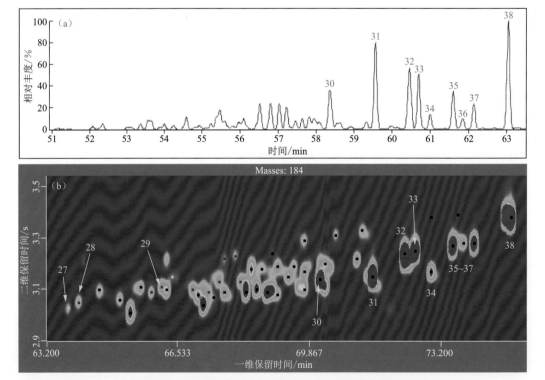

图 2-284　取代基碳数为 4 的萘系列化合物的 GC-MS 色谱图(a)和全二维点阵图(b)

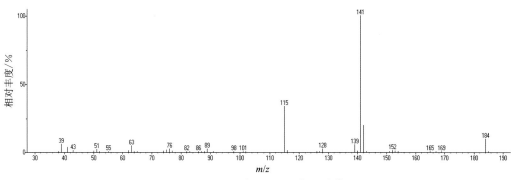

图 2-285 27 号峰 1-丁基萘的质谱图

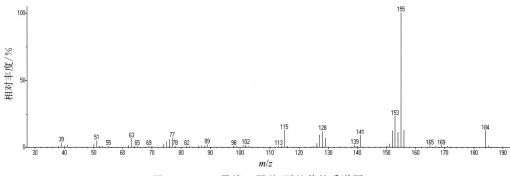

图 2-286 28 号峰 甲基-丙基萘的质谱图

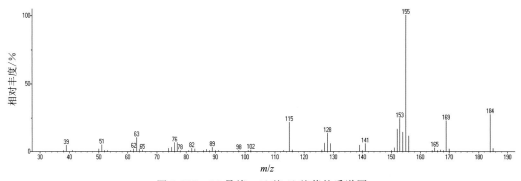

图 2-287 29 号峰 乙基-乙基萘的质谱图

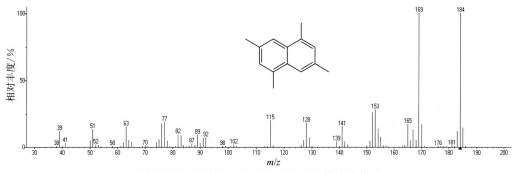

图 2-288 30 号峰 1,3,5,7-四甲基萘的质谱图

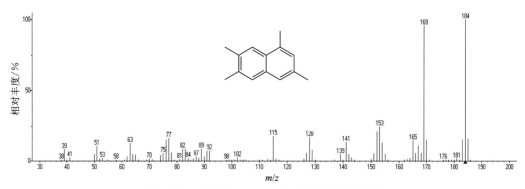

图 2-289　31 号峰　1,3,6,7-四甲基萘的质谱图

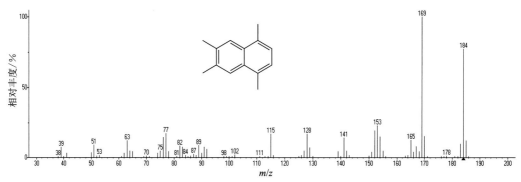

图 2-290　32 号峰　1,2,4,6-+1,2,4,7-+1,4,6,7-四甲基萘的质谱图

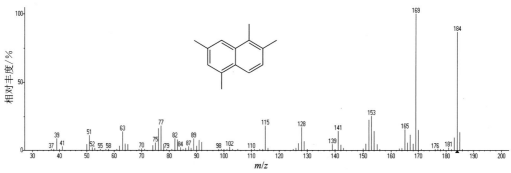

图 2-291　33 号峰　1,2,5,7-四甲基萘的质谱图

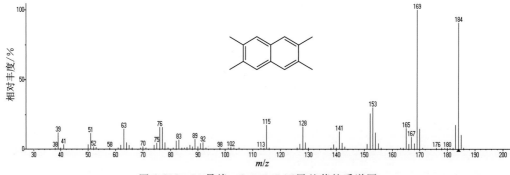

图 2-292　34 号峰　2,3,6,7-四甲基萘的质谱图

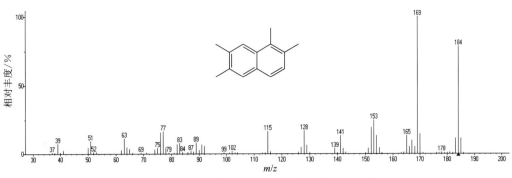

图 2-293　35 号峰　1,2,6,7-四甲基萘的质谱图

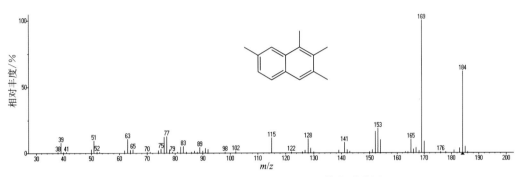

图 2-294　36 号峰　1,2,3,7-四甲基萘的质谱图

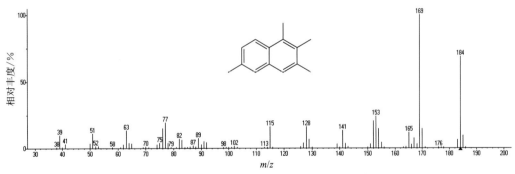

图 2-295　37 号峰　1,2,3,6-四甲基萘的质谱图

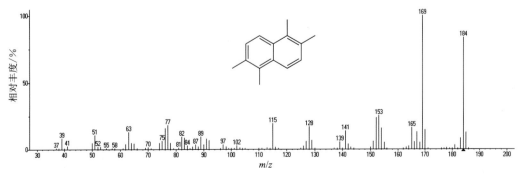

图 2-296　38 号峰　1,2,5,6-四甲基萘的质谱图

取代基碳数为 5 的萘系列化合物数量更多（如图 2-297b 上黑点标记所示），在全二维谱图中位于甲基-二苯并噻吩左下方。本节仅识别出取代基全是甲基的 4 个化合物，峰号

自 39 至 42,其他化合物的结构未知,如图 2-297 所示。各化合物的质谱图如图 2-298～图 2-301所示,鉴定结果见表 2-18。

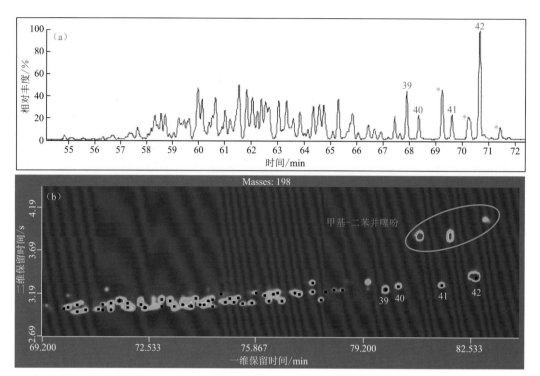

图 2-297　取代基碳数为 5 的萘系列化合物的 GC-MS 色谱图(a)和全二维点阵图(b)

图(a)中 * 标记的是甲基-二苯并噻吩,与图(b)中椭圆线圈出的化合物对应

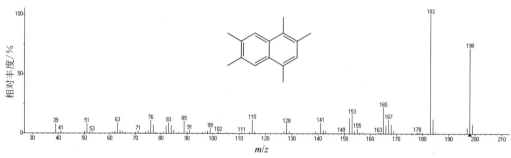

图 2-298　39 号峰　1,2,4,6,7-五甲基萘的质谱图

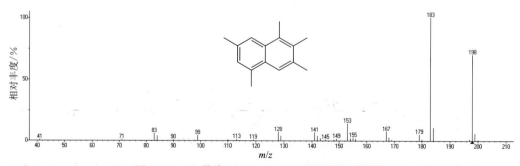

图 2-299　40 号峰　1,2,3,5,7-五甲基萘的质谱图

第 2 章 石油地质样品的全二维气相色谱图识别

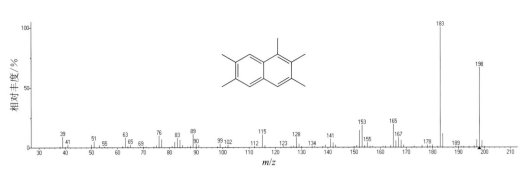

图 2-300　41 号峰　1,2,3,6,7-五甲基萘的质谱图

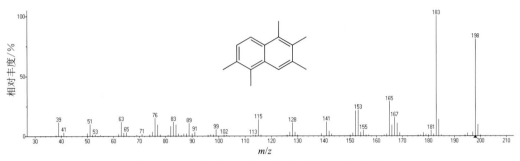

图 2-301　42 号峰　1,2,3,5,6-五甲基萘的质谱图

表 2-18　萘系列化合物鉴定结果表

峰号	一维保留时间 /min	二维保留时间 /s	特征离子 (m/z)	分子式	相对分子质量	名　称
1	39.73	3.24	128	$C_{10}H_8$	128	Naphthalene 萘
2	47.73	3.20	142	$C_{11}H_{10}$	142	2-Methyl-naphthalene 2-甲基萘
3	48.80	3.31	142	$C_{11}H_{10}$	142	1-Methyl-naphthalene 1-甲基萘
4	54.40	3.16	156	$C_{12}H_{12}$	156	2-Ethyl-naphthalene 2-乙基萘
5	54.53	3.26	156	$C_{12}H_{12}$	156	1-Ethyl-naphthalene 1-乙基萘
6	55.20	3.13	156	$C_{12}H_{12}$	156	2,6-Dimethyl-naphthalene 2,6-二甲基萘
7	55.33	3.10	156	$C_{12}H_{12}$	156	2,7-Dimethyl-naphthalene 2,7-二甲基萘
8	56.00	3.21	156	$C_{12}H_{12}$	156	1,7-Dimethyl-naphthalene 1,7-二甲基萘
9	56.13	3.24	156	$C_{12}H_{12}$	156	1,3-Dimethyl-naphthalene 1,3-二甲基萘

续表

峰 号	一维保留时间 /min	二维保留时间 /s	特征离子 (m/z)	分子式	相对分子质量	名　称
10	56.40	3.22	156	$C_{12}H_{12}$	156	1,6-Dimethyl-naphthalene 1,6-二甲基萘
11	57.33	3.31	156	$C_{12}H_{12}$	156	1,4-Dimethyl-naphthalene 1,4-二甲基萘
12	57.47	3.22	156	$C_{12}H_{12}$	156	2,3-Dimethyl-naphthalene 2,3-二甲基萘
13	57.60	3.32	156	$C_{12}H_{12}$	156	1,5-Dimethyl-naphthalene 1,5-二甲基萘
14	58.27	3.34	156	$C_{12}H_{12}$	156	1,2-Dimethyl-naphthalene 1,2-二甲基萘
15	60.13	3.19	170	$C_{13}H_{14}$	170	1-Propyl-naphthalene 1-丙基萘
16	60.53	3.10	170	$C_{13}H_{14}$	170	2-Propyl-naphthalene 2-丙基萘
17	62.80	3.14	170	$C_{13}H_{14}$	170	1,3,7-Trimethyl-naphthalene 1,3,7-三甲基萘
18	63.20	3.14	170	$C_{13}H_{14}$	170	1,3,6-Trimethyl-naphthalene 1,3,6-三甲基萘
19	64.13	3.22	170	$C_{13}H_{14}$	170	1,3,5-＋1,4,6-Trimethyl-naphthalene 1,3,5-＋1,4,6-三甲基萘
20	64.40	3.16	170	$C_{13}H_{14}$	170	2,3,6-Trimethyl-naphthalene 2,3,6-三甲基萘
21	65.07	3.24	170	$C_{13}H_{14}$	170	1,2,7-Trimethyl-naphthalene 1,2,7-三甲基萘[1]
22	65.20	3.25	170	$C_{13}H_{14}$	170	1,6,7-Trimethyl-naphthalene 1,6,7-三甲基萘[1]
23	65.33	3.25	170	$C_{13}H_{14}$	170	2,3,5-Trimethyl-naphthalene 2,3,5-三甲基萘[1]
24	66.13	3.34	170	$C_{13}H_{14}$	170	1,2,4-Trimethyl-naphthalene 1,2,4-三甲基萘
25	66.53	3.37	170	$C_{13}H_{14}$	170	1,2,5-Trimethyl-naphthalene 1,2,5-三甲基萘
26	67.60	3.38	170	$C_{13}H_{14}$	170	1,2,3-＋1,4,5-Trimethyl-naphthalene 1,2,3-＋1,4,5-三甲基萘
27	63.73	3.03	184	$C_{14}H_{16}$	184	1-Butyl-naphthalene 1-丁基萘
28	64.00	3.05	184	$C_{14}H_{16}$	184	Methyl-propyl-naphthalene 甲基-丙基萘
29	66.13	3.11	184	$C_{14}H_{16}$	184	Ethyl-ethyl-naphthalene 乙基-乙基萘

续表

峰 号	一维保留时间 /min	二维保留时间 /s	特征离子 (m/z)	分子式	相对分子质量	名 称
30	70.13	3.14	184	$C_{14}H_{16}$	184	1,3,5,7-Tetramethyl-naphthalene 1,3,5,7-四甲基萘
31	71.47	3.15	184	$C_{14}H_{16}$	184	1,3,6,7-Tetramethyl-naphthalene 1,3,6,7-四甲基萘
32	72.27	3.24	184	$C_{14}H_{16}$	184	1,2,4,6- + 1,2,4,7- + 1,4,6,7-Tetramethyl-naphthalene 1,2,4,6- + 1,2,4,7- + 1,4,6,7-四甲基萘
33	72.53	3.25	184	$C_{14}H_{16}$	184	1,2,5,7-Tetramethyl-naphthalene 1,2,5,7-四甲基萘
34	72.93	3.17	184	$C_{14}H_{16}$	184	2,3,6,7-Tetramethyl-naphthalene 2,3,6,7-四甲基萘
35	73.47	3.27	184	$C_{14}H_{16}$	184	1,2,6,7-Tetramethyl-naphthalene 1,2,6,7-四甲基萘
36	73.73	3.28	184	$C_{14}H_{16}$	184	1,2,3,7-Tetramethyl-naphthalene 1,2,3,7-四甲基萘
37	74.00	3.28	184	$C_{14}H_{16}$	184	1,2,3,6-Tetramethyl-naphthalene 1,2,3,6-四甲基萘[②]
38	74.93	3.38	184	$C_{14}H_{16}$	184	1,2,5,6-Tetramethyl-naphthalene 1,2,5,6-四甲基萘
39	79.87	3.25	198	$C_{15}H_{18}$	198	1,2,4,6,7-Pentamethyl-naphthalene 1,2,4,6,7-五甲基萘
40	80.27	3.29	198	$C_{15}H_{18}$	198	1,2,3,5,7-Pentamethyl-naphthalene 1,2,3,5,7-五甲基萘
41	81.60	3.30	198	$C_{15}H_{18}$	198	1,2,3,6,7-Pentamethyl-naphthalene 1,2,3,6,7-五甲基萘
42	82.67	3.39	198	$C_{15}H_{18}$	198	1,2,3,5,6-Pentamethyl-naphthalene 1,2,3,5,6-五甲基萘

注：① 据王培荣(1993),21~23号化合物分离效果不好,鉴定此处是1,2,7-+1,2,6-+2,3,5-甲基萘三个化合物共馏,而据苏焕华等(2010),鉴定此处是1,2,7-+1,6,7-+1,2,6-三甲基萘三个化合物共馏;
② 据苏焕华等(2010),鉴定此处是1,2,5,6-+1,2,3,5-四甲基萘两个化合物共馏。

与其他环状化合物具有长侧链的正烷基取代一样,萘也具有系列的长侧链正烷基取代化合物,其质谱检测的特征离子为m/z 141。传统的色质分析很难在一般的实验地质样品中检测到这些化合物,但同样的样品在全二维气相色谱-飞行时间质谱分析时很容易发现这些化合物,如图2-302(b)所示。同甲基萘和乙基萘一样,这些化合物成对出现,每一对化合物之间的沸点和极性都有差异,取代基在"1"位的化合物极性相对较强,取代基在"2"位的化合物沸点随碳链碳数的增加而增加,且大于对应"1"位的沸点增加速度。这些化合物的质谱图如图2-303~图2-324所示,鉴定结果见表2-19。

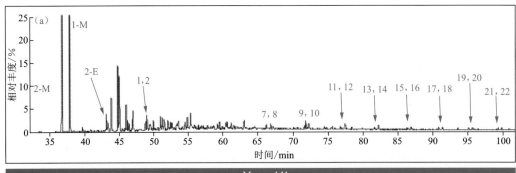

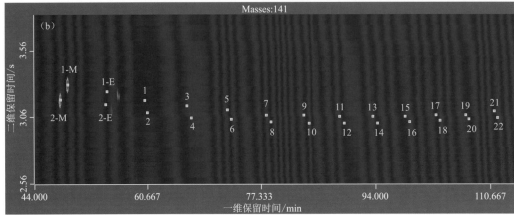

图 2-302　正烷基萘系列化合物的 GC-MS 色谱图(a)及全二维点阵图(b)

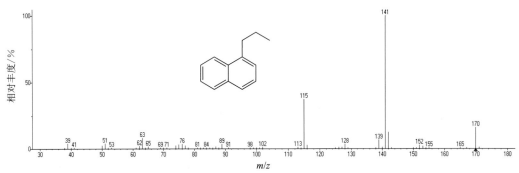

图 2-303　1 号峰　1-丙基萘质谱图

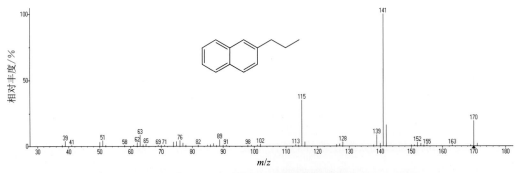

图 2-304　2 号峰　2-丙基萘质谱图

第 2 章 石油地质样品的全二维气相色谱图识别

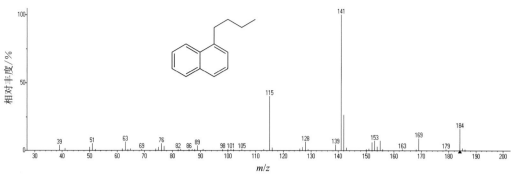

图 2-305　3 号峰　1-丁基萘质谱图

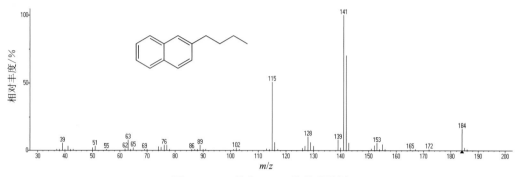

图 2-306　4 号峰　2-丁基萘质谱图

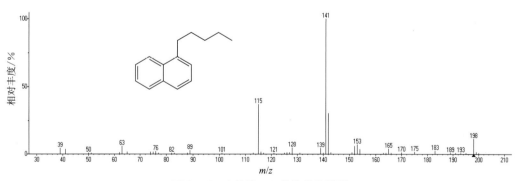

图 2-307　5 号峰　1-戊基萘质谱图

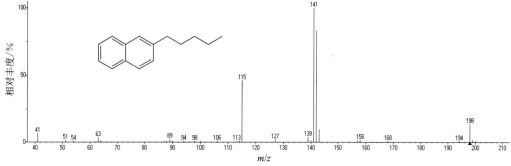

图 2-308　6 号峰　2-戊基萘质谱图

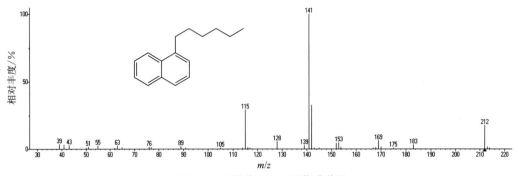

图 2-309　7 号峰　1-己基萘质谱图

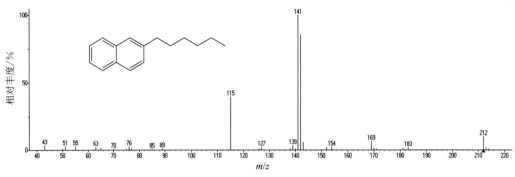

图 2-310　8 号峰　2-己基萘质谱图

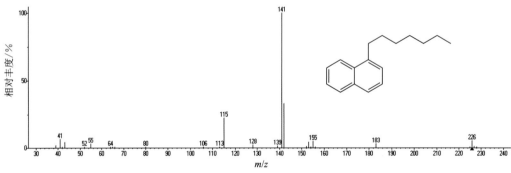

图 2-311　9 号峰　1-庚基萘质谱图

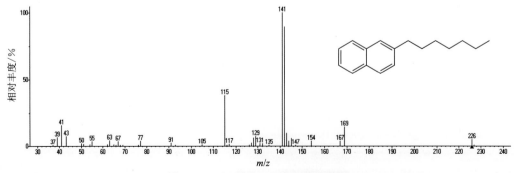

图 2-312　10 号峰　2-庚基萘质谱图

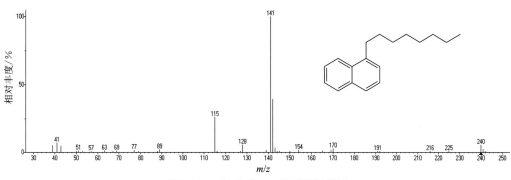

图 2-313　11 号峰　1-辛基萘质谱图

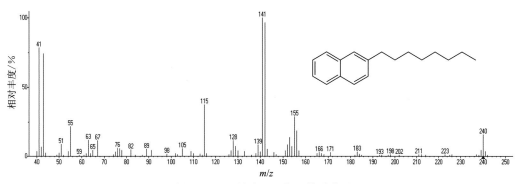

图 2-314　12 号峰　2-辛基萘质谱图

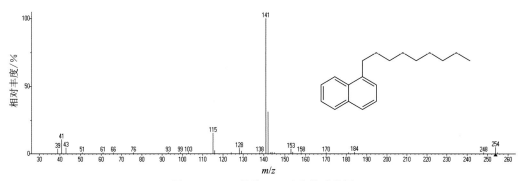

图 2-315　13 号峰　1-壬基萘质谱图

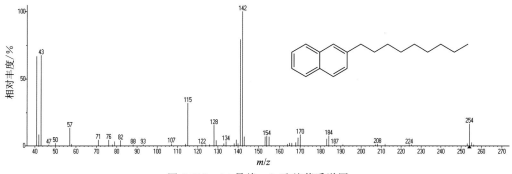

图 2-316　14 号峰　2-壬基萘质谱图

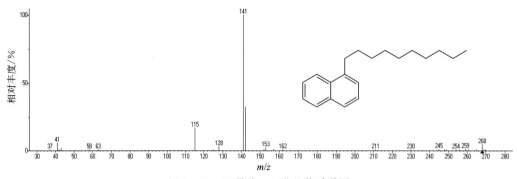

图 2-317　15 号峰　1-癸基萘质谱图

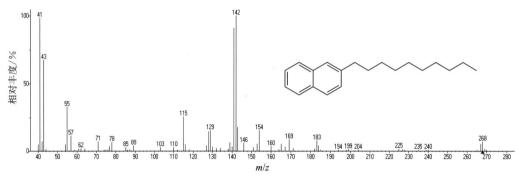

图 2-318　16 号峰　2-癸基萘质谱图

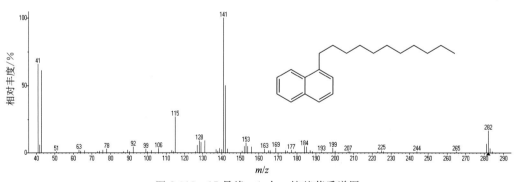

图 2-319　17 号峰　1-十一烷基萘质谱图

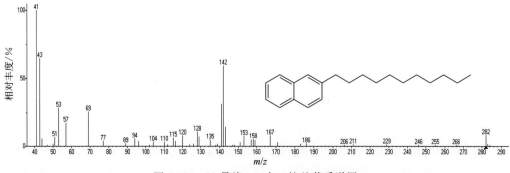

图 2-320　18 号峰　2-十一烷基萘质谱图

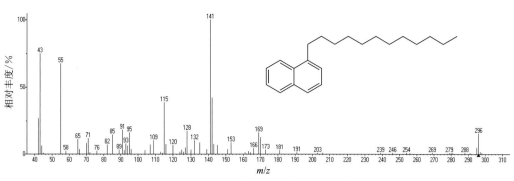

图 2-321　19 号峰　1-十二烷基萘质谱图

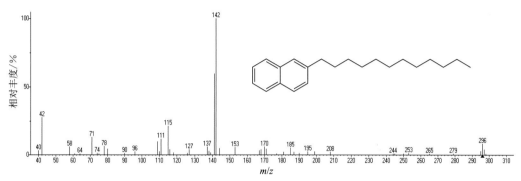

图 2-322　20 号峰　2-十二烷基萘质谱图

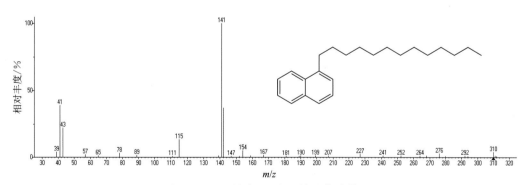

图 2-323　21 号峰　1-十三烷基萘质谱图

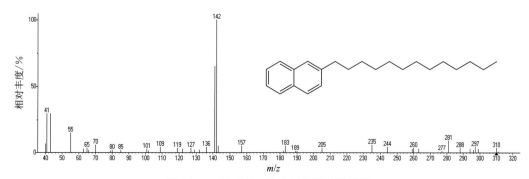

图 2-324　22 号峰　2-十三烷基萘质谱图

表 2-19　长链正烷基萘系列化合物鉴定结果表

峰　号	一维保留时间 /min	二维保留时间 /s	特征离子 (m/z)	分子式	相对分子质量	名　称
1	60.13	3.19	141	$C_{13}H_{14}$	170	1-Propyl-naphthalene 1-丙基萘
2	60.53	3.10	141	$C_{13}H_{14}$	170	2-Propyl-naphthalene 2-丙基萘
3	66.40	3.15	141	$C_{14}H_{16}$	184	1-Butyl-naphthalene 1-丁基萘
4	67.07	3.06	141	$C_{14}H_{16}$	184	2-Butyl-naphthalene 2-丁基萘
5	72.27	3.12	141	$C_{15}H_{18}$	198	1-Pentyl-naphthalene 1-戊基萘
6	72.93	3.05	141	$C_{15}H_{18}$	198	2-Pentyl-naphthalene 2-戊基萘
7	78.00	3.08	141	$C_{16}H_{20}$	212	1-Hexyl-naphthalene 1-己基萘
8	78.67	3.03	141	$C_{16}H_{20}$	212	2-Hexyl-naphthalene 2-己基萘
9	83.33	3.08	141	$C_{17}H_{22}$	226	1-Heptyl-naphthalene 1-庚基萘
10	84.13	3.02	141	$C_{17}H_{22}$	226	2-Heptyl-naphthalene 2-庚基萘
11	88.53	3.07	141	$C_{18}H_{24}$	240	1-Octyl-naphthalene 1-辛基萘
12	89.20	3.02	141	$C_{18}H_{24}$	240	2-Octyl-naphthalene 2-辛基萘
13	93.47	3.07	141	$C_{19}H_{26}$	254	1-Nonyl-naphthalene 1-壬基萘
14	94.13	3.02	141	$C_{19}H_{26}$	254	2-Nonyl-naphthalene 2-壬基萘
15	98.13	3.07	141	$C_{20}H_{28}$	268	1-Decyll-naphthalene 1-癸基萘
16	98.80	3.03	141	$C_{20}H_{28}$	268	2-Decyll-naphthalene 2-癸基萘
17	102.67	3.08	141	$C_{21}H_{30}$	282	1-Undecyl-naphthalene 1-十一烷基萘
18	103.20	3.04	141	$C_{21}H_{30}$	282	2-Undecyl-naphthalene 2-十一烷基萘
19	106.93	3.08	141	$C_{22}H_{32}$	296	1-Dodecyl-naphthalene 1-十二烷基萘
20	107.47	3.05	141	$C_{22}H_{32}$	296	2-Dodecyl-naphthalene 2-十二烷基萘

续表

峰 号	一维保留时间 /min	二维保留时间 /s	特征离子 (m/z)	分子式	相对分子质量	名 称
21	111.07	3.11	141	$C_{23}H_{34}$	310	1-Tridecyl-naphthalene 1-十三烷基萘
22	111.60	3.06	141	$C_{23}H_{34}$	310	2-Tridecyl-naphthalene 2-十三烷基萘

2.2.3 联苯、二氢化苊、氧芴系列化合物的全二维气相色谱图及化合物名称

芳烃馏分中的联苯和二氢化苊是同分异构体,其甲基取代的系列化合物与氧芴的相对分子质量相同,因此这 3 种系列化合物可用其分子离子作为特征离子同时进行质谱检测。图 2-325 是 m/z 154,m/z 168,m/z 182,m/z 196,m/z 210 的全二维点阵图(图 2-325a,b,c,d,e)和常规色质分析谱图(图 2-325a′,b′,c′,d′,e′)。比较全二维谱图和常规色质分析谱图可以看出:用常规色质分析时存在一些共馏峰和无法检出的化合物,它们都可以在全二维谱图上被完全分开。对比相关资料及质谱图,鉴定得到 20 个化合物,鉴定结果见表 2-20。

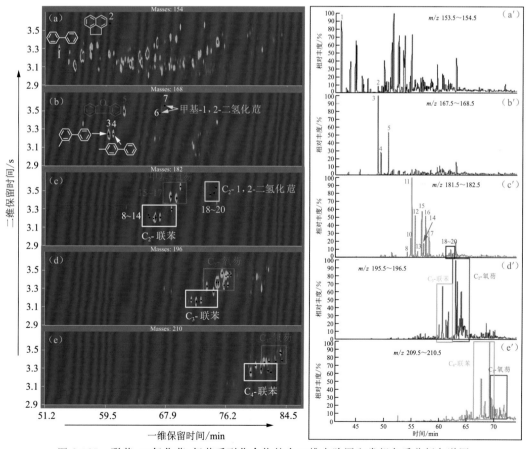

图 2-325 联苯、二氢化苊、氧芴系列化合物的全二维点阵图和常规色质分析色谱图

表 2-20　联苯、二氢化苊、氧芴系列化合物鉴定结果表

峰号	一维保留时间 /min	二维保留时间 /s	特征离子 (m/z)	分子式	相对分子质量	名　称
1	53.07	3.30	154	$C_{12}H_{10}$	154	Biphenyl 联苯
2	60.00	3.58	154	$C_{12}H_{10}$	154	1,2-Acenaphthene 1,2-二氢化苊
3	59.87	3.25	168	$C_{13}H_{12}$	168	3-Methyl-biphenyl 3-甲基联苯
4	60.53	3.25	168	$C_{13}H_{12}$	168	4-Methyl-biphenyl 4-甲基联苯
5	62.00	3.50	168	$C_{12}H_8O$	168	Dibenzofuran 氧芴
6	66.93	3.48	168	$C_{13}H_{12}$	168	Methyl-1,2-dihydro-acenaphthylene 甲基-1,2-二氢化苊
7	67.60	3.56	168	$C_{13}H_{12}$	168	Methyl-1,2-dihydro-acenaphthylene 甲基-1,2-二氢化苊
8	65.33	3.23	182	$C_{14}H_{14}$	182	Ethyl-biphenyl 乙基联苯
9	65.47	3.20	182	$C_{14}H_{14}$	182	Ethyl-biphenyl 乙基联苯
10	66.00	3.19	182	$C_{14}H_{14}$	182	3,5-Dimethyl-biphenyl 3,5-二甲基联苯
11	66.27	3.20	182	$C_{14}H_{14}$	182	3,3′-Dimethyl-biphenyl 3,3′-二甲基联苯
12	66.93	3.20	182	$C_{14}H_{14}$	182	3,4′-Dimethyl-biphenyl 3,4′-二甲基联苯
13	67.33	3.22	182	$C_{14}H_{14}$	182	4,4′-Dimethyl-biphenyl 4,4′-二甲基联苯
14	68.53	3.29	182	$C_{14}H_{14}$	182	3,4-Dimethyl-biphenyl 3,4-二甲基联苯
15	68.40	3.43	182	$C_{13}H_{10}O$	182	Methyl-dibenzofuran 甲基氧芴
16	69.20	3.41	182	$C_{13}H_{10}O$	182	Methyl-dibenzofuran 甲基氧芴
17	69.73	3.54	182	$C_{13}H_{10}O$	182	Methyl-dibenzofuran 甲基氧芴
18	73.87	3.45	182	$C_{14}H_{14}$	182	C_2-1,2-dihydro-acenaphthylene C_2-1,2-二氢化苊
19	74.40	3.47	182	$C_{14}H_{14}$	182	C_2-1,2-dihydro-acenaphthylene C_2-1,2-二氢化苊
20	74.67	3.53	182	$C_{14}H_{14}$	182	C_2-1,2-dihydro-acenaphthylene C_2-1,2-二氢化苊

2.2.4 芴系列化合物的全二维气相色谱图及化合物名称

芴系列化合物包括芴、甲基芴、C_2-芴、C_3-芴和C_4-芴,其质谱检测的特征离子为分子离子。图 2-326 是芴系列化合物的全二维点阵图(a,b,c,d,e)和常规色质分析色谱图(a′,b′,c′,d′,e′)。从图中可看出,在常规色质分析谱图中芴常与萘、联苯和氧芴共馏,消除了共馏峰的全二维谱图有利于化合物的识别和鉴定。芴系列化合物的名称见图内注。

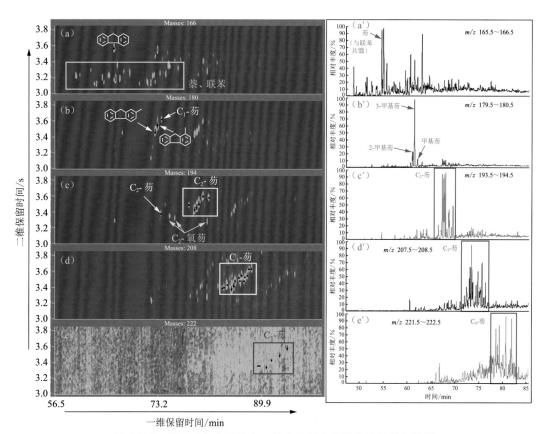

图 2-326 芴系列化合物的全二维点阵图和常规色质分析色谱图

2.2.5 菲系列化合物的全二维气相色谱图及化合物名称

菲系列化合物在全二维气相色谱图上处于萘系列化合物的右上方,一般包括菲、甲基菲以及取代基碳数从 2 到 4 的系列化合物,如图 2-327 所示。菲系列化合物的分子离子即 m/z 178,m/z 192,m/z 206,m/z 220,m/z 234 可分别用于菲、甲基菲及取代基碳数 2 至 4 的系列化合物的检测。由于消除了共馏峰的干扰,利用全二维分析可检测菲系列化合物的数量远高于常规色质分析,但一些化合物的具体构型和取代基位置不好确定。通过与相关资料比对,共鉴定出 43 个化合物,鉴定结果见表 2-21。

菲和甲基菲容易识别,如图 2-327 中 1~5 号峰所示,其他信息见表 2-21。

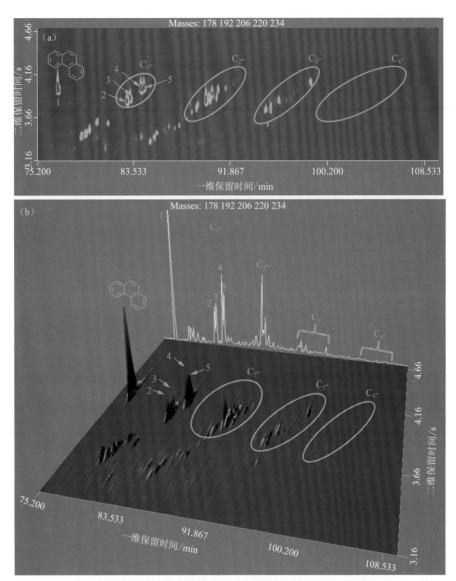

图 2-327　菲系列化合物的全二维点阵图(a)和三维立体图(b)

取代基碳数为 2 的菲系列在全二维分析谱图上可以检测到 14 个化合物,峰号自 6 至 19,如图 2-328 所示。从图中可以看出,传统色质检测时不能分离的化合物(图 2-328a)在全二维气相色谱-飞行时间质谱分析谱图(图 2-328b,c)上实现了较好的分离。7～10 号化合物在常规 GC-MS 上是共馏峰,在 GC×GC 上可以被分开,根据 GC×GC 的谱图特点判断 7,8,10 号化合物是乙基菲,9 号化合物是二甲基菲。由图 2-328 看出 3-甲基菲与 2-甲基菲的二维保留时间相近,说明它们的极性相近,而 9-甲基菲和 1-甲基菲的二维保留时间相近,说明它们的极性相近。根据 4 个甲基菲的排列方式可以判断 4 个乙基菲的排列方式。各化合物的鉴定结果见表 2-21。

取代基碳数为 3 的菲系列在全二维分析谱图上可以检测到 21 个化合物,峰号自 20 至 40,如图 2-329 所示。比较传统色质分析谱图(图 2-328a)与全二维气相色谱-飞行时间质谱分析(图 2-328b,c)谱图可以看出,全二维分析可以检测到更多的菲化合物,并且具有较好的分离度。其中的一些化合物通过质谱图(图 2-330～图 2-332 为例)和二维保留时间不能确定确切的结构,只能统一命名为 C_3-菲或三甲基菲。鉴定结果见表 2-21。

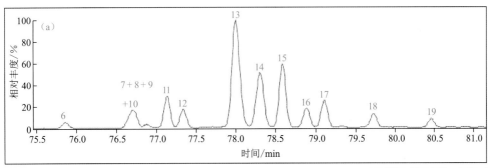

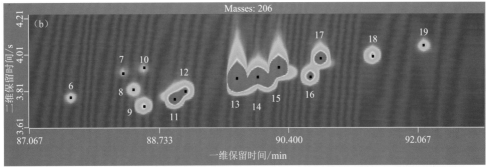

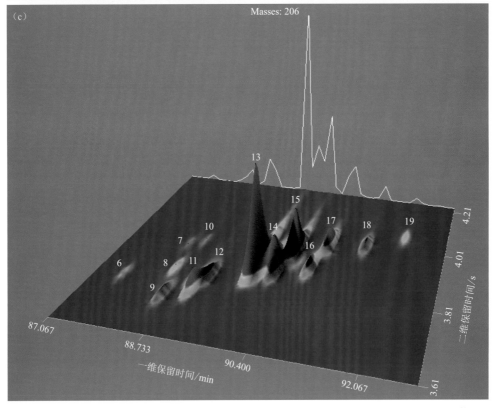

图 2-328　取代基碳数为 2 的菲系列化合物的 GC-MS 色谱图(a)、全二维点阵图(b)和三维立体图(c)

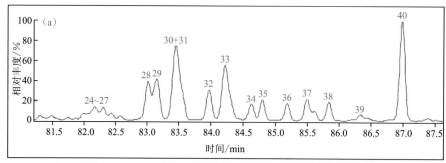

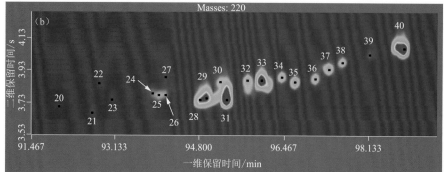

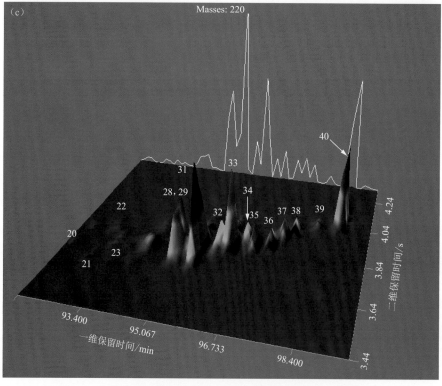

图 2-329 取代基碳数为 3 的菲系列化合物的 GC-MS 色谱图(a)、全二维点阵图(b)和三维立体图(c)

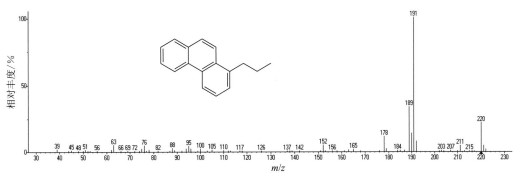

图 2-330　20 号峰　丙基菲质谱图（其中丙基取代基位置不确定）

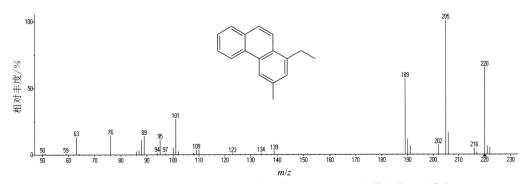

图 2-331　21 号峰　甲基-乙基菲质谱图（其中甲基和乙基取代基位置不确定）

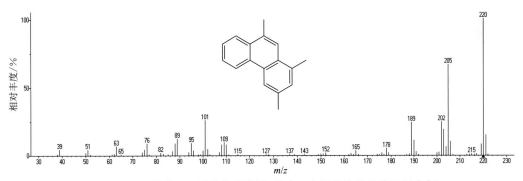

图 2-332　33 号峰　三甲基菲质谱图（其中三个甲基取代基位置不确定）

取代基碳数为 4 的菲系列化合物数量更多（如图 2-333b 中黑点标注所示），但由于每个化合物的结构难以确定，只能像一些取代基碳数为 3 的菲系列化合物那样给出笼统的名称，所以本节仅介绍其中的 3 个峰，峰号自 41 至 43，如图 2-333（b）所示，其他化合物的结构可能是丁基菲、甲基-丙基菲、二乙基菲、四甲基菲，具体结构不清。鉴定结果见表 2-21。

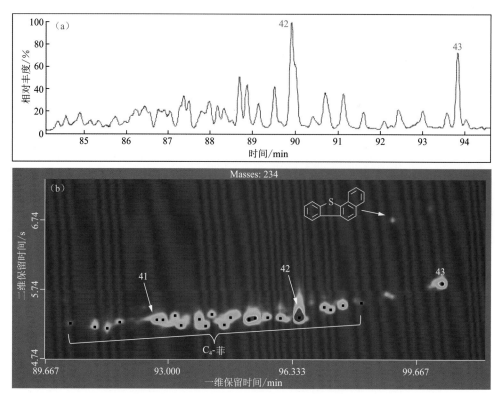

图 2-333 取代基碳数为 4 的菲系列化合物的 GC-MS 色谱图(a)全二维点阵图(b)

表 2-21 菲系列化合物鉴定结果表

峰 号	一维保留时间 /min	二维保留时间 /s	特征离子 (m/z)	分子式	相对分子质量	名　称
1	76.93	4.01	178	$C_{14}H_{10}$	178	Phenanthrene 菲
2	82.93	3.86	192	$C_{15}H_{12}$	192	3-Methyl-phenanthrene 3-甲基菲
3	83.20	3.90	192	$C_{15}H_{12}$	192	2-Methyl-phenanthrene 2-甲基菲
4	84.13	4.01	192	$C_{15}H_{12}$	192	9-Methyl-phenanthrene 9-甲基菲
5	84.40	4.02	192	$C_{15}H_{12}$	192	1-Methyl-phenanthrene 1-甲基菲
6	87.60	3.78	206	$C_{16}H_{14}$	206	3-Ethyl-phenanthrene 3-乙基菲
7	88.27	3.91	206	$C_{16}H_{14}$	206	9-Ethyl-phenanthrene 9-乙基菲
8	88.40	3.82	206	$C_{16}H_{14}$	206	2-Ethyl-phenanthrene 2-乙基菲
9	88.53	3.73	206	$C_{16}H_{14}$	206	3,6-Dimethyl-phenanthrene 3,6-二甲基菲

续表

峰 号	一维保留时间 /min	二维保留时间 /s	特征离子 (m/z)	分子式	相对分子质量	名 称
10	88.53	3.94	206	$C_{16}H_{14}$	206	1-Ethyl-phenanthrene 1-乙基菲
11	88.93	3.77	206	$C_{16}H_{14}$	206	2,6-＋3,5-Dimethyl-phenanthrene 2,6-＋3,5-二甲基菲
12	89.07	3.81	206	$C_{16}H_{14}$	206	2,7-Dimethyl-phenanthrene 2,7-二甲基菲
13	89.73	3.88	206	$C_{16}H_{14}$	206	2,10-＋1,3-＋3,9-＋3,10-dimethyl-phenanthrene 2,10-＋1,3-＋3,9-＋3,10-二甲基菲
14	90.00	3.89	206	$C_{16}H_{14}$	206	1,6＋2,9＋2,5-Dimethyl-phenanthrene 1,6＋2,9＋2,5-二甲基菲
15	90.27	3.94	206	$C_{16}H_{14}$	206	1,7-Dimethyl-phenanthrene 1,7-二甲基菲
16	90.67	3.89	206	$C_{16}H_{14}$	206	2,3-Dimethyl-phenanthrene 2,3-二甲基菲
17	90.80	3.99	206	$C_{16}H_{14}$	206	1,9-＋4,9-＋4,10-Dimethyl-phenanthrene 1,9-＋4,9-＋4,10-二甲基菲
18	91.47	4.00	206	$C_{16}H_{14}$	206	1,8-Dimethyl-phenanthrene 1,8-二甲基菲
19	92.13	4.06	206	$C_{16}H_{14}$	206	1,2-Dimethyl-phenanthrene 1,2-二甲基菲
20	92.00	3.71	220	$C_{17}H_{16}$	220	Propyl-phenanthrene 丙基菲
21	92.67	3.67	220	$C_{17}H_{16}$	220	Methyl-ethyl-phenanthrene 甲基-乙基菲
22	92.80	3.85	220	$C_{17}H_{16}$	220	Propyl-phenanthrene 丙基菲
23	93.07	3.75	220	$C_{17}H_{16}$	220	Propyl-phenanthrene 丙基菲
24	93.87	3.79	220	$C_{17}H_{16}$	220	C_3-Phenanthrene C_3-菲
25	94.00	3.78	220	$C_{17}H_{16}$	220	Methyl-ethyl-phenanthrene 甲基-乙基菲
26	94.13	3.78	220	$C_{17}H_{16}$	220	C_3-Phenanthrene C_3-菲
27	94.13	3.89	220	$C_{17}H_{16}$	220	C_3-Phenanthrene C_3-菲
28	94.80	3.75	220	$C_{17}H_{16}$	220	Trimethyl-phenanthrene 三甲基菲

续表

峰 号	一维保留时间 /min	二维保留时间 /s	特征离子 (m/z)	分子式	相对分子质量	名 称
29	94.93	3.76	220	$C_{17}H_{16}$	220	Trimethyl-phenanthrene 三甲基菲
30	95.20	3.86	220	$C_{17}H_{16}$	220	Methyl-ethyl-phenanthrene 甲基-乙基菲
31	95.33	3.75	220	$C_{17}H_{16}$	220	Trimethyl-phenanthrene 三甲基菲
32	95.73	3.87	220	$C_{17}H_{16}$	220	Trimethyl-phenanthrene 三甲基菲
33	96.00	3.87	220	$C_{17}H_{16}$	220	Trimethyl-phenanthrene 三甲基菲
34	96.40	3.89	220	$C_{17}H_{16}$	220	Trimethyl-phenanthrene 三甲基菲
35	96.67	3.86	220	$C_{17}H_{16}$	220	Trimethyl-phenanthrene 三甲基菲
36	97.07	3.88	220	$C_{17}H_{16}$	220	Trimethyl-phenanthrene 三甲基菲
37	97.34	3.94	220	$C_{17}H_{16}$	220	Trimethyl-phenanthrene 三甲基菲
38	97.60	3.98	220	$C_{17}H_{16}$	220	Trimethyl-phenanthrene 三甲基菲
39	98.13	4.03	220	$C_{17}H_{16}$	220	Trimethyl-phenanthrene 三甲基菲
40	98.80	4.07	220	$C_{17}H_{16}$	220	Trimethyl-phenanthrene 三甲基菲
41	92.67	5.31	234	$C_{18}H_{18}$	234	Retene 惹烯
42	96.50	5.35	234	$C_{18}H_{18}$	234	Tetramethyl-phenanthrene 四甲基菲
43	100.33	5.84	234	$C_{18}H_{18}$	234	Tetramethyl-phenanthrene 四甲基菲

2.2.6 芘和䓛系列化合物的全二维气相色谱图及化合物名称

芘和䓛同属4个环的芳香族化合物，也是同分异构体，其系列一般包括芘和䓛及其取代基碳数1～3的化合物。图2-334是芘和䓛系列化合物的全二维和常规色质分析谱图。在全二维气相色谱图中，该系列化合物处于三环芳烃的上方，含量低时在GC-MS上易与其他化合物共馏（图2-334a,b,c,d），全二维的分离度明显好于常规色质分析（图2-334a′,b′,c′,d′）。由于取代基碳数为2和3的系列化合物无相关参考资料，本节仅鉴定了芘和䓛及其甲基取代的9个化合物，结果见表2-22。

第2章 石油地质样品的全二维气相色谱图识别

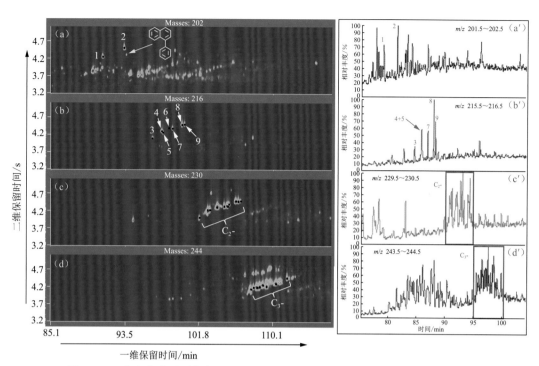

图 2-334　芘和荧蒽系列化合物的全二维点阵图（a～d）和 GC-MS 色谱图（a′～d′）

表 2-22　芘和荧蒽系列化合物鉴定结果表

峰号	一维保留时间 /min	二维保留时间 /s	特征离子 (m/z)	分子式	相对分子质量	名称
1	90.87	4.30	202	$C_{16}H_{10}$	202	Fluoranthene 荧蒽
2	93.27	4.55	202	$C_{16}H_{10}$	202	Pyrene 芘
3	96.47	4.15	216	$C_{17}H_{12}$	216	Benzo[a]fluorene 苯并[a]芴
4	97.53	4.35	216	$C_{17}H_{12}$	216	Benzo[b]fluorene + 4-methypyrene 苯并[b]芴+4-甲基芘
5	97.67	4.31	216	$C_{17}H_{12}$	216	2-Methylpyrene 2-甲基芘
6	98.33	4.44	216	$C_{17}H_{12}$	216	1-Methylpyrene 1-甲基芘
7	98.73	4.42	216	$C_{17}H_{12}$	216	Methylfluoranthene 甲基荧蒽
8	99.80	4.52	216	$C_{17}H_{12}$	216	Methylpyrene or methylfluoranthene 甲基芘或甲基荧蒽
9	100.07	4.54	216	$C_{17}H_{12}$	216	Methylpyrene or methylfluoranthene 甲基芘或甲基荧蒽

需要指出的是，在一些严重降解的石油地质样品芳烃馏分中，6号峰的位置多出了两个峰，如图 2-335(b)示，这两个峰在常规色质分析时共馏，如图 2-335(a)所示，根据出峰位

置和其质谱图(图 2-336～图 2-338)判断,其中 6-1 和 6-2 是甲基芘的同分异构体,6-3 结构未知。

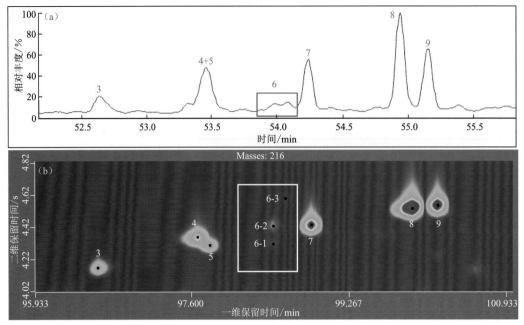

图 2-335 严重降解样品中甲基芘和甲基䓛化合物的 GC-MS 色谱图(a)和全二维点阵图(b)

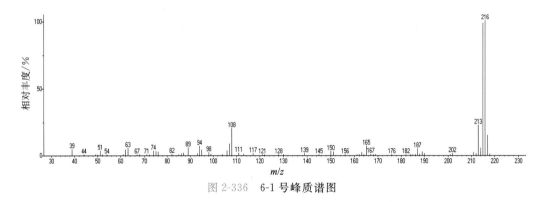

图 2-336 6-1 号峰质谱图

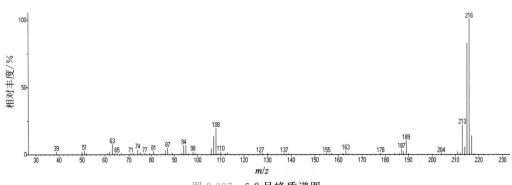

图 2-337 6-2 号峰质谱图

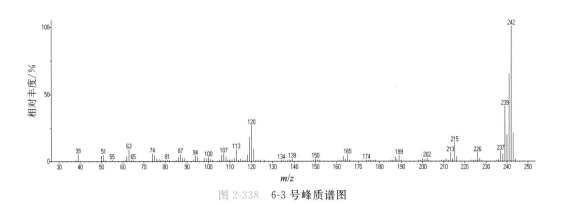

图 2-338　6-3 号峰质谱图

2.2.7　䓛系列化合物的全二维气相色谱图及化合物名称

䓛系列一般包括䓛及甲基䓛、取代基碳数为 2 的化合物等。图 2-339 是䓛系列化合物的全二维分析和常规色质分析谱图，在全二维气相色谱图中，标注了同为四环芳烃的芘和萤蒽与䓛系列化合物的相对位置，䓛在右上方，说明其极性较强。在全二维分析时不受其他峰的干扰（图 2-339a，b，c），检测到的化合物多，分离度也明显好于常规色质分析（图 2-339a′，b′，c′）。本节共鉴定了 9 个化合物，其中 4 号和 5 号峰在一维色谱上是共馏峰，但在二维色谱上因为该处有 2 个化合物，质谱图无法判定哪一个是 2-甲基䓛，暂定这两个化合物均为甲基䓛。其他化合物的鉴定结果见表 2-23。

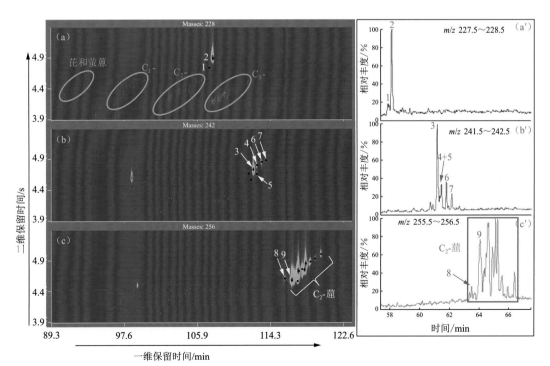

图 2-339　䓛系列化合物的全二维点阵图（a～c）和 GC-MS 色谱图（a′～c′）

图（a）～（c）中红色点标记的是芘和萤蒽类化合物

表 2-23 䓛系列化合物的鉴定结果表

峰 号	一维保留时间 /min	二维保留时间 /s	特征离子 (m/z)	分子式	相对分子质量	名 称
1	107.27	4.74	228	$C_{18}H_{12}$	228	Benzo[a]anthracene 苯并[a]蒽
2	107.67	4.89	228	$C_{18}H_{12}$	228	Chrysene 䓛
3	112.20	4.73	242	$C_{19}H_{14}$	242	3-Methylchrysene 3-甲基䓛
4	112.60	4.66	242	$C_{19}H_{14}$	242	Methylchrysene 甲基䓛
5	112.60	4.80	242	$C_{19}H_{14}$	242	Methylchrysene 甲基䓛
6	113.13	4.85	242	$C_{19}H_{14}$	242	6-Methylchrysene 6-甲基䓛
7	113.67	4.88	242	$C_{19}H_{14}$	242	1-Methylchrysene 1-甲基䓛
8	115.80	4.61	256	$C_{20}H_{16}$	256	Ethylchrysene 乙基䓛
9	116.60	4.58	256	$C_{20}H_{16}$	256	Dimethylchrysene 二甲基䓛

2.2.8 苝系列化合物的全二维气相色谱图及化合物名称

苝系列一般包括苝、甲基苝及取代基碳数为 2 的化合物等,属于典型的五环芳烃。其分子离子即为特征离子,分别为 m/z 252,m/z 266 及 m/z 280。不同相对分子质量的化合物都有对应的同分异构体共存,其结构和名称见表 2-24。

图 2-340 是苝系列化合物的全二维点阵图和常规色质分析谱图。从图中可以看出:常规色质分析含量较低的样品时,基线噪音较高,化合物之间的分离度也远逊色于全二维分析。由于甲基苝及 C_2 取代的苝同分异构体比较多,本节仅鉴定分子离子为 m/z 252 的化合物,鉴定结果见表 2-25。

表 2-24 苝系列化合物同分异构体的结构及名称

序 号	特征离子 (m/z)	分子离子	可能的化学结构
1	252	252	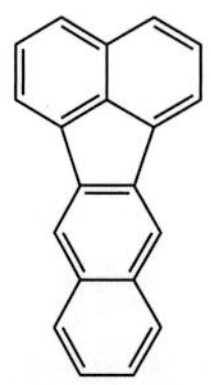 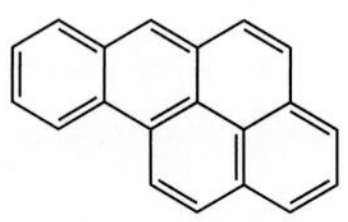 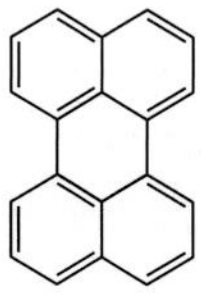 苯并[k]荧蒽 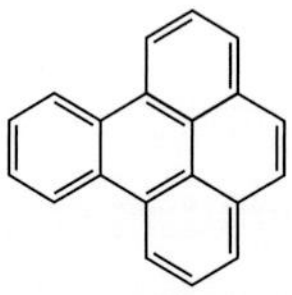　　苯并[e]苝　　苯并[a]苝　　苝

续表

序 号	特征离子 (m/z)	分子离子	可能的化学结构
2	266	266	茚并菲　二苯并芴　甲基苯并芘　甲基䓛
3	280	280	二甲基-苯并芘　甲基二苯并芴　甲基茚并菲

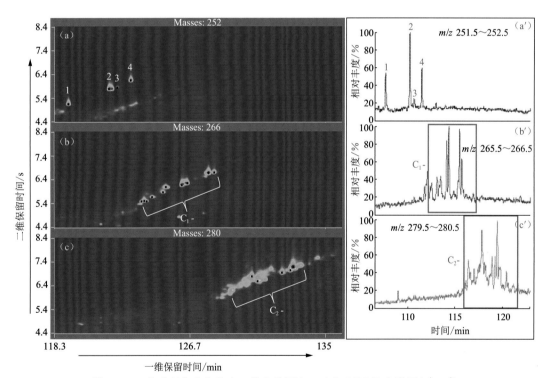

图 2-340　䓛系列化合物的全二维点阵图(a～c)和 GC-MS 色谱图(a′～c′)

表 2-25　**分子离子为 m/z 252 的化合物鉴定结果表**

峰 号	一维保留时间 /min	二维保留时间 /s	特征离子 (m/z)	分子式	相对 分子质量	名　　称
1	119.17	5.15	252	$C_{20}H_{12}$	252	Benzo[k]fluoranthene 苯并[k]荧蒽

续表

峰号	一维保留时间/min	二维保留时间/s	特征离子(m/z)	分子式	相对分子质量	名称
2	121.67	5.79	252	$C_{20}H_{12}$	252	Benzo[e]pyrene 苯并[e]芘
3	122.17	5.86	252	$C_{20}H_{12}$	252	Benzo[a]pyrene 苯并[a]芘
4	123.00	6.16	252	$C_{20}H_{12}$	252	Perylene 苝

2.2.9 单芳甾系列化合物的全二维气相色谱图及化合物名称

单芳甾和甾类属于同一物源的化合物，是甾类化合物在环境外力作用下的环芳构化产物。根据环的芳构化位置，单芳甾分为 A 环、B 环和 C 环 3 种，本节仅介绍最为常见的 C 环单芳甾系列化合物，简称单芳甾系列化合物。

单芳甾系列化合物包括单芳甾及其甲基单芳甾等化合物，质谱检测的特征离子为 m/z 253 及 m/z 267。图 2-341 是单芳甾系列化合物的常规色质分析和全二维气相色谱-飞行时间质谱分析的谱图，可以看出甲基单芳甾的数量不少于单芳甾化合物，但对其结构的研究较少，仅鉴定出两个化合物。全部鉴定的 1～17 号峰的化合物质谱图如图 2-340～图 2-358 所示，各化合物名称见表 2-26。

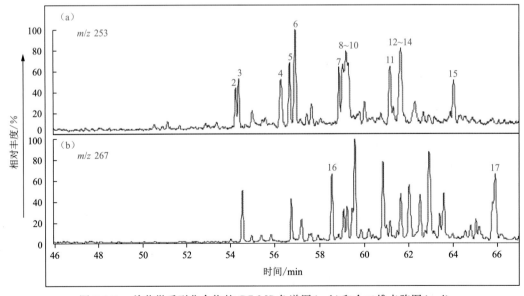

图 2-341 单芳甾系列化合物的 GC-MS 色谱图(a,b)和全二维点阵图(c,d)

第 2 章 石油地质样品的全二维气相色谱图识别

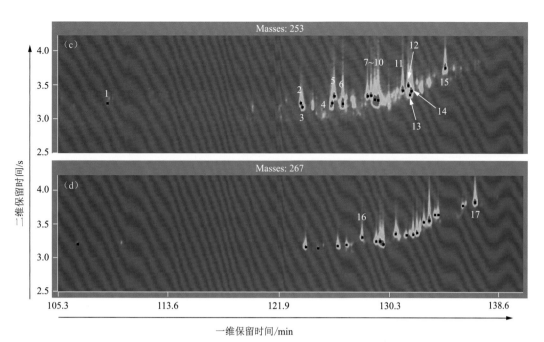

图 2-341(续) 单芳甾系列化合物的 GC-MS 色谱图(a,b)和全二维点阵图(c,d)

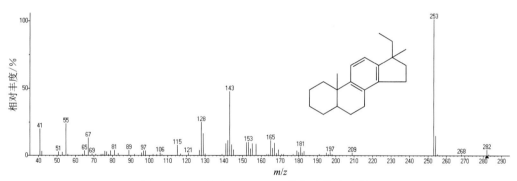

图 2-342 1 号峰 C_{21},C 环-单芳孕甾烷质谱图

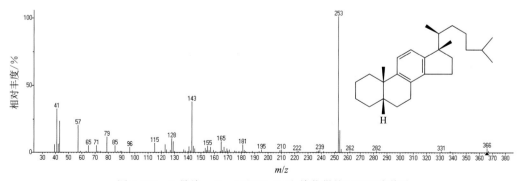

图 2-343 2 号峰 C_{27},5β(H)-C 环-单芳甾烷(20S)质谱图

· 169 ·

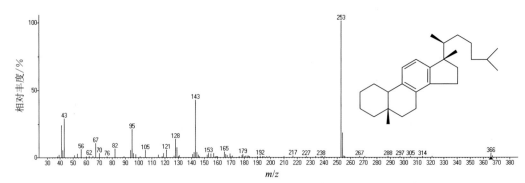

图 2-344　3 号峰　C_{27},5β-甲基-C 环-重排单芳甾烷(20S)质谱图

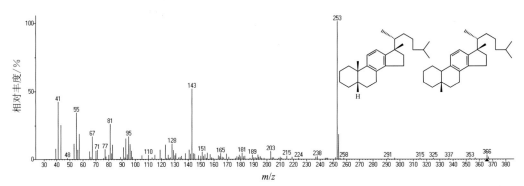

图 2-345　4 号峰　C_{27},5β(H)-C 环-单芳甾烷(20R)＋C_{27},5β-甲基-C 环-重排单芳甾烷(20R)质谱图

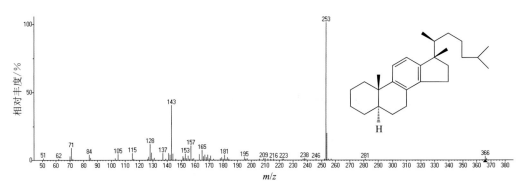

图 2-346　5 号峰　C_{27},5α(H)-C 环-单芳甾烷(20S)质谱图

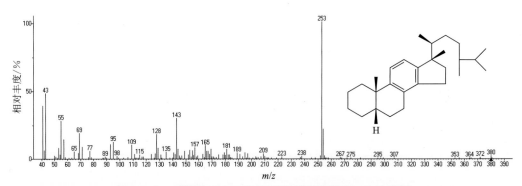

图 2-347　6 号峰　C_{28},5β(H)-C 环-单芳甾烷(20S)质谱图

第 2 章 石油地质样品的全二维气相色谱图识别

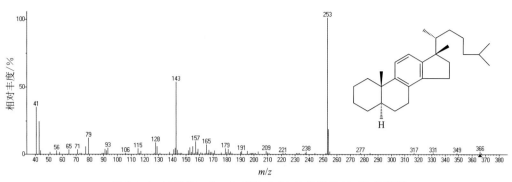

图 2-348　7 号峰　C_{27},5α(H)-C 环-单芳甾烷(20R)质谱图

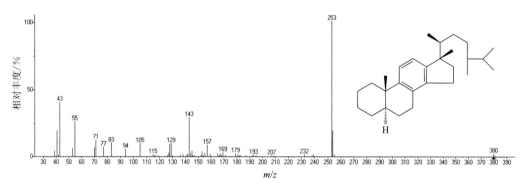

图 2-349　8 号峰　C_{28},5α(H)-C 环-单芳甾烷(20S)质谱图

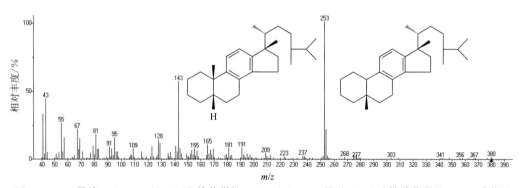

图 2-350　9 号峰　C_{28},5β(H)-C 环-单芳甾烷(20R)＋C_{28},5β-甲基-C 环-重排单芳甾烷(20R)质谱图

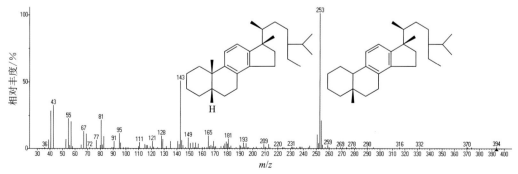

图 2-351　10 号峰　C_{29},5β(H)-C 环-单芳甾烷(20S)＋C_{29},5β-甲基-C 环-重排单芳甾烷(20S)质谱图

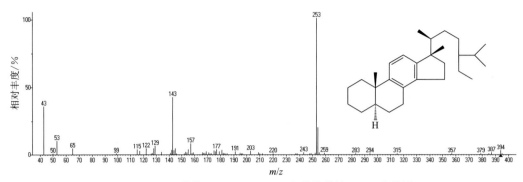

图 2-352　11 号峰　C_{29},5α(H)-C 环-单芳甾烷(20S)质谱图

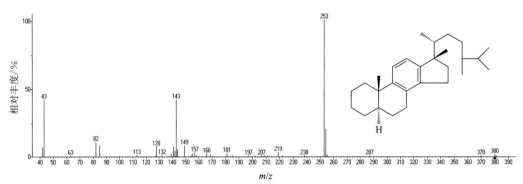

图 2-353　12 号峰　C_{28},5α(H)-C 环-单芳甾烷(20R)质谱图

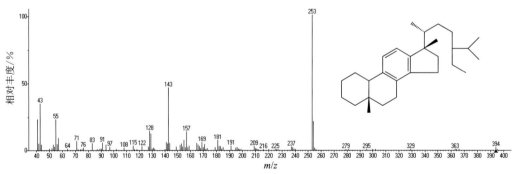

图 2-354　13 号峰　C_{29},5β-甲基-C 环-重排单芳甾烷(20R)质谱图

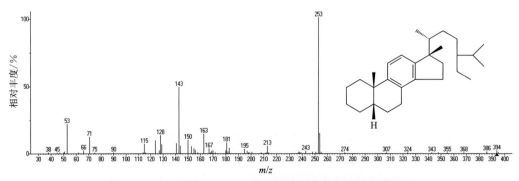

图 2-355　14 号峰　C_{29},5β(H)-C 环-单芳甾烷(20R)质谱图

第 2 章 石油地质样品的全二维气相色谱图识别

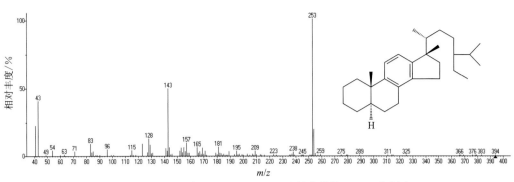

图 2-356　15 号峰　C_{29},$5\alpha(H)$-C 环-单芳甾烷(20R)质谱图

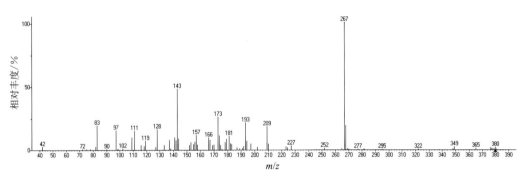

图 2-357　16 号峰　C_{28},甲基-C 环-单芳甾烷质谱图

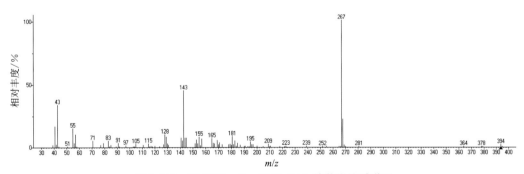

图 2-358　17 号峰　C_{29},甲基-C 环-单芳甾烷质谱图

表 2-26　C 环-单芳甾烷系列化合物鉴定结果表

峰号	一维保留时间/min	二维保留时间/s	特征离子(m/z)	分子式	相对分子质量	名　称
1	109.03	3.29	253	$C_{21}H_{30}$	282	C_{21},C-Ring-monoaromatic-pregnane C_{21},C 环-单芳孕甾烷
2	123.57	3.30	253	$C_{27}H_{42}$	366	C_{27},$5\beta(H)$-C-Ring-monoaromatic-sterane(20S) C_{27},$5\beta(H)$-C 环-单芳甾烷(20S)
3	123.70	3.24	253	$C_{27}H_{42}$	366	C_{27},5β-Methyl-C-Ring-diamonoaromatic-sterane(20S) C_{27},5β-甲基-C 环-重排单芳甾烷(20S)

续表

峰号	一维保留时间/min	二维保留时间/s	特征离子(m/z)	分子式	相对分子质量	名称
4	125.97	3.30	253	$C_{27}H_{42}$	366	C_{27},5β(H)-C-Ring-monoaromatic-sterane(20R)＋C_{27},5β-Methyl-C-Ring-diamonoaromatic-sterane(20R) C_{27},5β(H)-C环-单芳甾烷(20R)＋C_{27},5β-甲基-C环-重排单芳甾烷(20R)
5	126.10	3.40	253	$C_{27}H_{42}$	366	C_{27},5α(H)-C-Ring-monoaromatic-sterane(20S) C_{27},5α(H)-C环-单芳甾烷(20S)
6	126.77	3.29	191	$C_{28}H_{44}$	380	C_{28},5β(H)-C-Ring-monoaromatic-sterane(20S) C_{28},5β(H)-C环-单芳甾烷(20S)
7	128.63	3.40	253	$C_{27}H_{42}$	366	C_{27},5α(H)-C-Ring-monoaromatic-sterane(20R) C_{27},5α(H)-C环-单芳甾烷(20R)
8	128.90	3.41	253	$C_{28}H_{44}$	380	C_{28},5α(H)-C-Ring-monoaromatic-sterane(20S) C_{28},5α(H)-C环-单芳甾烷(20S)
9	129.17	3.35	213	$C_{28}H_{44}$	380	C_{28},5β(H)-C-Ring-monoaromatic-sterane(20R)＋C_{28},5β-Methyl-C-Ring-diamonoaromatic-sterane(20R) C_{28},5β(H)-C环-单芳甾烷(20R)＋C_{28},5β-甲基-C环-重排单芳甾烷(20R)
10	129.43	3.34	253	$C_{29}H_{46}$	394	C_{29},5β(H)-C-Ring-monoaromatic-sterane(20S)＋C_{29},5β-Methyl-C-Ring-diamonoaromatic-sterane(20S) C_{29},5β(H)-C环-单芳甾烷(20S)＋C_{29},5β-甲基-C环-重排单芳甾烷(20S)
11	131.30	3.49	253	$C_{29}H_{46}$	394	C_{29},5α(H)-C-Ring-monoaromatic-sterane(20S) C_{29},5α(H)-C环-单芳甾烷(20S)
12	131.70	3.56	253	$C_{28}H_{44}$	380	C_{28},5α(H)-C-Ring-monoaromatic-sterane(20R) C_{28},5α(H)-C环-单芳甾烷(20R)
13	131.83	3.42	253	$C_{29}H_{46}$	394	C_{29},5β-Methyl-C-Ring-diamonoaromatic-sterane(20R) C_{29},5β-甲基-C环-重排单芳甾烷(20R)
14	131.97	3.48	253	$C_{29}H_{46}$	394	C_{29},5β(H)-C-Ring-monoaromatic-sterane(20R) C_{29},5β(H)-C环-单芳甾烷(20R)

续表

峰 号	一维保留时间/min	二维保留时间/s	特征离子(m/z)	分子式	相对分子质量	名 称
15	134.50	3.82	253	$C_{29}H_{46}$	394	C_{29},5α(H)-C-Ring-monoaromatic-sterane(20R) C_{29},5α(H)-C 环-单芳甾烷(20R)
16	128.23	3.37	267	$C_{28}H_{44}$	380	C_{28},Methyl-C-Ring-monoaromatic-sterane C_{28},甲基-C 环-单芳甾烷
17	136.77	3.89	267	$C_{29}H_{46}$	394	C_{29},Methyl-C-Ring-monoaromatic-sterane C_{29},甲基-C 环-单芳甾烷

2.2.10 三芳甾系列化合物的全二维气相色谱图及化合物名称

单芳甾进一步演化,可使 A,B,C 3 个环都发生芳构化形成三芳甾系列化合物,因此三芳甾常和单芳甾一起作为成熟度的指标。

三芳甾系列化合物包括三芳甾及甲基三芳甾等化合物。三芳甾的质谱检测特征离子为 m/z 231,图 2-359 是三芳甾的常规色质分析和全二维气相色谱-飞行时间质谱分析的谱图,可以看出全二维分析消除了常规色质的共馏峰(峰号 2~4),能够检测到更多的化合物,全部鉴定的 1~13 号峰的化合物质谱图如图 2-360~图 2-372 所示,各化合物名称见表 2-27。

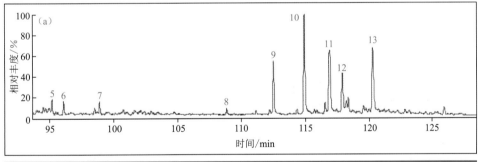

图 2-359 三芳甾的 GC-MS 色谱图(a)、全二维点阵图(b)和三维立体图(c)

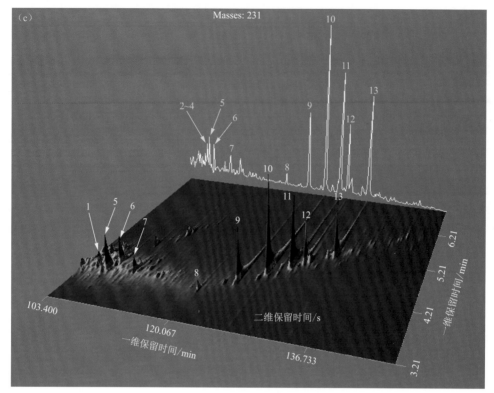

图 2-359(续)　三芳甾的 GC-MS 色谱图(a)、全二维点阵图(b)和三维立体图(c)

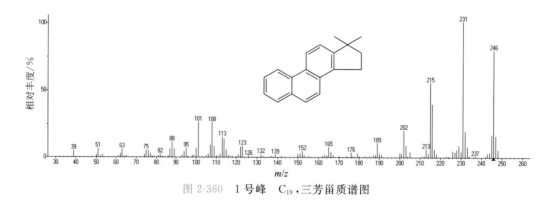

图 2-360　1 号峰　C_{19}，三芳甾质谱图

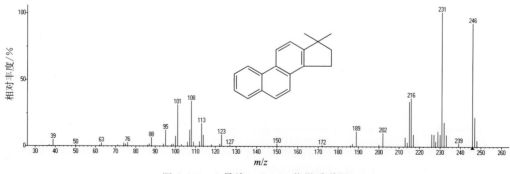

图 2-361　2 号峰　C_{19}，三芳甾质谱图

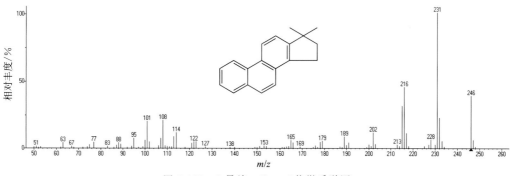

图 2-362　3 号峰　C_{19},三芳甾质谱图

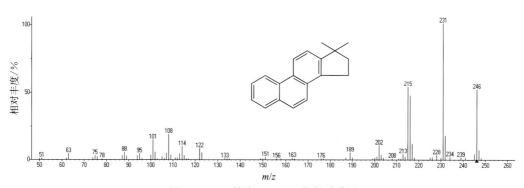

图 2-363　4 号峰　C_{19},三芳甾质谱图

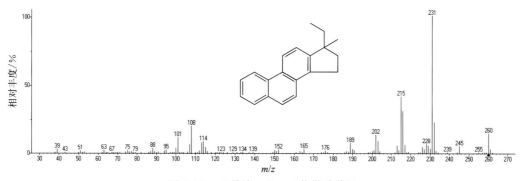

图 2-364　5 号峰　C_{20},三芳甾质谱图

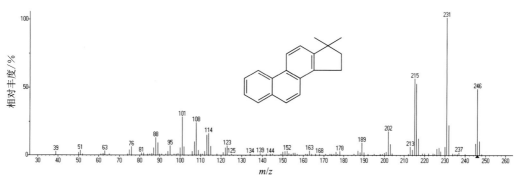

图 2-365　6 号峰　C_{19},三芳甾质谱图

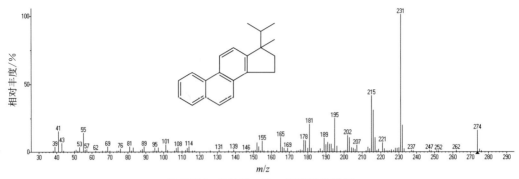

图 2-366　7 号峰　C_{21} 三芳甾质谱图

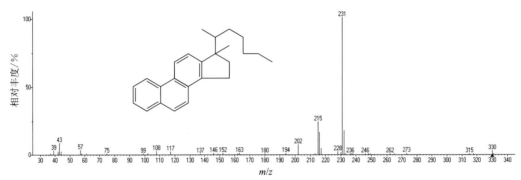

图 2-367　8 号峰　C_{25} 三芳甾质谱图

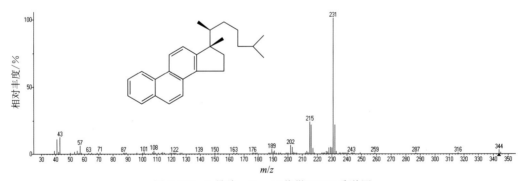

图 2-368　9 号峰　C_{26} 三芳甾(20S)质谱图

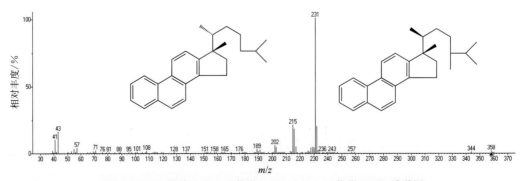

图 2-369　10 号峰　C_{26} 三芳甾(20R)＋C_{27} 三芳甾(20S)质谱图

第 2 章 石油地质样品的全二维气相色谱图识别

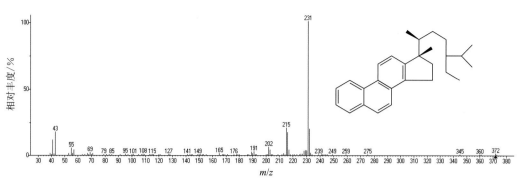

图 2-370　11 号峰　C_{28}，三芳甾（20S）质谱图

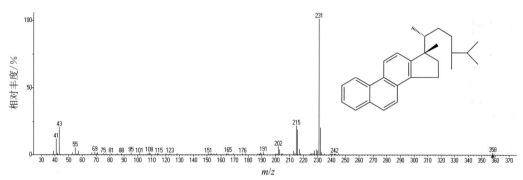

图 2-371　12 号峰　C_{27}，三芳甾（20R）质谱图

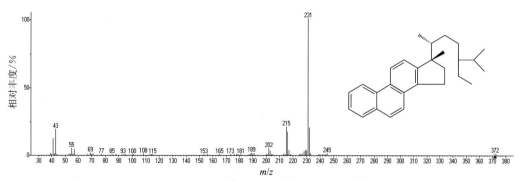

图 2-372　13 号峰　C_{28}，三芳甾（20R）质谱图

表 2-27　三芳甾化合物鉴定结果表

峰 号	一维保留时间 /min	二维保留时间 /s	特征离子 （m/z）	分子式	相对 分子质量	名　称
1	105.00	4.07	231	$C_{19}H_{18}$	246	C_{19}，Triaromatic-sterane C_{19}，三芳甾
2	106.60	3.89	231	$C_{19}H_{18}$	246	C_{19}，Triaromatic-sterane C_{19}，三芳甾
3	106.60	4.08	231	$C_{19}H_{18}$	246	C_{19}，Triaromatic-sterane C_{19}，三芳甾
4	106.60	4.21	231	$C_{19}H_{18}$	246	C_{19}，Triaromatic-sterane C_{19}，三芳甾

续表

峰 号	一维保留时间 /min	二维保留时间 /s	特征离子 (m/z)	分子式	相对分子质量	名 称
5	107.00	3.97	231	$C_{20}H_{20}$	260	C_{20},Triaromatic-sterane C_{20},三芳甾
6	107.80	4.16	231	$C_{19}H_{18}$	246	C_{19},Triaromatic-sterane C_{19},三芳甾
7	110.73	4.02	231	$C_{21}H_{22}$	274	C_{21},Triaromatic-sterane C_{21},三芳甾
8	120.47	3.89	231	$C_{25}H_{30}$	330	C_{25},Triaromatic-sterane C_{25},三芳甾
9	124.20	4.25	231	$C_{26}H_{32}$	344	C_{26},Triaromatic-sterane(20S) C_{26},三芳甾(20S)
10	127.00	4.54	231	$C_{27}H_{34}$	358	C_{26},Triaromatic-sterane(20R)+ C_{27},Triaromatic-sterane(20S) C_{26},三芳甾(20R)+C_{27},三芳甾(20S)
11	129.53	4.77	231	$C_{28}H_{36}$	372	C_{28},Triaromatic-sterane(20S) C_{28},三芳甾(20S)
12	130.73	4.91	231	$C_{27}H_{34}$	358	C_{27},Triaromatic-sterane(20R) C_{27},三芳甾(20R)
13	134.07	5.22	231	$C_{28}H_{36}$	372	C_{28},Triaromatic-sterane(20R) C_{28},三芳甾(20R)

同单芳甾一样,甲基三芳甾比三芳甾具有更多的化合物,如图 2-373 所示。甲基三芳甾的质谱检测特征离子为 m/z 245,在全二维谱图中化合物的分离度很高。参考《有机质谱在石油化学中的应用》(苏焕华等,2010)可以鉴定出 25 个化合物,相应的质谱图如图 2-374~图 2-398所示,各化合物名称见表 2-28。

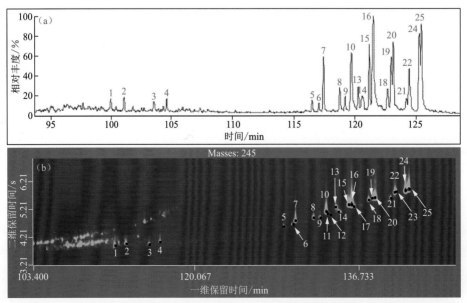

图 2-373 甲基三芳甾的 GC-MS 色谱图(a)、全二维点阵图(b)和三维立体图(c)

第 2 章 石油地质样品的全二维气相色谱图识别

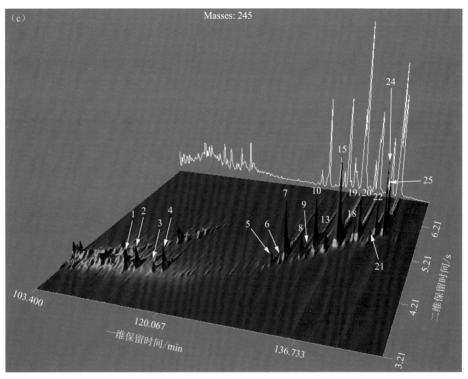

图 2-373（续） 甲基三芳甾的 GC-MS 色谱图(a)、全二维点阵图(b)和三维立体图(c)

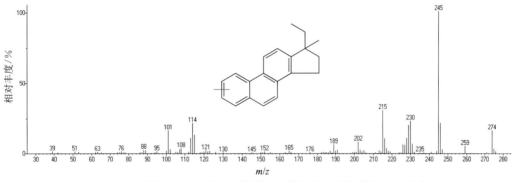

图 2-374 1 号峰 C_{21}, 甲基三芳甾烷质谱图（其中甲基的位置不确定）

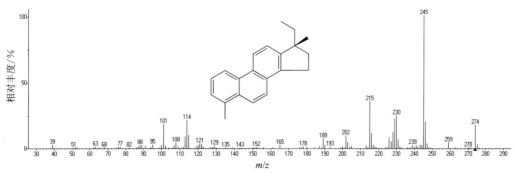

图 2-375 2 号峰 C_{21}, 4-甲基三芳甾烷质谱图

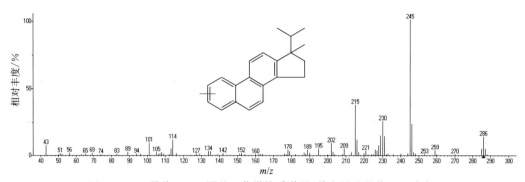

图 2-376　3 号峰　C_{22},甲基三芳甾烷质谱图(其中甲基的位置不确定)

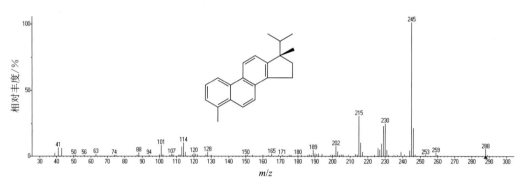

图 2-377　4 号峰　C_{22},4-甲基三芳甾烷质谱图

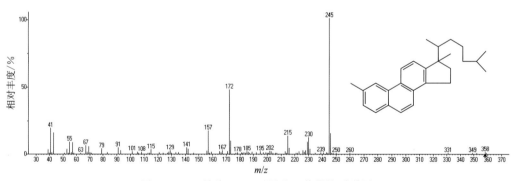

图 2-378　5 号峰　C_{27},2-甲基三芳甾烷质谱图

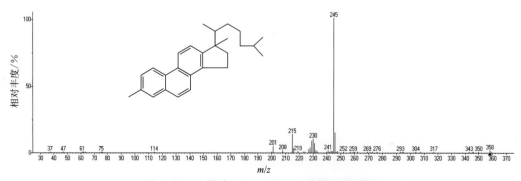

图 2-379　6 号峰　C_{27},3-甲基三芳甾烷质谱图

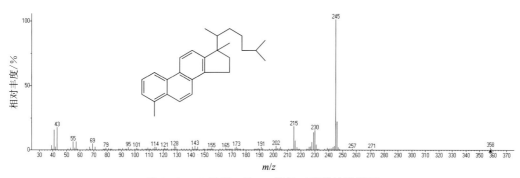

图 2-380　7 号峰　C_{27},4-甲基三芳甾烷质谱图

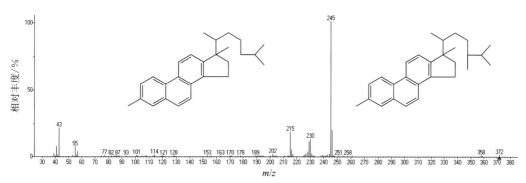

图 2-381　8 号峰　C_{27},3-甲基三芳甾烷＋C_{28},3,24-二甲基三芳甾烷质谱图

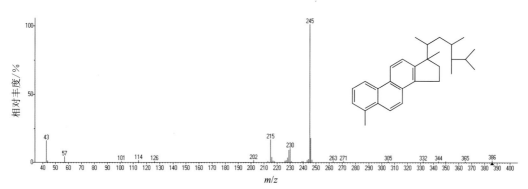

图 2-382　9 号峰　C_{29},4,23,24-三甲基三芳甲藻甾烷质谱图

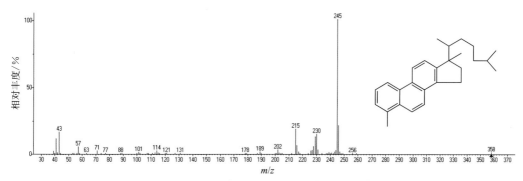

图 2-383　10 号峰　C_{27},4-甲基三芳甾烷质谱图

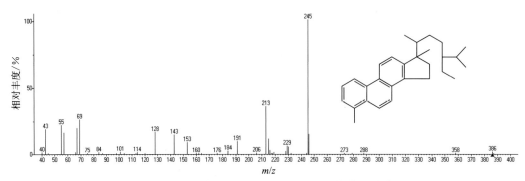

图 2-384　11 号峰　C_{29},4-甲基-24-乙基三芳甾烷质谱图

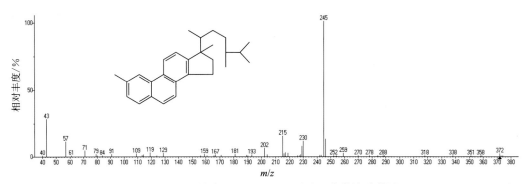

图 2-385　12 号峰　C_{28},2,24-二甲基三芳甾烷质谱图

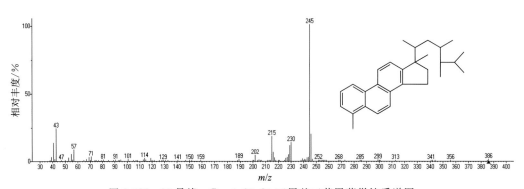

图 2-386　13 号峰　C_{29},4,23,24-三甲基三芳甲藻甾烷质谱图

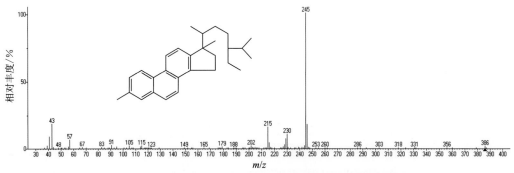

图 2-387　14 号峰　C_{29},3-甲基-24-乙基三芳甾烷质谱图

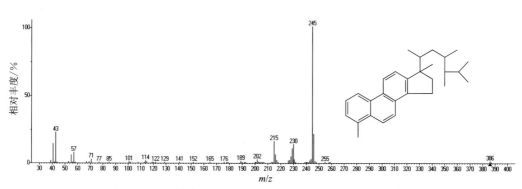

图 2-388　15 号峰　C_{29},4,23,24-三甲基三芳甲藻甾烷质谱图

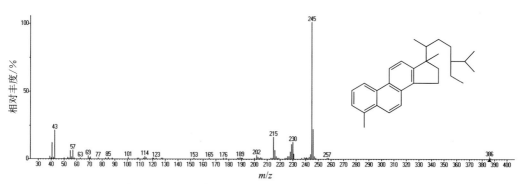

图 2-389　16 号峰　C_{29},4-甲基-24-乙基三芳甾烷质谱图

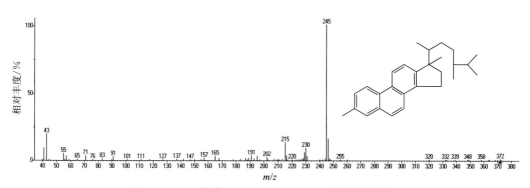

图 2-390　17 号峰　C_{28},3,24-二甲基三芳甾烷质谱图

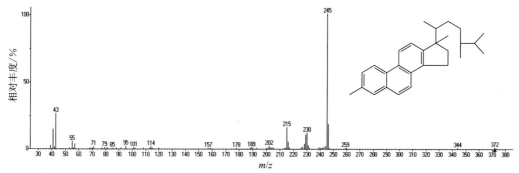

图 2-391　18 号峰　C_{28},3,24-二甲基三芳甾烷质谱图

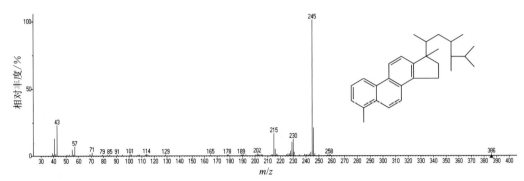

图 2-392　19 号峰　C_{29},4,23,24-三甲基三芳甲藻甾烷质谱图

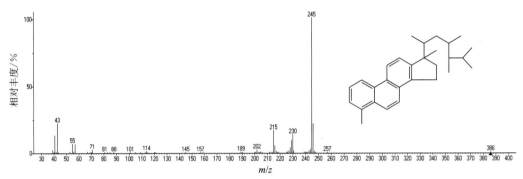

图 2-393　20 号峰　C_{29},4,23,24-三甲基三芳甲藻甾烷质谱图

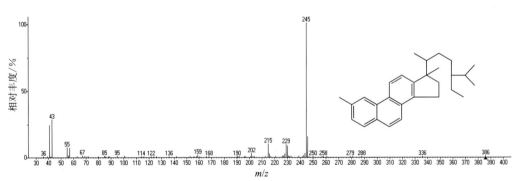

图 2-394　21 号峰　C_{29},2-甲基-24-乙基三芳甾烷质谱图

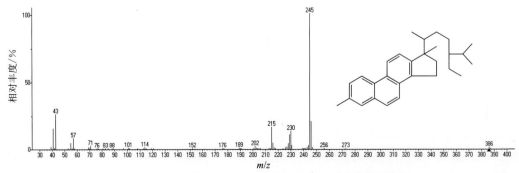

图 2-395　22 号峰　C_{29},3-甲基-24-乙基三芳甾烷质谱图

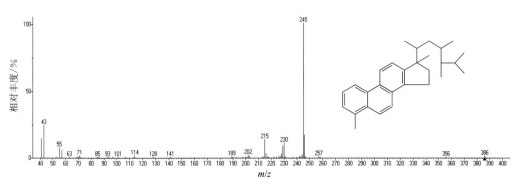

图 2-396　23 号峰　C_{29},4,23,24-三甲基三芳甲藻甾烷质谱图

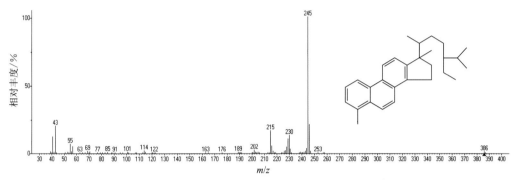

图 2-397　24 号峰　C_{29},4-甲基-24-乙基三芳甾烷质谱图

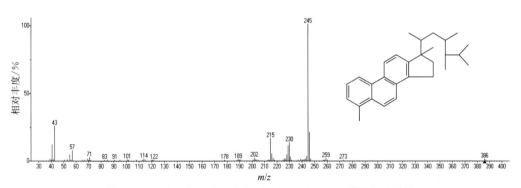

图 2-398　25 号峰　C_{29},4,23,24-三甲基三芳甲藻甾烷质谱图

表 2-28　甲基三芳甾烷系列化合物鉴定结果表

峰号	一维保留时间/min	二维保留时间/s	特征离子(m/z)	分子式	相对分子质量	名　称
1	111.80	3.93	245	$C_{21}H_{22}$	274	C_{21},Methyl- triaromatic-sterane C_{21},甲基三芳甾烷
2	113.00	3.99	245	$C_{21}H_{22}$	274	C_{21},4-Methyl- triaromatic-sterane C_{21},4-甲基三芳甾烷
3	115.40	3.98	245	$C_{22}H_{24}$	288	C_{22},Methyl- triaromatic-sterane C_{22},甲基三芳甾烷

续表

峰 号	一维保留时间 /min	二维保留时间 /s	特征离子 (m/z)	分子式	相对分子质量	名 称
4	116.47	4.05	245	$C_{22}H_{24}$	288	C_{22},4-Methyl- triaromatic-sterane C_{22},4-甲基三芳甾烷
5	129.00	4.60	245	$C_{27}H_{34}$	358	C_{27},2-Methyl- triaromatic-sterane C_{27},2-甲基三芳甾烷
6	129.93	4.68	245	$C_{27}H_{34}$	358	C_{27},3-Methyl- triaromatic-sterane C_{27},3-甲基三芳甾烷
7	130.20	4.81	245	$C_{27}H_{34}$	358	C_{27},4-Methyl- triaromatic-sterane C_{27},4-甲基三芳甾烷
8	132.07	4.91	245	$C_{28}H_{36}$	372	C_{27},3-Methyl- triaromatic-sterane+C_{28},3,24-Dimethyl-triaromatic-sterane C_{27},3-甲基三芳甾烷+C_{28},3,24-二甲基三芳甾烷
9	132.73	4.92	245	$C_{29}H_{38}$	386	C_{29},4,23,24-Trimethyl-triaromatic-dinosterane C_{29},4,23,24-三甲基三芳甲藻甾烷
10	133.40	5.13	245	$C_{27}H_{34}$	358	C_{27},4-Methyl-triaromatic-sterane C_{27},4-甲基三芳甾烷
11	133.53	5.10	245	$C_{29}H_{38}$	386	C_{29},4-Methyl-24-ethyl-triaromatic-sterane C_{29},4-甲基-24-乙基三芳甾烷
12	133.80	5.02	245	$C_{28}H_{36}$	372	C_{28},2,24-Dimethyl-triaromatic-sterane C_{28},2,24-二甲基三芳甾烷
13	134.33	5.30	245	$C_{29}H_{38}$	386	C_{29},4,23,24-Trimethyl-triaromatic-dinosterane C_{29},4,23,24-三甲基三芳甲藻甾烷
14	134.60	5.14	245	$C_{29}H_{38}$	386	C_{29},3-Methyl-24-ethyl-triaromatic-sterane C_{29},3-甲基-24-乙基三芳甾烷
15	135.67	5.38	245	$C_{29}H_{38}$	386	C_{29},4,23,24-Trimethyl-triaromatic-dinosterane C_{29},4,23,24-三甲基三芳甲藻甾烷
16	135.93	5.39	245	$C_{29}H_{38}$	386	C_{29},4-Methyl-24-ethyl-triaromatic-sterane C_{29},4-甲基-24-乙基三芳甾烷
17	136.20	5.33	245	$C_{28}H_{36}$	372	C_{28},3,24-Dimethyl-triaromatic-sterane C_{28},3,24-二甲基三芳甾烷

续表

峰 号	一维保留时间/min	二维保留时间/s	特征离子(m/z)	分子式	相对分子质量	名 称
18	137.67	5.56	245	$C_{28}H_{36}$	372	C_{28},3,24-Dimethyl-triaromatic-sterane C_{28},3,24-二甲基三芳甾烷
19	138.07	5.62	245	$C_{29}H_{38}$	386	C_{29},4,23,24-Trimethyl-triaromatic-dinosterane C_{29},4,23,24-三甲基三芳甲藻甾烷
20	138.33	5.63	245	$C_{29}H_{38}$	386	C_{29},4,23,24-Trimethyl-triaromatic-dinosterane C_{29},4,23,24-三甲基三芳甲藻甾烷
21	139.93	5.62	245	$C_{29}H_{38}$	386	C_{29},2-Methyl-24-ethyl-triaromatic-sterane C_{29},2-甲基-24-乙基三芳甾烷
22	140.33	5.82	245	$C_{29}H_{38}$	386	C_{29},3-Methyl-24-ethyl-triaromatic-sterane C_{29},3-甲基-24-乙基三芳甾烷
23	140.47	5.85	245	$C_{29}H_{38}$	386	C_{29},4,23,24-Trimethyl-triaromatic-dinosterane C_{29},4,23,24-三甲基三芳甲藻甾烷
24	141.40	5.89	245	$C_{29}H_{38}$	386	C_{29},4-Methyl-24-ethyl-triaromatic-sterane C_{29},4-甲基-24-乙基三芳甾烷
25	141.80	5.93	245	$C_{29}H_{38}$	386	C_{29},4,23,24-Trimethyl-triaromatic-dinosterane C_{29},4,23,24-三甲基三芳甲藻甾烷

2.2.11 单芳断藿烷系列化合物的全二维气相色谱图及化合物名称

藿烷类化合物的环也可被芳构化,形成单芳、双芳、三芳以及四芳等多种多样的芳构化藿烷。研究较多的是 D 环芳构化的单芳藿烷,此系列化合物的断裂一般发生在 8~14 位,因此本节介绍的单芳断藿烷系列化合物指的是 D 环-单芳-8,14-断藿烷,其质谱检测的特征离子为 m/z 365,另有 m/z 159 的特征离子碎片。图 2-399 是该系列化合物的常规色质分析色谱图(a)和全二维谱图(b,c)。在全二维气相色谱图上单芳断藿烷系列化合物分离度很高,可以鉴定出 14 个化合物,相应的质谱图如图 2-340~图 2-413 所示,各化合物名称见表 2-29。

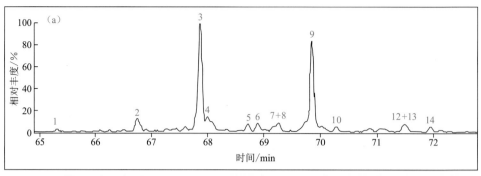

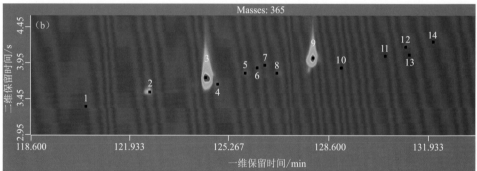

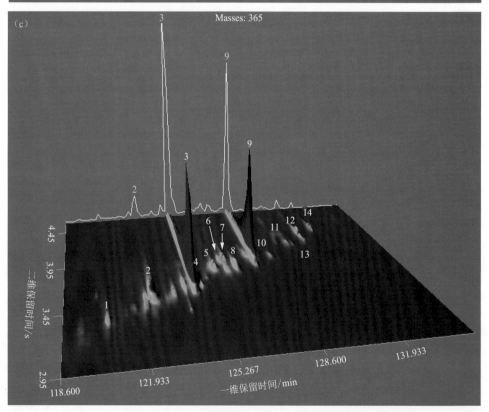

图 2-399　D 环-单芳-8,14 断-藿烷系列化合物的 GC-MS 色谱图(a)、全二维点阵图(b)和三维立体图(c)

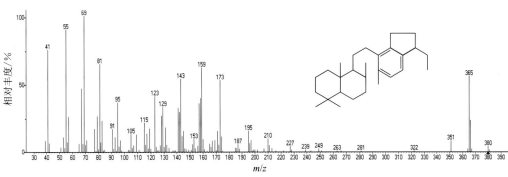

图 2-400　1 号峰　C_{28},单芳-8,14-断-藿烷质谱图

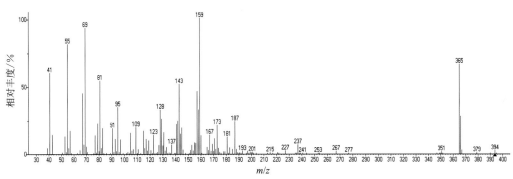

图 2-401　2 号峰　C_{29},单芳-8,14-断-藿烷质谱图

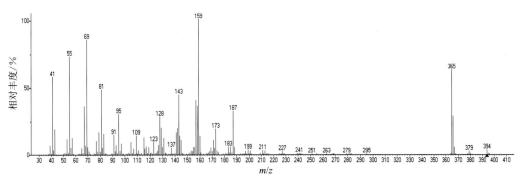

图 2-402　3 号峰　C_{29},单芳-8,14-断-藿烷质谱图

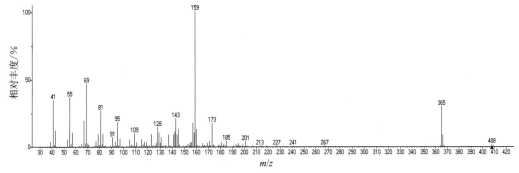

图 2-403　4 号峰　C_{30},单芳-8,14-断-藿烷质谱图

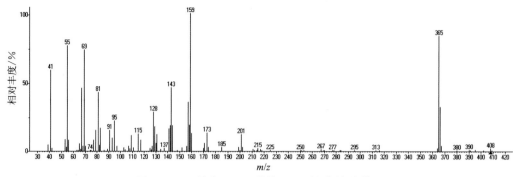

图 2-404　5号峰　C_{30},单芳-8,14-断-藿烷质谱图

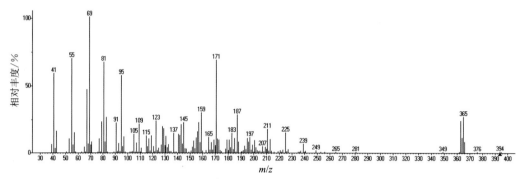

图 2-405　6号峰　C_{29},单芳-8,14-断-藿烷质谱图

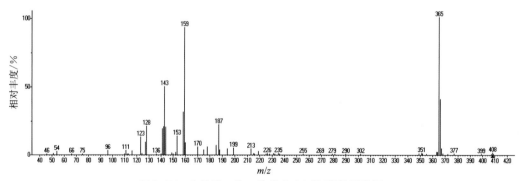

图 2-406　7号峰　C_{30},单芳-8,14-断-藿烷质谱图

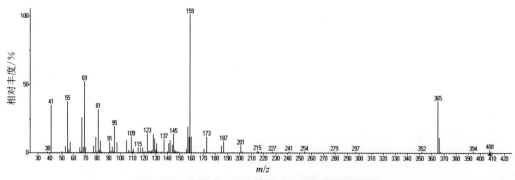

图 2-407　8号峰　C_{30},单芳-8,14-断-藿烷质谱图

第 2 章 石油地质样品的全二维气相色谱图识别

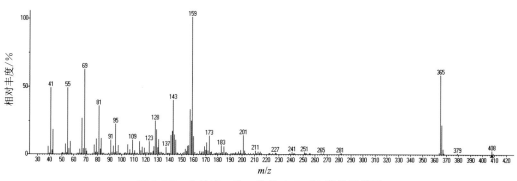

图 2-408　9 号峰　C_{30},单芳-8,14-断藿烷质谱图

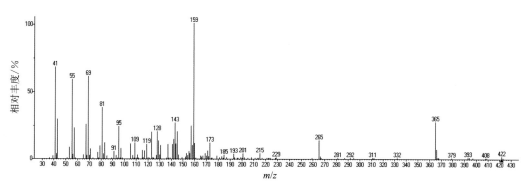

图 2-409　10 号峰　C_{31},单芳-8,14-断藿烷质谱图

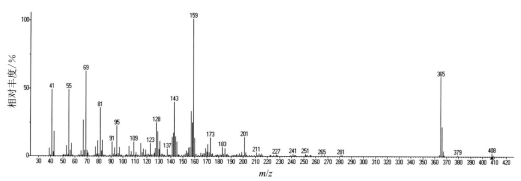

图 2-410　11 号峰　C_{30},单芳-8,14-断藿烷质谱图

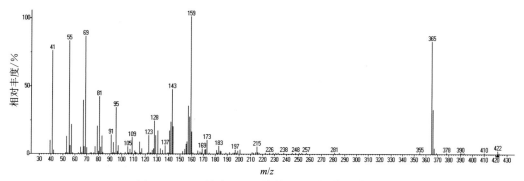

图 2-411　12 号峰　C_{31},单芳-8,14-断藿烷质谱图

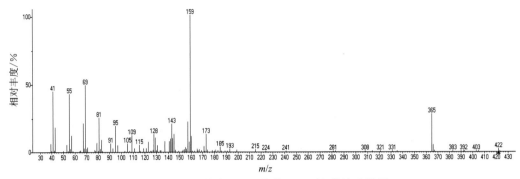

图 2-412　13 号峰　C_{31},单芳-8,14-断-藿烷质谱图

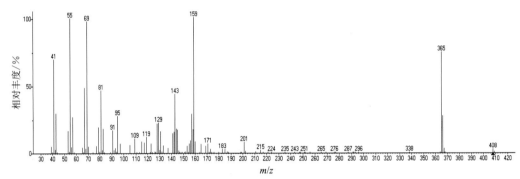

图 2-413　14 号峰　C_{30},单芳-8,14-断-藿烷质谱图

表 2-29　D 环-单芳-8,14-断-藿烷系列化合物鉴定结果表

峰号	一维保留时间 /min	二维保留时间 /s	特征离子 (m/z)	分子式	相对分子质量	名　称
1	120.47	3.36	365	$C_{28}H_{44}$	380	C_{28},Monoaromatic-8,14-seco-hopane C_{28},单芳-8,14-断-藿烷
2	122.60	3.55	365	$C_{29}H_{46}$	394	C_{29},Monoaromatic-8,14-seco-hopane C_{29},单芳-8,14-断-藿烷
3	124.47	3.75	365	$C_{29}H_{46}$	394	C_{29},Monoaromatic-8,14-seco-hopane C_{29},单芳-8,14-断-藿烷
4	124.87	3.66	365	$C_{30}H_{48}$	408	C_{30},Monoaromatic-8,14-seco-hopane C_{30},单芳-8,14-断-藿烷
5	125.80	3.82	365	$C_{30}H_{48}$	408	C_{30},Monoaromatic-8,14-seco-hopane C_{30},单芳-8,14-断-藿烷
6	126.20	3.89	365	$C_{29}H_{46}$	394	C_{29},Monoaromatic-8,14-seco-hopane C_{29},单芳-8,14-断-藿烷
7	126.47	3.92	365	$C_{30}H_{48}$	408	C_{30},Monoaromatic-8,14-seco-hopane C_{30},单芳-8,14-断-藿烷
8	126.87	3.82	365	$C_{30}H_{48}$	408	C_{30},Monoaromatic-8,14-seco-hopane C_{30},单芳-8,14-断-藿烷
9	128.07	4.02	365	$C_{30}H_{48}$	408	C_{30},Monoaromatic-8,14-seco-hopane C_{30},单芳-8,14-断-藿烷

续表

峰号	一维保留时间 /min	二维保留时间 /s	特征离子 (m/z)	分子式	相对分子质量	名 称
10	129.00	3.89	365	$C_{31}H_{50}$	422	C_{31},Monoaromatic-8,14-seco-hopane C_{31},单芳-8,14-断-藿烷
11	130.47	4.05	365	$C_{30}H_{48}$	408	C_{30},Monoaromatic-8,14-seco-hopane C_{30},单芳-8,14-断-藿烷
12	131.13	4.18	365	$C_{31}H_{50}$	422	C_{31},Monoaromatic-8,14-seco-hopane C_{31},单芳-8,14-断-藿烷
13	131.27	4.07	365	$C_{31}H_{50}$	422	C_{31},Monoaromatic-8,14-seco-hopane C_{31},单芳-8,14-断-藿烷
14	132.07	4.25	365	$C_{30}H_{48}$	408	C_{30},Monoaromatic-8,14-seco-hopane C_{30},单芳-8,14-断-藿烷

2.2.12 单芳-断-降藿烯系列化合物的全二维气相色谱图及化合物名称

单芳-断-降藿烯系列化合物指的是 D 环-单芳-8,14-断-藿烷降解一个甲基在 E 环形成双键的一系列化合物,其质谱检测的特征离子是 m/z 363,图 2-414 是该系列化合物的 GC-MS 色谱图(a)、全二维点阵图(b)和三维立体图(c)。在全二维气相色谱图上鉴定出 5 个化合物,其中 3 号化合物与 3′号化合物(C_{29},单芳-8,14-断藿烷)在常规色质分析谱图上是共馏峰,但在全二维分析谱图中可以被分开,相应的质谱图如图 2-415~图 2-419 所示,各化合物鉴定结果见表 2-30。

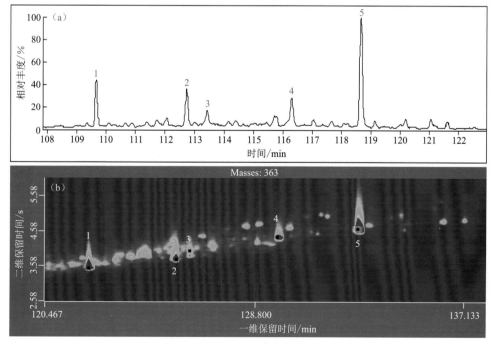

图 2-414 D 环-单芳-8,14-断-降藿烯系列化合物的 GC-MS 色谱图(a)、全二维点阵图(b)和三维立体图(c)

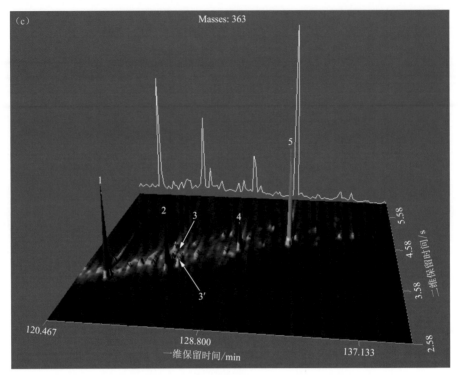

图 2-414(续)　D 环-单芳-8,14-断-降霍烯系列化合物的 GC-MS 色谱图(a)、全二维点阵图(b)和三维立体图(c)

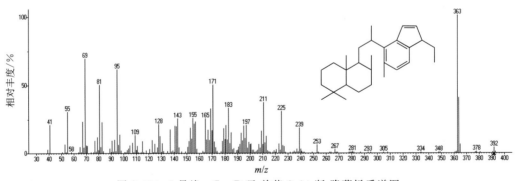

图 2-415　1 号峰　C_{29},D 环-单芳-8,14-断-降霍烯质谱图

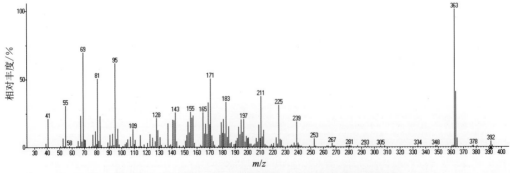

图 2-416　2 号峰　C_{29},D 环-单芳-8,14-断-降霍烯质谱图

第 2 章 石油地质样品的全二维气相色谱图识别

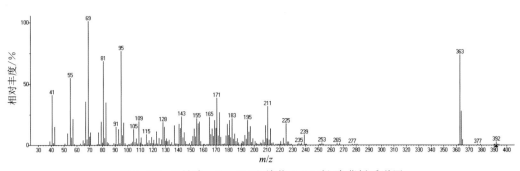

图 2-417　3 号峰　C_{29}，D 环-单芳-8,14-断-降藿烯质谱图

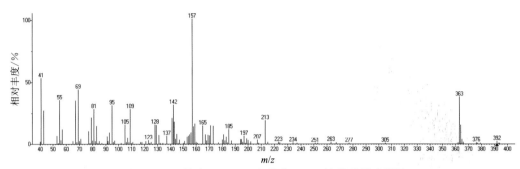

图 2-418　4 号峰　C_{29}，D 环-单芳-8,14-断-降藿烯质谱图

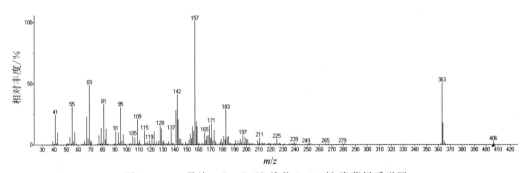

图 2-419　5 号峰　C_{30}，D 环-单芳-8,14-断-降藿烯质谱图

表 2-30　D 环-单芳-8,14-断-降藿烯系列化合物鉴定结果表

峰号	一维保留时间/min	二维保留时间/s	特征离子（m/z）	分子式	相对分子质量	名　称
1	122.20	3.55	363	$C_{29}H_{44}$	392	C_{29}，D-Ring-monoaromatic-8,14-seco-norhopene C_{29}，D 环-单芳-8,14-断-降藿烯
2	125.67	3.79	363	$C_{29}H_{44}$	392	C_{29}，D-Ring-monoaromatic-8,14-seco-norhopene C_{29}，D 环-单芳-8,14-断-降藿烯
3	126.20	4.03	363	$C_{29}H_{44}$	392	C_{29}，D-Ring-monoaromatic-8,14-seco-norhopene C_{29}，D 环-单芳-8,14-断-降藿烯
4	129.67	4.42	157	$C_{29}H_{44}$	392	C_{29}，D-Ring-monoaromatic-8,14-seco-norhopene C_{29}，D 环-单芳-8,14-断-降藿烯
5	132.87	4.65	157	$C_{30}H_{46}$	406	C_{30}，D-Ring-monoaromatic-8,14-seco-norhopene C_{30}，D 环-单芳-8,14-断-降藿烯

2.2.13 单芳-断-降藿烷系列化合物的全二维气相色谱图及化合物名称

单芳-断-降藿烷系列化合物指的是 D 环-单芳-8,14-断-藿烷降解一个甲基形成的一系列化合物,其质谱检测的特征离子是 m/z 351,碎片离子 m/z 145 相对丰度也较高。图 2-420 是该系列化合物的 GC-MS 色谱图(a)、全二维点阵图(b)和三维立体图(c)。单芳-断-降藿烷系列化合物鉴定出 6 个化合物,对应的质谱图如图 2-421～图 2-426 所示,各化合物鉴定结果见表 2-31。

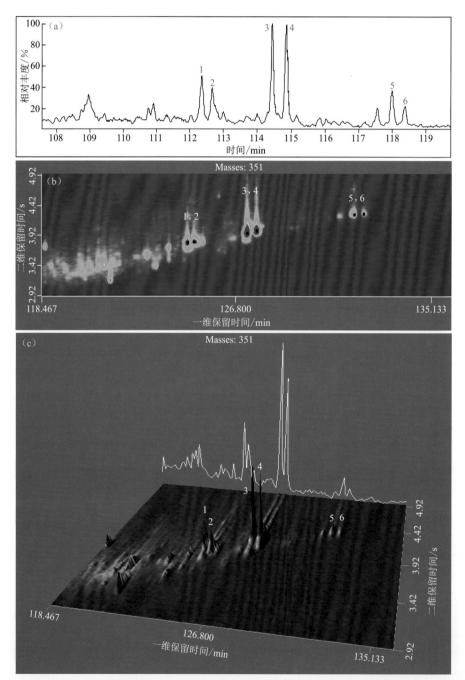

图 2-420　D 环-单芳-8,14-断-降藿烷系列化合物的 GC-MS 色谱图(a)、全二维点阵图(b)和三维立体图(c)

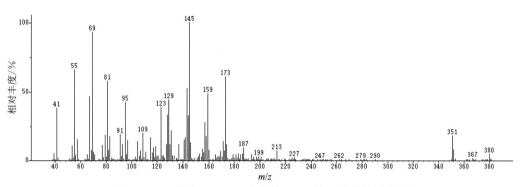

图 2-421　1 号峰　C_{28}，D 环-单芳-8,14-断-双降藿烷质谱图

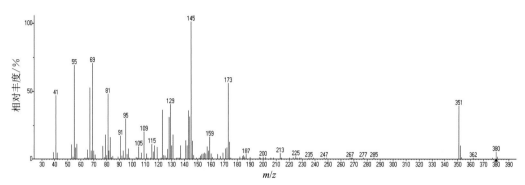

图 2-422　2 号峰　C_{28}，D 环-单芳-8,14-断-双降藿烷质谱图

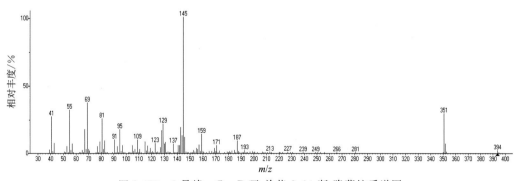

图 2-423　3 号峰　C_{29}，D 环-单芳-8,14-断-降藿烷质谱图

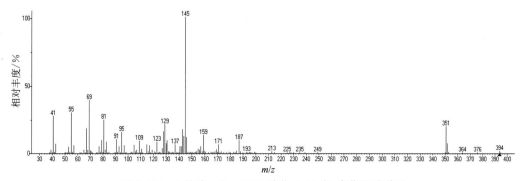

图 2-424　4 号峰　C_{29}，D 环-单芳-8,14-断-降藿烷质谱图

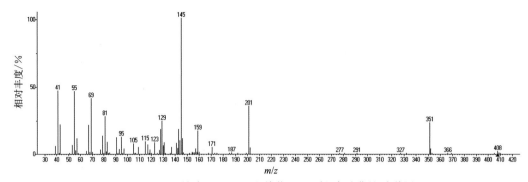

图 2-425　5 号峰　C_{30}，D 环-单芳-8,14-断-降升藿烷质谱图

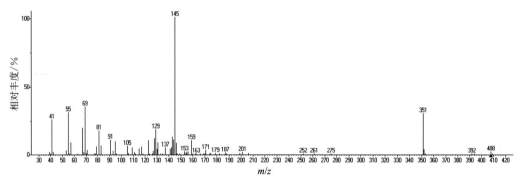

图 2-426　6 号峰　C_{30}，D 环-单芳-8,14-断-降升藿烷质谱图

表 2-31　D 环-单芳-8,14-断-降藿烷系列化合物鉴定结果表

峰号	一维保留时间 /min	二维保留时间 /s	特征离子 （m/z）	分子式	相对分子质量	名　称
1	124.73	3.81	145	$C_{28}H_{44}$	380	C_{28}，D-Ring-monoaromatic-8,14-seco-bisnorhopane C_{28}，D 环-单芳-8,14-断-双降藿烷
2	125.00	3.82	145	$C_{28}H_{44}$	380	C_{28}，D-Ring-monoaromatic-8,14-seco-bisnorhopane C_{28}，D 环-单芳-8,14-断-双降藿烷
3	127.27	3.97	145	$C_{29}H_{46}$	394	C_{29}，D-Ring-monoaromatic-8,14-seco-norhopane C_{29}，D 环-单芳-8,14-断-降藿烷
4	127.67	4.00	145	$C_{29}H_{46}$	394	C_{29}，D-Ring- monoaromatic-8,14-seco-norhopane C_{29}，D 环-单芳-8,14-断-降藿烷
5	131.80	4.28	145	$C_{30}H_{48}$	408	C_{30}，D-Ring-monoaromatic-8,14-seco-norhomohopane C_{30}，D 环-单芳-8,14-断-降升藿烷
6	132.20	4.29	145	$C_{30}H_{48}$	408	C_{30}，D-Ring-monoaromatic-8,14-seco-norhomohopane C_{30}，D 环-单芳-8,14-断-降升藿烷

2.2.14 芳构化的五环三萜烷的全二维气相色谱图及化合物名称

不同程度的芳构化不仅发生在藿烷类化合物上,非藿烷类化合物(如奥利烷、羽扇烷等)以及尚不能准确定名的五环三萜烷化合物同样也会发生不同程度的芳构化,简便起见,将它们统称芳构化的五环三萜烷。

根据芳构化的环数将芳构化的五环三萜烷分别称为单芳、双芳、三芳、四芳以及五芳-五环三萜烷,它们在全二维气相色谱图上的分布如图2-427所示。

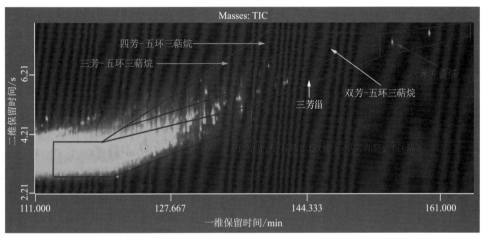

图 2-427　不同芳构化的五环三萜烷在全二维气相色谱图上的分布特征

1) 双芳-五环三萜烷的全二维气相色谱图及化合物名称

双芳-五环三萜烷多指 A 和 B 环发生芳构化的化合物,这些化合物的碳数从 26 到 32,可能的同分异构体及化学结构见表 2-32。总碳数相同、化学结构及特征离子不同的化合物具体名称待定,暂统一以碳数笼统命名。

表 2-32　A,B 环-双芳-五环三萜烷系列化合物的同分异构体

序号	化合物名称	特征离子 (m/z)	分子离子	可能的化学结构
1	C_{26},A,B-Ring-diaromatic-pentacyclotriterpane C_{26},A,B 环-双芳-五环三萜烷	181	346	
2	C_{27},A,B-Ring-diaromatic-pentacyclotriterpane C_{27},A,B 环-双芳-五环三萜烷	195	360	

续表

序号	化合物名称	特征离子 (m/z)	分子离子	可能的化学结构
3	C_{28}, A, B-Ring-diaromatic-pentacyclotriterpane C_{28}, A, B 环-双芳-五环三萜烷	195	374	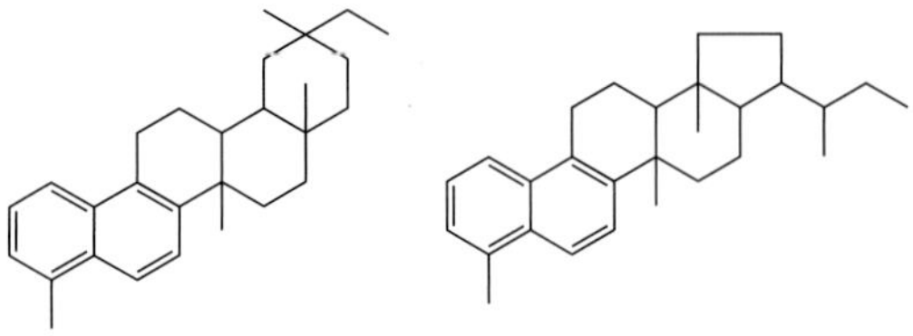
4	C_{30}, A, B-Ring-diaromatic-pentacyclotriterpane C_{30}, A, B 环-双芳-五环三萜烷	195	402	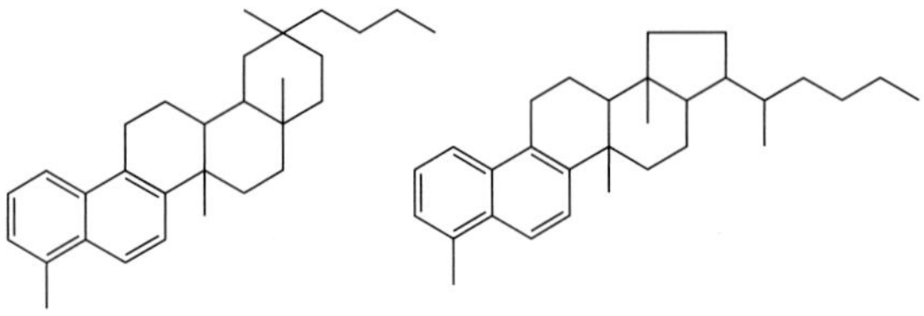
5	C_{28}, A, B-Ring-diaromatic-pentacyclotriterpane C_{28}, A, B 环-双芳-五环三萜烷	209	374	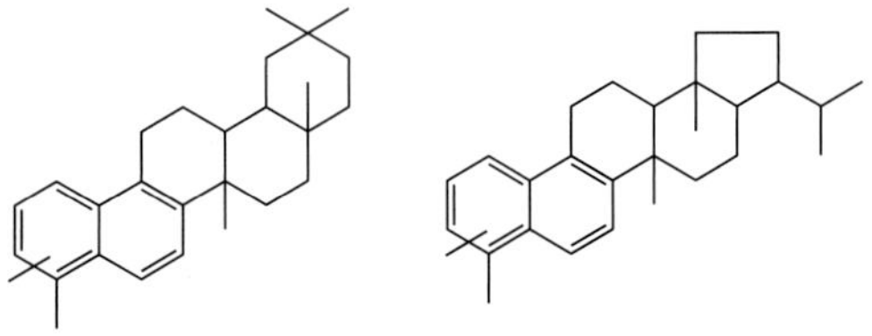
6	C_{31}, A, B-Ring-diaromatic-pentacyclotriterpane C_{31}, A, B 环-双芳-五环三萜烷	209	416	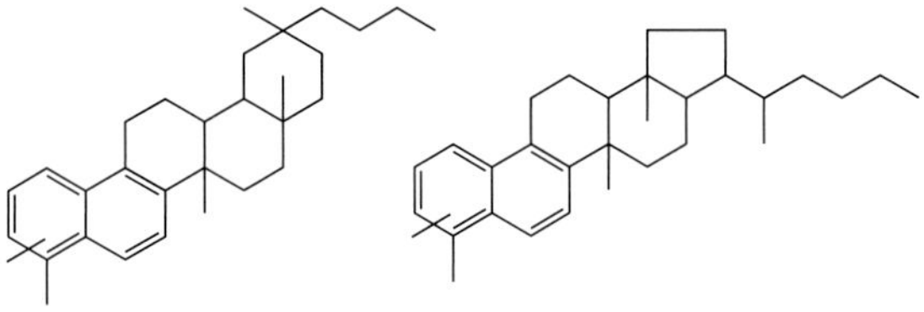
7	C_{29}, A, B-Ring-diaromatic-pentacyclotriterpane C_{29}, A, B 环-双芳-五环三萜烷	223	388	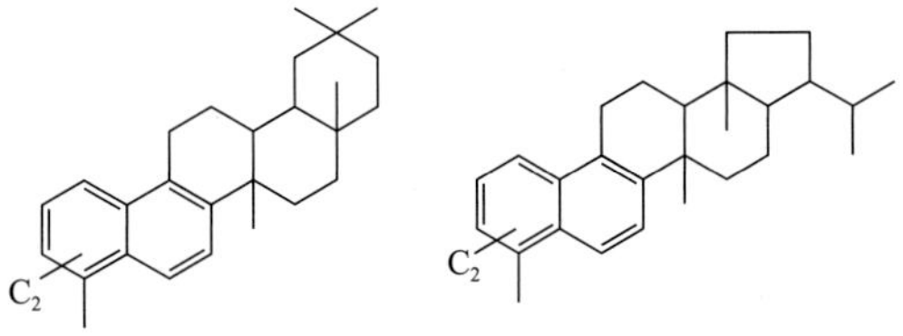
8	C_{32}, A, B-Ring-diaromatic-pentacyclotriterpane C_{32}, A, B 环-双芳-五环三萜烷	223	430	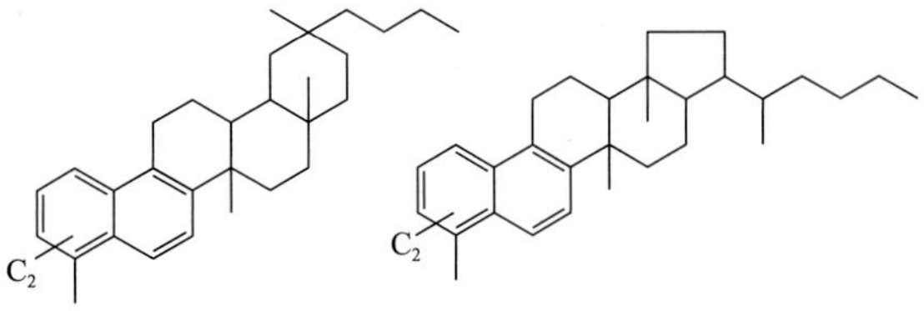

图 2-428 是特征离子为 m/z 195 的 A, B 环-双芳-五环三萜烷的 GC-MS 色谱图(a)和全二维分析谱图(b, c)。该系列共检测到 11 个化合物，其中 1~8 号峰同为 C_{27}, A, B 环-双芳-五环三萜烷，质谱图以 8 号峰为例，如图 2-429 所示；9~10 号峰为 C_{28}, A, B 环-双芳-五环三萜烷，质谱图以 9 号峰为例，如图 2-430 所示；11 号峰为 C_{30}, A, B 环-双芳-五环三萜烷，质谱图如图 2-431 所示。各化合物的二维保留特征及鉴定结果见表 2-33。

第 2 章 石油地质样品的全二维气相色谱图识别

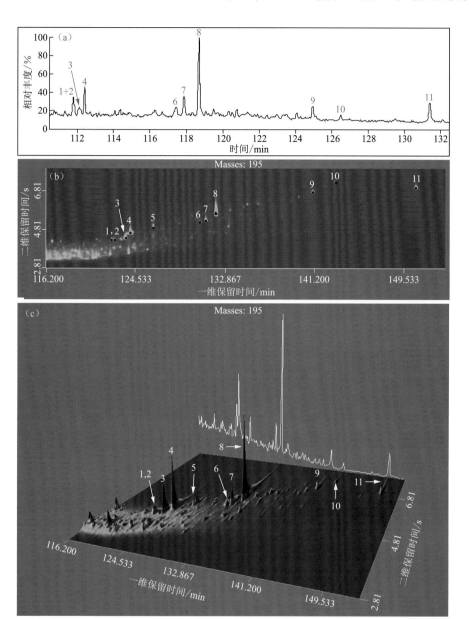

图 2-428 m/z 195 的 A,B 环-双芳-五环三萜烷的 GC-MS 色谱图(a)、全二维点阵图(b)和三维立体图(c)

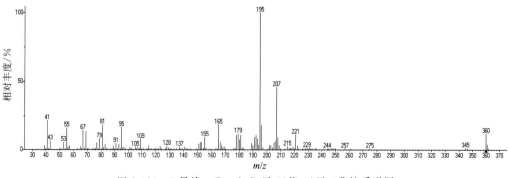

图 2-429 8 号峰 C_{27},A,B 环-双芳-五环三萜烷质谱图

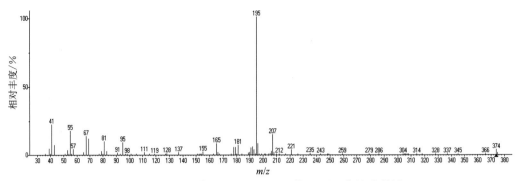

图 2-430　10 号峰　C_{28},A,B 环-双芳-五环三萜烷质谱图

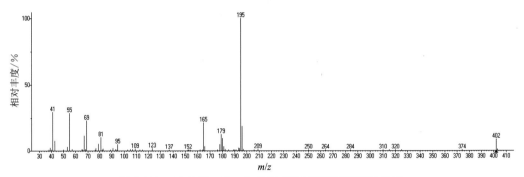

图 2-431　11 号峰　C_{30},A,B 环-双芳-五环三萜烷质谱图

表 2-33　m/z 195 的 A,B 环-双芳-五环三萜烷鉴定结果表

峰号	一维保留时间 /min	二维保留时间 /s	特征离子 (m/z)	分子式	相对分子质量	名　称
1	122.33	4.28	195	$C_{27}H_{36}$	360	C_{27},A,B-Ring-diaromatic-pentacyclotriterpane C_{27},A,B 环-双芳-五环三萜烷
2	122.60	4.29	195	$C_{27}H_{36}$	360	C_{27},A,B-Ring-diaromatic-pentacyclotriterpane C_{27},A,B 环-双芳-五环三萜烷
3	123.67	4.39	195	$C_{27}H_{36}$	360	C_{27},A,B-Ring-diaromatic-pentacyclotriterpane C_{27},A,B 环-双芳-五环三萜烷
4	124.07	4.66	195	$C_{27}H_{36}$	360	C_{27},A,B-Ring-diaromatic-pentacyclotriterpane C_{27},A,B 环-双芳-五环三萜烷
5	126.20	4.93	195	$C_{26}H_{34}$	346	C_{27},A,B-Ring-diaromatic-pentacyclotriterpane C_{27},A,B 环-双芳-五环三萜烷
6	130.47	5.20	195	$C_{27}H_{36}$	360	C_{27},A,B-Ring-diaromatic-pentacyclotriterpane C_{27},A,B 环-双芳-五环三萜烷

续表

峰 号	一维保留时间/min	二维保留时间/s	特征离子(m/z)	分子式	相对分子质量	名 称
7	131.00	5.30	195	$C_{27}H_{36}$	360	C_{27},A,B-Ring-diaromatic-pentacyclotriterpane C_{27},A,B环-双芳-五环三萜烷
8	131.93	5.66	195	$C_{27}H_{36}$	360	C_{27},A,B-Ring-diaromatic-pentacyclotriterpane C_{27},A,B环-双芳-五环三萜烷
9	141.00	6.77	195	$C_{28}H_{38}$	374	C_{28},A,B-Ring-diaromatic-pentacyclotriterpane C_{28},A,B环-双芳-五环三萜烷
10	143.13	7.28	195	$C_{28}H_{38}$	374	C_{28},A,B-Ring-diaromatic-pentacyclotriterpane C_{28},A,B环-双芳-五环三萜烷
11	150.60	7.01	195	$C_{30}H_{42}$	402	C_{30},A,B-Ring-diaromatic-pentacyclotriterpane C_{30},A,B环-双芳-五环三萜烷

图 2-432 是特征离子为 m/z 209 的 A,B 环-双芳-五环三萜烷的 GC-MS 色谱图(a)、全二维点阵图(b)和三维立体图(c)。该系列共检测到 6 个化合物,其中 1~5 号峰同为 C_{28},A,B 环-双芳-五环三萜烷,质谱图以 5 号峰为例,如图 2-433 所示;6 号峰为 C_{31},A,B 环-双芳-五环三萜烷,质谱图如图 2-434 所示。各化合物的二维保留特征及鉴定结果见表 2-34。

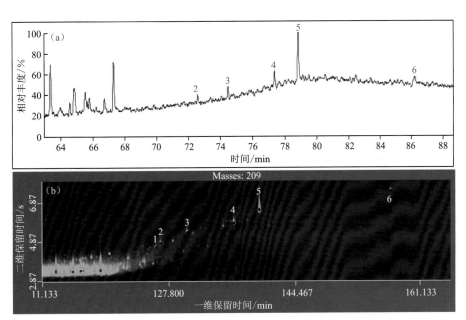

图 2-432 m/z 209 的 A,B 环-双芳-五环三萜烷的 GC-MS 色谱图(a)、全二维点阵图(b)和三维立体图(c)

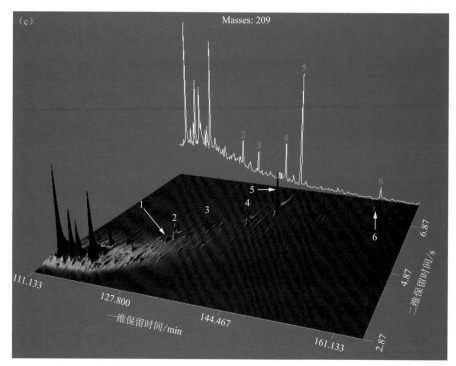

图 2-432（续） m/z 209 的 A,B 环-双芳-五环三萜烷的 GC-MS 色谱图(a)、全二维点阵图(b)和三维立体图(c)

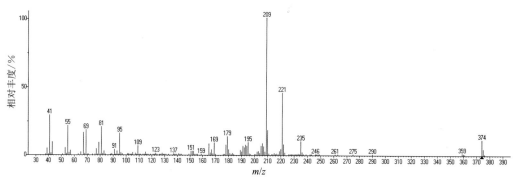

图 2-433 5 号峰 C_{28},A,B 环-双芳-五环三萜烷质谱图

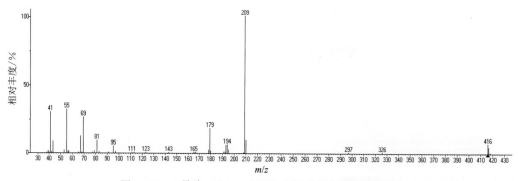

图 2-434 6 号峰 C_{31},A,B 环-双芳-五环三萜烷质谱图

表 2-34　m/z 209 的 A,B 环-双芳-五环三萜烷鉴定结果表

峰号	一维保留时间 /min	二维保留时间 /s	特征离子 (m/z)	分子式	相对 分子质量	名　称
1	126.07	4.63	209	$C_{28}H_{38}$	374	C_{28},A,B-Ring-diaromatic-pentacyclotriterpane C_{28},A,B 环-双芳-五环三萜烷
2	126.60	4.82	209	$C_{28}H_{38}$	374	C_{28},A,B-Ring-diaromatic-pentacyclotriterpane C_{28},A,B 环-双芳-五环三萜烷
3	130.20	5.39	209	$C_{28}H_{38}$	374	C_{28},A,B-Ring-diaromatic-pentacyclotriterpane C_{28},A,B 环-双芳-五环三萜烷
4	136.33	5.88	209	$C_{28}H_{38}$	374	C_{28},A,B-Ring-diaromatic-pentacyclotriterpane C_{28},A,B 环-双芳-五环三萜烷
5	139.67	6.53	209	$C_{28}H_{38}$	374	C_{28},A,B-Ring-diaromatic-pentacyclotriterpane C_{28},A,B 环-双芳-五环三萜烷
6	157.13	7.50	209	$C_{31}H_{44}$	416	C_{31},A,B-Ring-diaromatic-pentacyclotriterpane C_{31},A,B 环-双芳-五环三萜烷

图 2-435 是特征离子为 m/z 223 的 A,B 环-双芳-五环三萜烷的 GC-MS 色谱图(a)、全二维点阵图(b)和三维立体图(c)。该系列共检测到 6 个化合物,其中 1～3 号峰同为 C_{29},A,B 环-双芳-五环三萜烷,质谱图以 3 号峰为例,如图 2-436 所示;4～6 号峰为 C_{32},A,B 环-双芳-五环三萜烷,质谱图以 6 号峰为例,如图 2-437 所示。各化合物的二维保留特征及鉴定结果见表 2-35。

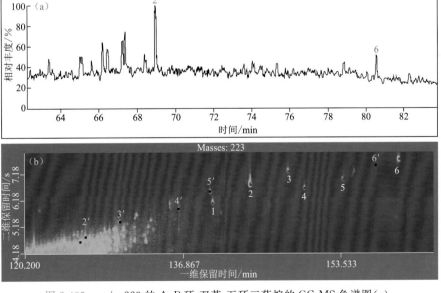

图 2-435　m/z 223 的 A,B 环-双芳-五环三萜烷的 GC-MS 色谱图(a)、
全二维点阵图(b)和三维立体图(c)

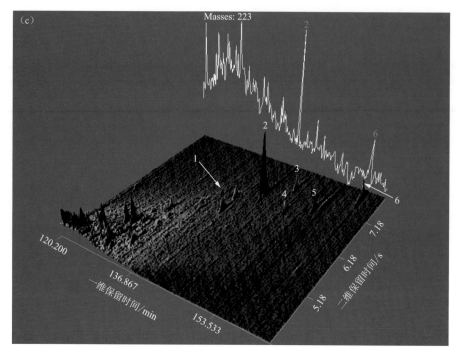

图 2-435(续) m/z 223 的 A,B 环-双芳-五环三萜烷的 GC-MS 色谱图(a)、
全二维点阵图(b)和三维立体图(c)

图中 1~6 标记了特征离子是 m/z 223 的化合物,1′~6′ 标记了特征离子是 m/z 209 的化合物

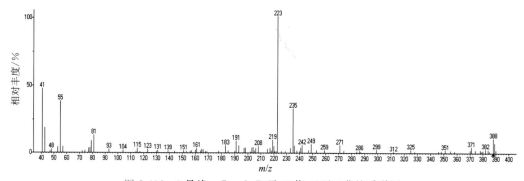

图 2-436　3 号峰　C_{29},A,B 环-双芳-五环三萜烷质谱图

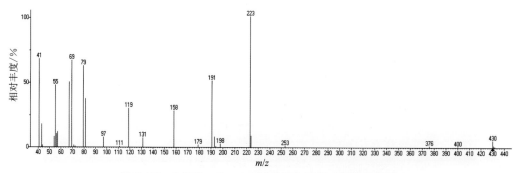

图 2-437　6 号峰　C_{32},A,B 环-双芳-五环三萜烷质谱图

表 2-35　m/z 223 的 A,B 环-双芳-五环三萜烷鉴定结果表

峰号	一维保留时间/min	二维保留时间/s	特征离子(m/z)	分子式	相对分子质量	名称
1	139.93	6.09	223	$C_{29}H_{40}$	388	C_{29},A,B-Ring-diaromatic-pentacyclotriterpane C_{29},A,B 环-双芳-五环三萜烷
2	143.93	6.78	223	$C_{29}H_{40}$	388	C_{29},A,B-Ring-diaromatic-pentacyclotriterpane C_{29},A,B 环-双芳-五环三萜烷
3	147.93	7.29	223	$C_{29}H_{40}$	388	C_{29},A,B-Ring-diaromatic-pentacyclotriterpane C_{29},A,B 环-双芳-五环三萜烷
4	149.67	6.66	223	$C_{32}H_{46}$	430	C_{32},A,B-Ring-diaromatic-pentacyclotriterpane C_{32},A,B 环-双芳-五环三萜烷
5	153.80	6.97	223	$C_{32}H_{46}$	430	C_{32},A,B-Ring-diaromatic-pentacyclotriterpane C_{32},A,B 环-双芳-五环三萜烷
6	159.67	7.74	223	$C_{32}H_{46}$	430	C_{32},A,B-Ring-diaromatic-pentacyclotriterpane C_{32},A,B 环-双芳-五环三萜烷

2) 三芳-五环三萜烷的全二维气相色谱图及化合物名称

三芳-五环三萜烷系列化合物有两个系列，一个系列是 A,B,C 环发生了芳构化的五环三萜烷，另一个系列是 B,C,D 环发生了芳构化的五环三萜烷。A,B,C 环-三芳-五环三萜烷(或烯)化合物用其分子离子或者特征离子 m/z 255，m/z 257 在质谱上进行检测，B,C,D 环-三芳-五环三萜烷化合物用其特征离子 m/z 299 在质谱上进行检测。图 2-438 是三芳-五环三萜烷的 GC-MS 色谱图(a)、全二维点阵图(b)和三维立体图(c)。共检测到两个系列的 10 个化合物，其质谱图如图 2-439～图 2-448 所示。各化合物的二维保留特征及鉴定结果见表 2-36。

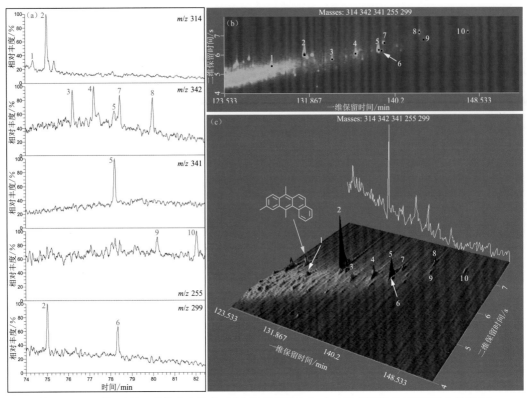

图 2-438 三芳-五环三萜烷化合物的 GC-MS 色谱图(a)、全二维点阵图(b)和三维立体图(c)

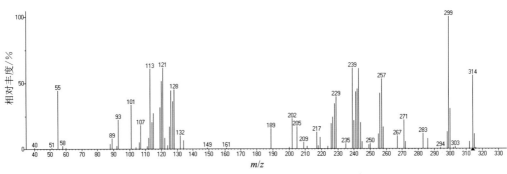

图 2-439 1号峰 C_{24},B,C,D环-三芳-五环三萜烷质谱图

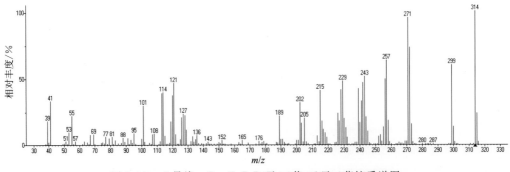

图 2-440 2号峰 C_{24},B,C,D环-三芳-五环三萜烷质谱图

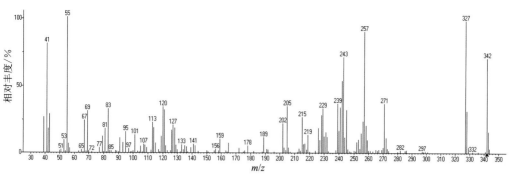

图 2-441　3 号峰　C_{26}，A,B,C 环-三芳-五环三萜烷质谱图

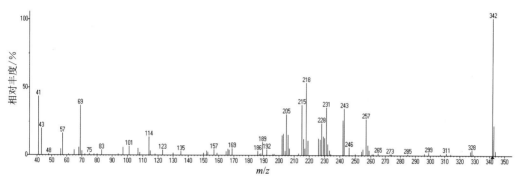

图 2-442　4 号峰　C_{26}，A,B,C 环-三芳-五环三萜烷质谱图

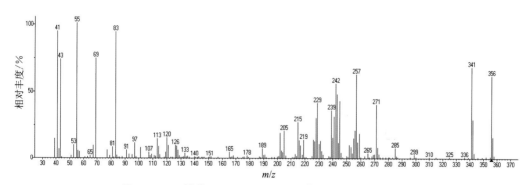

图 2-443　5 号峰　C_{27}，A,B,C 环-三芳-五环三萜烷质谱图

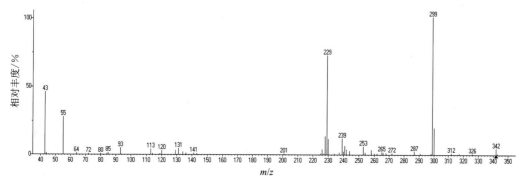

图 2-444　6 号峰　C_{26}，B,C,D 环-三芳-五环三萜烷质谱图

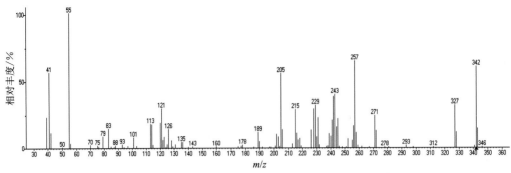

图 2-445　7 号峰　C_{26}，A，B，C 环-三芳-五环三萜烷质谱图

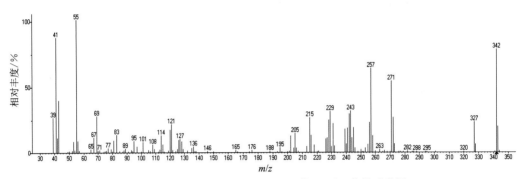

图 2-446　8 号峰　C_{26}，A，B，C 环-三芳-五环三萜烷质谱图

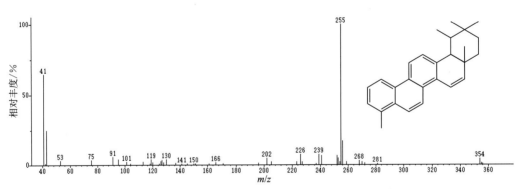

图 2-447　9 号峰　C_{27}，A，B，C 环-三芳-五环三萜烯质谱图

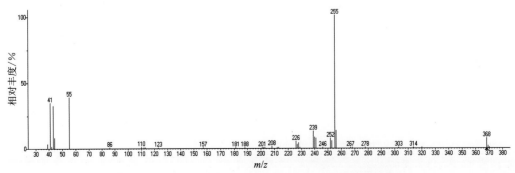

图 2-448　10 号峰　C_{28}，A，B，C 环-三芳-五环三萜烯质谱图

表 2-36　三芳-五环三萜烷鉴定结果表

峰号	一维保留时间 /min	二维保留时间 /s	特征离子 (m/z)	分子式	相对分子质量	名称
1	128.20	5.43	299	$C_{24}H_{26}$	314	C_{24}, B,C,D-Ring-triaromatic pentacyclotriterpane C_{24}, B,C,D 环-三芳-五环三萜烷
2	131.40	6.01	314	$C_{24}H_{26}$	314	C_{24}, B,C,D-Ring-triaromatic pentacyclotriterpane C_{24}, B,C,D 环-三芳-五环三萜烷
3	134.07	5.78	327	$C_{26}H_{30}$	342	C_{26}, A,B,C-Ring-triaromatic pentacyclotriterpane C_{26}, A,B,C 环-三芳-五环三萜烷
4	136.33	6.08	342	$C_{26}H_{30}$	342	C_{26}, A,B,C-Ring-triaromatic pentacyclotriterpane C_{26}, A,B,C 环-三芳-五环三萜烷
5	138.60	6.25	356	$C_{27}H_{32}$	356	C_{27}, A,B,C-Ring-triaromatic pentacyclotriterpane C_{27}, A,B,C 环-三芳-五环三萜烷
6	138.87	6.22	299	$C_{26}H_{30}$	342	C_{26}, B,C,D-Ring-triaromatic pentacyclotriterpane C_{26}, B,C,D 环-三芳-五环三萜烷
7	139.13	6.62	342	$C_{26}H_{30}$	342	C_{26}, A,B,C-Ring-triaromatic pentacyclotriterpane C_{26}, A,B,C 环-三芳-五环三萜烷
8	142.60	7.17	342	$C_{26}H_{30}$	342	C_{26}, A,B,C-Ring-triaromatic pentacyclotriterpane C_{26}, A,B,C 环-三芳-五环三萜烷
9	143.00	6.79	255	$C_{27}H_{30}$	354	C_{27}, A,B,C-Ring-triaromatic pentacyclotriterpene C_{27}, A,B,C 环-三芳-五环三萜烯
10	147.40	7.17	255	$C_{28}H_{32}$	368	C_{28}, A,B,C-Ring-triaromatic pentacyclotriterpene C_{28}, A,B,C 环-三芳-五环三萜烯

3) 四芳-五环三萜烷的全二维气相色谱图及化合物名称

四芳-五环三萜烷系列化合物是 A,B,C,D 环发生了芳构化的五环三萜烷系列,图 2-449 是四芳-五环三萜烷的 GC-MS 色谱图(a)、全二维点阵图(b)和三维立体图(c)。共检测到 5 个化合物,其质谱图如图 2-450～图 2-454 所示。各化合物的二维保留特征及鉴定结果见表 2-37。

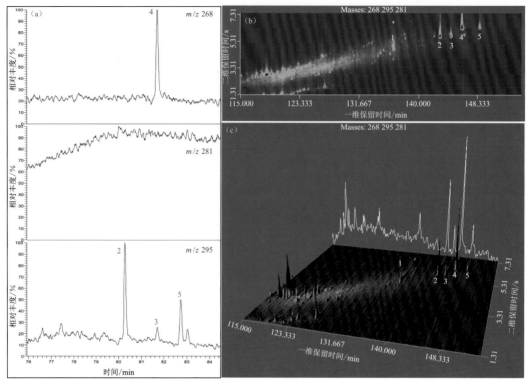

图 2-449　四芳-五环三萜烷化合物的 GC-MS 色谱图(a)、全二维点阵图(b)和三维立体图(c)

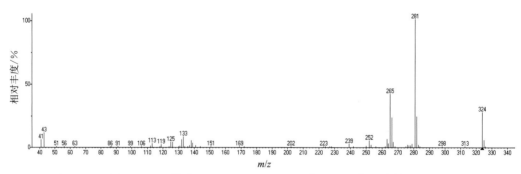

图 2-450　1号峰　C_{25},A,B,C,D环-四芳-五环三萜烷质谱图

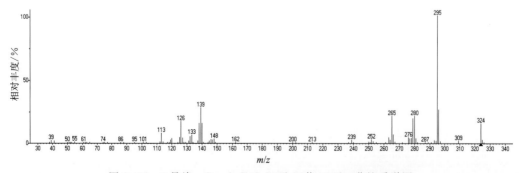

图 2-451　2号峰　C_{25},A,B,C,D环-四芳-五环三萜烷质谱图

第 2 章 石油地质样品的全二维气相色谱图识别

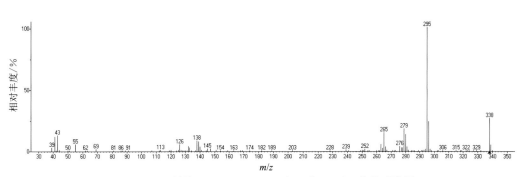

图 2-452 3 号峰 C_{26},A,B,C,D 环-四芳-五环三萜烷质谱图

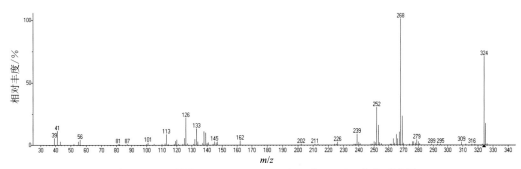

图 2-453 4 号峰 C_{25},A,B,C,D 环-四芳-五环三萜烷质谱图

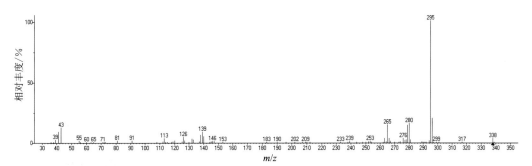

图 2-454 5 号峰 C_{26},A,B,C,D 环-四芳-五环三萜烷质谱图

表 2-37 四芳-五环三萜烷鉴定结果表

峰号	一维保留时间 /min	二维保留时间 /s	特征离子 (m/z)	分子式	相对分子质量	名 称
1	136.47	6.77	281	$C_{25}H_{24}$	324	C_{25},A,B,C,D-Ring-tetraaromatic-pentacyclotriterpane C_{25},A,B,C,D 环-四芳-五环三萜烷
2	143.00	7.73	295	$C_{25}H_{24}$	324	C_{25},A,B,C,D-Ring-tetraaromatic-pentacyclotriterpane C_{25},A,B,C,D 环-四芳-五环三萜烷
3	144.47	7.75	295	$C_{26}H_{26}$	338	C_{26},A,B,C,D-Ring-tetraaromatic-pentacyclotriterpane C_{26},A,B,C,D 环-四芳-五环三萜烷

续表

峰号	一维保留时间/min	二维保留时间/s	特征离子(m/z)	分子式	相对分子质量	名称
4	146.20	0.42	268	$C_{25}H_{24}$	324	C_{25},A,B,C,D-Ring-tetraaromatic-pentacyclotriterpane C_{25},A,B,C,D环-四芳-五环三萜烷
5	148.73	0.38	295	$C_{26}H_{26}$	338	C_{26},A,B,C,D-Ring-tetraaromatic-pentacyclotriterpane C_{26},A,B,C,D环-四芳-五环三萜烷

4）五芳-五环三萜烷的全二维气相色谱图及化合物名称

5个环全部被芳构化的五环三萜烷目前仅鉴定出1个化合物，其GC-MS色谱图和全二维点阵图如图2-455所示，质谱图如图2-456所示，分子离子即特征离子，其化合物鉴定结果见表2-38。

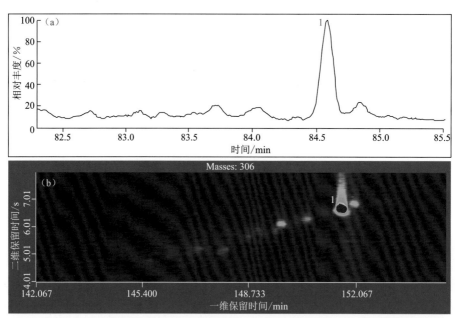

图2-455 五芳-五环三萜烷化合物的GC-MS色谱图(a)和全二维点阵图(b)

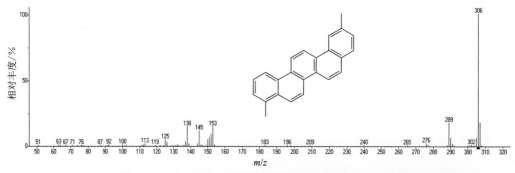

图2-456 1号峰 C_{24},A,B,C,D,E环-五芳-五环三萜烷(2,9-二甲基䓛)质谱图

表 2-38 五芳五环三萜烷鉴定结果表

峰号	一维保留时间/min	二维保留时间/s	特征离子（m/z）	分子式	相对分子质量	名称
1	151.80	2.66	306	$C_{24}H_{18}$	306	C_{24},A,B,C,D,E-Ring-pentaaromatic pentacyclotriterpane C_{24},A,B,C,D,E 环-五芳-五环三萜烷 （2,9-二甲基䓛）

2.2.15 苯并藿烷的全二维气相色谱图及化合物名称

苯并藿烷是原核生物细胞膜中的 C_{35} 四羟基藿烷在成岩早期阶段经过脱水、缩合并芳构化而形成的化合物。其色质分析的特征离子同检测藿烷类化合物一样，都是 m/z 191，在全二维分析谱图上的分布区域如图 2-427 所示。图 2-457 是苯并藿烷化合物的 GC-MS 色谱图(a)、全二维点阵图(b)和三维立体图(c)，共检测到 3 个化合物，其质谱图如图 2-458～图 2-460 所示，其鉴定结果见表 2-39。

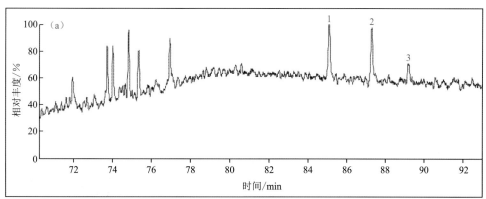

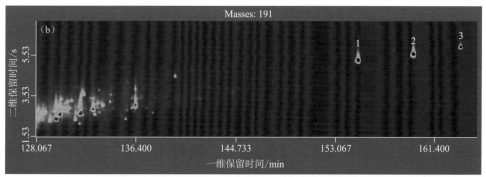

图 2-457 苯并藿烷化合物的 GC-MS 色谱图(a)、全二维点阵图(b)和三维立体图(c)

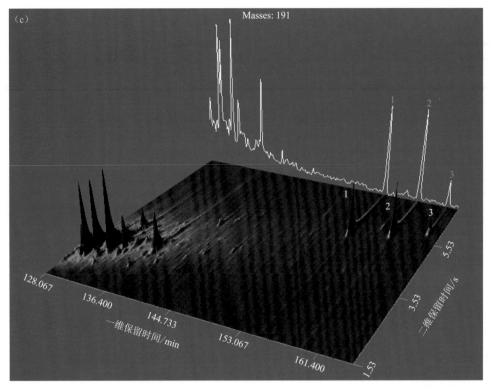

图 2-457（续） 苯并藿烷化合物的 GC-MS 色谱图（a）、全二维点阵图（b）和三维立体图（c）

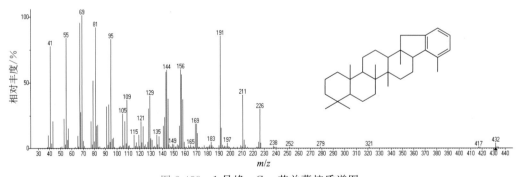

图 2-458　1 号峰　C_{32}，苯并藿烷质谱图

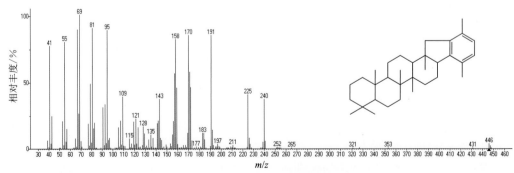

图 2-459　2 号峰　C_{33}，苯并藿烷质谱图

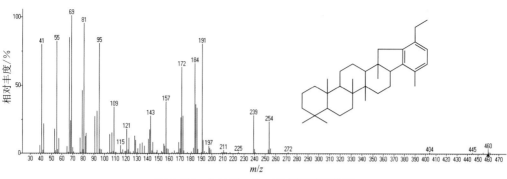

图 2-460 3号峰 C_{34},苯并藿烷质谱图

表 2-39 苯并藿烷鉴定结果表

峰 号	一维保留时间/min	二维保留时间/s	特征离子(m/z)	分子式	相对分子质量	名 称
1	154.87	7.33	191	$C_{32}H_{48}$	432	C_{32},Benzohopane C_{32},苯并藿烷
2	159.53	7.64	191	$C_{33}H_{50}$	446	C_{33},Benzohopane C_{33},苯并藿烷
3	163.53	8.00	191	$C_{34}H_{52}$	460	C_{34},Benzohopane C_{34},苯并藿烷

2.3 >> 杂原子化合物的全二维气相色谱图特征及鉴定

杂原子化合物是指含有 S,N,O 原子的化合物。石油地质样品中这类化合物的种类和含量都比较丰富,但能够被全二维气相色谱检测的化合物有限。结合石油地质应用的具体情况,本节仅介绍部分含硫和含氮化合物。

2.3.1 含硫化合物的全二维气相色谱图及化合物名称

石油样品中含硫化合物种类较多,能够被全二维气相色谱检测的有硫醇、硫酚、烷基四氢化噻吩、硫代单金刚烷、硫代双金刚烷、烷基噻吩、苯并噻吩、二苯并噻吩、菲并噻吩以及苯并萘并噻吩等。图 2-461 是塔里木油田凝析油直接进样检测所得到的含硫化合物的全二维分析谱图,该图直观展示了不同种类的含硫化合物在全二维谱图上的分布特征。

1) 烷基四氢化噻吩系列的全二维气相色谱图及化合物名称

烷基四氢化噻吩系列有两类,一类是一个烷基作为取代基,其质谱检测的特征离子是 m/z 87,另一类是环上有一个甲基和一个长链烷基作为取代基,其质谱检测的特征离子是 m/z 101,其中甲基的取代位置未知。图 2-462 是烷基四氢化噻吩的全二维气相色谱-飞行时间质谱分析图谱,图中此类化合物在二维方向上的出峰位置在甲基苯和萘系列化合物之间,图中标记的 1~22 号峰为鉴定的烷基四氢化噻吩,图 2-463~图 2-476 是部分化合物对应峰号的质谱图,由于有些化合物的烷基位置尚无法确定,表 2-40 中的化合物名称都没有烷基位置的标记,但单取代基的四氢化噻吩一般在"2"位是最稳定的。从图 2-462 中可以看出,在特

征离子 m/z 87 下,烷基四氢化噻吩与烷基苯、萘等化合物共馏,在特征离子 m/z 101 下,甲基-烷基四氢化噻吩与萘类化合物共馏,用常规色质分析无法直接检测该类化合物,但用全二维气相色谱可以。

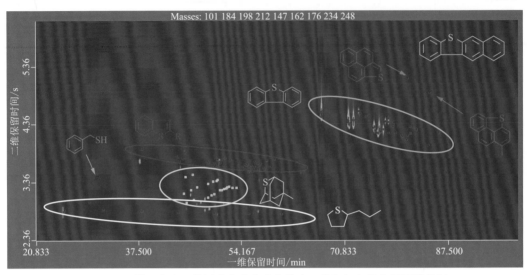

图 2-461　塔里木油田凝析油含硫化合物的全二维气相色谱分析图谱特征

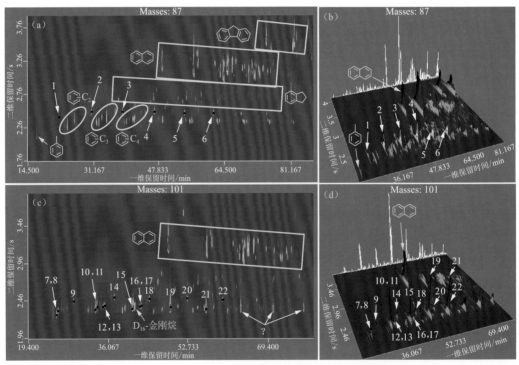

图 2-462　烷基四氢化噻吩的全二维气相色谱-飞行时间质谱分析(a,c 全二维点阵图,b,d 三维立体图)

根据全二维谱图的族分离特点和"瓦片效应",在同一升温梯度下,长侧链取代的同系列化合物,其二维保留时间相近,随着取代基碳数的增加,在一维保留时间上等间距分布。图 2-462(c)中标记为"?"的化合物在一维保留时间上与 19 号和 21 号化合物等间距排列,二维保留时间上与 19 号和 21 号化合物相似,因此推测这 3 个化合物可能是长链的甲基四氢化噻吩系列。但由于附近化合物的干扰,这 3 个化合物的质谱图未能表现出长链的甲基四氢化噻吩的特征,因此对这 3 个化合物只能推测其为长链取代的甲基四氢化噻吩系列化合物。

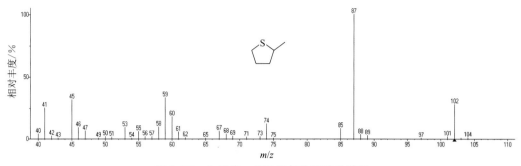

图 2-463　1 号峰　甲基四氢化噻吩质谱图

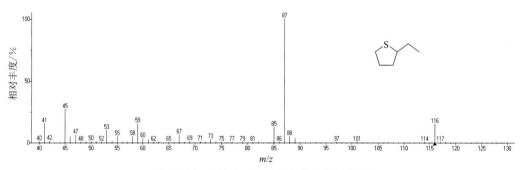

图 2-464　2 号峰　乙基四氢化噻吩质谱图

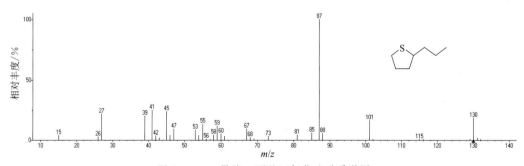

图 2-465　3 号峰　丙基四氢化噻吩质谱图

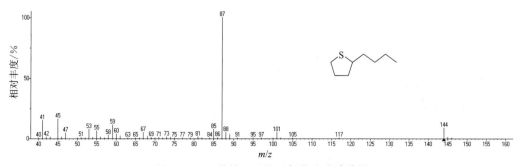

图 2-466　4 号峰　丁基四氢化噻吩质谱图

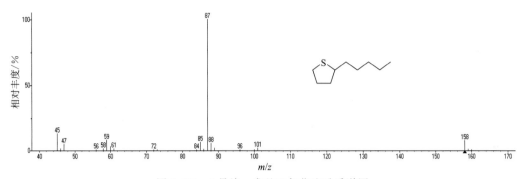

图 2-467　5 号峰　戊基四氢化噻吩质谱图

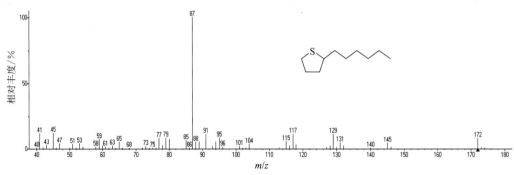

图 2-468　6 号峰　己基四氢化噻吩质谱图

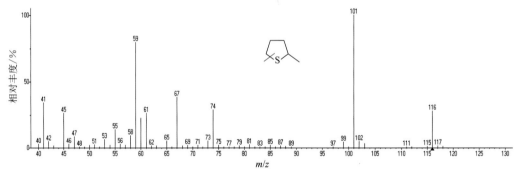

图 2-469　7 号峰　二甲基四氢化噻吩质谱图（其中甲基取代的位置未知）

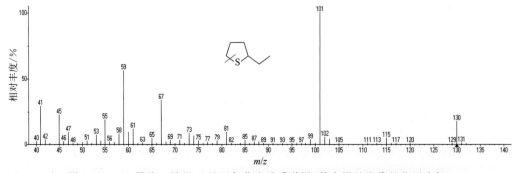

图 2-470　11 号峰　甲基-乙基四氢化噻吩质谱图（其中甲基取代的位置未知）

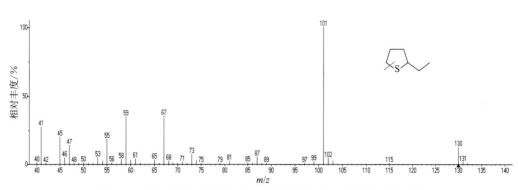

图 2-471　14 号峰　甲基-乙基四氢化噻吩质谱图（其中甲基取代的位置未知）

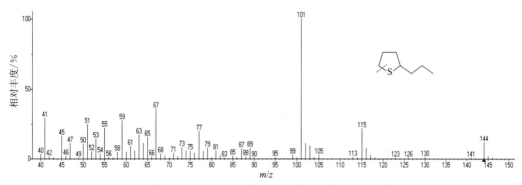

图 2-472　15 号峰　甲基-丙基四氢化噻吩质谱图（其中甲基取代的位置未知）

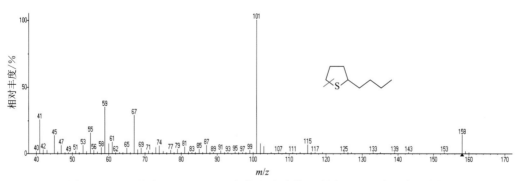

图 2-473　19 号峰　甲基-丁基四氢化噻吩质谱图（其中甲基取代的位置未知）

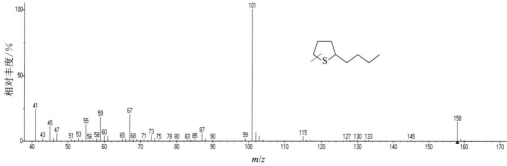

图 2-474　20 号峰　甲基-丁基四氢化噻吩质谱图（其中甲基取代的位置未知）

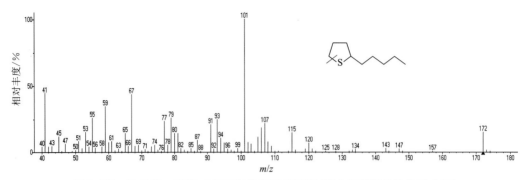

图 2-475　21 号峰　甲基-戊基四氢化噻吩质谱图（其中甲基取代的位置未知）

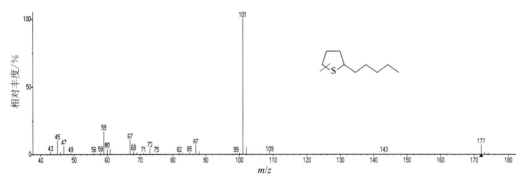

图 2-476　22 号峰　甲基-戊基四氢化噻吩质谱图（其中甲基取代的位置未知）

表 2-40　烷基四氢化噻吩鉴定结果表

峰号	一维保留时间 /min	二维保留时间 /s	特征离子 (m/z)	分子式	相对分子质量	名　称
1	22.30	2.42	87	$C_5H_{10}S$	102	Tetrahydrothiophene,methyl- 甲基四氢化噻吩
2	30.50	2.50	87	$C_6H_{12}S$	116	Tetrahydrothiophene,ethyl- 乙基四氢化噻吩
3	38.50	2.50	87	$C_7H_{14}S$	130	Tetrahydrothiophene,propyl- 丙基四氢化噻吩
4	46.70	2.50	87	$C_8H_{16}S$	144	Tetrahydrothiophene,butyl- 丁基四氢化噻吩
5	54.40	2.48	87	$C_9H_{18}S$	158	Tetrahydrothiophene,pentyl- 戊基四氢化噻吩
6	61.70	2.48	87	$C_{10}H_{20}S$	172	Tetrahydrothiophene,hexyl- 己基四氢化噻吩
7	25.20	2.28	101	$C_6H_{12}S$	116	Tetrahydrothiophene,dimethyl- 二甲基四氢化噻吩
8	25.50	2.33	101	$C_6H_{12}S$	116	Tetrahydrothiophene,dimethyl- 二甲基四氢化噻吩
9	28.70	2.46	101	$C_6H_{12}S$	116	Tetrahydrothiophene,dimethyl- 二甲基四氢化噻吩

续表

峰 号	一维保留时间 /min	二维保留时间 /s	特征离子 (m/z)	分子式	相对分子质量	名 称
10	33.30	2.35	101	$C_7H_{14}S$	130	Tetrahydrothiophene,methyl-ethyl- 甲基-乙基四氢化噻吩
11	33.60	2.38	101	$C_7H_{14}S$	130	Tetrahydrothiophene,methyl-ethyl- 甲基-乙基四氢化噻吩
12	34.90	2.38	101	$C_7H_{14}S$	130	Tetrahydrothiophene,methyl-ethyl- 甲基-乙基四氢化噻吩
13	35.30	2.40	101	$C_7H_{14}S$	130	Tetrahydrothiophene,methyl-ethyl- 甲基-乙基四氢化噻吩
14	37.30	2.51	101	$C_7H_{14}S$	130	Tetrahydrothiophene,methyl-ethyl- 甲基-乙基四氢化噻吩
15	41.10	2.36	101	$C_8H_{16}S$	144	Tetrahydrothiophene,methyl-propyl- 甲基-丙基四氢化噻吩
16	42.70	2.39	101	$C_8H_{16}S$	144	Tetrahydrothiophene,methyl-propyl- 甲基-丙基四氢化噻吩
17	43.00	2.40	101	$C_8H_{16}S$	144	Tetrahydrothiophene,methyl-propyl- 甲基-丙基四氢化噻吩
18	44.90	2.49	101	$C_8H_{16}S$	144	Tetrahydrothiophene,methyl-propyl- 甲基-丙基四氢化噻吩
19	49.10	2.38	101	$C_9H_{18}S$	158	Tetrahydrothiophene,methyl-butyl- 甲基-丁基四氢化噻吩
20	52.60	2.50	101	$C_9H_{18}S$	158	Tetrahydrothiophene,methyl-butyl- 甲基-丁基四氢化噻吩
21	56.60	2.36	101	$C_{10}H_{20}S$	172	Tetrahydrothiophene,methyl-pentyl- 甲基-戊基四氢化噻吩
22	60.00	2.48	101	$C_{10}H_{20}S$	172	Tetrahydrothiophene,methyl-pentyl- 甲基-戊基四氢化噻吩

2) 烷基噻吩系列的全二维气相色谱图及化合物名称

理论上噻吩环上可以有 4 个取代基,但实际检测到的烷基噻吩取代基不超过 3 个,可以用特征离子 m/z 97, m/z 111, m/z 125, m/z 139 在质谱上检测。图 2-477 是烷基噻吩的全二维气相色谱-飞行时间质谱分析图谱,图中此类化合物在二维方向上的出峰位置位于单环烷烃的上方,碳数相同的烷基噻吩具有明显的族分离特性,在常规色质分析谱图上烷基噻吩和单环烷烃是共馏峰,由于烷基噻吩的含量低,一般不易被检出。图 2-478 中标记出 38 个烷基噻吩,其中特征离子是 m/z 97 的噻吩类化合物与烷基苯的出峰位置十分接近。图 2-479～图 2-488 是部分化合物对应峰号的质谱图,各化合物的鉴定结果见表 2-41。

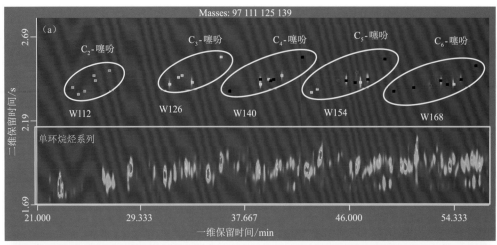

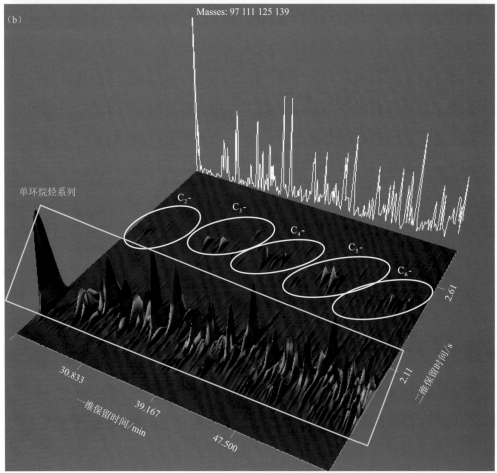

图 2-477 烷基噻吩的全二维气相色谱-飞行时间质谱分析（a 全二维点阵图，b 三维立体图）
其中"W112"表示该类化合物的相对分子质量是 112，其他同理
图(a)中灰色点标记特征离子是 m/z 97 的化合物，黄色点标记特征离子是 m/z 111 的化合物，
黑色点标记特征离子是 m/z 125 的化合物，红色点标记特征离子是 m/z 139 的化合物

第 2 章 石油地质样品的全二维气相色谱图识别

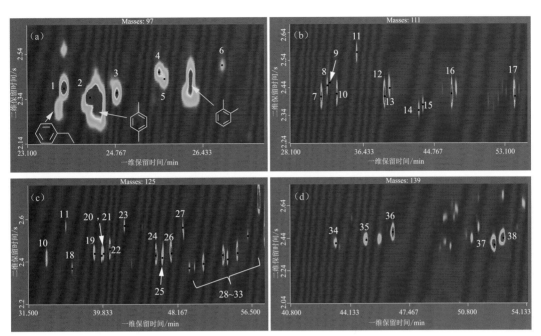

图 2-478 不同取代基的噻吩类化合物的全二维点阵图

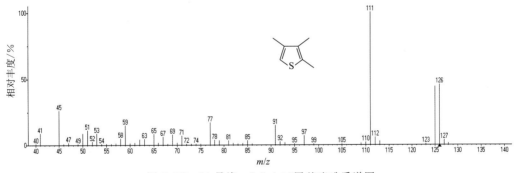

图 2-479 11 号峰 2,3,4-三甲基噻吩质谱图

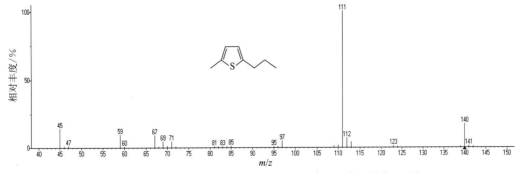

图 2-480 12 号峰 甲基-丙基噻吩质谱图（其中取代基的位置未知）

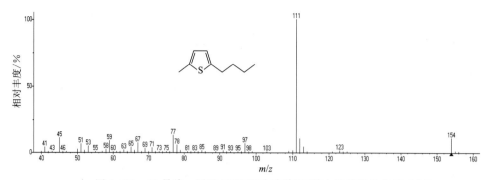

图 2-481　16 号峰　甲基-丁基噻吩质谱图（其中取代基的位置未知）

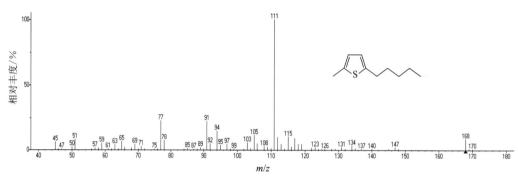

图 2-482　17 号峰　甲基-戊基噻吩质谱图（其中取代基的位置未知）

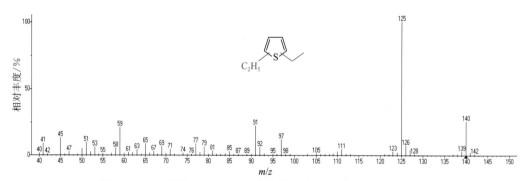

图 2-483　22 号峰　C_2-乙基噻吩质谱图（其中取代基的位置未知）

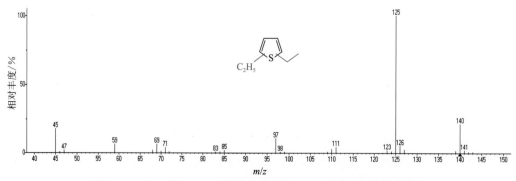

图 2-484　23 号峰　C_2-乙基噻吩质谱图（其中取代基的位置未知）

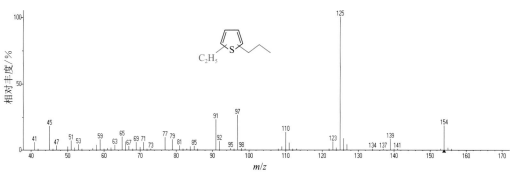

图 2-485　24 号峰　C_2-丙基噻吩质谱图（其中取代基的位置未知）

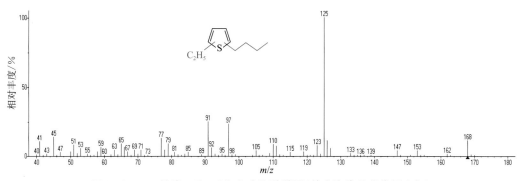

图 2-486　30 号峰　C_2-丁基噻吩的质谱图（其中取代基的位置未知）

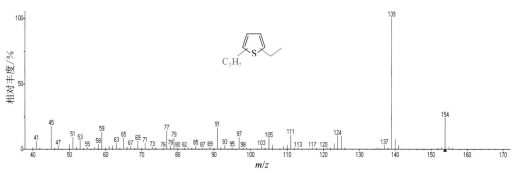

图 2-487　36 号峰　C_3-乙基噻吩质谱图（其中取代基的位置未知）

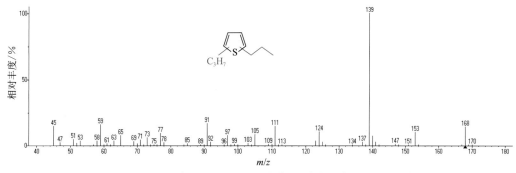

图 2-488　38 号峰　C_3-丙基噻吩质谱图（其中取代基的位置未知）

表 2-41 烷基噻吩鉴定结果表

峰 号	一维保留时间 /min	二维保留时间 /s	特征离子 (m/z)	分子式	相对分子质量	名 称
1	23.80	2.39	97	C_6H_8S	112	Thiophene, 2-ethyl- 2-乙基噻吩
2	24.30	2.35	97	C_6H_8S	112	Thiophene, 2,5-dimethyl- 2,5-二甲基噻吩
3	24.80	2.37	97	C_6H_8S	112	Thiophene, 2,4-dimethyl- 2,4-二甲基噻吩
4	25.60	2.46	97	C_6H_8S	112	Thiophene, 2,3-dimethyl- 2,3-二甲基噻吩
5	25.70	2.43	97	C_6H_8S	112	Thiophene, dimethyl- 二甲基噻吩
6	26.80	2.49	97	C_6H_8S	112	Thiophene, 3,4-dimethyl- 3,4-二甲基噻吩
7	31.60	2.41	111	$C_7H_{10}S$	126	Thiophene, 2-ethyl-5-methyl- 2-乙基-5-甲基噻吩
8	32.30	2.45	111	$C_7H_{10}S$	126	Thiophene, 3-ethyl-5-methyl- 3-乙基-5-甲基噻吩
9	32.60	2.46	111	$C_7H_{10}S$	126	Thiophene, methyl-ethyl- 甲基-乙基噻吩
10	33.40	2.42	111	$C_7H_{10}S$	126	Thiophene, 2,3,5-trimethyl- 2,3,5-三甲基噻吩
11	35.70	2.57	111	$C_7H_{10}S$	126	Thiophene, 2,3,4-trimethyl- 2,3,4-三甲基噻吩
12	38.90	2.41	111	$C_8H_{12}S$	140	Thiophene, methyl-propyl- 甲基-丙基噻吩
13	39.50	2.44	111	$C_8H_{12}S$	140	Thiophene, methyl-propyl- 甲基-丙基噻吩
14	42.90	2.36	111	$C_9H_{14}S$	154	Thiophene, methyl-butyl- 甲基-丁基噻吩
15	43.40	2.38	111	$C_9H_{14}S$	154	Thiophene, methyl-butyl- 甲基-丁基噻吩
16	46.80	2.42	111	$C_9H_{14}S$	154	Thiophene, methyl-butyl- 甲基-丁基噻吩
17	54.20	2.41	111	$C_{10}H_{16}S$	168	Thiophene, methyl-pentyl- 甲基-戊基噻吩
18	36.40	2.37	125	$C_8H_{12}S$	140	Thiophene, C_2-ethyl- C_2-乙基噻吩
19	38.90	2.44	125	$C_8H_{12}S$	140	Thiophene, C_2-ethyl- C_2-乙基噻吩
20	39.70	2.43	125	$C_8H_{12}S$	140	Thiophene, C_2-ethyl- C_2-乙基噻吩

续表

峰　号	一维保留时间 /min	二维保留时间 /s	特征离子 (m/z)	分子式	相对分子质量	名　　称
21	39.90	2.44	125	$C_8H_{12}S$	140	Thiophene, C_2-ethyl- C_2-乙基噻吩
22	40.60	2.46	125	$C_8H_{12}S$	140	Thiophene, C_2-ethyl- C_2-乙基噻吩
23	42.20	2.57	125	$C_8H_{12}S$	140	Thiophene, C_2-ethyl- C_2-乙基噻吩
24	45.70	2.43	125	$C_9H_{14}S$	154	Thiophene, C_2-propyl- C_2-丙基噻吩
25	46.40	2.42	125	$C_9H_{14}S$	154	Thiophene, C_2-propyl- C_2-丙基噻吩
26	47.40	2.44	125	$C_9H_{14}S$	154	Thiophene, C_2-propyl- C_2-丙基噻吩
27	48.70	2.56	125	$C_9H_{14}S$	154	Thiophene, C_2-propyl- C_2-丙基噻吩
28	49.40	2.37	125	$C_{10}H_{16}S$	168	Thiophene, C_2-butyl- C_2-丁基噻吩
29	51.00	2.39	125	$C_{10}H_{16}S$	168	Thiophene, C_2-butyl- C_2-丁基噻吩
30	53.20	2.43	125	$C_{10}H_{16}S$	168	Thiophene, C_2-butyl- C_2-丁基噻吩
31	53.7	2.41	125	$C_{10}H_{16}S$	168	Thiophene, C_2-butyl- C_2-丁基噻吩
32	54.80	2.44	125	$C_{10}H_{16}S$	168	Thiophene, C_2-butyl- C_2-丁基噻吩
33	56.00	2.52	125	$C_{10}H_{16}S$	168	Thiophene, C_2-butyl- C_2-丁基噻吩
34	43.30	2.41	139	$C_9H_{14}S$	154	Thiophene, C_3-ethyl- C_3-乙基噻吩
35	45.00	2.43	139	$C_9H_{14}S$	154	Thiophene, C_3-ethyl- C_3-乙基噻吩
36	46.50	2.479	139	$C_9H_{14}S$	154	Thiophene, C_3-ethyl- C_3-乙基噻吩
37	52.30	2.40	139	$C_{10}H_{16}S$	168	Thiophene, C_3-propyl- C_3-丙基噻吩
38	52.80	2.44	139	$C_{10}H_{16}S$	168	Thiophene, C_3-propyl- C_3-丙基噻吩

3) 苯并噻吩系列的全二维气相色谱图及化合物名称

苯并噻吩系列化合物在全二维分析谱图上位于单金刚烷系列化合物和苯系列化合物的上方,芴系列化合物的下方,包括苯并噻吩、甲基苯并噻吩、C_2 和 C_3 及 C_4 取代的苯并噻吩,如图 2-489 所示。图中标记出两类化合物,其中黑色点标记的是苯并噻吩系列,红色点标记的是二氢化苯并噻吩。这两类化合物的化学结构相似,在全二维分析谱图上位置也相近。用常规的 GC-MS 分析这两类化合物时,由于其他化合物的干扰而不易被检出,但用 GC×GC 分析时就完全不受影响。

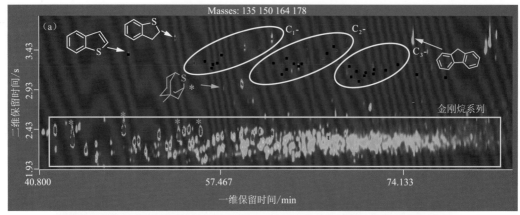

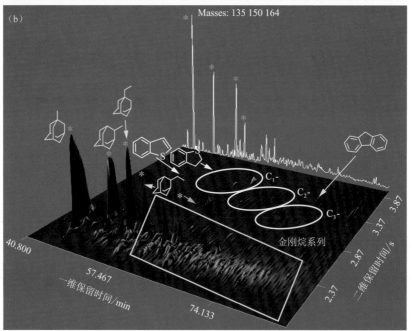

图 2-489 苯并噻吩系列化合物的全二维点阵图(a)和三维立体图(b)

图中 * 标记的是 4 个金刚烷化合物,其结构式如图(b)中所示

苯并噻吩系列化合物可用其特征离子 m/z 134, m/z 148, m/z 162, m/z 176, m/z 190 等进行检测,从苯并噻吩到 C_4 取代的苯并噻吩可检测到 36 个化合物,如图 2-490 所示,典型化合物的质谱图如图 2-491~图 2-498 所示,各化合物的鉴定结果见表 2-42。从图 2-490 中可以看出,2 号化合物和 C_5-苯、5 号化合物和 C_2-硫代金刚烷在常规 GC-MS 上是共馏峰(图 b,b′),7~34 号化合物在其特征离子下也与一些化合物是共馏峰。因此,用常规 GC-MS 检测苯并噻吩化合物时,易受共馏峰的干扰,得到的定量结果不准确。

第 2 章 石油地质样品的全二维气相色谱图识别

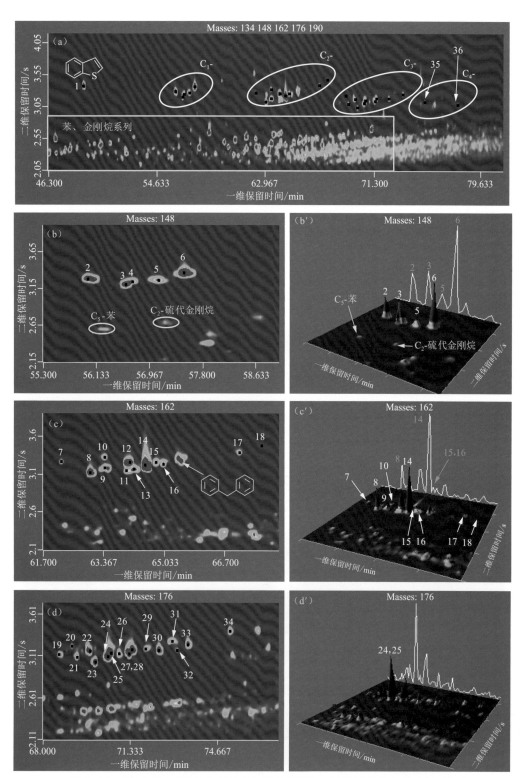

图 2-490 苯并噻吩系列化合物的全二维点阵图（a～b）和三维立体图（b′～d′）

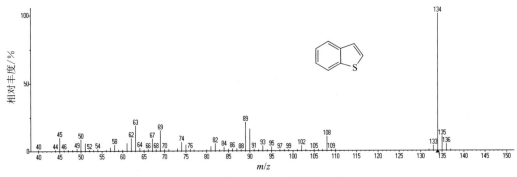

图 2-491　1 号峰　苯并[b]噻吩质谱图

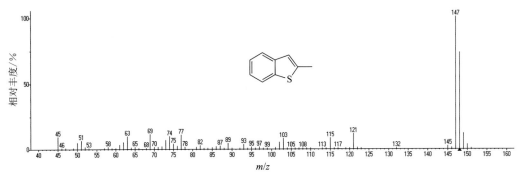

图 2-492　6 号峰　甲基苯并[b]噻吩质谱图（其中甲基取代的位置未知）

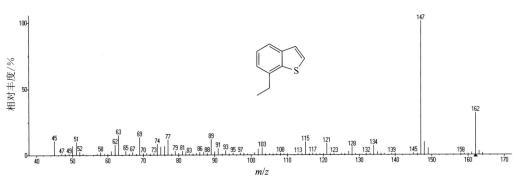

图 2-493　7 号峰　乙基苯并[b]噻吩质谱图（其中甲基取代的位置未知）

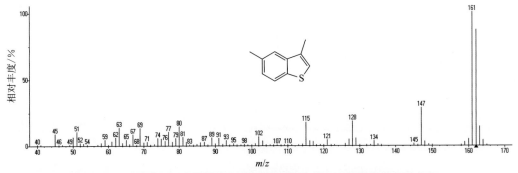

图 2-494　8 号峰　二甲基苯并[b]噻吩质谱图（其中甲基取代的位置未知）

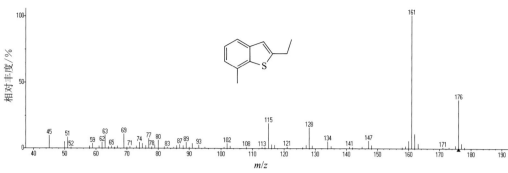

图 2-495　19 号峰　甲基-乙基苯并[b]噻吩质谱图（其中取代的位置未知）

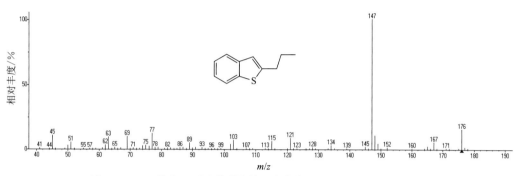

图 2-496　20 号峰　丙基苯并[b]噻吩质谱图（其中取代的位置未知）

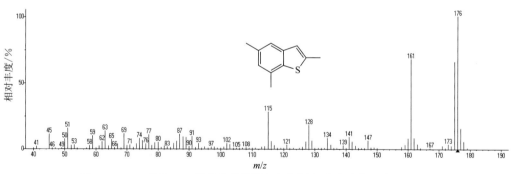

图 2-497　24 号峰　三甲基苯并[b]噻吩质谱图（其中取代的位置未知）

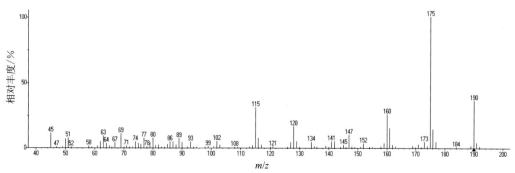

图 2-498　35 号峰　C_4-苯并[b]噻吩质谱图

表 2-42 苯并噻吩系列化合物鉴定结果表

峰号	一维保留时间 /min	二维保留时间 /s	特征离子 (m/z)	分子式	相对分子质量	名　称
1	49.00	3.38	134	C_8H_6S	134	Benzo[b]thiophene 苯并[b]噻吩
2	56.00	3.29	147	C_9H_8S	148	Benzo[b]thiophene,methyl- 甲基苯并[b]噻吩
3	56.60	3.21	147	C_9H_8S	148	Benzo[b]thiophene,methyl- 甲基苯并[b]噻吩
4	56.70	3.25	147	C_9H_8S	148	Benzo[b]thiophene,methyl- 甲基苯并[b]噻吩
5	57.10	3.27	147	C_9H_8S	148	Benzo[b]thiophene,methyl- 甲基苯并[b]噻吩
6	57.50	3.37	147	C_9H_8S	148	Benzo[b]thiophene,methyl- 甲基苯并[b]噻吩
7	62.20	3.27	147	$C_{10}H_{10}S$	162	Benzo[b]thiophene,ethyl- 乙基苯并[b]噻吩
8	63.00	3.13	161	$C_{10}H_{10}S$	162	Benzo[b]thiophene,dimethyl- 二甲基苯并[b]噻吩
9	63.40	3.19	161	$C_{10}H_{10}S$	162	Benzo[b]thiophene,dimethyl- 二甲基苯并[b]噻吩
10	63.40	3.33	147	$C_{10}H_{10}S$	162	Benzo[b]thiophene,ethyl- 乙基苯并[b]噻吩
11	64.00	3.15	161	$C_{10}H_{10}S$	162	Benzo[b]thiophene,dimethyl- 二甲基苯并[b]噻吩
12	64.10	3.26	161	$C_{10}H_{10}S$	162	Benzo[b]thiophene,dimethyl- 二甲基苯并[b]噻吩
13	64.20	3.17	161	$C_{10}H_{10}S$	162	Benzo[b]thiophene,dimethyl- 二甲基苯并[b]噻吩
14	64.50	3.23	161	$C_{10}H_{10}S$	162	Benzo[b]thiophene,dimethyl- 二甲基苯并[b]噻吩
15	64.80	3.26	162	$C_{10}H_{10}S$	162	Benzo[b]thiophene,dimethyl- 二甲基苯并[b]噻吩
16	65.00	3.24	161	$C_{10}H_{10}S$	162	Benzo[b]thiophene,dimethyl- 二甲基苯并[b]噻吩
17	67.10	3.39	161	$C_{10}H_{10}S$	162	Benzo[b]thiophene,dimethyl- 二甲基苯并[b]噻吩
18	67.70	3.48	161	$C_{10}H_{10}S$	162	Benzo[b]thiophene,dimethyl- 二甲基苯并[b]噻吩
19	68.60	3.13	161	$C_{11}H_{12}S$	176	Benzo[b]thiophene,methyl-ethyl- 甲基-乙基苯并[b]噻吩
20	69.10	3.23	147	$C_{11}H_{12}S$	176	Benzo[b]thiophene,propyl- 丙基苯并[b]噻吩

续表

峰 号	一维保留时间 /min	二维保留时间 /s	特征离子 (m/z)	分子式	相对分子质量	名 称
21	69.30	3.10	147	$C_{11}H_{12}S$	176	Benzo[b]thiophene,propyl- 丙基苯并[b]噻吩
22	69.80	3.17	161	$C_{11}H_{12}S$	176	Benzo[b]thiophene,methyl-ethyl- 甲基-乙基苯并[b]噻吩
23	70.00	3.04	176	$C_{11}H_{12}S$	176	Benzo[b]thiophene,trimethyl- 三甲基苯并[b]噻吩
24	70.50	3.11	176	$C_{11}H_{12}S$	176	Benzo[b]thiophene,trimethyl- 三甲基苯并[b]噻吩
25	70.60	3.15	161	$C_{11}H_{12}S$	176	Benzo[b]thiophene,methyl-ethyl- 甲基-乙基苯并[b]噻吩
26	70.90	3.14	176	$C_{11}H_{12}S$	176	Benzo[b]thiophene,trimethyl- 三甲基苯并[b]噻吩
27	71.30	3.14	176	$C_{11}H_{12}S$	176	Benzo[b]thiophene,trimethyl- 三甲基苯并[b]噻吩
28	71.40	3.19	176	$C_{11}H_{12}S$	176	Benzo[b]thiophene,trimethyl- 三甲基苯并[b]噻吩
29	71.90	3.20	176	$C_{11}H_{12}S$	176	Benzo[b]thiophene,trimethyl- 三甲基苯并[b]噻吩
30	72.40	3.18	176	$C_{11}H_{12}S$	176	Benzo[b]thiophene,trimethyl- 三甲基苯并[b]噻吩
31	72.90	3.29	176	$C_{11}H_{12}S$	176	Benzo[b]thiophene,methyl-ethyl- 甲基-乙基苯并[b]噻吩
32	73.10	3.18	176	$C_{11}H_{12}S$	176	Benzo[b]thiophene,trimethyl- 三甲基苯并[b]噻吩
33	73.50	3.25	176	$C_{11}H_{12}S$	176	Benzo[b]thiophene,trimethyl- 三甲基苯并[b]噻吩
34	75.10	3.41	161	$C_{11}H_{12}S$	176	Benzo[b]thiophene,trimethyl- 三甲基苯并[b]噻吩
35	75.20	3.13	175	$C_{12}H_{14}S$	190	Benzo[b]thiophene,C_4- C_4-苯并[b]噻吩
36	77.80	3.09	175	$C_{12}H_{14}S$	190	Benzo[b]thiophene,C_4- C_4-苯并[b]噻吩

4) 长链烷基苯并噻吩系列的全二维气相色谱图及化合物名称

利用特征离子 m/z 147，m/z 161，m/z 175 可以检测到苯并噻吩环上取代基碳数更多的系列化合物，且取代基为正构烷基，这些化合物是长链烷基苯并噻吩系列，包括正烷基-苯并噻吩、正烷基-甲基苯并噻吩和正烷基-C_2-苯并噻吩，但是取代基的位置无法确定。与长侧链取代基的其他环状化合物（如十氢化萘）相似，长链烷基苯并噻吩中同碳数的化合物也是成对出现的，或互为同分异构体，如图 2-499 所示。在图 2-499 中共检测到 3 个类别的 26 个化合物,编号自 37 至 62,各化合物的质谱图如图 2-500~图 2-525 所示,各化合物鉴定结果见表 2-43。

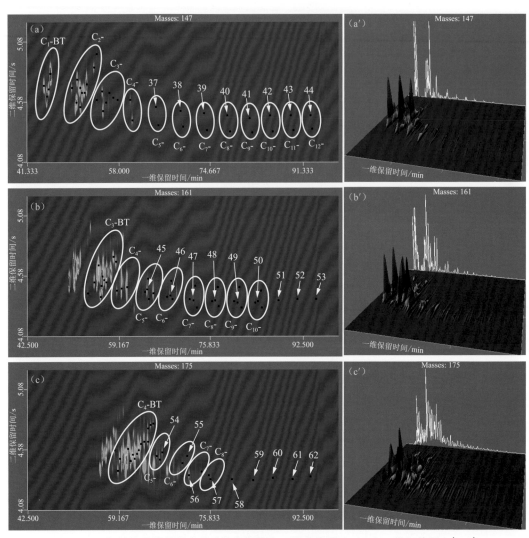

图 2-499　长链烷基苯并噻吩系列化合物的全二维点阵图(a~c)和三维立体图(a′~c′)

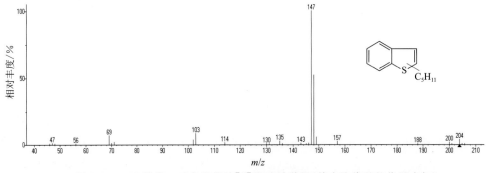

图 2-500　37 号峰　正戊基苯并[b]噻吩质谱图(其中取代基的位置未知)

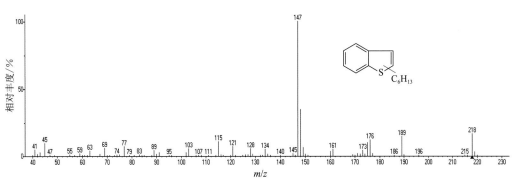

图 2-501　38 号峰　正己基苯并[b]噻吩质谱图（其中取代基的位置未知）

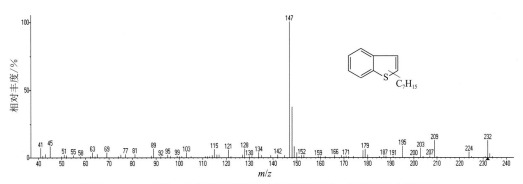

图 2-502　39 号峰　正庚基苯并[b]噻吩质谱图（其中取代基的位置未知）

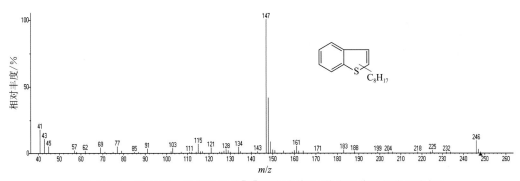

图 2-503　40 号峰　正辛基苯并[b]噻吩质谱图（其中取代基的位置未知）

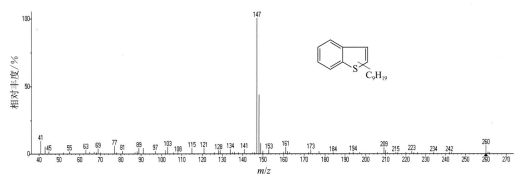

图 2-504　41 号峰　正壬基苯并[b]噻吩质谱图（其中取代基的位置未知）

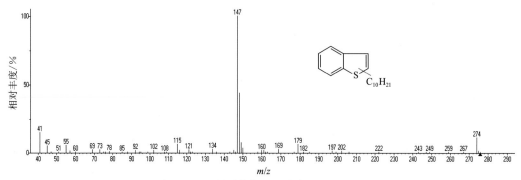

图 2-505　42 号峰　正癸基苯并[b]噻吩质谱图（其中取代基的位置未知）

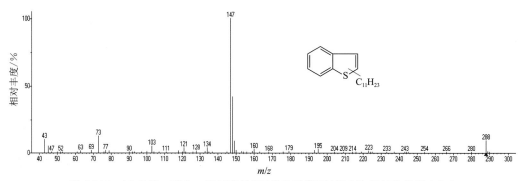

图 2-506　43 号峰　正十一烷基苯并[b]噻吩质谱图（其中取代基的位置未知）

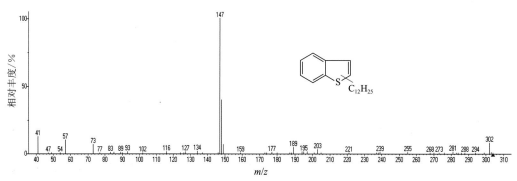

图 2-507　44 号峰　正十二烷基苯并[b]噻吩质谱图（其中取代基的位置未知）

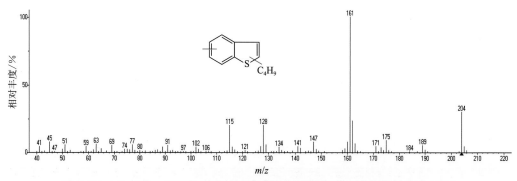

图 2-508　45 号峰　正丁基-甲基-苯并[b]噻吩质谱图（其中取代基的位置未知）

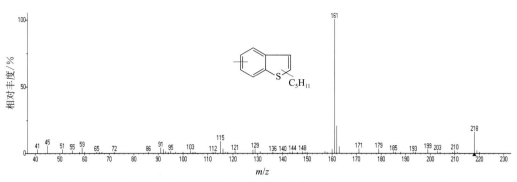

图 2-509　46 号峰　正戊基-甲基-苯并[b]噻吩质谱图（其中取代基的位置未知）

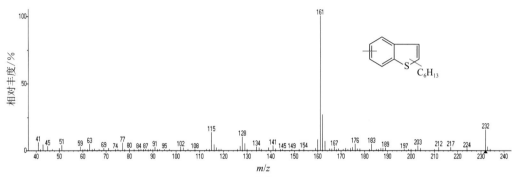

图 2-510　47 号峰　正己基-甲基-苯并[b]噻吩质谱图（其中取代基的位置未知）

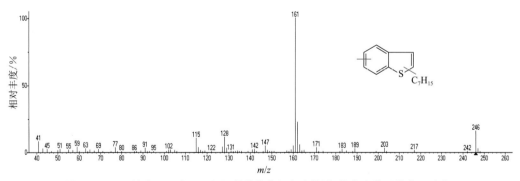

图 2-511　48 号峰　正庚基-甲基-苯并[b]噻吩质谱图（其中取代基的位置未知）

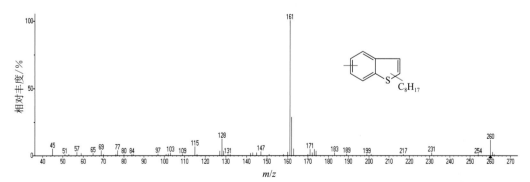

图 2-512　49 号峰　正辛基-甲基-苯并[b]噻吩质谱图（其中取代基的位置未知）

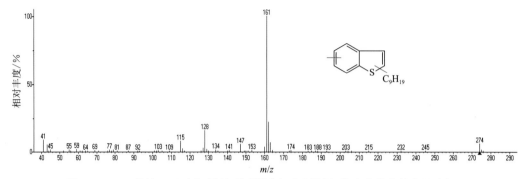

图 2-513　50 号峰　正壬基-甲基-苯并[b]噻吩质谱图（其中取代基的位置未知）

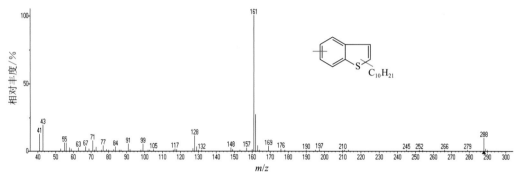

图 2-514　51 号峰　正癸基-甲基-苯并[b]噻吩质谱图（其中取代基的位置未知）

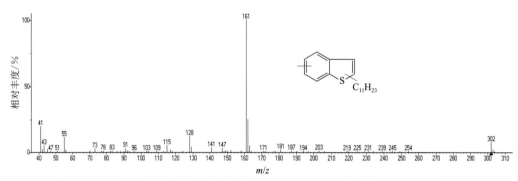

图 2-515　52 号峰　正十一烷基-甲基-苯并[b]噻吩质谱图（其中取代基的位置未知）

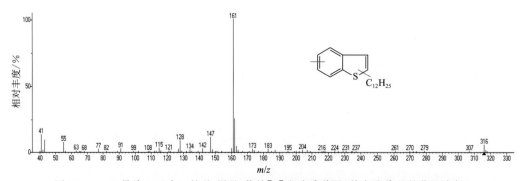

图 2-516　53 号峰　正十二烷基-甲基-苯并[b]噻吩质谱图（其中取代基的位置未知）

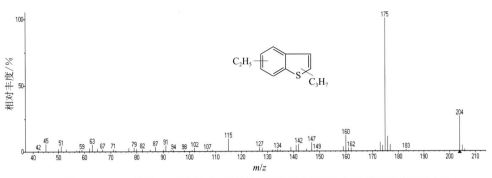

图 2-517　54 号峰　正丙基-C_2-苯并[b]噻吩质谱图（其中取代基的位置未知）

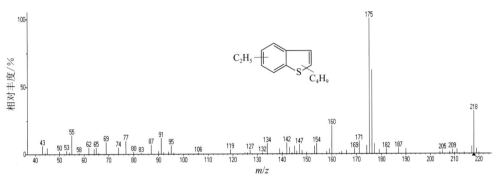

图 2-518　55 号峰　正丁基-C_2-苯并[b]噻吩质谱图（其中取代基的位置未知）

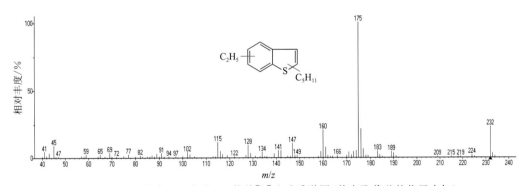

图 2-519　56 号峰　正戊基-C_2-苯并[b]噻吩质谱图（其中取代基的位置未知）

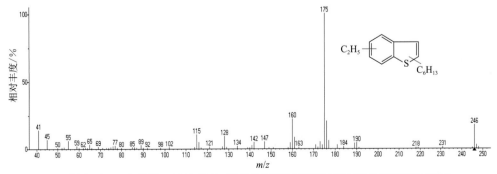

图 2-520　57 号峰　正己基-C_2-苯并[b]噻吩质谱图（其中取代基的位置未知）

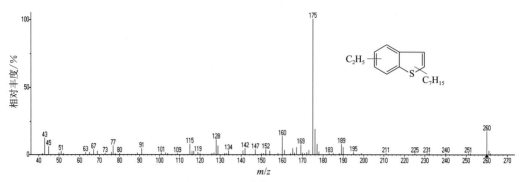

图 2-521　58号峰　正庚基-C_2-苯并[b]噻吩质谱图（其中取代基的位置未知）

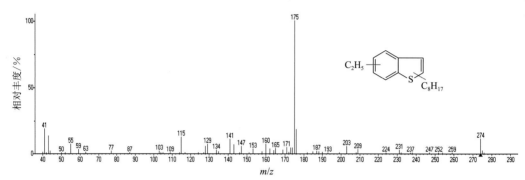

图 2-522　59号峰　正辛基-C_2-苯并[b]噻吩质谱图（其中取代基的位置未知）

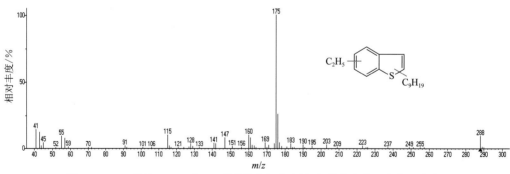

图 2-523　60号峰　正壬基-C_2-苯并[b]噻吩质谱图（其中取代基的位置未知）

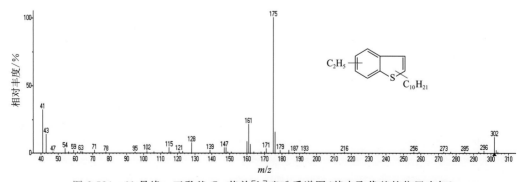

图 2-524　61号峰　正癸基-C_2-苯并[b]噻吩质谱图（其中取代基的位置未知）

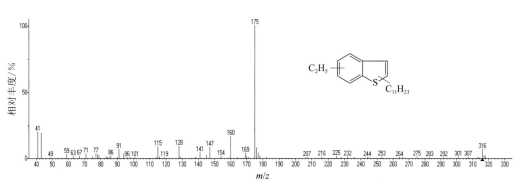

图 2-525　60号峰　正十一烷基-C_2-苯并[b]噻吩质谱图(其中取代基的位置未知)

表 2-43　长(正构)烷基苯并噻吩系列化合物鉴定结果表

峰号	一维保留时间 /min	二维保留时间 /s	特征离子 (m/z)	分子式	相对分子质量	名　称
37	64.67	4.55	147	$C_{13}H_{16}S$	204	Benzo[b]thiophene, n-C_5- 正戊基苯并[b]噻吩
38	69.17	4.51	147	$C_{14}H_{18}S$	218	Benzo[b]thiophene, n-C_6- 正己基苯并[b]噻吩
39	73.50	4.49	147	$C_{15}H_{20}S$	232	Benzo[b]thiophene, n-C_7- 正庚基苯并[b]噻吩
40	77.67	4.47	147	$C_{16}H_{22}S$	246	Benzo[b]thiophene, n-C_8- 正辛基苯并[b]噻吩
41	81.67	4.45	147	$C_{17}H_{24}S$	260	Benzo[b]thiophene, n-C_9- 正壬基苯并[b]噻吩
42	85.33	4.47	147	$C_{18}H_{26}S$	274	Benzo[b]thiophene, n-C_{10}- 正癸基苯并[b]噻吩
43	89.00	4.47	147	$C_{19}H_{28}S$	288	Benzo[b]thiophene, n-C_{11}- 正十一烷基苯并[b]噻吩
44	92.50	4.47	147	$C_{20}H_{30}S$	302	Benzo[b]thiophene, n-C_{12}- 正十二烷基苯并[b]噻吩
45	64.17	4.47	161	$C_{13}H_{16}S$	204	Benzo[b]thiophene, methyl-n-C_4- 正丁基-甲基-苯并[b]噻吩
46	68.50	4.42	161	$C_{14}H_{18}S$	218	Benzo[b]thiophene, methyl-n-C_5- 正戊基-甲基-苯并[b]噻吩
47	72.67	4.38	161	$C_{15}H_{20}S$	232	Benzo[b]thiophene, methyl-n-C_6- 正己基-甲基-苯并[b]噻吩
48	76.67	4.38	161	$C_{16}H_{22}S$	246	Benzo[b]thiophene, methyl-n-C_7- 正庚基-甲基-苯并[b]噻吩

续表

峰 号	一维保留时间 /min	二维保留时间 /s	特征离子 (m/z)	分子式	相对分子质量	名 称
49	80.67	4.36	161	$C_{17}H_{24}S$	260	Benzo[b]thiophene,methyl-n-C_8-正辛基-甲基-苯并[b]噻吩
50	84.33	4.37	161	$C_{18}H_{26}S$	274	Benzo[b]thiophene,methyl-n-C_9-正壬基-甲基-苯并[b]噻吩
51	88.00	4.38	161	$C_{19}H_{28}S$	288	Benzo[b]thiophene,methyl-n-C_{10}-正癸基-甲基-苯并[b]噻吩
52	91.33	4.39	161	$C_{20}H_{30}S$	302	Benzo[b]thiophene,methyl-n-C_{11}-正十一烷基-甲基-苯并[b]噻吩
53	94.67	4.39	161	$C_{21}H_{32}S$	316	Benzo[b]thiophene,methyl-n-C_{12}-正十二烷基-甲基-苯并[b]噻吩
54	67.00	4.56	175	$C_{13}H_{16}S$	204	Benzo[b]thiophene,C_2-n-C_3-正丙基-C_2-苯并[b]噻吩
55	71.50	4.53	175	$C_{14}H_{18}S$	218	Benzo[b]thiophene,C_2-n-C_4-正丁基-C_2-苯并[b]噻吩
56	71.83	4.37	175	$C_{15}H_{20}S$	232	Benzo[b]thiophene,C_2-n-C_5-正戊基-C_2-苯并[b]噻吩
57	75.83	4.34	175	$C_{16}H_{22}S$	246	Benzo[b]thiophene,C_2-n-C_6-正己基-C_2-苯并[b]噻吩
58	79.67	4.34	175	$C_{17}H_{24}S$	260	Benzo[b]thiophene,C_2-n-C_7-正庚基-C_2-苯并[b]噻吩
59	83.50	4.33	175	$C_{18}H_{26}S$	274	Benzo[b]thiophene,C_2-n-C_8-正辛基-C_2-苯并[b]噻吩
60	87.00	4.35	175	$C_{19}H_{28}S$	288	Benzo[b]thiophene,C_2-n-C_9-正壬基-C_2-苯并[b]噻吩
61	90.50	4.34	175	$C_{20}H_{30}S$	302	Benzo[b]thiophene,C_2-n-C_{10}-正癸基-C_2-苯并[b]噻吩
62	93.67	4.36	175	$C_{21}H_{32}S$	316	Benzo[b]thiophene,C_2-n-C_{11}-正十一烷基-C_2-苯并[b]噻吩

5）二氢化苯并噻吩的全二维气相色谱图及化合物名称

在苯并噻吩和单金刚烷的中间区域，利用 m/z 136，m/z 150，m/z 164 特征离子可以检测到 2,3-二氢化苯并噻吩、甲基-2,3-二氢化苯并噻吩以及 C_2-2,3-二氢化苯并噻吩系列化合物 11 个，如图 2-526 所示。从图中可以看出，二氢化苯并噻吩类化合物在特征离子下的共馏峰更多，用常规 GC-MS 不易被检出。典型化合物的质谱图如图 2-527～图 2-531 所示，各化合物的鉴定结果见表 2-44。

第 2 章 石油地质样品的全二维气相色谱图识别

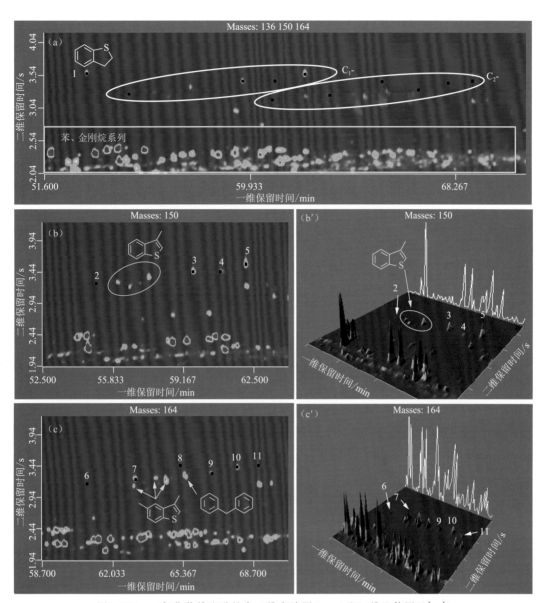

图 2-526 二氢化苯并噻吩的全二维点阵图(a~c)和三维立体图(b′,c′)

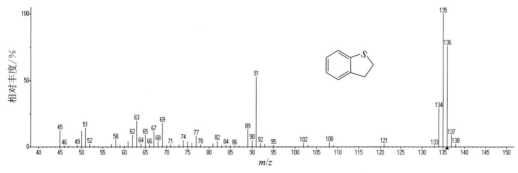

图 2-527 1 号峰 2,3-二氢化苯并噻吩质谱图

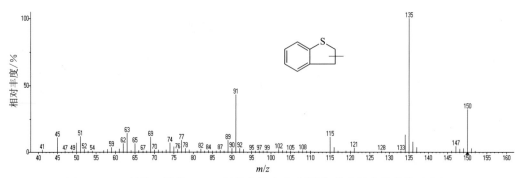

图 2-528　2 号峰　甲基-2,3-二氢化苯并噻吩质谱图(其中取代基的位置未知)

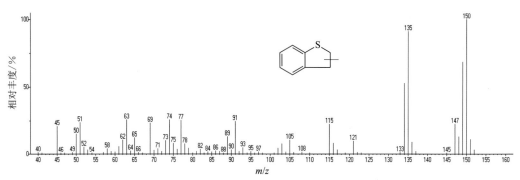

图 2-529　3 号峰　甲基-2,3-二氢化苯并噻吩质谱图(其中取代基的位置未知)

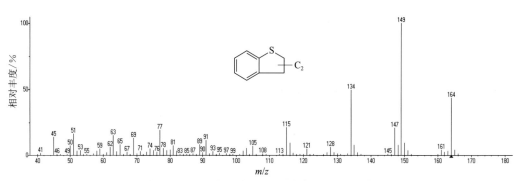

图 2-530　7 号峰　C_2-2,3-二氢化苯并噻吩质谱图(其中取代基的位置未知)

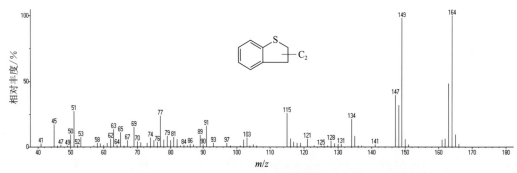

图 2-531　9 号峰　C_2-2,3-二氢化苯并噻吩质谱图(其中取代基的位置未知)

表 2-44 二氢化苯并噻吩系列化合物鉴定结果表

峰 号	一维保留时间/min	二维保留时间/s	特征离子(m/z)	分子式	相对分子质量	名 称
1	53.30	3.58	135	C_8H_8S	136	Benzo-thiophene,2,3-dihydro- 2,3-二氢化苯并噻吩
2	55.00	3.27	135	$C_9H_{10}S$	150	Benzo-thiophene, 2,3-dihydro-methyl- 甲基-2,3-二氢化苯并噻吩
3	59.60	3.46	135	$C_9H_{10}S$	150	Benzo-thiophene, 2,3-dihydro-methyl- 甲基-2,3-二氢化苯并噻吩
4	60.90	3.46	135	$C_9H_{10}S$	150	Benzo-thiophene, 2,3-dihydro-methyl- 甲基-2,3-二氢化苯并噻吩
5	62.10	3.57	135	$C_9H_{10}S$	150	Benzo-thiophene, 2,3-dihydro-methyl- 甲基-2,3-二氢化苯并噻吩
6	60.80	3.17	149	$C_{10}H_{12}S$	164	Benzo-thiophene, 2,3-dihydro-dimethyl- C_2-2,3-二氢化苯并噻吩
7	63.10	3.24	149	$C_{10}H_{12}S$	164	Benzo-thiophene, 2,3-dihydro-dimethyl- C_2-2,3-二氢化苯并噻吩
8	65.20	3.45	149	$C_{10}H_{12}S$	164	Benzo-thiophene, 2,3-dihydro-dimethyl- C_2-2,3-二氢化苯并噻吩
9	66.70	3.33	149	$C_{10}H_{12}S$	164	Benzo-thiophene, 2,3-dihydro-dimethyl- C_2-2,3-二氢化苯并噻吩
10	67.90	3.43	149	$C_{10}H_{12}S$	164	Benzo-thiophene, 2,3-dihydro-dimethyl- C_2-2,3-二氢化苯并噻吩
11	68.90	3.46	149	$C_{10}H_{12}S$	164	Benzo-thiophene, 2,3-dihydro-dimethyl- C_2-2,3-二氢化苯并噻吩

6）二苯并噻吩系列的全二维气相色谱图及化合物名称

二苯并噻吩系列化合物蕴藏着丰富的地球化学信息，其化合物的比值是油气地球化学工作者最常用的参数之一，利用其特征离子 m/z 184，m/z 198，m/z 212，m/z 226，m/z 240 可以检测二苯并噻吩、甲基-二苯并噻吩、C_2-二苯并噻吩、C_3-二苯并噻吩以及 C_4-二苯并噻吩。在全二维气相色谱-飞行时间质谱仪上可以检测到二苯并噻吩系列化合物 52 个，如图 2-532 所示，远多于常规色质分析检测到的化合物，如图 2-533 所示。

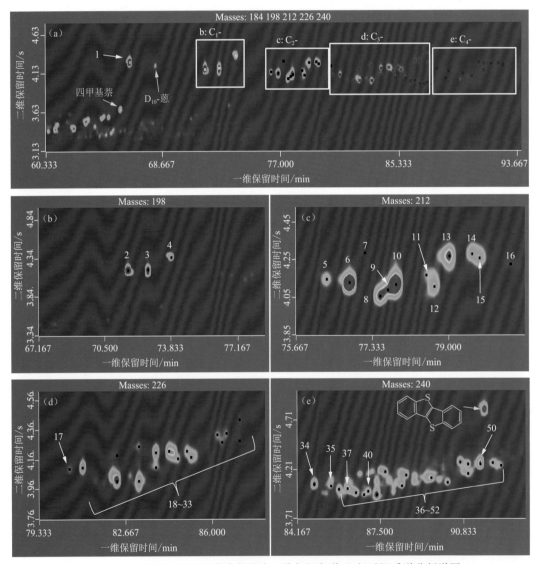

图 2-532　二苯并噻吩系列化合物的全二维气相色谱-飞行时间质谱分析谱图

利用全二维分析检测到的二苯并噻吩系列化合物，有的可以根据标准图谱比对定性，无标准图谱的化合物可以根据其质谱图（图 2-534～图 2-541）推测出其取代基的个数及每个取代基的碳数，但取代基在环上的位置无法准确判断。各化合物的鉴定结果见表 2-45。

第 2 章　石油地质样品的全二维气相色谱图识别

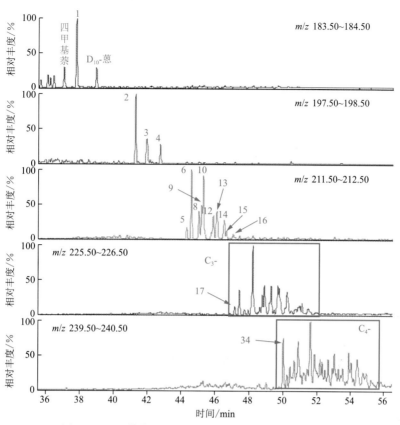

图 2-533　二苯并噻吩系列化合物的常规色质分析谱图

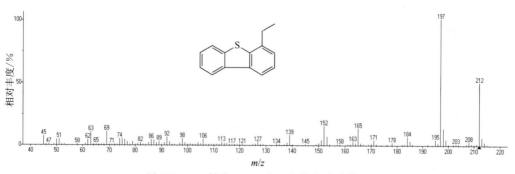

图 2-534　5 号峰　4-乙基二苯并噻吩质谱图

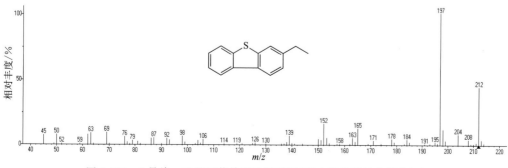

图 2-535　7 号峰　乙基二苯并噻吩质谱图（其中取代基的位置未知）

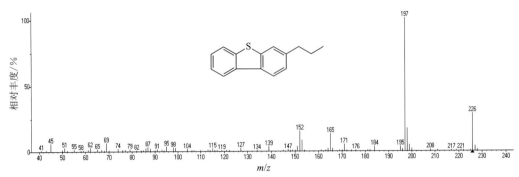

图 2-536　17 号峰　丙基二苯并噻吩质谱图（其中取代基的位置未知）

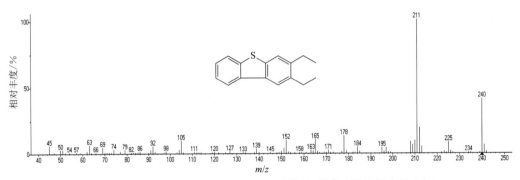

图 2-537　34 号峰　二乙基二苯并噻吩质谱图（其中取代基的位置未知）

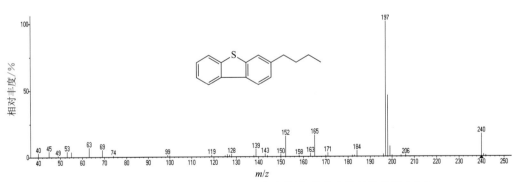

图 2-538　35 号峰　丁基二苯并噻吩质谱图（其中取代基的位置未知）

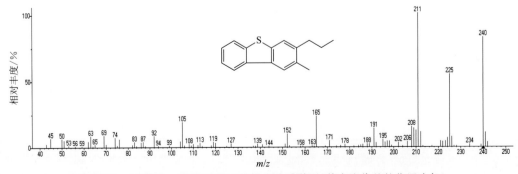

图 2-539　37 号峰　甲基-丙基二苯并噻吩质谱图（其中取代基的位置未知）

第 2 章 石油地质样品的全二维气相色谱图识别

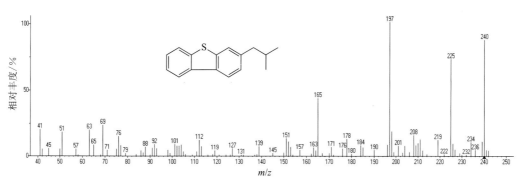

图 2-540 40 号峰 异丁基二苯并噻吩质谱图(其中取代基的位置未知)

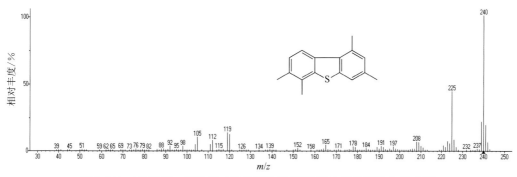

图 2-541 50 号峰 四甲基二苯并噻吩质谱图(其中取代基的位置未知)

表 2-45 二苯并噻吩系列化合物鉴定结果表

峰 号	一维保留时间/min	二维保留时间/s	特征离子(m/z)	分子式	相对分子质量	名 称
1	66.33	4.28	184	$C_{12}H_8S$	184	Dibenzothiophene 二苯并噻吩
2	71.67	4.19	198	$C_{13}H_{10}S$	198	Dibenzothiophene, 4-methyl- 4-甲基二苯并噻吩
3	72.67	4.20	198	$C_{13}H_{10}S$	198	Dibenzothiophene, 3-methyl- 3-甲基二苯并噻吩
4	73.83	4.36	198	$C_{13}H_{10}S$	198	Dibenzothiophene, 1-methyl- 1-甲基二苯并噻吩
5	76.33	4.15	197	$C_{14}H_{12}S$	212	Dibenzothiophene, 4-ethyl- 4-乙基二苯并噻吩
6	76.83	4.13	212	$C_{14}H_{12}S$	212	Dibenzothiophene, 4,6-dimethyl- 4,6-二甲基二苯并噻吩
7	77.17	4.29	197	$C_{14}H_{12}S$	212	Dibenzothiophene, ethyl- 乙基二苯并噻吩
8	77.50	4.06	212	$C_{14}H_{12}S$	212	Dibenzothiophene, 2,4-dimethyl- 2,4-二甲基二苯并噻吩
9	77.67	4.09	212	$C_{14}H_{12}S$	212	Dibenzothiophene, 2,6-dimethyl- 2,6-二甲基二苯并噻吩
10	77.83	4.12	212	$C_{14}H_{12}S$	212	Dibenzothiophene, 3,6-dimethyl- 3,6-二甲基二苯并噻吩

续表

峰 号	一维保留时间/min	二维保留时间/s	特征离子（m/z）	分子式	相对分子质量	名 称
11	78.50	4.17	212	$C_{14}H_{12}S$	212	Dibenzothiophene, 2,8-dimethyl- 2,8-二甲基二苯并噻吩
12	78.67	4.11	212	$C_{14}H_{12}S$	212	Dibenzothiophene, 2,7- ＋3,7-dimethyl- 2,7- ＋3,7-二甲基二苯并噻吩
13	79.00	4.27	212	$C_{14}H_{12}S$	212	Dibenzothiophene, 1,4- ＋1,6-dimethyl- 1,4- ＋1,6-二甲基二苯并噻吩
14	79.50	4.28	212	$C_{14}H_{12}S$	212	Dibenzothiophene, 1,3- ＋3,4-dimethyl- 1,3- ＋3,4-二甲基二苯并噻吩
15	79.67	4.26	212	$C_{14}H_{12}S$	212	Dibenzothiophene, 1,7-dimethyl- 1,7-二甲基二苯并噻吩
16	80.33	4.24	212	$C_{14}H_{12}S$	212	Dibenzothiophene, 2,3- ＋1,9-dimethyl- 2,3- ＋1,9-二甲基二苯并噻吩
17	80.50	4.10	197	$C_{15}H_{14}S$	226	Dibenzothiophene, propyl- 丙基二苯并噻吩
18	81.00	4.11	226	$C_{15}H_{14}S$	226	Dibenzothiophene, 4-ethyl-6-methyl- 4-乙基-6-甲基二苯并噻吩
19	82.33	4.02	226	$C_{15}H_{14}S$	226	Dibenzothiophene, 2,4,6-trimethyl- 2,4,6-三甲基二苯并噻吩
20	82.33	4.19	226	$C_{15}H_{14}S$	226	Dibenzothiophene, trimethyl- 三甲基二苯并噻吩
21	83.17	4.02	226	$C_{15}H_{14}S$	226	Dibenzothiophene, trimethyl- 三甲基二苯并噻吩
22	83.17	4.25	226	$C_{15}H_{14}S$	226	Dibenzothiophene, trimethyl- 三甲基二苯并噻吩
23	83.83	4.11	226	$C_{15}H_{14}S$	226	Dibenzothiophene, trimethyl- 三甲基二苯并噻吩
24	83.83	4.21	226	$C_{15}H_{14}S$	226	Dibenzothiophene, trimethyl- 三甲基二苯并噻吩
25	84.33	4.22	226	$C_{15}H_{14}S$	226	Dibenzothiophene, trimethyl- 三甲基二苯并噻吩
26	84.67	4.17	226	$C_{15}H_{14}S$	226	Dibenzothiophene, trimethyl- 三甲基二苯并噻吩
27	85.00	4.22	226	$C_{15}H_{14}S$	226	Dibenzothiophene, trimethyl- 三甲基二苯并噻吩
28	85.17	4.18	226	$C_{15}H_{14}S$	226	Dibenzothiophene, trimethyl- 三甲基二苯并噻吩
29	86.17	4.33	226	$C_{15}H_{14}S$	226	Dibenzothiophene, trimethyl- 三甲基二苯并噻吩
30	86.33	4.28	226	$C_{15}H_{14}S$	226	Dibenzothiophene, trimethyl- 三甲基二苯并噻吩
31	86.50	4.34	226	$C_{15}H_{14}S$	226	Dibenzothiophene, trimethyl- 三甲基二苯并噻吩
32	87.00	4.29	226	$C_{15}H_{14}S$	226	Dibenzothiophene, trimethyl- 三甲基二苯并噻吩

续表

峰 号	一维保留时间/min	二维保留时间/s	特征离子（m/z）	分子式	相对分子质量	名　称
33	87.00	4.43	226	$C_{15}H_{14}S$	226	Dibenzothiophene, trimethyl- 三甲基二苯并噻吩
34	84.83	4.06	211	$C_{16}H_{16}S$	240	Dibenzothiophene, diethyl- 二乙基二苯并噻吩
35	85.50	4.05	197	$C_{16}H_{16}S$	240	Dibenzothiophene, butyl- 丁基二苯并噻吩
36	85.83	4.01	240	$C_{16}H_{16}S$	240	Dibenzothiophene, tetramethyl- 四甲基二苯并噻吩
37	86.17	4.02	240	$C_{16}H_{16}S$	240	Dibenzothiophene, methyl-propyl- 甲基-丙基二苯并噻吩
38	86.50	3.98	240	$C_{16}H_{16}S$	240	Dibenzothiophene, tetramethyl- 四甲基二苯并噻吩
39	86.83	3.96	240	$C_{16}H_{16}S$	240	Dibenzothiophene, tetramethyl- 四甲基二苯并噻吩
40	87.00	3.99	240	$C_{16}H_{16}S$	240	Dibenzothiophene, isobutyl- 异丁基二苯并噻吩
41	87.33	3.96	240	$C_{16}H_{16}S$	240	Dibenzothiophene, tetramethyl- 四甲基二苯并噻吩
42	87.67	4.19	240	$C_{16}H_{16}S$	240	Dibenzothiophene, tetramethyl- 四甲基二苯并噻吩
43	88.33	4.05	240	$C_{16}H_{16}S$	240	Dibenzothiophene, tetramethyl- 四甲基二苯并噻吩
44	88.50	4.16	240	$C_{16}H_{16}S$	240	Dibenzothiophene, tetramethyl- 四甲基二苯并噻吩
45	89.50	4.13	240	$C_{16}H_{16}S$	240	Dibenzothiophene, tetramethyl- 四甲基二苯并噻吩
46	90.17	4.08	240	$C_{16}H_{16}S$	240	Dibenzothiophene, tetramethyl- 四甲基二苯并噻吩
47	90.67	4.29	240	$C_{16}H_{16}S$	240	Dibenzothiophene, tetramethyl- 四甲基二苯并噻吩
48	91.00	4.17	240	$C_{16}H_{16}S$	240	Dibenzothiophene, tetramethyl- 四甲基二苯并噻吩
49	91.00	4.27	240	$C_{16}H_{16}S$	240	Dibenzothiophene, tetramethyl- 四甲基二苯并噻吩
50	91.50	4.27	240	$C_{16}H_{16}S$	240	Dibenzothiophene, tetramethyl- 四甲基二苯并噻吩
51	92.00	4.28	240	$C_{16}H_{16}S$	240	Dibenzothiophene, tetramethyl- 四甲基二苯并噻吩
52	92.33	4.25	240	$C_{16}H_{16}S$	240	Dibenzothiophene, tetramethyl- 四甲基二苯并噻吩

7）菲并噻吩系列的全二维气相色谱图及化合物名称

菲并噻吩系列化合物在全二维分析谱图上位于 C_3-二苯并噻吩及 C_4-二苯并噻吩系列化合物的上方，苯并萘并噻吩的下方，如图 2-542 所示。利用其特征离子 m/z 208，m/z 222，m/z 236，m/z 250 可以分别检测到菲并噻吩、甲基菲并噻吩及 C_2 和 C_3 取代基的系列化合物 16 个。菲并噻吩、甲基菲并噻吩与烷基二苯并噻吩在常规 GC-MS 上是共馏峰（如图 2-543a 和 b 所示），C_3-菲并噻吩与甲基苯并萘并噻吩在常规 GC-MS 上是共馏峰（如图 2-543d 所示）。由此可见，用常规 GC-MS 检测菲并噻吩类化合物时，要注意共馏峰的干扰。各个化合物的鉴定依据质谱图（图 2-544～图 2-547）碎片离子推断，结果见表 2-46。

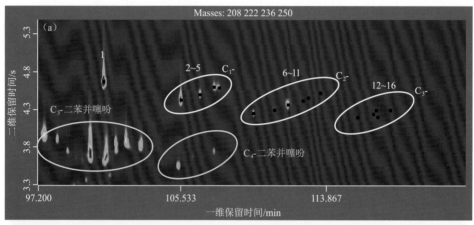

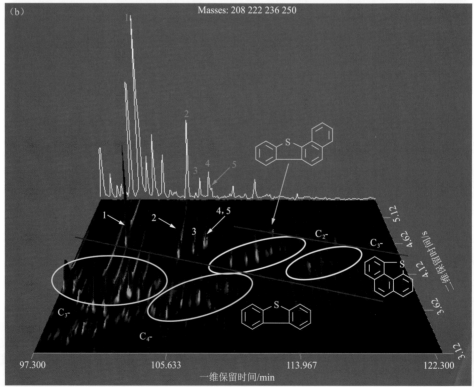

图 2-542　菲并噻吩系列化合物在全二维分析谱图上的分布

第 2 章　石油地质样品的全二维气相色谱图识别

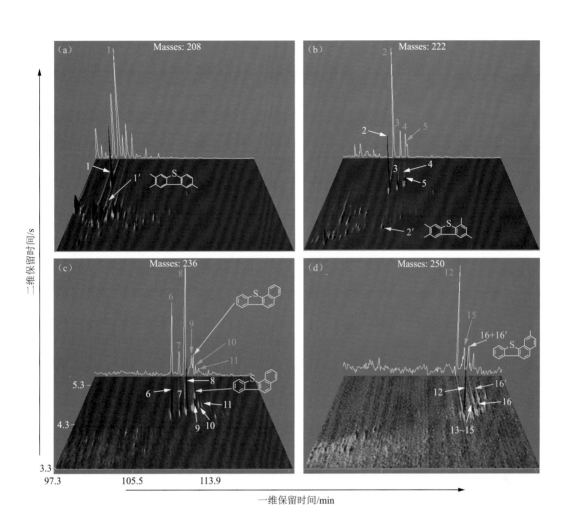

图 2-543　菲并噻吩系列化合物在全二维分析谱图上的识别和鉴定
图(a)中 1 号化合物和 1′号化合物(三甲基二苯并噻吩)是共馏峰,
图(b)中 2 号化合物和 2′号化合物(四甲基二苯并噻吩)是共馏峰,
图(d)中 16 号化合物和 16′号化合物(甲基苯并萘并噻吩)是共馏峰

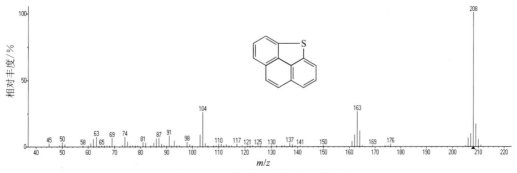

图 2-544　1 号峰　菲并噻吩质谱图

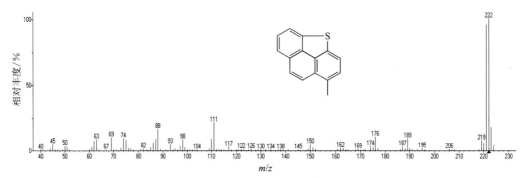

图 2-545　4 号峰　甲基菲并噻吩质谱图（其中取代基的位置未知）

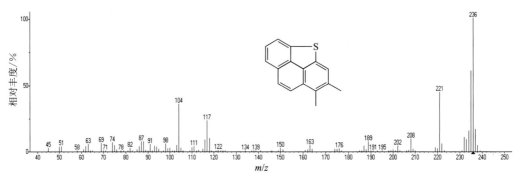

图 2-546　8 号峰　C_2-菲并噻吩质谱图（其中取代基的位置未知）

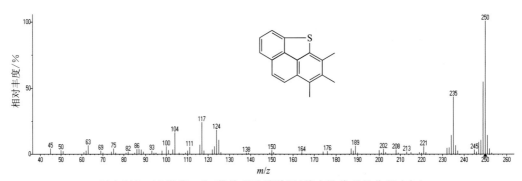

图 2-547　12 号峰　C_3-菲并噻吩质谱图（其中取代基的位置未知）

表 2-46　菲并噻吩系列化合物鉴定结果表

峰号	一维保留时间 /min	二维保留时间 /s	特征离子 （m/z）	分子式	相对分子质量	名称
1	101.00	4.68	208	$C_{14}H_8S$	208	Phenanthro-thiophene 菲并噻吩
2	105.50	4.44	222	$C_{15}H_{10}S$	222	Phenanthro-thiophene, methyl- 甲基菲并噻吩
3	106.60	4.51	222	$C_{15}H_{10}S$	222	Phenanthro-thiophene, methyl- 甲基菲并噻吩

续表

峰 号	一维保留时间 /min	二维保留时间 /s	特征离子 (m/z)	分子式	相对分子质量	名 称
4	107.40	4.61	222	$C_{15}H_{10}S$	222	Phenanthro-thiophene, methyl- 甲基菲并噻吩
5	107.70	4.60	222	$C_{15}H_{10}S$	222	Phenanthro-thiophene, methyl- 甲基菲并噻吩
6	109.70	4.26	236	$C_{16}H_{12}S$	236	Phenanthro-thiophene, C_2- C_2-菲并噻吩
7	110.90	4.30	236	$C_{16}H_{12}S$	236	Phenanthro-thiophene, C_2- C_2-菲并噻吩
8	111.70	4.37	236	$C_{16}H_{12}S$	236	Phenanthro-thiophene, C_2- C_2-菲并噻吩
9	112.50	4.43	236	$C_{16}H_{12}S$	236	Phenanthro-thiophene, C_2- C_2-菲并噻吩
10	112.80	4.47	236	$C_{16}H_{12}S$	236	Phenanthro-thiophene, C_2- C_2-菲并噻吩
11	113.50	4.53	236	$C_{16}H_{12}S$	236	Phenanthro-thiophene, C_2- C_2-菲并噻吩
12	115.60	4.21	250	$C_{17}H_{14}S$	250	Phenanthro-thiophene, C_3- C_3-菲并噻吩
13	116.50	4.25	250	$C_{17}H_{14}S$	250	Phenanthro-thiophene, C_3- C_3-菲并噻吩
14	116.70	4.30	250	$C_{17}H_{14}S$	250	Phenanthro-thiophene, C_3- C_3-菲并噻吩
15	116.80	4.21	250	$C_{17}H_{14}S$	250	Phenanthro-thiophene, C_3- C_3-菲并噻吩
16	117.50	4.30	250	$C_{17}H_{14}S$	250	Phenanthro-thiophene, C_3- C_3-菲并噻吩

8) 苯并萘并噻吩的全二维气相色谱图及化合物名称

苯并萘并噻吩系列化合物包括苯并萘并噻吩的 3 个同分异构体以及取代基分别为甲基、C_2、C_3 的化合物，利用其特征离子 m/z 234，m/z 248，m/z 262，m/z 276 在全二维分析谱图上可以识别出 29 个化合物，如图 2-548 所示。苯并萘并噻吩类化合物在用常规 GC-MS 检测时常与 C_4-菲、C_5-菲共馏，因此用 GC×MS 分析可以检测到更多的化合物。典型化合物的质谱图如图 2-549～图 2-555 所示，各化合物的鉴定结果见表 2-47。

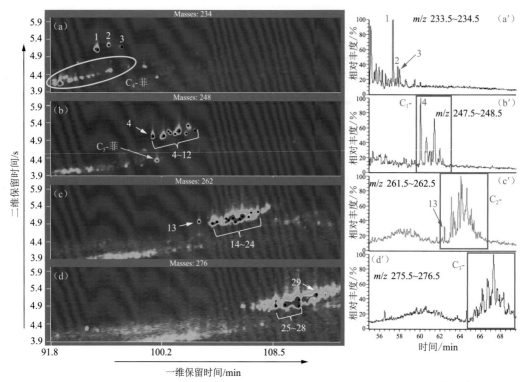

图 2-548　苯并萘并噻吩系列化合物的全二维点阵图(a～d)和 GC-MS 色谱图(a′～d′)

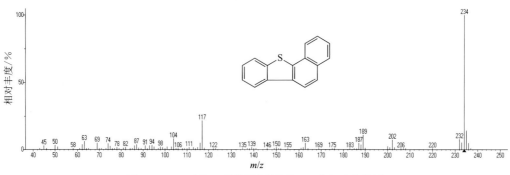

图 2-549　1 号峰　苯并[b]萘并[2,1-d]噻吩质谱图

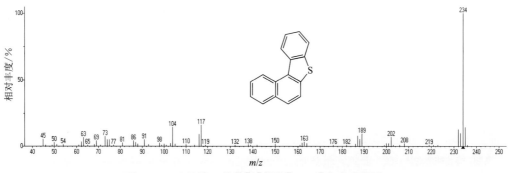

图 2-550　2 号峰　苯并[b]萘并[1,2-d]噻吩质谱图

图 2-551　3 号峰　苯并[b]萘并[2,3-d]噻吩质谱图

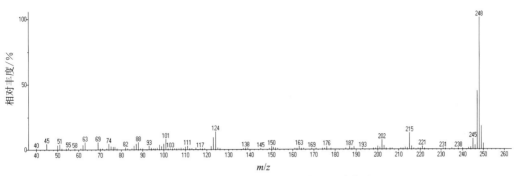

图 2-552　4 号峰　甲基苯并萘并噻吩质谱图

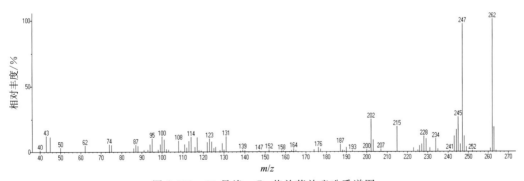

图 2-553　13 号峰　C_2-苯并萘并噻吩质谱图

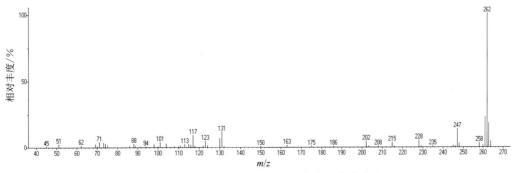

图 2-554　14 号峰　二甲基苯并萘并噻吩质谱图

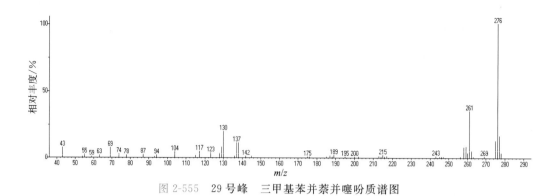

图 2-555　29号峰　三甲基苯并萘并噻吩质谱图

表 2-47　菲并噻吩系列化合物鉴定结果表

峰　号	一维保留时间 /min	二维保留时间 /s	特征离子 （m/z）	分子式	相对 分子质量	名　称
1	95.33	5.09	234	$C_{16}H_{10}S$	234	Benzo[b]naphtho[2,1-d]thiophene 苯并[b]萘并[2,1-d]噻吩
2	96.17	5.23	234	$C_{16}H_{10}S$	234	Benzo[b]naphtho[1,2-d]thiophene 苯并[b]萘并[1,2-d]噻吩
3	97.17	5.17	234	$C_{16}H_{10}S$	234	Benzo[b]naphtho[2,3-d]thiophene 苯并[b]萘并[2,3-d]噻吩
4	99.50	5.00	248	$C_{17}H_{12}S$	248	Benzo-naphtho-thiophene,methyl- 甲基苯并萘并噻吩
5	100.33	5.02	248	$C_{17}H_{12}S$	248	Benzo-naphtho-thiophene,methyl- 甲基苯并萘并噻吩
6	100.67	5.15	248	$C_{17}H_{12}S$	248	Benzo-naphtho-thiophene,methyl- 甲基苯并萘并噻吩
7	100.83	5.10	248	$C_{17}H_{12}S$	248	Benzo-naphtho-thiophene,methyl- 甲基苯并萘并噻吩
8	101.17	5.08	248	$C_{17}H_{12}S$	248	Benzo-naphtho-thiophene,methyl- 甲基苯并萘并噻吩
9	101.50	5.16	248	$C_{17}H_{12}S$	248	Benzo-naphtho-thiophene,methyl- 甲基苯并萘并噻吩
10	102.00	5.09	248	$C_{17}H_{12}S$	248	Benzo-naphtho-thiophene,methyl- 甲基苯并萘并噻吩
11	102.17	5.31	248	$C_{17}H_{12}S$	248	Benzo-naphtho-thiophene,methyl- 甲基苯并萘并噻吩
12	102.67	5.18	248	$C_{17}H_{12}S$	248	Benzo-naphtho-thiophene,methyl- 甲基苯并萘并噻吩
13	103.00	4.96	262	$C_{18}H_{14}S$	262	Benzo-naphtho-thiophene,C_2- C_2-苯并萘并噻吩
14	104.00	4.91	262	$C_{18}H_{14}S$	262	Benzo-naphtho-thiophene,dimethyl- 二甲基苯并萘并噻吩

续表

峰 号	一维保留时间/min	二维保留时间/s	特征离子(m/z)	分子式	相对分子质量	名 称
15	104.33	4.95	262	$C_{18}H_{14}S$	262	Benzo-naphtho-thiophene,dimethyl-二甲基苯并萘并噻吩
16	105.00	5.05	262	$C_{18}H_{14}S$	262	Benzo-naphtho-thiophene,dimethyl-二甲基苯并萘并噻吩
17	105.17	5.00	262	$C_{18}H_{14}S$	262	Benzo-naphtho-thiophene,dimethyl-二甲基苯并萘并噻吩
18	105.50	5.06	262	$C_{18}H_{14}S$	262	Benzo-naphtho-thiophene,dimethyl-二甲基苯并萘并噻吩
19	105.67	4.95	262	$C_{18}H_{14}S$	262	Benzo-naphtho-thiophene,dimethyl-二甲基苯并萘并噻吩
20	106.00	5.03	262	$C_{18}H_{14}S$	262	Benzo-naphtho-thiophene,dimethyl-二甲基苯并萘并噻吩
21	106.17	5.11	262	$C_{18}H_{14}S$	262	Benzo-naphtho-thiophene,dimethyl-二甲基苯并萘并噻吩
22	106.67	5.10	262	$C_{18}H_{14}S$	262	Benzo-naphtho-thiophene,dimethyl-二甲基苯并萘并噻吩
23	106.83	5.22	262	$C_{18}H_{14}S$	262	Benzo-naphtho-thiophene,dimethyl-二甲基苯并萘并噻吩
24	107.33	5.25	262	$C_{18}H_{14}S$	262	Benzo-naphtho-thiophene,dimethyl-二甲基苯并萘并噻吩
25	108.67	4.90	276	$C_{19}H_{16}S$	276	Benzo-naphtho-thiophene,trimethyl-三甲基苯并萘并噻吩
26	109.33	4.93	276	$C_{19}H_{16}S$	276	Benzo-naphtho-thiophene,trimethyl-三甲基苯并萘并噻吩
27	110.33	4.91	276	$C_{19}H_{16}S$	276	Benzo-naphtho-thiophene,trimethyl-三甲基苯并萘并噻吩
28	110.33	5.08	276	$C_{19}H_{16}S$	276	Benzo-naphtho-thiophene,trimethyl-三甲基苯并萘并噻吩
29	111.67	5.22	276	$C_{19}H_{16}S$	276	Benzo-naphtho-thiophene,trimethyl-三甲基苯并萘并噻吩

2.3.2 含氮化合物的全二维气相色谱图及化合物名称

石油地质样品中含氮化合物的种类和数量不亚于含硫化合物的种类和数量,但目前应用较多的是中性含氮化合物,本节仅讨论最常用的咔唑和苯并咔唑类化合物在全二维分析谱图上的分布特征及识别鉴定。

1) 咔唑类化合物的全二维气相色谱图及化合物名称

咔唑类化合物一般包括咔唑和 $C_1 \sim C_5$ 不同取代基的咔唑,其质谱分析检测的特征离子分别是 $m/z\ 167, m/z\ 181, m/z\ 195, m/z\ 209, m/z\ 223$ 及 $m/z\ 237$,取代基碳数相同的咔唑族分离特征明显,如图 2-556 所示。从图 2-556 可以看出,咔唑类化合物的相对丰度很低,易受其他化合物的干扰,用常规色质分析无法直接检测芳烃中的含氮化合物,但用 GC×GC-TOFMS 可以。理论上讲,随着取代基碳数的增加,化合物的个数也随之增加,但由于高碳数化合物含量的降低,可检测的化合物数量并没有增加,其中取代基为 C_5 的咔唑仅识别出 2 个化合物,如图 2-556(a)所示;取代基为 C_2 和 C_3 的咔唑识别出的化合物数量最多,如图 2-557 和图 2-558 所示。对照图 2-557 和图 2-558 可以看出,当全二维气相色谱图的同一区域的化合物含量接近,且绝对量不太高时,三维立体图的化合物更容易被识别。

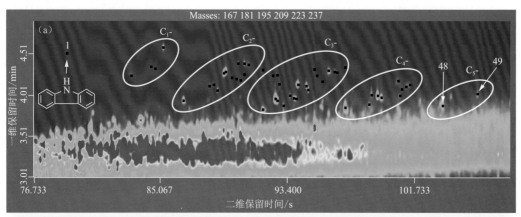

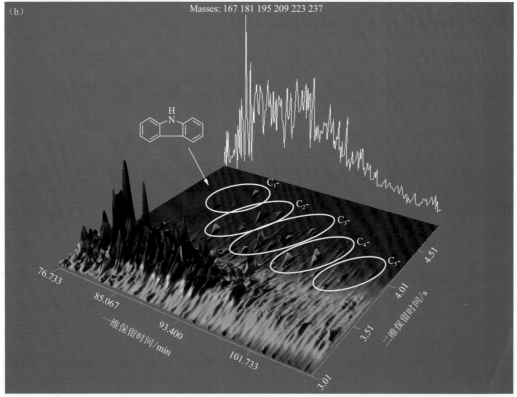

图 2-556　咔唑类化合物的全二维谱图特征(a 全二维点阵图,b 三维立体图)

第 2 章 石油地质样品的全二维气相色谱图识别

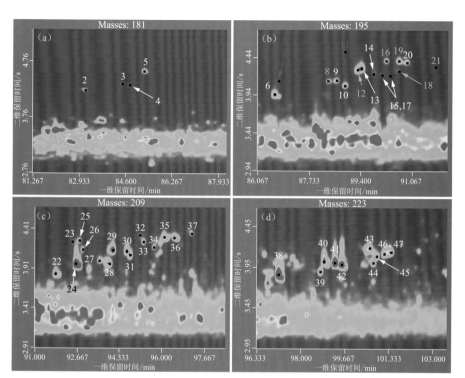

图 2-557 $C_1 \sim C_4$ 取代的咔唑类化合物的全二维谱图特征（全二维点阵图）

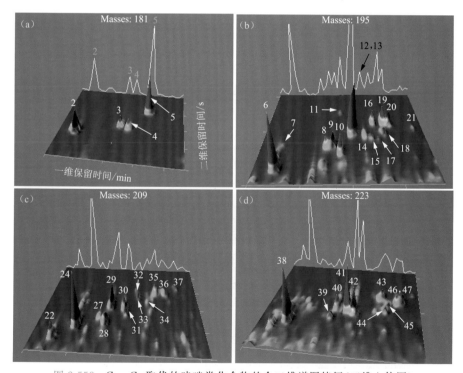

图 2-558 $C_1 \sim C_4$ 取代的咔唑类化合物的全二维谱图特征（三维立体图）

利用全二维气相色谱-飞行时间质谱可以检测到咔唑类化合物 49 个，典型化合物的质谱图如图 2-559～图 2-570 所示。部分化合物根据相关文献和质谱图可以推断出取代基的类型和位置,但多数无法判断,只能推断出取代基的碳数。各化合物的鉴定结果见表 2-48。

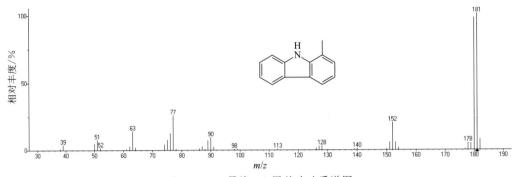

图 2-559　2 号峰　1-甲基咔唑质谱图

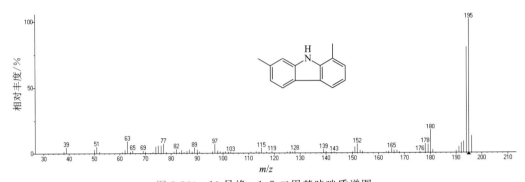

图 2-560　10 号峰　1,7-二甲基咔唑质谱图

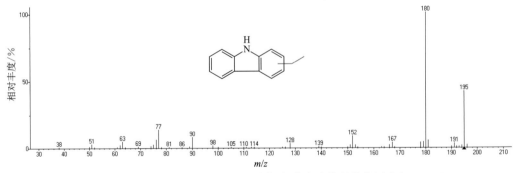

图 2-561　11 号峰　乙基咔唑质谱图(其中取代基的位置未知)

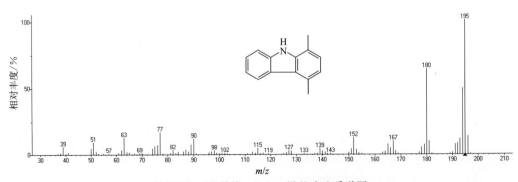

图 2-562　12 号峰　1,4-二甲基咔唑质谱图

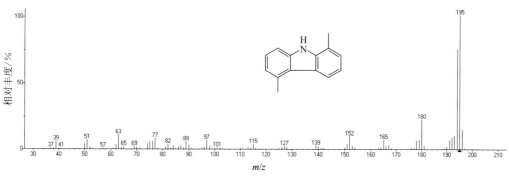

图 2-563　13 号峰　1,5-二甲基咔唑质谱图

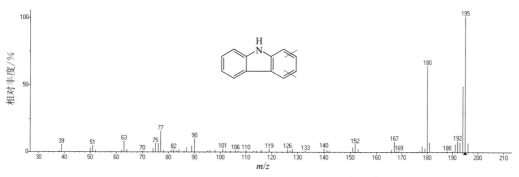

图 2-564　18 号峰　二甲基咔唑质谱图（其中取代基的位置未知）

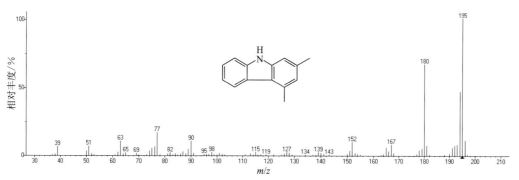

图 2-565　19 号峰　2,4-二甲基咔唑质谱图

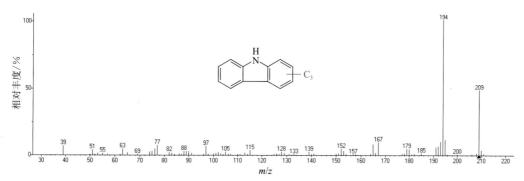

图 2-566　23 号峰　C_3-咔唑质谱图（其中取代基的位置未知）

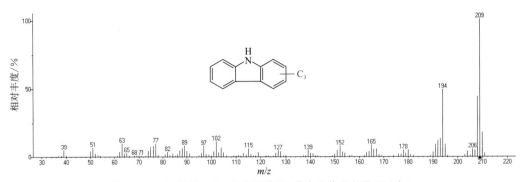

图 2-567　24 号峰　C_3-咔唑质谱图（其中取代基的位置未知）

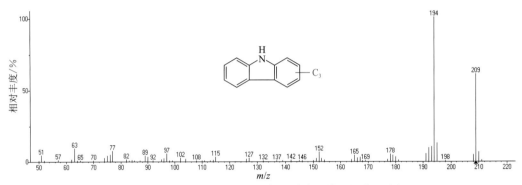

图 2-568　25 号峰　C_3-咔唑质谱图（其中取代基的位置未知）

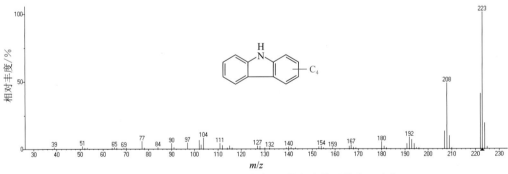

图 2-569　38 号峰　C_4-咔唑质谱图（其中取代基的位置未知）

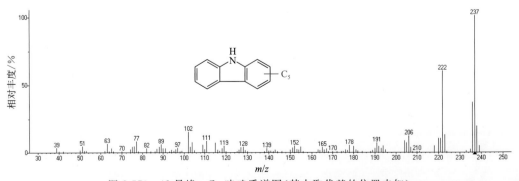

图 2-570　48 号峰　C_5-咔唑质谱图（其中取代基的位置未知）

表 2-48　咔唑类化合物鉴定结果表*

峰　号	一维保留时间 /min	二维保留时间 /s	特征离子 (m/z)	分子式	相对分子质量	名　称
1	78.87	4.52	167	$C_{12}H_9N$	167	Carbazole 咔唑
2	83.13	4.25	181	$C_{13}H_{11}N$	181	Carbazole, 1-Methyl- 1-甲基咔唑
3	84.47	4.36	181	$C_{13}H_{11}N$	181	Carbazole, 3-Methyl- 3-甲基咔唑
4	84.73	4.34	181	$C_{13}H_{11}N$	181	Carbazole, 2-Methyl- 2-甲基咔唑
5	85.27	4.59	181	$C_{13}H_{11}N$	181	Carbazole, 4-Methyl- 4-甲基咔唑
6	86.60	3.95	195	$C_{14}H_{13}N$	195	Carbazole, 1,8-dimethyl- 1,8-二甲基咔唑
7	86.73	4.12	180	$C_{14}H_{13}N$	195	Carbazole, 1-ethyl- 1-乙基咔唑
8	88.33	4.13	195	$C_{14}H_{13}N$	195	Carbazole, 1,3-dimethyl- 1,3-二甲基咔唑
9	88.60	4.14	195	$C_{14}H_{13}N$	195	Carbazole, 1,6-dimethyl- 1,6-二甲基咔唑
10	88.87	4.07	195	$C_{14}H_{13}N$	195	Carbazole, 1,7-dimethyl- 1,7-二甲基咔唑
11	88.87	4.52	180	$C_{14}H_{13}N$	195	Carbazole, ethyl- 乙基咔唑
12	89.27	4.28	195	$C_{14}H_{13}N$	195	Carbazole, 1,4-dimethyl- 1,4-二甲基咔唑①
13	89.40	4.30	195	$C_{14}H_{13}N$	195	Carbazole, 1,5-dimethyl- 1,5-二甲基咔唑②
14	89.80	4.23	195	$C_{14}H_{13}N$	195	Carbazole, 2,6-dimethyl- 2,6-二甲基咔唑
15	90.07	4.21	195	$C_{14}H_{13}N$	195	Carbazole, 2,7-dimethyl- 2,7-二甲基咔唑
16	90.20	4.40	195	$C_{14}H_{13}N$	195	Carbazole, 1,2-dimethyl- 1,2-二甲基咔唑
17	90.33	4.20	195	$C_{14}H_{13}N$	195	Carbazole, dimethyl- 二甲基咔唑
18	90.60	4.27	195	$C_{14}H_{13}N$	195	Carbazole, dimethyl- 二甲基咔唑

续表

峰 号	一维保留时间 /min	二维保留时间 /s	特征离子 (m/z)	分子式	相对分子质量	名　称
19	90.60	4.40	195	$C_{14}H_{13}N$	195	Carbazole, 2,4-dimethyl- 2,4-二甲基咔唑
20	90.87	4.38	195	$C_{14}H_{13}N$	195	Carbazole, 2,5-dimethyl- 2,5-二甲基咔唑
21	91.80	4.32	195	$C_{14}H_{13}N$	195	Carbazole, 2,3-dimethyl- 2,3-二甲基咔唑
22	91.80	3.84	209	$C_{15}H_{15}N$	209	Carbazole, C_3- C_3-咔唑
23	92.47	4.25	194	$C_{15}H_{15}N$	209	Carbazole, C_3- C_3-咔唑
24	92.60	3.97	209	$C_{15}H_{15}N$	209	Carbazole, C_3- C_3-咔唑
25	92.73	4.26	194	$C_{15}H_{15}N$	209	Carbazole, C_3- C_3-咔唑
26	92.87	4.15	194	$C_{15}H_{15}N$	209	Carbazole, C_3- C_3-咔唑
27	93.53	4.01	209	$C_{15}H_{15}N$	209	Carbazole, C_3- C_3-咔唑
28	93.93	3.97	209	$C_{15}H_{15}N$	209	Carbazole, C_3- C_3-咔唑
29	94.07	4.14	209	$C_{15}H_{15}N$	209	Carbazole, C_3- C_3-咔唑
30	94.60	4.13	209	$C_{15}H_{15}N$	209	Carbazole, C_3- C_3-咔唑
31	94.73	4.08	209	$C_{15}H_{15}N$	209	Carbazole, C_3- C_3-咔唑
32	95.13	4.33	209	$C_{15}H_{15}N$	209	Carbazole, C_3- C_3-咔唑
33	95.27	4.25	209	$C_{15}H_{15}N$	209	Carbazole, C_3- C_3-咔唑
34	95.67	4.18	194	$C_{15}H_{15}N$	209	Carbazole, C_3- C_3-咔唑
35	96.07	4.31	209	$C_{15}H_{15}N$	209	Carbazole, C_3- C_3-咔唑
36	96.47	4.30	209	$C_{15}H_{15}N$	209	Carbazole, C_3- C_3-咔唑
37	97.13	4.35	209	$C_{15}H_{15}N$	209	Carbazole, C_3- C_3-咔唑

续表

峰 号	一维保留时间 /min	二维保留时间 /s	特征离子 (m/z)	分子式	相对分子质量	名 称
38	97.13	3.86	223	$C_{16}H_{17}N$	223	Carbazole, C_4-C_4-咔唑
39	98.73	3.90	223	$C_{16}H_{17}N$	223	Carbazole, C_4-C_4-咔唑
40	98.87	4.02	223	$C_{16}H_{17}N$	223	Carbazole, C_4-C_4-咔唑
41	99.27	4.01	223	$C_{16}H_{17}N$	223	Carbazole, C_4-C_4-咔唑
42	99.53	3.99	223	$C_{16}H_{17}N$	223	Carbazole, C4-C4-咔唑
43	100.60	4.19	223	$C_{16}H_{17}N$	223	Carbazole, C_4-C_4-咔唑
44	100.73	4.01	223	$C_{16}H_{17}N$	223	Carbazole, C_4-C_4-咔唑
45	100.87	4.09	223	$C_{16}H_{17}N$	223	Carbazole, C_4-C_4-咔唑
46	101.13	4.12	223	$C_{16}H_{17}N$	223	Carbazole, C_4-C_4-咔唑
47	101.40	4.14	223	$C_{16}H_{17}N$	223	Carbazole, C_4-C_4-咔唑
48	103.53	3.89	237	$C_{17}H_{19}N$	237	Carbazole, C_5-C_5-咔唑
49	105.80	4.03	237	$C_{17}H_{19}N$	237	Carbazole, C_5-C_5-咔唑

* 注：王培荣的《非烃地球化学和应用》一书中认为①处是 1,4-二甲基咔唑和 4-乙基咔唑共馏，②处是 1,5-二甲基咔唑和 3-乙基咔唑共馏。

2) 苯并咔唑类化合物的全二维气相色谱图及化合物名称

苯并咔唑类化合物除了油气地球化学研究中常用的苯并[a]咔唑和苯并[c]咔唑外,还包括甲基和 C_2,C_3 取代基的众多苯并咔唑化合物,它们在全二维分析谱图上位于三环芳烃的上方、三芳甾烷系列化合物的左侧,族分离特征和"瓦片效应"明显,易于识别,如图 2-571 所示。

苯并咔唑类化合物质谱检测的特征离子分别是 m/z 217,m/z 231,m/z 245 和 m/z 259,其中 m/z 217 用以检测苯并[a]咔唑和苯并[c]咔唑,其他特征离子分布用以检测 C_1~C_3 取代基的苯并咔唑。含有取代基的苯并咔唑共检测到 29 个化合物,如图 2-572 所示。碳数相同的化合物的质谱图基本一致（典型化合物质谱图如图 2-573~图 2-577 所示),但无法确定取代基的类型和位置。各化合物鉴定结果见表 2-49。

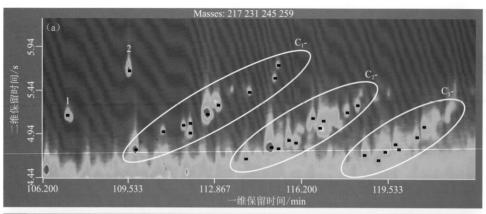
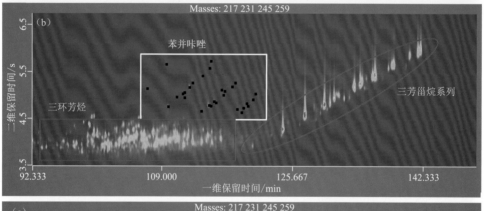
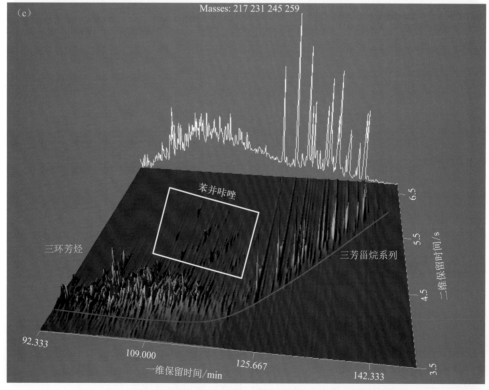

图 2-571 苯并咔唑类化合物的全二维点阵图(a,b)和三维立体图(c)

第 2 章 石油地质样品的全二维气相色谱图识别

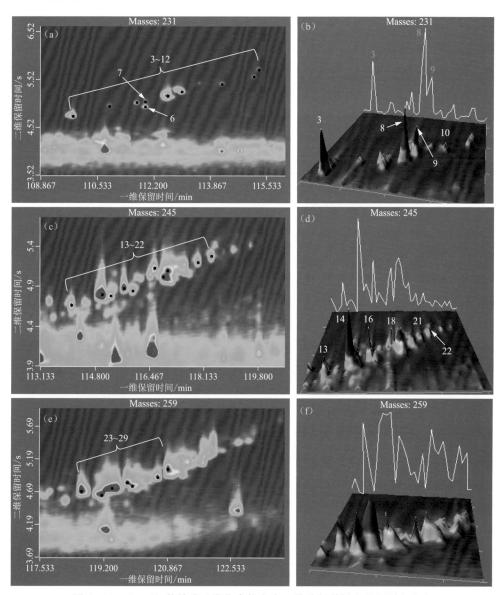

图 2-572 $C_1 \sim C_3$ 苯并咔唑类化合物在全二维分析谱图上的识别和鉴定

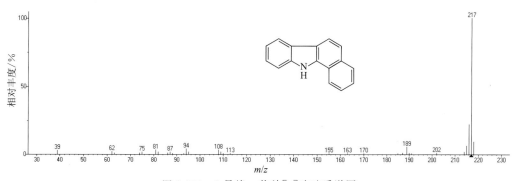

图 2-573 1 号峰 苯并[a]咔唑质谱图

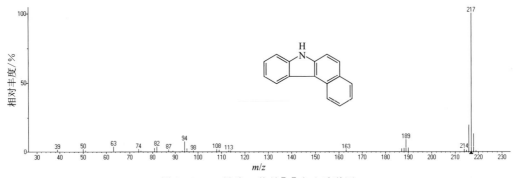

图 2-574　2 号峰　苯并[c]咔唑质谱图

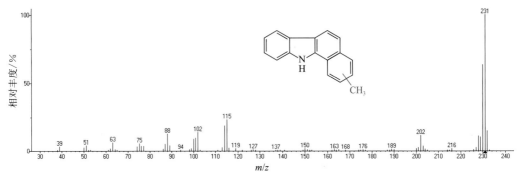

图 2-575　8 号峰　甲基苯并咔唑质谱图（其中苯并咔唑结构和取代基的位置未知）

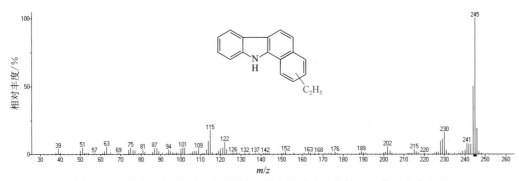

图 2-576　14 号峰　C_2-苯并咔唑质谱图（其中苯并咔唑结构和取代基的位置未知）

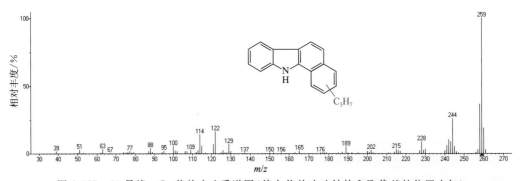

图 2-577　25 号峰　C_3-苯并咔唑质谱图（其中苯并咔唑结构和取代基的位置未知）

表 2-49 苯并咔唑类化合物鉴定结果表

峰 号	一维保留时间 /min	二维保留时间 /s	特征离子 (m/z)	分子式	相对分子质量	名 称
1	107.13	5.16	217	$C_{16}H_{11}N$	217	Benzo[a]carbazole 苯并[a]咔唑
2	109.53	5.67	217	$C_{16}H_{11}N$	217	Benzo[c]carbazole 苯并[c]咔唑
3	109.80	4.77	231	$C_{17}H_{13}N$	231	Benzo-carbazole, methyl- 甲基苯并咔唑
4	110.87	4.981	231	$C_{17}H_{13}N$	231	Benzo-carbazole, methyl- 甲基苯并咔唑
5	111.67	5.049	231	$C_{17}H_{13}N$	231	Benzo-carbazole, methyl- 甲基苯并咔唑
6	111.93	4.96	231	$C_{17}H_{13}N$	231	Benzo-carbazole, methyl- 甲基苯并咔唑
7	111.93	5.07	231	$C_{17}H_{13}N$	231	Benzo-carbazole, methyl- 甲基苯并咔唑
8	112.60	5.18	231	$C_{17}H_{13}N$	231	Benzo-carbazole, methyl- 甲基苯并咔唑
9	113.00	5.28	231	$C_{17}H_{13}N$	231	Benzo-carbazole, methyl- 甲基苯并咔唑
10	114.20	5.431	231	$C_{17}H_{13}N$	231	Benzo-carbazole, methyl- 甲基苯并咔唑
11	115.13	5.59	231	$C_{17}H_{13}N$	231	Benzo-carbazole, methyl- 甲基苯并咔唑
12	115.27	5.74	231	$C_{17}H_{13}N$	231	Benzo-carbazole, methyl- 甲基苯并咔唑
13	114.07	4.67	245	$C_{18}H_{15}N$	245	Benzo-carbazole, C_2- C_2-苯并咔唑
14	115.00	4.8	245	$C_{18}H_{15}N$	245	Benzo-carbazole, C_2- C_2-苯并咔唑
15	115.27	4.79	245	$C_{18}H_{15}N$	245	Benzo-carbazole, C_2- C_2-苯并咔唑
16	115.67	4.88	245	$C_{18}H_{15}N$	245	Benzo-carbazole, C_2- C_2-苯并咔唑
17	115.93	4.85	245	$C_{18}H_{15}N$	245	Benzo-carbazole, C_2- C_2-苯并咔唑
18	116.6	5.13	245	$C_{18}H_{15}N$	245	Benzo-carbazole, C_2- C_2-苯并咔唑
19	116.87	5.02	245	$C_{18}H_{15}N$	245	Benzo-carbazole, C_2- C_2-苯并咔唑
20	117.00	5.10	245	$C_{18}H_{15}N$	245	Benzo-carbazole, C_2- C_2-苯并咔唑

续表

峰 号	一维保留时间 /min	二维保留时间 /s	特征离子 (m/z)	分子式	相对分子质量	名 称
21	117.93	5.201	245	$C_{18}H_{15}N$	245	Benzo-carbazole，C_2-C_2-苯并咔唑
22	118.33	5.28	245	$C_{18}H_{15}N$	245	Benzo-carbazole，C_2-C_2-苯并咔唑
23	118.60	4.71	259	$C_{19}H_{17}N$	259	Benzo-carbazole，C_3-C_3-苯并咔唑
24	119.13	4.65	259	$C_{19}H_{17}N$	259	Benzo-carbazole，C_3-C_3-苯并咔唑
25	119.40	4.75	259	$C_{19}H_{17}N$	259	Benzo-carbazole，C_3-C_3-苯并咔唑
26	119.80	4.84	244	$C_{19}H_{17}N$	259	Benzo-carbazole，C_3-C_3-苯并咔唑
27	119.93	4.78	259	$C_{19}H_{17}N$	259	Benzo-carbazole，C_3-C_3-苯并咔唑
28	120.60	4.92	259	$C_{19}H_{17}N$	259	Benzo-carbazole，C_3-C_3-苯并咔唑
29	120.87	5.04	259	$C_{19}H_{17}N$	259	Benzo-carbazole，C_3-C_3-苯并咔唑

第 3 章

全二维气相色谱的石油地质应用

石油地质实验技术是油气勘探开发生产中十分重要和不可缺少的技术之一,石油地质实验是石油地质研究与油气勘探决策的重要基础性工作。每个大油田、大气田的发现及开发都与石油地质实验提供的大量基础数据和信息密切相关。从这个意义上来说,石油地质的发展离不开石油地质实验技术的进步,而石油地质实验技术的进步依赖于分析技术的创新。既然全二维气相色谱仪(GC×GC)被色谱界称为"革命性"的仪器,那么它所能解决的一定都是难题,而且利用全二维气相色谱分析地质样品带来的认识也必然是全新的。

中国石油勘探开发研究院油气地球化学重点实验室自 2007 年引进该仪器以来,连续 10 年开发仪器的功能,建立了系列新技术、新方法,解决了一些长期想解决却无法解决的难题,并取得了多项创新性成果,为石油地质研究人员提供了大量的技术服务。

本章通过汇集作者近年来在国内外发表的论文和相关专利,介绍全二维气相色谱和飞行时间质谱/氢火焰离子化检测器所建立的方法及其石油地质应用,主要内容包括以下几个方面。

3.1 ›› 石油地质样品的全二维气相色谱图识别

石油地质样品的化合物种类繁多,在全二维气相色谱服务于石油地质研究之前,采用传统的色谱图谱对全二维分析谱图中化合物的识别是一项量大且繁琐的基础性工作,需要耗费大量的时间和精力,而 GC×GC 的图谱完全不同于传统的色谱图谱,可以识别出很多过去不能认知的化合物,尤其是有效识别了很多传统色谱中的共馏峰,其具体内容已在第 2 章中介绍,附录中的论文 1 和论文 2 只是其中的部分工作。

3.2 ›› 饱和烃和芳烃同时分析方法及其全二维谱图特征

石油地质样品中的饱和烃和芳烃是开展油气研究最重要的两类组分,常规的色谱无法达到二者同时分析的要求,因此采用一维气相色谱分析时需要借助更多的前处理手段把它们分开,但长时间的样品前处理过程会造成一些组分的损失,所以建立饱和烃和芳烃同时检测的分析方法不仅减少了复杂的样品分离过程,而且使样品的信息保持得更完整。

饱和烃和芳烃同时检测的实验方法如下。

(1) 样品处理:将 20 mg 原油或者岩石氯仿可溶有机质转入到 3 g 的细硅胶柱上,先用 10 mL 二氯甲烷淋洗后,再用氮吹仪浓缩至 2 mL 的样品瓶中待检。

(2) 依照表 3-1 的实验条件分流进样进行检测。

表 3-1 饱和烃和芳烃同时分析实验条件

一维柱系统	Petro,50 m×0.2 mm×0.5 μm
二维柱系统	DB-17HT,3 m×0.1 mm×0.1 μm
一维柱子升温程序	80 ℃(0.2 min) $\xrightarrow{2\ ℃/min}$ 310 ℃
二维柱子升温程序	90 ℃(0.2 min) $\xrightarrow{2\ ℃/min}$ 330 ℃
进样口温度/℃	300
进样量/μL	1
进样模式	Split 50∶1
载气	He 流速为 1.8 mL/min
调制器温度	比一维炉温高 30 ℃
调制周期	10 s,其中 2.5 s 为热吹时间
传输线温度/℃	280
电离能量/eV	−70
检测器电压/V	1 475
采集速率/(谱图·s^{-1})	100
质量扫描范围/amu	40~520
离子源温度/℃	240
采集延迟时间/min	13

图 1-2 和图 1-3 分别是 LG7 样品饱和烃和芳烃同时分析的 GC×GC-TOFMS 二维点阵图和三维立体图。由图可以看出:全二维气相色谱-飞行时间质谱可以同时分析饱和烃和芳烃组分,各化合物的组成特征直观可视,石油地质样品的全貌一目了然,且该方法的图像进一步处理后可直接进行油源对比研究。

从图 1-2 和图 1-3 还可以看出:整个谱图被分成了不同的区域,代表不同极性的不同族的化合物。其中最下方一族是极性最弱的烷烃(包括直链和支链),在一维方向随保留时间增加碳数增加,在它的上方依次是环烷烃(包括烷基环戊烷、烷基环己烷、双环及多环烷烃)、单环芳烃(苯系列)、双环芳烃(萘、联苯等系列)、三环芳烃(二苯并噻吩、菲等系列)、四环芳烃(芘、荧蒽等系列)、五环芳烃(苯并荧蒽、苉等系列)、三环萜烷、甾烷、藿烷、三芳甾系列。

生物标志化合物一直是石油地质研究的重点。用全二维气相色谱分析,可以在一张图上同时展示饱和烃、芳烃组分中的生物标志化合物,使这些化合物的相对含量可以直观地进行比较。图 3-1 就是饱和烃和芳烃组分同时分析得到的部分生物标志化合物的全二维点阵图。图中关注的生物标志化合物按照极性依次排序,分别是三环萜烷、甾烷、四环萜烷和单芳甾、藿烷、伽马蜡烷和三芳甾烃系列化合物。

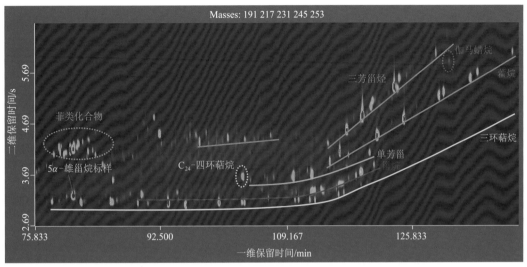

图 3-1　饱和烃和芳烃组分同时分析的生物标志化合物的全二维点阵图

在对原油样品的饱和烃和芳烃同时进行分析时,要注意除去其中的水。

3.3 » 地球化学参数校准及应用

GC-MS 是石油地质研究样品中定性、定量分析化合物的常用手段,根据 GC-MS 分析结果计算出的地球化学参数被广泛应用于地球化学领域。但由于柱容量和分离能力等的限制,一些生物标志化合物在 GC-MS 上未能得到很好的分离,含量低的时候易受共馏峰和背景噪音的干扰,不能得到正确的地球化学参数值,从而影响地球化学的研究结果。采用 GC×GC-TOFMS 和 GC-MS 分析相同的样品,并利用分析数据计算一些常用的地球化学参数,讨论两台仪器结果偏差的原因,建立地球化学参数校准因子,可以更准确地应用地球化学参数。

从计算的 23 个常用地球化学参数的对比结果上看,有 10 个参数的偏差小于 5%,说明这些地球化学参数在两台仪器上的分析结果一致。在剩下的 13 个参数中,Ts/Tm、Ts/(Ts+Tm)、乙基萘比和二甲基萘比 4 个参数由于共馏峰的影响,GC-MS 得到的结果不准确。甲基菲比、甲基菲指数、1,2,5-三甲基萘/菲 3 个参数由于分离情况不完全导致 GC-MS 结果不准确。伽马蜡烷和 C_{29}-甾烷含量低时易受其他物质和基线噪音的干扰无法检出或者积分结果不准确,特别是在经历生物降解的样品中,随着降解程度的增加,两类化合物受共馏峰的影响增大,两台仪器的偏差增大。此外,在 GC-MS 上不易得到的与金刚烷类化合物有关的地球化学参数,在 GC×GC-TOFMS 上可以得到。

具体内容见附录论文 3。

3.4 新的系列化合物的发现与结构鉴定

用 GC×GC-TOFMS 分析石油样品，一次进样能采集到所有出峰化合物的信息，通过对样品的数据处理分析，可以发现一些新的化合物。例如在分析 ND1 井凝析油样品时，发现该样品中的长侧链取代化合物中除了常规的长侧链取代环己烷、环戊烷、甲基环己烷和苯系列外，还存在一个成对出现的长侧链取代化合物（图 3-2）。通过解析这个系列化合物的质谱信息，推断该系列化合物是长侧链取代的十氢化萘，取代基数量为 $C_1 \sim C_{26}$。这个系列化合物之前从未被发现，后用合成的标样证实了推断的正确性，并区分出了不同同分异构体的分布位置。

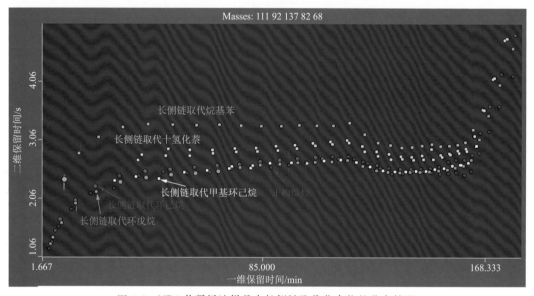

图 3-2　ND1 井凝析油样品中长侧链取代化合物的分布情况

ND1 井有丰富的金刚烷类化合物信息，用 GC×GC-TOFMS 分析，共检测出单金刚烷 25 个，双金刚烷 16 个，三金刚烷 3 个。其中单金刚烷类化合物除了常规 GC-MS 检测的 17 个化合物，还检测出 5 个 C_3-金刚烷、2 个 C_4-金刚烷和 1,2,3,5,7-五甲基金刚烷的存在。双金刚烷类化合物除了常规检测的 9 个化合物，还检测出 4 个 C_2-双金刚烷、3 个 C_3-双金刚烷的存在。除此之外，在 ND1 井中还发现一系列新的化合物，经推断这类化合物应该是 $C_3 \sim C_9$-金刚烷，长侧链取代的位置为 2 位。

ND1 井凝析油是中国目前发现的储层温度最高的油样，也是渤海湾盆地最深的工业油气井，它的发现为重新认识渤海湾盆地深层油气资源提供了重要参考。通过运用全二维气相色谱-飞行时间质谱仪等方法，发现正构烷烃含量很丰富，从 $n\text{-}C_3 \sim n\text{-}C_{38}$ 均能检测到，且长侧链取代化合物也保存完整，结合金刚烷系列化合物的分布特点，可以确定该油藏原油已发生了裂解，但是裂解程度并不高，说明液相石油在储层温度 200 ℃ 的地质条件下是可以存在的，液相石油的消失温度可能要超过传统认知。

与该部分相关的内容见附录论文 4 和论文 5。

3.5 凝析油的族组分定量分析方法

凝析油中轻烃含量高,包含丰富的地球化学信息,但由于其挥发性强,常规的石油样品分析方法无法满足凝析油样品的分析要求。棒薄层色谱无法得到凝析油样品的族组分定量信息;气相色谱-氢火焰离子化检测器(GC-FID)只能得到少数高含量化合物如正构烷烃等的定量结果,剩下大部分的化合物由于相互干扰而难以进行定量分析。采用全二维气相色谱分析凝析油样品,可以直接进样分析,避免了轻组分的挥发损失。与飞行时间质谱联用,能得到主要化合物的定性结果;与 FID 检测器联用,可以通过软件的分类功能得到样品的族组分定量结果(图 3-3)。目前运用该方法已成功得到四川盆地 22 个凝析油样品的定量结果,而且方法重复性好,7 次实验重复性小于 3%,满足复杂体系的分析要求。

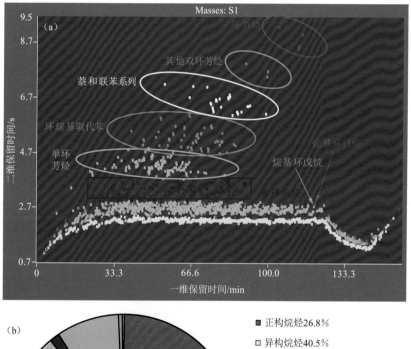

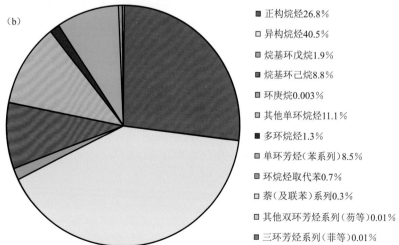

图 3-3 潼南 101# 凝析油样品的全二维谱图及其族组分定量结果

凝析油中 C_8 之前的化合物含量高、组分丰富，含有丰富的地球化学信息。用常规色谱方法定量这部分化合物，受共馏峰的影响只能识别出 54 个化合物，而用全二维气相色谱能检测出 67 个化合物，其中包含在常规色谱上的 9 对共馏峰和不易被检测到的环庚烷化合物（图 3-4）。

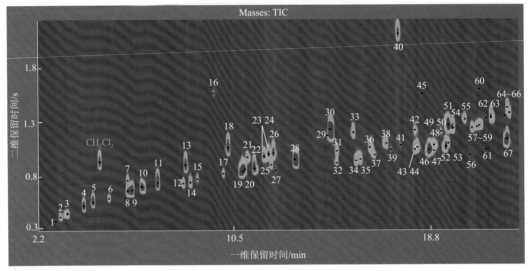

图 3-4　凝析油样品 $n\text{-}C_3 \sim n\text{-}C_8$ 色谱段化合物的全二维点阵图

用全二维气相色谱分析凝析油无需对样品进行前处理，简化了方法，避免了轻组分的损失，而且全二维气相色谱图可以清晰地展现样品的组分分布。结合氢火焰离子化检测器分析，不仅可以获得常规色谱、色谱-质谱无法得到的轻组分共馏化合物、长侧链取代环烷烃类化合物和高含量金刚烷类化合物的直接定量结果，还可以得到凝析油的族组分定量信息。全二维气相色谱对凝析油的定量分析解决了轻质油组分定量难的问题，为研究凝析油的次生蚀变、成熟度变化、沉积环境以及油源对比等提供了新的方法。一些在传统色谱分析中无法定量的化合物或族组分有可能成为今后石油地质研究的新参数。

本部分内容已经成为石油行业的标准，更详细的信息参见附录论文 6。

3.6 » 金刚烷类化合物的定量分析方法

金刚烷作为识别高成熟油裂解产物的新指标，得到国际学术界的认可和肯定，并被推广应用。国外 Wei Zhibin 等采用 GC-MS 内标法定量金刚烷类化合物，选取 6 个不同取代基的金刚烷氘代化合物作为内标，定量不同取代基的金刚烷类化合物。而国内由于内标化合物的缺乏，目前只选用 D_{16}-金刚烷一种内标物在 GC-MS 上定量所有的金刚烷类化合物，这样得到的定量结果与国外的有很大差别。用 GC×GC-FID 分析金刚烷类化合物，由于 FID 检测器对所有碳氢化合物的响应均一致，因此用一个 D_{16}-金刚烷作内标就可以定量出所有的金刚烷类化合物。

单金刚烷、C_1-金刚烷、C_2-金刚烷和双金刚烷的极性与单环芳烃接近。当样品中金刚烷类化合物含量较高时，可以不受芳烃组分的影响直接在原油中就能检测出来（如一些凝析油样品);当样品中金刚烷类化合物含量低时,芳烃组分的存在会影响它的定量结果。因此,要定量样品中低含量的金刚烷类化合物,需要先用前处理的手段去除芳烃组分,然后再用 GC×GC-FID 分析即可。用 GC×GC-FID 定量金刚烷类化合物的方法检测限低,几 mg/kg 的金刚烷类化合物就可以定量,且重复性好,7 次分析的重复性小于 5%,满足分析要求。

用国内 GC-MS 方法和 GC×GC-FID 方法分别定量 6 个样品的饱和烃组分,计算出常规 26 个金刚烷类化合物的含量。对比两台仪器的结果可以看出,在定量的 26 个金刚烷类化合物中,只有单金刚烷的定量结果 6 个样品在两台仪器上的偏差均小于 5%,其他 25 个金刚烷类化合物都差别很大,说明在 GC-MS 的选择离子模式下只用一个标样定量不同取代基的金刚烷类化合物的方法是不准确的。

用 GC×GC-FID 定量石油样品中金刚烷类化合物的方法操作简单、重复性好,值得推广。

方法可靠性论证实验等具体内容见附录专利 1。

3.7 » 生物标志化合物的定量分析方法

石油样品中的生物标志化合物是用来判定油气性质,进行油源对比,判识沉积环境等的重要指标。常用的生物标志化合物有甾烷、藿烷、三芳甾烷和标志性的芳烃等化合物。因为标样的缺乏和共馏峰的干扰,传统的分析方法如 GC 和 GC-MS 无法得到这些生物标志化合物的绝对定量结果,只能得到"半定量"的结果,数据不够精确。采用 GC×GC-FID 定量生物标志化合物,其中 GC×GC 可以消除共馏峰的干扰,再结合 FID 检测器,用 1 个标准样品就能得到所有生物标志化合物更加精确的定量数据。

在用 GC×GC-FID 分析样品的饱和烃组分时,藿烷类化合物与 GC 上共馏的甾烷、三环萜烷等化合物由于极性的差异很容易被分开,且在全二维谱图上极易分辨,定量结果较易获得。甾烷类化合物由于沸点相近,在一维色谱上不易被分开,在二维色谱上也存在同样的问题。通过对实验方法的摸索,在高温区域采用较慢的升温程序和较高的载气流速,可得到一个很好的分离效果。

芳烃化合物由于极性范围广,在 GC 上共馏的化合物在二维色谱上能被很好地分开,较易得到芳烃组分中所需要的各生物标志化合物的定量结果。

方法可靠性论证实验等具体内容见附录论文 7 和专利 2。

3.8 » 生排烃模拟实验液态烃全组分定量分析方法

化学成因法是新领域、新区块油气勘探开发中资源量的主要评价方法,在油气勘探的发展过程中起着关键的决策支持作用。热模拟生烃实验是化学成因法的主要手段之一,即利用有机质热演化的时间-温度补偿原理,在高温高压条件下短时间内进行热解生烃模拟,以再现地质过程的低温长时间有机质热演化过程。目前,热模拟生烃实验的气态产物收集和定量方法较成熟,所得气体成分组成特征与实际油气田天然气的组成特征较吻合。但模拟实验得出的液态烃组成往往与原油的组成特征差别较大,在原油中大量存在的

$C_6 \sim C_{13}$ 化合物在模拟实验产物分析的气相色谱图中几乎看不到,也就是说,有较大一部分的烃类组分组成由于分析方法的原因没有被认知,其含量长期以来也未能计入生烃总量,这会对资源量的准确评估带来严重影响。利用全二维气相色谱解决了生排烃过程中全组分定量的难题,使化学成因法在技术上的缺陷得以弥补。

生排烃热模拟实验产物的全组分定量难点在于对液态烃中轻组分的分析。较强的挥发性使得轻组分在除水和恒重过程中损失殆尽,而传统的气相色谱由于分离度的原因,大量的化合物干扰标样,同样无法得到轻烃的定量数据。利用全二维气相色谱的分离能力,通过溶剂及轻烃挥发前后的两次定量分析可以得到轻烃的定量结果,结合其他的常规分析方法可以得到模拟产物的全组分定量分析数据。分析流程如图 3-5 所示,结果表明下花园地区的烃源岩轻烃总量占所生成液态烃的 40% 以上。

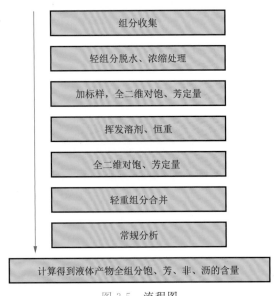

图 3-5 流程图

详细内容见附录专利 3。

3.9 » 稠油中"不可识别未知混合物"的成分解析

稠油因赋存状态、流动性差、被隔离等因素导致其直接开采效率极低,被称为储层中的难动用资源。原生稠油所占稠油的比例十分有限,绝大部分稠油为次生稠油,即由正常原油经历热液蚀变或遭受生物降解形成的。这类原油在世界各地十分普遍,如在我国的塔里木、渤海湾、准噶尔、松辽等盆地生物降解油超过原油产量的 20%。受分析条件的限制,长期以来人们对稠油成分中化合物的认识水平远不及人们对正常原油的认识。常规的气相色谱-质谱是分析石油样品的有效手段,但用它分析稠油时,由于峰容量和分辨率的限制,导致色谱基线抬高而形成一个大鼓包,被称为"基线鼓包"。由于长期以来鼓包中的化合物无法识别,因此被称为"不可分辨的复杂混合物"(unresolved complex mixture, UCM)。剖析"基线鼓包"中化合物的分子组成一直是分析工作者的追求。国外有学者尝试用化学方法如分子筛、尿素络合和四氧化钌氧化法来处理 UCM,再结合 GC-MS 分析化合物组成,但这些方法所分析的 UCM 组成还不到 10%。还有些学者采用多种仪器相结合的方法进行分析,如 ^{1}H 和 ^{13}C FT-NMR、FT-IR、UV、TLC、EI-MS、CI-MS、GC-MSMS 等技术都被采用过,但得到的也是部分化合物的信息。对于饱和烃组分,浓度高的化合物会在"基线鼓包"上面出峰,但有时一个峰可能是几个化合物的共馏峰;浓度低的化合物会被抬起的"基线鼓包"掩盖而无法被检出。一般"基线鼓包"中不可分辨的化合物含量占饱和烃总量的 30%~80% 以上,因此,有效解析"基线鼓包"中的化合物分子组成不仅是次生蚀变原油研究中的基础科学问题,而且对稠油成因机理的认识、开发方案的进步、石油炼制方案的制订等有重要指导意义。

利用 GC×GC-TOFMS 对辽河油田严重生物降解原油进行分析,并通过大量繁杂的质谱解析发现饱和烃"基线鼓包"是由大量的含量极低、性质接近的环状化合物组成的,芳烃"基线鼓包"主要是带由支链烷烃和环烷烃取代基的单环、双环、三环、四环、五环和芳构化的生物标志化合物,这些化合物按照取代基碳数的增加依次在第一维色谱上规律排列,同时按照芳环个数和芳环上取代的环烷烃个数增加依次在第二维色谱上规律排列。"基线鼓包"中化合物的总量超过了饱和烃总量的 70%,超过芳烃总量的 90%。

详细内容见附录论文 8 和论文 9。

3.10 » 石油地质应用

全二维气相色谱具有强大的柱容量和分离能力,分析石油地质样品时可以得到更多的化合物信息,结合全二维气相色谱的准确定量结果,石油地质工作者更容易发现新的规律和认识,促进油气勘探、开发的发展。

原油的稳定性是目前国际上研究的一个热点。迄今为止,纯粹油的裂解生气模拟实验做得比较多,主要认为原油在温度高于 160 ℃ 时或埋深超过 6 000 m 时开始裂解成气,随后液相石油逐渐消失。而随着向深层油气勘探的进展,一些油藏在远大于此温度或此深度条件下依然以油相大量存在,说明原油的稳定性可能比预期的要高很多。地质条件下深层高温储集层中的液相石油随着勘探的深入愈来愈多地被发现,一些超深层油气井的发现也将引领各油田向深层开展大规模的油气勘探工作。全二维气相色谱可以直接分析超深层油气藏中的凝析油样品,得到的金刚烷类化合物的定性、定量结果可用来衡量超深层油藏的原油裂解程度,其他单体化合物和族组分的定性、定量结果可用于油气来源与成因研究,试图为认识流体成因与相态提供依据,并为深层油气勘探提供参考。

中国渤海湾盆地冀中坳陷 ND1 井在埋深 5 641~6 027 m、储层温度 190~201 ℃ 的雾迷山组潜山储集层中获得了高产油气流,主要以凝析油为主。运用 GC×GC-TOFMS 和 GC×GC-FID 对该凝析油进行分析,在原油中发现了保存较为完整的长链化合物和金刚烷系列化合物,这说明原油的热稳定性能比传统认识要高很多,反映出渤海湾盆地深层油气的勘探潜力较大。

最近在塔里木盆地库车坳陷西部博孜地区埋深超过 7 000 m 的白垩系砂岩储集层中获得高产工业油气流,这是我国目前发现的最深陆相来源的工业油流;但是该区地温梯度低,实测的储层温度为 130.6 ℃,这无疑为研究深层原油的稳定性提供了得天独厚的条件。运用 GC×GC-TOFMS 对原油化合物进行分析鉴定,运用 GC×GC-FID 进行原油族组分定量，数据表明该凝析油已处于高成熟裂解过程中。

而在塔里木盆地塔北南斜坡富源区块完钻的井深 771 m 的 FY1 井中，运用 GC×GC-TOFMS 对原有的化合物进行深入分析，发现金刚烷类化合物种类少且含量很低，未发现双金刚烷等代表高温高度裂解的化合物，说明该油为正常成熟度原油，因此可预测 7 700 m 以下深层还存在液态石油的勘探潜力，为引领塔里木盆地向深层开展大规模的石油勘探工作提供了理论依据。

此外，全二维气相色谱在检测原油中的含硫化合物如四氢噻吩、噻吩、苯并噻吩、二苯并噻吩、菲并噻吩、苯并萘并噻吩等时优势明显，其中检测出的硫代金刚烷类化合物被认为是烃类与硫酸盐发生 TSR 反应的标志性产物。含硫化合物的检测可以识别深层油气藏是否发生了 TSR 以及蚀变强度，并进一步预测硫化氢的分布，也可为钻探预警，以防范因硫化氢而导致的各种安全生产事故。

目前已在塔里木盆地 ZS1C 井的 6 861~6 944 m 获日产气 158 545 m³，这是塔里木盆地盐下白云岩层首次获得工业油气流。运用 GC×GC 等分析手段，在 ZS1C 凝析油中检测到十分丰富的金刚烷类化合物、硫代单金刚烷和硫代双金刚烷类化合物，以及大量的含硫化合物，并开展了硫同位素、碳同位素等测试，证实该区发生过强烈的 TSR 作用，凝析油可能是深层油藏发生 TSR 作用后的残余物。

更详细的石油地质应用见附录论文 5、论文 10~论文 13。

附 录

论文 1 2010 年发表于《质谱学报》

全二维气相色谱-飞行时间质谱对饱和烃分析的图谱识别及特征

王汇彤,翁 娜,张水昌,陈建平,魏彩云

摘 要 采用全二维气相色谱-飞行时间质谱分析方法对典型石油样品中饱和烃组分进行了定性分析,依据极性大小和环数多少的分布特征,解析了点阵图谱中的烷烃、单环烷烃、双环烷烃、单金刚烷和双金刚烷系列、三环萜烷类、甾烷类和藿烷类等生物标志化合物的识别;讨论了典型生物标志化合物单金刚烷、三环萜烷和藿烷类的全二维点阵谱图特征;检测到过去 GC/MS 分析中常被忽视的 $C_{31} \sim C_{35}$ 三环萜烷,为石油地质实验和研究提供了参考依据。全二维气相色谱-飞行时间质谱相比于 GC/MS 灵敏度更高、峰容量更大,适合复杂混合物体系的分析,对石油样品的分析有很好的应用前景。

关键词 全二维气相色谱 飞行时间质谱 饱和烃 三环萜烷 石油地质

气相色谱作为混合物的分离工具,已在石油样品分析领域中得到了广泛应用。但普通的一维色谱仪有峰容量不够、分辨率低等特点,在分析复杂的石油样品时组分峰重叠严重,使得定性定量结果不准确。增加柱长可提高分离度,但也增加了分析时间,因此通过增加柱长来提高分辨率的方法是十分有限的。为了解决这个问题,国内外学者做了大量的研究,一些新的技术也被应用到这个领域。Gough 等利用分子筛的方法,Pool 等利用去卷积的方法,Blomberg 等利用多维色谱的方法分析原油样品,这些方法在一定程度上得到了更完整的化合物信息。

全二维气相色谱(GC×GC)是 20 世纪 90 年代发展起来的分离复杂混合物的一种全新手段。它是把分离机理不同而又相互独立的两支色谱柱,通过一个调制器以串联方式连接成二维气相色谱柱系统。与通常的一维气相色谱相比,全二维气相色谱具有分辨率高、峰容量大、灵敏度好、分析速度快以及定性更有规律可循等特点。由于调制器的捕集、聚焦、再分配作用,单个化合物被分割成若干个碎片峰通过检测器检测。数据处理时需要把这些峰碎片重新组合成一个峰,最理想的方法是借助于质谱对碎片的识别。数据采集系统会采集到每一个碎片的质谱信息,通过软件的比对把谱图相似的碎片峰合在一起定性定量。

由于采集速度和容量的限制,传统的四极杆质谱不能满足 GC×GC 的分析要求。飞行时间质谱(TOFMS)是 20 世纪 90 年代以来应用最广的质谱分析技术之一。它是利用动能相同而质-荷比不同的离子在恒定电场中运动,经过恒定距离所需时间不同的原理对物质成分或结构进行测定的一种分析方法。飞行时间质谱分析技术的优点在于理论上对测定对象没有质量范围的限制,具有极快的响应速度以及较高的灵敏度。它采集频率可调,每秒能产生

1~500个谱的谱图,能够精确处理GC×GC得到的碎片峰,并得到较为精准的质谱信息。全二维气相色谱和飞行时间质谱的联用更加适合复杂体系的定性定量。

国外已将GC×GC,GC×GC-TOFMS应用于石油炼制样品中的轻质和中等馏分分析。对于石油地质样品,利用GC×GC分离和保留时间定性石油中的部分生物标志化合物,利用GC×GC-TOFMS分析石油中的基线鼓包(unresolved complex mixture,UCM)的组成已有报道。

国内近年来利用GC×GC技术进行分析测试成果显著。在烟草、酿酒、食品、炼油产品和环境监测上已有报道,但未见将GC×GC-TOFMS技术应用在石油地质研究上。本工作的意义是在建立原油样品分析方法的基础上,利用GC×GC-TOFMS技术分析原油样品的饱和烃组分,通过与饱和烃标样的质谱图谱对比和其全二维谱图特征解析,对饱和烃的全二维气相色谱图特征和各化合物的分布进行描述,进一步探讨全二维气相色谱-飞行时间质谱在石油地质试验中的优越性。

1 实验部分

1.1 仪器与试剂

全二维气相色谱-飞行时间质谱仪(GC×GC-TOFMS):美国LECO公司产品,GC×GC系统由配有氢火焰离子化检测器(FID)的Agilent7890气相色谱仪和双喷口热调制器组成;飞行时间质谱仪:美国LECO公司的Pegasus Ⅳ,系为Chroma TOF软件;氮吹仪:美国进口。

正己烷(重蒸,色谱纯):购自北京化工厂;细硅胶(100~200目,200 ℃下活化4 h)。

1.2 样品

原油样品:取自塔里木盆地轮古和轮南地区的10个油样,井号和深度列于表1。

表1 样品信息

井号	深度/m	井号	深度/m
LG35	6 155~6 165	LG40	5 339.5~5 346
LG35	6 198~6 460	LN14	5 256~5 266
LG4	5 270~5 295.48	LN14	5 274.15~5 363
LG18	5 462.35~5 546.8	LN631	5 885.02~5 990
LG7	5 165~5 175	LN8	5 167.23~5 230

饱和烃的制备:① 取60 mg左右原油样品,用适量正己烷(重蒸过的分析纯)溶解;② 取6 g细硅胶转入玻璃柱中,震荡压实,少许正己烷淋洗后,将原油样品溶液分几次全部转入玻璃柱中;③ 当柱子下端溶液流出时,分批加入10 mL正己烷,收集饱和烃馏分,用氮吹仪吹至约1 mL,转至进样瓶中。

1.3 GC×GC-TOFMS实验条件

1.3.1 GC×GC条件

本实验的色谱柱均为美国J & W Scientific公司产品,具体条件列于表2。

表 2　GC×GC 实验条件

一维柱系统	Petro,50 m ×0.2 mm ×0.5 μm
二维柱系统	DB-17HT ,3 m ×0.1 mm ×0.1 μm
一维柱子升温程序	80 ℃(0.2 min) $\xrightarrow{2\ ℃/min}$ 310 ℃(25 min)
二维柱子升温程序	100 ℃(0.2 min) $\xrightarrow{2\ ℃/min}$ 330 ℃(25 min)
进样口温度	300 ℃
进样量	1 μL
进样模式	Split 40∶1
载　气	He 流速1.8 mL/min
调制器温度	比一维炉温高 30 ℃
调制周期	10 s,其中 2.5 s 热吹时间
传输线温度	280 ℃

1.3.2　TOFMS 条件

TOFMS 实验条件列于表 3。

表 3　TOFMS 实验条件

电离能量/eV	−70
检测器电压/V	1 475
采集速率/(谱图·s^{-1})	100
质量扫描范围/amu	40～520
离子源温度/℃	240
采集延迟时间/min	11

1.4　标样

将 5α-雄甾烷、$C_{24}D_{50}$、D_{16}-单金刚烷作为标样加到样品的饱和烃组分中,以便于样品中化合物的定性。

2　实验方法

2.1　气相色谱和质谱方法

与一维气相色谱方法的建立类似,二维气相色谱方法的建立需要考虑的参数主要包括柱系统的选择、升温程序、调制周期、热吹时间及载气流速等。

由于石油地质样品中生物标志化合物含量较高,正构烷烃含量较大,因此一维柱选用柱膜较厚的非极性色谱柱 Petro(50 m×0.2 mm×0.5 μm),二维柱选用较长的中等极性的 DB-17HT(3 m×0.1 mm×0.1 μm)。

用全二维气相色谱-飞行时间质谱分析石油类样品,甾烷、藿烷类物质二维出峰时间较晚,易与二维柱子的柱流失重合。为了避免这种现象,本工作选用较大的调制周期 10 s(一般为 6～8 s)以满足分析要求。热吹时间对高沸点化合物的分离很重要,过小的热吹时间会使后面的藿烷等化合物拖尾,因此采用 2.5 s 热吹时间。为了保证 R 和 S 型生物标志化合物的良好分离并缩短分析时间,本实验采用较大的载气流速 1.8 mL/min。

从二维柱子流出的峰都很窄,一般为 100 ms,要保证每个峰至少有 7～10 张谱图的采集才能被很好的定性,因此设定质谱的采集速率为 100 谱图/s。

2.2 数据处理方法

本实验所用的是 Chroma TOF 数据处理软件,NIST 05 谱库。被测物从第一维柱子出来,经过调制器的捕集、聚焦,再传送进入第二维柱子,这个过程通常一个组要被分割成 3～4 个碎片峰进入检测器鉴定,每一个碎片峰 TOFMS 都会给出定性信息。数据处理方法中参数的设定就是让软件在一定的一维峰宽(一般设为调制周期的 4 倍)、二维峰宽(一般为 0.1 s)和信噪比内,通过质谱图的比对,把相似度在 700 以上峰的碎片合并起来计算峰面积。软件会自动完成计算基线、峰查找、谱库比对和峰面积积分等工作。由于 NIST 05 谱库中几乎没有生物标志化合物的图谱,因此,各个化合物的定性和识别需要通过保留时间和一些石油地质专业图谱集的谱图比对完成。

3 结果与讨论

3.1 饱和烃组分的全二维点阵谱图特征

在 GC×GC 中,第一维和第二维色谱柱分别是非极性柱和极性或中等极性柱,可使化合物在一维上按照沸点、二维上按照极性分离。由于两根柱子固定相极性的改变,以及线性程序升温方法的共同作用,使得两维间交叉信息减少,实现真正的正交分离。由于化合物的极性不同,GC×GC 谱图被明显分割成不同的区带,每一个区带代表一族物质,这就是 GC×GC 的族分离特性,饱和烃组分的全二维点阵谱图具有族分离特性。

LG7 样品饱和烃组分的全二维点阵图示于图 1。从图中可以看出:整个谱图被分成了不同的区域,代表不同极性的不同族化合物。最下方一族是极性最弱的烷烃(包括直链和支链),在一维方向随保留时间增加碳数增加,姥姣烷、植烷在 C_{17},C_{18} 处特征明显,该组各化合物易于识别,大多数可被软件直接鉴定。在它的上方依次是烷基环戊烷,烷基环己

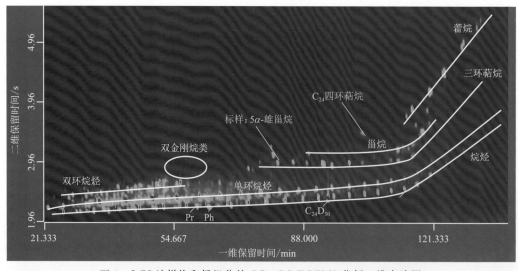

图 1　LG7 油样饱和烃组分的 GC×GC-TOFMS 分析二维点阵图

烷,双环烷烃(十氢化萘、单金刚烷系列),三环萜烷类,甾烷类和藿烷类。如表 4 所示,每一族化合物都有不同的结构特征和特征离子,这些化合物很难由软件自动准确识别或者无法识别,需要与相关出版物的标准图谱比对方可定性识别。

表 4　饱和烃组分中不同族化合物的结构特征和特征质量数

特征离子 (m/z)	名　称	基本结构	特征离子 (m/z)	名　称	基本结构
57,71,85,99,…	正构烷烃		187,188,201,215	双金刚烷类	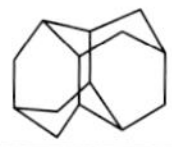
57,M-15,M-43	异构烷烃	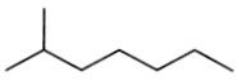	191	三环萜烷类	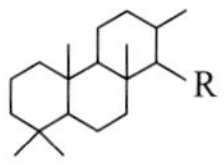
83,82	长侧链取代环己烷	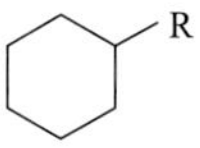	191	四环萜烷	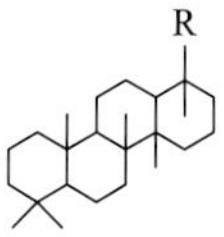
138,152,166	十氢化萘系列	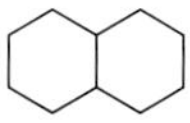	191	藿烷类	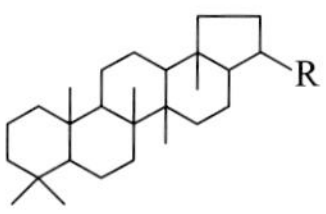
123	二环倍半萜烷	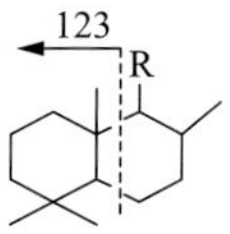	191	伽马蜡烷	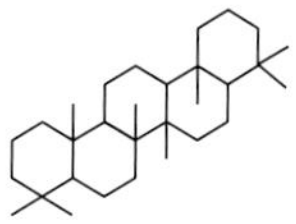
135,136,149,163,177	单金刚烷类	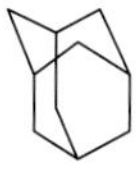	217	甾烷类	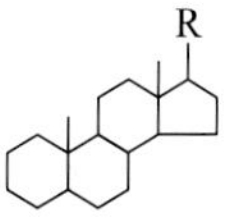

用 GC×GC 分析样品得到的化合物谱图除了有族分离的特性,还有"瓦片效应",即同族带上相同碳数取代基的化合物呈线性排列,不同碳数取代基的同族化合物呈瓦片状排列。双环烷烃系列在其选择离子下的全二维点阵图示于图 2。从图 2 可以清楚地看出:十氢化萘及其不同碳数取代基的物质在全二维轮廓图上呈瓦片状排列。

3.2　典型生物标志化合物的谱图特征

3.2.1　金刚烷类化合物的谱图特征

近年来,国内外很多学者利用金刚烷研究原油的成熟度、裂解程度、生物降解程度以及油源对比等。采用 GC/MS 方法检测原油中的金刚烷化合物也早有报道,Wei 等利用 D_3-1-单金刚烷、D_4-单金刚烷、D_3-1-甲基双金刚烷、D_5-2-乙基双金刚烷、D_4-双金刚烷、D_4-三金刚烷 6 个标样定性原油中的金刚烷类化合物。国内由于缺乏金刚烷类化合物的标样,尤其在分析较低含量的金刚烷样品时,定性较困难。全二维气相色谱由于其高灵敏度可以检测到信噪比较高的该类化合物,根据 GC×GC 的族分离和"瓦片效应"推测出峰位置,结合 TOFMS 得到的质谱图可对金刚烷类化合物定性。

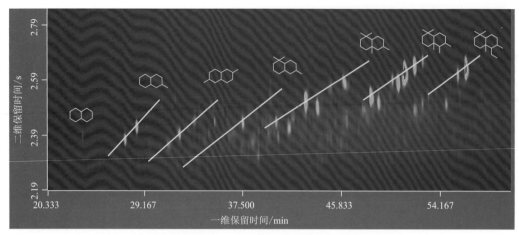

图 2　LG7 油样中十氢化萘类化合物的全二维点阵图

样品中检测到的金刚烷类物质全二维点阵图示于图 3。各个化合物的定性结果列于表 5。可以看出：无论是何种形式的金刚烷化合物都在一张图谱上显示，根据其点的亮度和颜色可以直观地反映不同金刚烷的分布和组成关系。

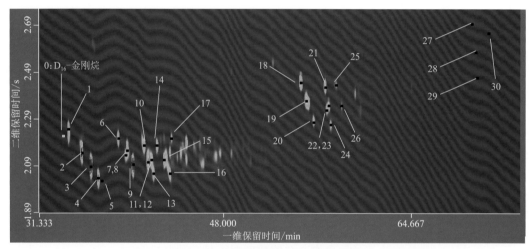

图 3　金刚烷类化合物全二维点阵图

表 5　金刚烷类化合物鉴定表

峰号	名　称	一维保留时间/min	二维保留时间/s	$M^+(m/z)$	基峰(m/z)
0	D_{16}-单金刚烷	33.500	2.22	152	152
1	金刚烷	34.000	2.25	136	$136(M^+)$
2	1-甲基金刚烷	35.167	2.15	150	$135(M-CH_3)$
3	1,3-二甲基金刚烷	36.000	2.09	164	$149(M-CH_3)$
4	1,3,5-三甲基金刚烷	36.667	2.04	178	$163(M-CH_3)$
5	1,3,5,7-四甲基金刚烷	37.000	2.03	192	$177(M-CH_3)$
6	2-甲基金刚烷	38.500	2.21	150	$135(M-CH_3)$
7	顺式-1,4-二甲基金刚烷	39.167	2.14	164	$149(M-CH_3)$

续表

峰号	名 称	一维保留时间/min	二维保留时间/s	$M^+(m/z)$	基峰(m/z)
8	反式-1,4-二甲基金刚烷	39.333	2.16	164	149(M-CH_3)
9	1,3,6-三甲基金刚烷	39.833	2.10	178	163(M-CH_3)
10	1,2-二甲基金刚烷	40.833	2.18	164	149(M-CH_3)
11	顺式-1,3,4-三甲基金刚烷	41.167	2.11	178	163(M-CH_3)
12	反式-1,3,4-三甲基金刚烷	41.500	2.12	178	163(M-CH_3)
13	1,2,5,7-四甲基金刚烷	41.667	2.06	192	177(M-CH_3)
14	1-乙基金刚烷	42.000	2.18	164	135(M-C_2H_5)
15	1-乙基 3-甲基金刚烷	42.667	2.12	178	149(M-C_2H_5)
16	1-乙基-3,5-二甲基金刚烷	43.167	2.06	192	163(M-C_2H_5)
17	2-乙基金刚烷	43.333	2.21	164	135(M-C_2H_5)
18	双金刚烷	54.833	2.45	188	188(M^+)
19	4-甲基双金刚烷	55.333	2.37	202	187(M-CH_3)
20	4,9-二甲基双金刚烷	56.000	2.28	216	201(M-CH_3)
21	1-甲基双金刚烷	57.000	2.43	202	187(M-CH_3)
22	1,2-+2,4-二甲基双金刚烷	57.167	2.33	216	201(M-CH_3)
23	4,8-二甲基双金刚烷	57.333	2.35	216	201(M-CH_3)
24	三甲基双金刚烷	57.500	2.27	230	215(M-CH_3)
25	3-甲基双金刚烷	58.000	2.44	202	187(M-CH_3)
26	3,4-二甲基双金刚烷	58.500	2.35	216	201(M-CH_3)
27	三金刚烷	70.000	2.699	240	240(M^+)
28	9-甲基三金刚烷	70.333	2.58	254	239(M-CH_3)
29	二甲基三金刚烷	70.500	2.47	268	253(M-CH_3)
30	5-甲基三金刚烷	71.500	2.66	254	239(M-CH_3)

3.2.2 三环萜烷、藿烷类化合物的谱图特征

三环萜烷、藿烷类化合物参数可以用来进行油源对比、母源和沉积环境的研究,是石油地质样品中较为重要的生物标志化合物。采用 GC/MS 技术对其分析基本能满足要求,但一些"共馏"化合物会造成定量偏差,一些新的化合物往往会被掩盖。

LG7 样品饱和烃组分在 m/z 191 下的全二维点阵图示于图 4。从图 4 可以看出:随着环数的增加,三环萜烷、四环萜烷、藿烷、伽马蜡烷在二维上的保留时间增加,特征离子 m/z 191 的化合物被分成了 4 个族。在 GC/MS 上有共馏的化合物在全二维上被彻底分开,各化合物的定性结果列于表 6。在三环萜烷一族中可以发现:在 C_{29} 三环萜烷峰的后面,有一些规律的、成对出峰的物质,它们都有较强的离子碎片 m/z 191。根据它们在保留时间上与藿烷的关系,结合 TOFMS 得到的质谱图比对,可以判断这些物质是成对出现的 $C_{30} \sim C_{35}$ 三环萜烷,其质谱图谱示于图 5。C_{30} 以后出现的三环萜烷在一维色谱上往往会被同时出峰的藿烷类物质所掩盖而无法检测到,同时也影响藿烷相关参数的准确应用。

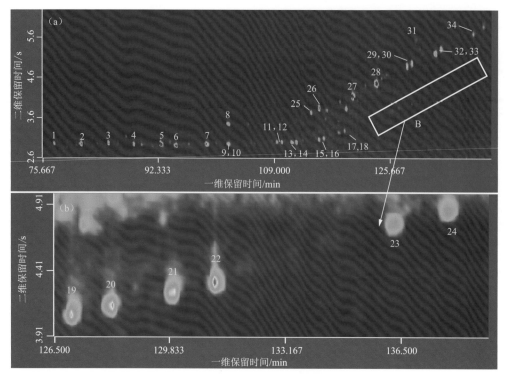

图 4　m/z 191 的全二维点阵图

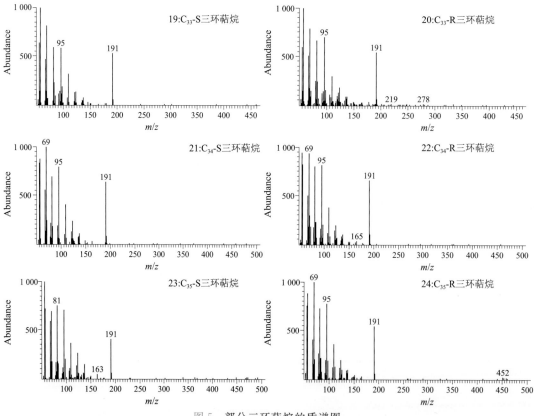

图 5　部分三环萜烷的质谱图

表6 金刚烷类化合物鉴定表

峰号	名称	一维保留时间 /min	二维保留时间 /s	M^+ (m/z)	基峰 (m/z)	分子式
1	C_{19}三环萜烷	78.000	3.17	262	191	$C_{19}H_{34}$
2	C_{20}三环萜烷	82.000	3.14	276	191	$C_{20}H_{36}$
3	C_{21}三环萜烷	85.833	3.17	290	191	$C_{21}H_{38}$
4	C_{22}三环萜烷	89.500	3.14	304	191	$C_{22}H_{40}$
5	C_{23}三环萜烷	93.500	3.14	318	191	$C_{23}H_{42}$
6	C_{24}三环萜烷	95.833	3.10	332	191	$C_{24}H_{44}$
7	C_{25}三环萜烷	100.167	3.15	346	191	$C_{25}H_{46}$
8	C_{24}四环萜烷	103.333	3.65	330	191	$C_{24}H_{42}$
9	C_{26}三环萜烷	103.167	3.15	360	191	$C_{26}H_{48}$
10	C_{26}三环萜烷	103.500	3.14	360	191	$C_{26}H_{48}$
11	C_{28}三环萜烷	110.167	3.20	388	191	$C_{28}H_{52}$
12	C_{28}三环萜烷	110.667	3.20	388	191	$C_{28}H_{52}$
13	C_{29}三环萜烷	112.167	3.19	402	191	$C_{29}H_{54}$
14	C_{29}三环萜烷	112.833	3.19	402	191	$C_{29}H_{54}$
15	C_{30}三环萜烷	116.000	3.30	416	191	$C_{30}H_{56}$
16	C_{30}三环萜烷	116.667	3.34	416	191	$C_{30}H_{56}$
17	C_{31}三环萜烷	119.000	3.49	430	191	$C_{31}H_{58}$
18	C_{31}三环萜烷	119.833	3.54	430	191	$C_{31}H_{58}$
19	C_{33}三环萜烷	127.000	4.07	458	191	$C_{33}H_{62}$
20	C_{33}三环萜烷	128.167	4.14	458	191	$C_{33}H_{62}$
21	C_{34}三环萜烷	131.167	4.32	472	191	$C_{34}H_{64}$
22	C_{34}三环萜烷	130.000	4.24	472	191	$C_{34}H_{64}$
23	C_{35}三环萜烷	136.167	4.75	486	191	$C_{35}H_{66}$
24	C_{35}三环萜烷	137.833	4.85	486	191	$C_{35}H_{66}$
25	$18\alpha(H)$-22,29,30-三降藿烷(Ts)	114.833	3.95	370	191	$C_{27}H_{46}$
26	$17\alpha(H)$-22,29,30-三降藿烷(Tm)	116.167	4.13	370	191	$C_{27}H_{46}$
27	$17\alpha(H),21\beta(H)$-30-降藿烷	121.000	4.45	398	191	$C_{29}H_{50}$
28	$17\alpha(H),21\beta(H)$-藿烷	124.667	4.78	412	191	$C_{30}H_{52}$
29	$17\alpha(H),21\beta(H)$-29-升藿烷(22S)	129.333	5.24	426	191	$C_{31}H_{54}$
30	$17\alpha(H),21\beta(H)$-29-升藿烷(22R)	129.833	5.31	426	191	$C_{31}H_{54}$
31	伽马蜡烷	130.667	5.92	412	191	$C_{30}H_{52}$
32	$17\alpha(H),21\beta(H)$-29-二升藿烷(22S)	133.500	5.58	440	191	$C_{32}H_{56}$
33	$17\alpha(H),21\beta(H)$-29-二升藿烷(22R)	134.500	5.70	440	191	$C_{32}H_{56}$
34	$17\alpha(H),21\beta(H)$-29-三升藿烷(22S)	139.167	6.10	454	191	$C_{33}H_{58}$

4 结 论

全二维气相色谱-飞行时间质谱相比于 GC/MS 灵敏度更高、峰容量更大,非常适合复杂混合物体系的分析。石油地质实验工作者正是基于此,开始应用该技术于石油地质样品的分析。相信随着该项技术的初步完善和方法的普及,会有更多、更准确的地球化学参数应用到油气勘探中,同时也会有更多的未知化合物被地球化学家认知。全二维气相色谱-飞行时间质谱对石油样品的分析有很好的应用前景。

论文 2 2010 年发表于《科学通报》

全二维气相色谱-飞行时间质谱对原油芳烃分析的图谱识别

王汇彤,翁 娜,张水昌,朱光有,陈建平,魏彩云

摘 要 全二维气相色谱-飞行时间质谱是 20 世纪 90 年代商品化的仪器,具有峰容量大、分辨率及灵敏度高等特点,但利用该仪器解决石油地质方面的问题国内外才刚刚起步。通过对柱系统的选择以及对升温程序、调制周期、热吹时间、载气流速、采集频率、数据处理等实验参数的优化,建立了全二维气相色谱-飞行时间质谱对石油样品中芳烃组分的分析方法。芳烃组分中极性不同的化合物在二维谱图上呈区带分布,每一区带中同系物根据取代基个数的不同呈瓦片状排列,根据其二维图谱特征和各化合物的质谱谱图,对各化合物进行了定性鉴定,化合物依据极性大小和环数多少在同一谱图上的不同空间展布,直观反映芳烃中各化合物的分布特征。一些在普通色谱-质谱图谱上重叠的化合物峰在二维谱图上被完全分开,有利于化合物的定量分析。一些含量很低的杂环原子芳香族化合物由于其极性的差异可以与干扰的芳烃化合物分开,利用该方法可以很清晰地被识别和鉴定。该方法的建立和图谱识别为石油地质实验工作者提供了利用全二维气相色谱-飞行时间质谱开展工作的参考依据。

关键词 全二维气相色谱 飞行时间质谱 芳烃 杂环化合物 石油地质实验

芳烃是岩石和原油中重要的组成部分之一,它们包含数百种化合物,具有丰富的地球化学信息。一些芳烃化合物与菲的峰强度比值可以用来反映陆生植物的相关信息,芳环稠合噻吩化合物与盐湖相地层有关,短链烷基二苯并噻吩与菲的比值是很好的盐湖相地层指示物等。目前在石油地质实验中主要是应用一维气相色谱分析原油中的芳烃组分。芳烃化合物种类繁多,有相同芳环数的化合物挥发性差别不大,一维色谱仪由于峰容量和分辨率等较低,有时无法满足分析要求,易造成组分峰重叠严重,使得定性定量结果不准确。增加柱长可提高分离度,但相应的分析时间也会增加,因此通过提高柱长来提高分辨率的方法是十分有限的。利用分子筛的方法、去卷积的方法和多维色谱的方法在原油样品分析方面取得了一定进展,这些方法在一定程度上得到了较完整的石油样品化合物信息。

全二维气相色谱(GC×GC)是 20 世纪 90 年代发展起来的分离复杂混合物的一种全新手段。它是由分离机理不同而又相互独立的两支色谱柱通过一个调制器以串联方式连接起来的二维气相色谱柱系统。与通常的一维气相色谱相比,全二维气相色谱具有分辨率高、灵敏度好、峰容量大、分析速度快以及定性更有规律可循等特点。由于调制器的捕集、聚焦、再分配作用,单个化合物被分割成若干个碎片峰通过检测器检测。数据处理时需要把这些峰碎片重新组合起来成为一个峰,最理想的方法是借助于质谱对碎片的识别。数据采集系统会采集到每一个碎片的质谱信息,通过软件的比对把谱图相似的碎片峰合在一起定性定量。

由于采集速度和容量的限制,传统的四极杆质谱不能满足 GC×GC 的分析要求。飞行时间质谱(TOFMS)是 20 世纪 90 年代以来应用最广的质谱分析技术之一。它是利用动能相

同而质-荷比不同的离子在恒定电场中运动,经过恒定距离所需的时间不同的原理对物质成分或结构进行测定的一种分析方法。飞行时间质谱分析技术的优点在于理论上对测定对象没有质量范围限制、极快的响应速度以及较高的灵敏度。它采用高采集频率,每秒能产生1~500个谱的谱图,能够精确处理GC×GC得到的碎片峰,并得到较为精准的质谱信息。全二维气相色谱和飞行时间质谱的联用更加适合复杂体系的定性定量。

国外学者已将GC×GC,GC×GC-TOFMS应用到石油炼制样品中的轻质和中等馏分分析领域。在石油地质样品上,利用GC×GC分析石油中的部分生物标志化合物,利用GC×GC-TOFMS分析石油中的基线鼓包UCMs(unresolved complex mixtures)的组成已有报道。

国内近年来在烟草、酿酒、食品、炼油产品和环境监测等领域利用GC×GC技术进行分析测试成果显著,但在石油地质研究上GC×GC-TOFMS技术的应用目前仍是空白。本文是在建立原油样品分析方法的基础上,利用GC×GC-TOFMS技术分析原油样品的芳烃组分,通过与标准质谱图谱对比和其全二维谱图特征解析,对芳烃的全二维气相色谱图特征和各化合物的分布进行了描述,探讨了全二维气相色谱-飞行时间质谱在石油地质样品分析中的应用。

1 实验部分

(1) 仪器与试剂。全二维气相色谱-飞行时间质谱仪(GC×GC-TOFMS,美国LECO公司),GC×GC系统由配有氢火焰离子化检测器(FID)的Agilent7890气相色谱仪和双喷口热调制器组成;飞行时间质谱仪为美国LECO公司的Pegasus Ⅳ,系统为Chroma TOF软件;氮吹仪(美国进口)。正己烷(分析纯)和二氯甲烷(分析纯),购自北京化工厂,使用前进一步提纯;细硅胶(100~200目,200 ℃下活化4 h)。

(2) 样品。本文所用原油样品取自塔里木盆地轮南地区轮古7井,该油位于井下5 165~5 175 m的奥陶系储层,外观为黑色,密度是0.860 7 g/cm^3,属正常原油。芳烃制备过程如下:① 取原油样品60 mg左右,用适量正己烷溶解。② 取细硅胶6 g转入玻璃柱中,震荡压实。少许正己烷淋洗后,将原油样品溶液分几次全部转入玻璃柱中。③ 当柱子下端溶液流出时,分批加10 mL正己烷,收集饱和烃馏分;加20 mL二氯甲烷淋洗,收集芳烃馏分。过柱子同时用紫外灯照射,通过观察荧光来控制加入溶剂的量。用氮吹仪吹至约1 mL转至进样瓶。

(3) GC×GC/TOFMS实验条件。GC×GC具体条件如表1所示。实验采用的色谱柱均为美国J & W Scientific公司产品。TOFMS实验条件如表2所示。

表1 GC×GC实验条件

物理量	参数值
一维柱系统	Petro,50 m×0.2 mm×0.5 μm
二维柱系统	DB-17HT,3 m×0.1 mm×0.1 μm
一维柱子升温程序	80 ℃(0.2 min) $\xrightarrow{2\ ℃/min}$ 310 ℃(25 min)
二维柱子升温程序	90 ℃(0.2 min) $\xrightarrow{2\ ℃/min}$ 320 ℃(25 min)

续表

物理量	参数值
进样口温度	300 ℃
进样量	1 μL
进样模式	Split 50∶1
载　气	He 流速为 1.8 mL/min
调制器温度	比一维炉温高 30 ℃
调制周期	10 s,其中 2.5 s 热吹时间
传输线温度	280 ℃

表 2　TOFMS 实验条件

物理量	参数值
电离能量/eV	−70
检测器电压/V	1 475
采集速率/(谱图·s^{-1})	100
质量扫描范围/aum	40～520
离子源温度/℃	240
采集延迟时间/min	13

（4）标样。将氘代蒽作为标样加到样品的芳烃组分中以便于样品中化合物的定性。

2　实验方法

2.1　气相色谱方法和质谱方法

与一维气相色谱方法的建立类似,二维气相色谱方法的建立需要考虑的参数主要包括柱系统的选择、升温程序、调制周期、热吹时间及载气流速等。

由于石油地质样品中化合物种类繁多,芳烃化合物中烷基苯类化合物从甲苯到 C_{19} 取代苯都能检测到,说明样品中各化合物的沸点范围大,因此一维柱选用柱膜较厚且较长的非极性色谱柱 Petro(50 m×0.2 mm×0.5 μm)。由于芳烃组分中化合物极性和沸点分布广,二维柱选用较长的中等极性且耐高温低流失的色谱柱 DB-17HT(3 m×0.1 mm×0.1 μm)。为了提高一维柱子和二维柱子的分离效果,我们缩小两维柱系统的温差,使芳烃同族的化合物在一维柱子上分离得更好。

为了能用全二维气相色谱-飞行时间质谱检测到五环芳烃类化合物,我们采用较大的调制周期。五环芳烃类化合物的极性强,在二维柱子上的出峰时间晚,易与二维柱子的柱流失重合。为了避免这个现象,我们选用 10 s 的调制周期以满足分析要求。此调制周期对一维柱子的分辨效果影响不大,因为飞行时间质谱做检测器主要是依靠各碎片的质谱特征来合峰,不会影响化合物的检测。热吹时间对高沸点化合物的分离很重要,过小的热吹时间会使后面的化合物拖尾,因此采用 2.5 s 的热吹时间。考虑到实验时间的问题,我们采用较大的载气流速(1.8 mL/min)。

从二维柱子流出的峰都很窄,一般为 100 ms,要保证每个峰至少有 7～10 张谱图的采集才能被很好地定性,我们设定质谱的采集速率是 100 谱图/s。由于加入的氘代蒽标样的溶剂是甲苯,因此质谱采集的延迟时间采用 13 min。

2.2 数据处理方法

本文所用仪器所带的数据处理软件是 Chroma TOF 软件,NIST 05 谱库。被测物从第一维柱子出来,经过调制器的捕集、聚焦,再传送进入第二维柱子,这个过程中通常一个组分要被分割成 3～4 个碎片峰进入检测器被鉴定,每一个碎片峰 TOFMS 都会给出定性信息。数据处理方法中参数的设定就是让软件在一定的一维峰宽(一般设为调制周期的 4 倍)、二维峰宽(一般为 0.1 s)和信噪比内,通过质谱图的比对,把相似度在 700(参数可设)以上的峰的碎片合并起来计算峰面积。软件会根据设定自动完成计算基线、峰查找、谱库比对和峰面积积分等工作,地球化学各参数与一般色谱-质谱分析方法一样通过峰面积计算求得。

飞行时间质谱采用 EI 源电离,其采集的质谱图谱各碎片峰与四级杆 EI 电离源的碎片峰一致,因此各个化合物的定性及识别和普通色谱/质谱一样,也是通过保留时间和 NIST 05 谱库对比,结合一些石油地质专业图谱集的谱图进行。

3 结果与讨论

3.1 芳烃组分的全二维点阵谱图特征

在 GC×GC 中,第一维和第二维色谱柱分别是非极性柱和极性或中等极性柱,可使化合物在一维上按照沸点、二维上按照极性分离。由于两根柱子固定相极性的改变加上线性程序升温方法的共同作用,两维间交叉信息减少,实现真正的正交分离。由于化合物的极性不同,GC×GC 谱图被明显分割成不同的区带,每一个区带代表一族物质,这就是 GC×GC 的族分离特性。芳烃化合物具有种类多、极性范围大的特点,因此 GC×GC 的族分离特性能够更好地把它们分开。

图 1 是 LG7 样品芳烃组分的全二维点阵图。从图中可以看出,整个谱图被分成了不同的区域,代表不同极性的不同族的化合物。最下方一族是相对极性最弱的苯系列化合物(长侧链和支链取代苯),在一维方向随保留时间增加,取代基碳数增加,由于这类化合物特征离子明显,大多数可被软件直接鉴定。在它上方依次是环基取代苯系列,双

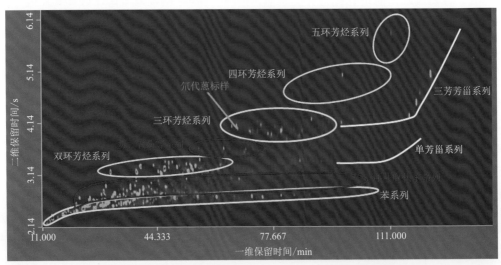

图 1 LG7 油样芳烃组分的 GC×GC-TOFMS 分析二维点阵图

环芳烃系列，芴、氧芴等系列，三环芳烃系列，四环芳烃系列，五环芳烃系列，单芳甾和三芳甾烃系列。如表3所示，每一族化合物都有不同的结构特征和特征离子，这些化合物中有些软件很难准确识别或者无法识别，需要与相关出版物的标准图谱比对方可定性识别。

表3 芳烃组分中不同族化合物的结构特征和特征质量数

特征离子(m/z)	名 称	基本结构	特征离子(m/z)	名 称	基本结构
106,120,134	C_2～C_4取代苯		178,192,206,220,234	菲系列	
91	长侧链烷基取代苯		252	五环芳烃系列	
184,198,212,226	二苯并噻吩系列		234,248,262	苯并萘并噻吩系列	
128,142,156,170,184	萘系列		160,174,188	环己基-苯系列	
154,168,182,196	联苯、二氢苊、氧芴系列		117,132,131	1,2-二氢化茚系列	
166,180 194	芴系列		147,161,175	长链烷基苯并噻吩系列	
133	1-烷基-2,3,6-三甲基苯		231,245	三芳甾烃系列	
202,216,230	荧蒽和芘系列		253	单芳甾系列	
228,242,256	䓛系列				

用GC×GC分析样品得到的化合物谱图除了有族分离的特性，还有"瓦片效应"，即芳环带上相同碳数取代基的化合物呈线性排列，不同碳数取代基的同族化合物呈瓦片状排列。图2是双环、三环、四环和五环芳烃系列在其选择离子下的全二维点阵图。从图2中可以看出，不同碳数取代基的化合物在全二维轮廓图上呈瓦片状排列。

3.2 典型化合物的谱图特征

由于GC×GC的族分离和"瓦片效应"，一些以往在一维色谱上分不开的峰，在二维色谱上可以很好地被分开。在陆相和海相原油中，联苯类物质和氧芴类物质含量差异大。但这两种物质是分子离子相同、特征离子相同且沸点相近的两类物质，在一维色谱上有些是"共馏峰"，在GC/MS的选择离子下也无法分开。但在GC×GC上，由于极性的差异它们被很好地分开，因此定性定量更为准确。图3就是在联苯和氧芴的选择离子m/z 154，168，182，196下的全二维3D图和色谱图，其中二甲基联苯和甲基二苯并呋喃由于特征离子和分子离子都是m/z 182，在GC/MS上无法分开，但在GC×GC上就可以。定性如表

4所示。

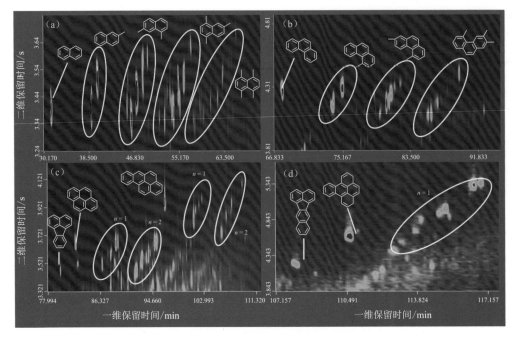

图 2　LG7 油样中二环、三环、四环和五环芳烃化合物的全二维点阵图
(a) 二环；(b) 三环；(c) 四环；(d) 五环；$n=1$ 表示芳环上有一个取代基的化合物，
$n=2$ 表示芳环上有两个取代基的化合物

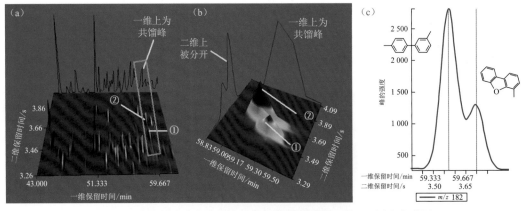

图 3　联苯及氧芴类化合物在其特征离子下的全二维 3D 图和色谱图
(a)和(b) 全二维 3D 图；(c) 色谱图；标记①为二甲基联苯，②为甲基二苯并呋喃

表 4　图 3 中标志化合物的定性峰表

峰 号	名　称	一维保留时间/min	二维保留时间/s
1	二甲基联苯	59.333 3	3.58
2	甲基二苯并呋喃	59.333 3	3.72

原油中有一部分含氧、氮、硫原子的杂环原子化合物，黄弟藩等认为这些物质是生物

标志烃类的先驱物,比生物标志烃类更接近生源,有更为直接的古环境意义。但由于它们极性大、含量低,难以分离鉴定,使得在研究上还有一定的困难。在芳烃组分中有少量的杂环原子化合物,它们的沸点往往与结构相近的芳环类化合物接近,在一维色谱上易重叠出峰。以往人们把研究重点放在这些化合物的分离鉴定上,也摸索出了很好的方法,但对于含量很低的样品,检测仍存在一定困难。用 GC×GC 分析原油样品,不用进行特殊的前处理也能检测到含量较低的部分杂环芳烃化合物。它们由于极性的差异,在族分离的特性下与芳烃类化合物被很好地分开。

图 4 是在原油样品 LG7 的芳烃组分中检测到的长链烷基苯并噻吩系列化合物,它们的特征离子分别是 m/z 147,161,175,根据质量色谱图和与谱库比对的质谱图,共检测出相似度在 850 以上的化合物 31 种。

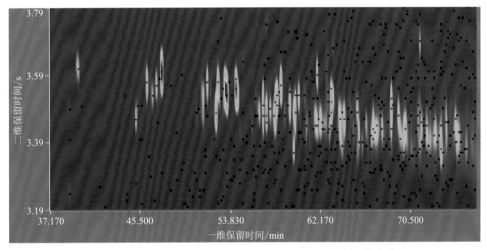

图 4　LG7 样品在选择离子 m/z 147,161,175 下的全二维点阵图

除此之外,LG7 样品中还检测出氧芴、硫芴、咔唑、苯并萘并噻吩和烷基苯并咔唑这几类杂原子芳烃化合物。以往含氮化合物在芳烃组分中不易检测到,因为其会被含量很高的菲类物质所掩盖。全二维气相色谱-飞行时间质谱能在芳烃组分中一次性检测到含氮的杂环芳烃化合物,对于含氮化合物的定性和定量有很大的帮助。

4　结　论

全二维气相色谱-飞行时间质谱相比于 GC-MS 有灵敏度更高、峰容量大的优点。原油中的芳烃各组分根据族分离特性和"瓦片效应"在全二维色谱上被很好地分开,极性的差异使得很多以往在一维色谱上共馏的化合物在二维色谱上得到了很好的分离。相对含量较低的化合物也能同时被检测,因此全二维气相色谱-飞行时间质谱非常适合复杂混合物体系和痕量物质的分析。目前,石油地质实验工作者已开始将该技术应用于石油地质样品的分析,相信随着该项技术的完善和方法的普及,会有更多、更准确的地球化学参数应用到油气勘探中,同时也会有更多的未知化合物被地球化学家认知,从而解决勘探生产中的问题,全二维气相色谱-飞行时间质谱在石油地质行业中有很好的应用前景。

论文 3　2011 年发表于《中国科学：地球科学》

全二维气相色谱-飞行时间质谱与
常规色质分析的地球化学参数对比

王汇彤，翁　娜，张水昌，朱光有，魏彩云

摘　要　利用全二维气相色谱-飞行时间质谱和气相色谱-质谱分别对原油的饱和烃和芳烃组分进行分析，比较了一些常用的油气地球化学参数在两种仪器上的差异。结果表明，在计算的 23 个常用油气地球化学参数中，有 10 个参数的偏差在 5% 以内，说明部分油气地球化学参数在两种不同的仪器上分析结果基本一致。但对于伽马蜡烷/$\alpha\beta$-藿烷、Ts/Tm、2-乙基萘/1-乙基萘(ENR)、(2,6-二甲基萘+2,7-二甲基萘)/1,5-二甲基萘(DNR)等参数，两种仪器的结果偏差较大；一些低含量的化合物在常规的色谱-质谱上检测不到，无法获得相关的地球化学参数。产生上述问题的原因在于常规色质在分析时受色谱柱分离能力和柱容量的限制，化合物易形成"共馏峰"，影响峰面积积分结果，低含量的化合物易受基线噪音和其他物质的干扰而无法检出。全二维气相色谱-飞行时间质谱在有效消除"共馏峰"及分离能力上要强于常规色质分析，能得到相对更真实的基线，在化合物的峰面积积分结果上更为精确，从而得到更客观的油气地球化学参数。因此，全二维气相色谱-飞行时间质谱可能成为油气地球化学研究的有效工具之一。

关键词　全二维气相色谱　飞行时间质谱　气相色谱-质谱　地球化学参数

气相色谱-质谱联用仪(GC-MS)在石油地质研究中应用广泛，是原油和岩石可溶有机质中生物标志化合物等有机化合物定性、定量分析的常用手段。分析结果可用于判断油气性质(类型与成熟度)、进行油气源对比、判识沉积环境、划分地层、确定生油层和储油层亲缘关系等。但由于常规气相色谱的峰容量和分离能力有限，在分析化合物种类多于 250 的复杂样品时，会形成一个基线鼓包 UCM(Unresolved Complex Mixture)，无法满足对化合物精准分析的要求。与其联用的四级杆质谱仪提供的质谱信息较为简单，且受干扰离子影响化合物的质谱图与标准谱库的谱图差别较大，无法为未知化合物的判定提供有效依据。因此，GC-MS 主要采用选择离子的方式对已知的化合物进行定性、定量分析。

为了弥补 GC-MS 分析的不足，人们从色谱和质谱两个方向对仪器进行了改进。在色谱方面，传统多维色谱(如 GC+GC)拓展了一维色谱的分离能力，可以分析复杂体系中感兴趣的馏分信息。在质谱方面，串联质谱(如气相色谱-质谱/质谱)或三级四极质谱的应用弥补了 GC-MS 在质谱检测方面的不足。它们可以减少 GC 共馏峰的干扰，提高信噪比，从而得到较为精确的质谱信息。尽管用串联质谱可以鉴别出石油复杂混合物中的单体化合物和化合物群族，但是它和传统多维色谱都无法做到一次进样就得到样品的全部信息。

全二维气相色谱(GC×GC)是多维色谱的一种，但不同于常规的二维色谱。二维色谱(GC+GC)是两根色谱柱简单连接，而全二维气相色谱的柱系统是由分离机理不同而又相互独立的两支色谱柱通过一个调制器连接而成。调制器的作用就是捕获第一根色谱柱分出的馏分，经过聚焦后以脉冲的方式进入第二根柱子进行分离，因此用全二维气相色谱分析复杂样品，能够一次得到样品的主要信息。全二维气相色谱具有分辨率高、灵敏度好、峰容量大以及定性更有规律可循等特点。用全二维气相色谱分析石油样品,样品不用经过前处理可以直接进样,较之一维色谱质谱大大节省了时间。在分离能力上,它比普通 GC 至少高出一个数量级,较于 GC-MS 有更强的分离能力。由于两根柱子不同的分离效果,以往在一维色谱上按照沸点分不开的物质,在二维色谱柱上可以按照极性很好地被分开,因此,全二维气相色谱能够更好地解决"共馏峰"问题,很多学者已经在这方面做了研究。与它匹配的飞行时间质谱(TOFMS)有极快的响应速度,较高的灵敏度,高采集频率(每秒产生 1~500 幅谱图)和对测定对象没有质量范围限制等特点,能得到化合物较为清晰的质谱信息,为未知化合物的定性提供了有力依据。全二维气相色谱与飞行时间质谱的联用更加适合复杂体系的定性定量。

随着全二维气相色谱-飞行时间质谱在不同领域应用的开展,有人对它的分析结果能否与过去常规的色质分析结果一致或者可比产生了疑问,不同领域的分析人员对此作过研究。比如,Blumberg 等和 Schoenmakers 等对比了 GC×GC 和 GC 的分析能力,认为GC×GC 在分析复杂样品上更有优势;Zhu 等利用 GC×GC-TOFMS 和 GC-MS 在分析烟用香精成分,发现 GC×GC 在分析复杂基质中的痕量物质上比 GC-MS 更有优势,并用GC×GC-FID 对样品进行了定量分析。对于石油地质样品,全二维气相色谱-飞行时间质谱与常规的色谱/质谱分析结果的比较还没有报道。

本文在建立饱和烃、芳烃的全二维气相色谱-飞行时间质谱分析方法和图谱鉴别的基础上,用全二维气相色谱-飞行时间质谱仪和色谱/质谱仪器分别对原油的饱和烃及芳烃组分进行了分析,并利用这些分析数据计算一些常用的地球化学参数,讨论了两台仪器结果偏差的原因,进一步探讨了全二维气相色谱-飞行时间质谱在石油地质实验中的应用问题。

1 实验部分

1.1 GC-MS 实验条件

DSQ Ⅱ四级杆气质联用仪(GC-MS)(Thermo Fisher Scientific 公司),色谱柱为美国 J&W Scientific 公司产品,GC-MS 实验条件如表 1 所示。

表 1 GC-MS 实验条件

	饱和烃分析方法	芳烃分析方法
色谱柱	HP-5MS 60 m×0.25 mm×0.25 μm	HP-5MS 60 m×0.25 mm×0.25 μm
升温程序	100 ℃(5 min) $\xrightarrow{4\ ℃/min}$ 220 ℃ $\xrightarrow{2\ ℃/min}$ 320 ℃(15 min)	100 ℃(5 min) $\xrightarrow{3\ ℃/min}$ 320 ℃(20 min)
进样口温度/℃	300	300
进样量/μL	1	1
进样模式	不分流	不分流
载气	He 气,1 mL/min	He 气,1 mL/min
传输线温度/℃	280	280
电离能量/eV	−70	−70
检测器电压/V	1 324	1 324

续表

	饱和烃分析方法	芳烃分析方法
质量方式及扫描范围	选择离子扫描	选择离子扫描
离子源温度/℃	250	250
采集延迟时间/min	8	8

1.2 GC×GC-TOFMS 实验条件

全二维气相色谱-飞行时间质谱仪(GC×GC-TOFMS)(美国 LECO 公司),GC×GC 系统由配有氢火焰离子化检测器(FID)的 Agilent7890 气相色谱仪和双喷口热调制器组成;飞行时间质谱仪为美国 LECO 公司的 Pegasus Ⅳ,数据处理系统为 Chroma TOF 软件。色谱柱均为美国 J & W Scientific 公司产品,GC×GC-TOFMS 实验条件如表2所示。

表2　GC×GC-TOFMS 实验条件

	饱和烃分析方法	芳烃分析方法
一维柱系统	Petro,50 m×0.2 mm×0.5 μm	Petro,50 m×0.2 mm×0.5 μm
二维柱系统	DB-17HT,3 m×0.1 mm×0.1 μm	DB-17HT,3 m×0.1 mm×0.1 μm
一维柱子升温程序	80 ℃(0.2 min) $\xrightarrow{2\ ℃/min}$ 310 ℃(25 min)	80 ℃(0.2 min) $\xrightarrow{2\ ℃/min}$ 310 ℃(25 min)
二维柱子升温程序	100 ℃(0.2 min) $\xrightarrow{2\ ℃/min}$ 330 ℃(25 min)	90 ℃(0.2 min) $\xrightarrow{2\ ℃/min}$ 320 ℃(25 min)
进样口温度/℃	300	300
进样量/L	1	1
进样模式	Split 40∶1	Split 50∶1
载　气	He,流速为 1.8 mL/min	He,流速为 1.8 mL/min
调制器温度	比一维炉温高 30 ℃	比一维炉温高 30 ℃
调制周期	10 s,其中 2.5 s 热吹时间	10 s,其中 2.5 s 热吹时间
传输线温度/℃	280	280
电离能量/eV	−70	−70
检测器电压/V	1 475	1 475
采集速率/(谱图·s^{-1})	100	100
质量扫描范围/amu	40～520	40～520
离子源温度/℃	240	240
采集延迟时间/min	11	13

1.3 样品及试剂

正己烷(分析纯)和二氯甲烷(分析纯),购自北京化工厂,使用前进一步提纯;细硅胶(100～200 目,200 ℃下活化 4 h)。

本文共选取 6 个样品进行分析。一个是取自塔里木盆地轮南地区轮古 7 井的 LG7 原油,一个是取自玉门油田柳 43 井的 O43 原油,剩下 4 个原油取自辽河盆地,相应编号为 C1～C4,其中 C1 和 C2 为未降解原油,C3 为轻度降解原油,C4 为中度降解原油。

饱和烃、芳烃制备：① 取原油样品 60 mg 左右，用适量正己烷（重蒸过的分析纯）溶解。② 取细硅胶 6 g 转入玻璃柱中，震荡压实。少许正己烷淋洗后，将原油样品溶液分几次全部转入玻璃柱中。③ 当柱子下端溶液流出时，分批加 10 mL 正己烷，收集饱和烃馏分；加 20 mL 二氯甲烷淋洗，收集芳烃馏分。过柱子同时用紫外灯照射，通过观察荧光来控制加入溶剂的量。用氮吹仪吹至约 1 mL 转至进样瓶。

2 结果与讨论

2.1 GC-MS 和 GC×GC-TOFMS 结果比较

用 GC-MS 和 GC×GC-TOFMS 分析 6 个样品的饱和烃、芳烃组分，用峰面积积分法得到化合物的峰高、峰面积信息，并计算一些常用的地球化学参数。表 3 列出的是用 GC-MS 和 GC×GC-TOFMS 分别分析 6 个样品，经过计算得到的地球化学参数信息。根据表 3 的数据，表 4 列出了这两种方法的对比结果。

表 3　用 GC-MS 和 GC×GC-TOFMS 分析得到的地球化学参数结果

地球化学参数	GC-MS						GC×GC-TOFMS					
样品号	O43	LG7	C1	C2	C3	C4	O43	LG7	C1	C2	C3	C4
Ts/Tm	0.30	0.30	1.03	1.23	0.97	0.95	0.31	0.32	1.13	1.31	1.06	1.07
Ts/(Ts+Tm)	0.23	0.26	0.50	0.51	0.49	0.49	0.23	0.26	0.52	0.58	0.51	0.52
Ts/$C_{30}17\alpha(H)$-藿烷	0.04	0.21	0.11	0.15	0.11	0.11	0.03	0.21	0.16	0.16	0.12	0.12
$\alpha\alpha\alpha$-C_{29}甾烷 20S/(20S+20R)	0.48	0.50	0.34	0.38	0.33	0.40	0.49	0.53	0.25	0.32	0.30	0.35
$C_{29}\beta\beta/(\beta\beta+\alpha\alpha)$	0.41	0.54	0.35	0.40	0.35	0.39	0.43	0.57	0.37	0.39	0.32	0.39
碳优势指数 CPI	1.24	1.01	1.19	1.16	1.20	1.21	1.17	0.98	1.20	1.20	1.14	1.17
奇偶优势 OEP	1.14	0.99	1.12	1.09	1.15	1.14	1.14	1.00	1.18	1.11	1.19	1.12
MPⅠ	0.49	0.45	0.46	0.45	0.44	0.42	0.46	0.42	0.43	0.43	0.42	0.39
MPⅡ	0.27	0.26	0.24	0.23	0.23	0.22	0.26	0.24	0.22	0.21	0.22	0.20
甲基萘比 MNR	1.82	1.46	1.18	1.14	1.29	1.28	1.65	1.33	1.08	1.13	1.36	1.17
乙基萘比 ENR	4.58	2.80	1.94	1.11	1.75	2.76	3.56	2.25	1.25	0.98	2.15	2.10
二甲基萘比 DNR	7.64	4.98	4.07	5.12	4.44	4.22	5.82	4.34	4.99	3.53	4.81	3.38
甲基菲比 MPR	1.00	1.14	1.06	1.02	0.99	1.32	0.94	1.04	0.98	0.94	0.91	1.13
甲基菲指数 MPI	0.62	0.60	0.72	0.66	0.69	0.71	0.58	0.57	0.57	0.55	0.56	0.57
MDI	—	—	—	—	—	—	0.49	0.42	0.40	0.34	0.48	
MAI	—	—	—	—	—	—	0.53	0.54	0.55	0.60	0.49	0.56
DMAI	—	—	—	—	—	—	0.47	0.30	0.43	0.47	0.38	0.44
伽马蜡烷/$\alpha\beta$-藿烷	0.25	0.08	0.01	0.02	0.01	0.03	0.23	0.06	0.01	0.02	0.01	0.02
Pr/Ph	0.68	0.64	1.83	2.24	1.64	1.67	0.70	0.71	1.64	2.37	1.60	1.55
1,2,5-三甲基萘/菲	0.77	0.39	1.79	0.97	1.46	1.41	0.71	0.41	2.04	0.94	1.67	1.44
二苯并噻吩/菲	0.04	0.67	0.11	0.19	0.11	0.16	0.04	0.69	0.10	0.18	0.10	0.15
三芳甾烷 $C_{28}/(C_{26}\sim C_{28})$	0.59	0.77	0.47	0.50	0.47	0.42	0.63	0.82	0.46	0.46	0.43	0.41

续表

地球化学参数	GC-MS						GC×GC-TOFMS					
样品号	O43	LG7	C1	C2	C3	C4	O43	LG7	C1	C2	C3	C4
$\alpha\alpha\alpha$(20R)甾烷 C_{28}/C_{29}	0.57	0.42	0.47	0.48	0.53	0.41	0.57	0.44	0.51	0.50	0.52	0.43

注：表 3 中 MPⅠ和 MPⅡ是与芳烃有关的地球化学参数。其中 MPⅠ=(3-甲基菲+2-甲基菲)/(3-甲基菲+2-甲基菲+9-甲基菲+1-甲基菲)；MPⅡ=2-甲基菲/(3-甲基菲+2-甲基菲+9-甲基菲+1-甲基菲)；MNR=2-甲基萘/1-甲基萘；其他参数公式如下：ENR=2-乙基萘/1-乙基萘；DNR=(2,6-二甲基萘+2,7-二甲基萘)/1,5-二甲基萘；MPR=2-甲基菲/1-甲基菲；MPI=1.5×(2-甲基菲+3-甲基菲)/(菲+1-甲基菲+9-甲基菲)；MDI=4-甲基双金刚烷/(1-+3-+4-甲基双金刚烷)；MAI=1-甲基金刚烷/(1-+2-甲基金刚烷)；DMAI=1,3-二甲基金刚烷/(1,2-+1,3-二甲基金刚烷)。"—"代表未检出。

表 4 采用 GC-MS 和 GC×GC-TOFMS 两台仪器得到地球化学参数的结果对比

地球化学意义	样品号	O43	LG7	C1	C2	C3	C4
	地球化学参数	Δ					
成熟度判识	Ts/Tm/%	1.63	3.39	4.64	3.25	4.08	5.60
	Ts/(Ts+Tm)/%	0.65	0.78	2.54	6.43	2.03	2.79
	Ts/$C_{30}17\alpha$(H)-藿烷/%	1.45	1.20	4.48	4.14	1.86	4.89
	$\alpha\alpha\alpha$-C_{29}甾烷 20S/(20+20R)/%	1.03	2.72	15.12	8.01	4.71	6.56
	C_{29}甾烷异构化 $\beta\beta/(\beta\beta+\alpha\alpha)$/%	2.72	2.71	1.69	1.62	4.18	0.20
	碳优势指数 CPI/%	2.53	1.51	0.78	1.73	2.41	1.65
	奇偶优势 OEP/%	0.13	0.30	2.79	0.84	1.85	0.78
	MPⅠ/%	3.17	4.27	3.07	2.01	3.07	3.44
	MPⅡ/%	2.43	2.73	4.25	4.63	3.97	4.90
	甲基萘比 MNR/%	4.92	4.56	4.30	0.49	2.66	4.51
	乙基萘比 ENR/%	12.54	10.78	5.00	6.42	10.21	13.63
	二甲基萘比 DNR/%	13.53	6.87	10.13	18.39	4.03	11.12
	甲基菲比 MPR/%	3.10	4.63	3.58	3.87	4.06	7.59
	甲基菲指数 MPI/%	4.54	1.10	11.84	5.09	10.73	10.70
	MDI	—	—	—	—	—	—
	MAI	—	—	—	—	—	—
	DMAI	—	—	—	—	—	—
沉积环境、物源判识	伽马蜡烷/$\alpha\beta$-藿烷/%	4.76	11.59	10.09	17.10	9.70	32.95
	Pr/Ph/%	1.52	4.90	5.44	2.87	1.51	3.53
	1,2,5-三甲基萘/菲/%	3.93	2.53	6.53	1.27	6.67	1.22
	二苯并噻吩/菲/%	1.37	1.10	1.79	2.15	2.53	1.62
	三芳甾烷 $C_{28}/(C_{26}\sim C_{28})$/%	2.96	3.14	1.51	4.68	3.66	1.90
	$\alpha\alpha\alpha$(20R)甾烷 C_{28}/C_{29}/%	0.61	1.86	4.97	2.51	0.86	2.96

注：表 4 中 Δ 表示两个数的偏差，计算公式为 $\Delta=|A_1-A_2|/(A_1+A_2)$，其中 A_1 表示 GC-MS 的分析结果，A_2 表示 GC×GC-TOFMS 的分析结果。

针对 GC×GC-TOFMS 这种新型仪器,我们对它的重复性进行了考察。选取 O43 样品的饱和烃组分,在相同条件下用 GC×GC-TOFMS 分析三次,计算它们的相对标准偏差(RSD),结果列于表 5。从中可以看出相对标准偏差≤3.34%,说明 GC×GC-TOFMS 适合复杂混合物体系的分析。

表 5　用 GC×GC-TOFMS 分析 O43 样品三次的重复性结果

地球化学参数	第一次	第二次	第三次	重复性/%
Ts/Tm	0.30	0.32	0.32	1.38
Ts/(Ts+Tm)	0.22	0.26	0.26	2.11
Ts/$C_{30}17\alpha(H)$-藿烷	0.03	0.04	0.04	0.41
$\alpha\alpha\alpha$-C_{29}甾烷 20S/(20S+20R)	0.50	0.48	0.47	1.92
C_{29}甾烷 $\beta\beta/(\beta\beta+\alpha\alpha)$	0.41	0.42	0.42	0.54
碳优势指数 CPI	1.17	1.17	1.17	0.36
奇偶优势 OEP	1.22	1.19	1.17	2.55
MAI	0.52	0.55	0.52	1.75
DMAI	0.44	0.46	0.51	3.34
伽马蜡烷/$\alpha\beta$-藿烷	0.20	0.19	0.20	0.59
Pr/Ph	0.78	0.79	0.79	0.83
$\alpha\alpha\alpha$(20R)甾烷 C_{28}/C_{29}	0.57	0.59	0.58	1.12

2.2　结果讨论

根据表 4,我们对 23 个常用地化参数的对比结果作如下讨论。

(1) 在分析的 6 个样品中,有 10 个参数在两台仪器上的计算结果偏差均小于 5%,属于仪器分析允许的误差范围,说明两种仪器在这些地化参数分析结果上一致。其中碳优势指数 CPI 和奇偶优势 OEP 相比于其他地球化学参数需要计算得更加精确。它们是用来估计原油成熟度的重要指标,当 OEP 取 1.20 和 1.0 时,尽管这两个数值的偏差只有 9%,但它在地球化学上的意义就完全不同。OEP<1.20 时有机质进入生油门限,OEP 接近 1.0 时进入成熟阶段。在本次实验中,分别用两台不同仪器的分析结果计算 6 个原油样品的 CPI 和 OEP 数值,两台仪器的偏差均小于 3%,说明用 GC×GC-TOFMS 计算出的 OEP 和 CPI 值完全可以满足地化应用的要求。

(2) 23 个常用的地球化学参数中有 13 个参数在两种仪器上的比较结果偏差较大,可分为以下 4 种情况讨论。

① 在两台仪器上均能检出,但由于"共馏峰"影响,一些化合物在 GC-MS 上共馏,但在 GC×GCTO-FMS 上可以分开,用 GC-MS 得到的峰面积积分结果不准确。如 Ts/Tm,Ts/(Ts+Tm),乙基萘比 ENR 和二甲基萘比 DNR。

Peters 等就曾提出 Ts/Tm 及 Ts/(Ts+Tm)在应用时需谨慎。在常规色质 m/z 191 质量色谱图上,Tm 和 Ts 往往与三环或四环萜烷同时析出,从而导致假的 Ts/Tm 及 Ts/(Ts+Tm)比值。在本文分析的 6 个样品中,Tm 与 C_{30}-三环萜烷在一维气相色谱-质谱上是"共馏峰",如图1(c)所示,但在全二维气相色谱上,由于 Tm 和 C_{30}-三环萜烷的结构不同,一个是五环三萜烷,一个是三环萜烷,极性的差异使它们在第二根极性色谱柱上可以很好地被分开,如图1(a)所示,出峰互不影响。由此看出用 GC×GC-TOFMS 分析得到的 Ts/Tm 及 Ts/(Ts+Tm) 值比 GC-MS 更为精确。虽然用串联质谱检测由 m/z 370 到 m/z 191 的变化可以排除三环萜烷的干扰,但需要借助前处理手段,延长了样品分析时间。相比之下 GC×GC-TOFMS 一次进样就可以得到全部结果,更为简便。

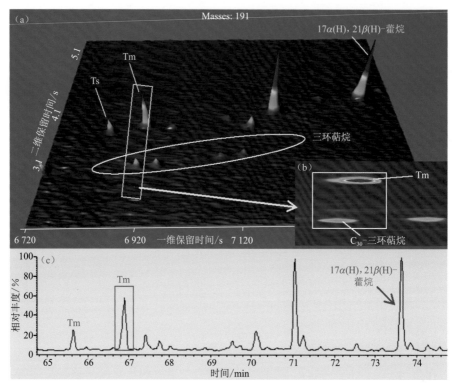

图1 GC×GC-TOFMS 和 GC-MS 分析 LG7 饱和烃组分在 m/z 191 下的谱图
(a) m/z 191 下的全二维 3D 图;(b) m/z 191 下的全二维轮廓图;(c) m/z 191 下 GC-MS 的色谱图;
从图(b)可以明显看出,Tm 和 C_{30}-三环萜烷在一维色谱柱上是共馏峰;从(c)图中可以看出,
在 GC-MS 色谱图上 Tm 出峰位置只有一个峰

用常规 GC-MS 分析芳烃组分时,C_2-萘的出峰效果不好,组分共馏现象严重。图2(c)所示的是 GC-MS 在选择离子 m/z 156 下得到的色谱图,在图上 2-乙基萘和 1-乙基萘因为没有完全分开而互相重叠,1,4-二甲基萘与 2,3-二甲基萘在 GC-MS 上是完全重叠的出峰,且与 1,5-二甲基萘的峰部分重叠,严重影响了各个化合物的峰面积积分结果。

图2(a)是 GC×GC 在 m/z 156 下得到的全二维 3D 图,从图中可以明显地看出 2-乙基萘和 1-乙基萘由于极性的差异被完全分开。在 GC/MS 上重叠的 1,4-二甲基萘、2,3-二甲基萘和 1,5-二甲基萘在全二维上可以被分开。因此,用 GC×GC 得到的积分结果计算出的乙基萘比 ENR 和二甲基萘比 DNR 相比于 GC-MS 更为可靠。

② 在两种仪器上均能检出,但受常规色谱分离能力的影响,一些化合物在 GC-MS 上分离情况不好,含量低的化合物易受其他峰的干扰而影响出峰效果,从而使积分结果不准

确。如甲基菲比、甲基菲指数、1,2,5-三甲基萘/菲、$\alpha\alpha\alpha$-C_{29}甾烷 $20S/(20S+20R)$和伽马蜡烷/$\alpha\beta$-藿烷。

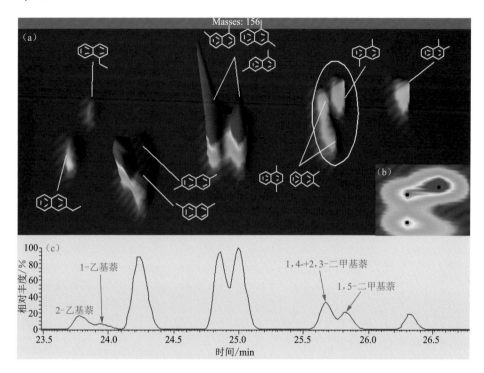

图 2　GC×GC-TOFMS 和 GC-MS 分析 LG7 饱和烃组分中 C_2-萘类物质的谱图
(a) m/z 156 下的全二维 3D 图；(b) m/z 156 下的全二维轮廓图；(c) m/z 156 下 GC/MS 的色谱图；
图(b)反映的是 1,4-二甲基萘、2,3-二甲基萘和 1,5-二甲基萘的分离情况,图中黑点代表一个化合物

在这些经常使用的地球化学参数中甲基菲比 MPR、甲基菲指数 MPI 和 1,2,5-三甲基萘/菲这三个参数在 4 个样品中两台仪器的结果偏差较大,其原因与色谱的分辨能力有关。一维色谱由于柱容量的限制,一些性质相近的化合物不能达到分离要求,如甲基菲类化合物。当 3-甲基菲与 2-甲基菲、9-甲基菲与 1-甲基菲的分离度不高于 85% 时,会使积分结果不准确,进而影响相应的参数计算。从图 3(b)可以看出,GC-MS 中的 3-甲基菲与 2-甲基菲、9-甲基菲与 1-甲基菲的分离度未达到 85%,而 GC×GC-TOFMS 的分离效果能满足要求。此外,由于 GC×GC 能有效地消除共馏峰的干扰,得到相对正确基线,因此 GC×GC-TOFMS 得到的积分结果更可靠。

在取自辽河盆地的 4 个样品 C_1～C_4 中,$\alpha\alpha\alpha$-C_{29}甾烷 $20S/(20S+20R)$参数用 GC×GC-TOFMS 分析的结果与 GC-MS 相差较大,原因在于这 4 个样品中甾烷含量低,用 GC-MS 分析时易受其他物质和基线噪音干扰使得积分结果存在误差。用 GC×GC 分析,由于两根柱子极性的差异可以把沸点相近而极性不同的化合物在二维柱子上分开,使它们之间互不干扰。此外,有国外学者指出 GC×GC 的 S/N 值(信噪比)要高出普通 GC 4～10倍,可以有效减少噪音对目标物质的干扰,搭配高采集频率的飞行时间质谱(100 Hz),使它在分析复杂基质的痕量物质时更有优势。

伽马蜡烷属于难挥发物质,出峰温度相对较高,气相色谱的毛细管柱随温度的升高会

有柱流失现象。温度越高,柱流失现象越明显,在一维色谱上会使基线抬高,含量低的物质容易被掩盖或干扰。柱流失物质与萜烷类物质都有 m/z 191 的碎片离子,因此在选择离子 m/z 191 下,萜烷类化合物受柱流失影响较大。

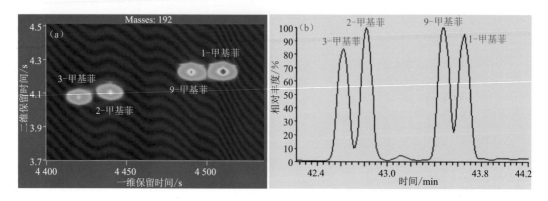

图 3　GC×GC 和 GC-MS 分析 LG7 饱和烃组分中甲基菲类物质的谱图
(a) 全二维轮廓图;(b) GC-MS 在 m/z 192 下的选择离子流图。从(a)图中可以看出,3-甲基菲和 2-甲基菲、9-甲基菲和 1-甲基菲能够完全分离;从(b)图中可以看出,在 GC-MS 色谱图上 3-甲基菲和 2-甲基菲、9-甲基菲和 1-甲基菲的分离度未达到 85%

　　相比于一维色谱,全二维气相色谱在高温时也有柱流失现象,但由于柱流失物质与一般物质的极性差异大,它们在二维柱子上要早于或晚于样品中的组分出峰,因此不会影响待测组分。在一维色谱中,由于峰容量和分离度的限制共馏峰现象严重,这就使得用 GC-MS 分析得到的基线相对不够真实。全二维气相色谱大大改善了分离度,共馏的化合物得到了进一步的分离,因此全二维得到更为真实的基线,对于出峰物质的峰面积积分结果也更为准确。

　　在我们所选取的 6 个样品中,O43 样品的伽马蜡烷含量较高,两台仪器得到的伽马蜡烷与 $\alpha\beta$-藿烷的比值分别为 0.25 和 0.23,偏差为 4.76%,是仪器分析允许的偏差范围。其他 5 个样品中伽马蜡烷含量低,用 GC-MS 定量时干扰物质多。如图 4 所示,LG7 样品饱和烃组分中伽马蜡烷的含量低,在用 GC-MS 分析得到的色谱图(图 4b)中,伽马蜡烷的峰形不好,与周围物质没有被分开,得到的积分结果不够准确。用 GC×GC 分析时,伽马蜡烷与周围萜烷类物质在结构和极性上的差异使得它们的出峰互不干扰,因此能得到更为精确的化合物定量信息。辽河盆地的 4 个样品是不同降解程度的原油,从未降解到中度降解,原油成分越来越复杂,用 GC-MS 分析时,会形成一个基线鼓包物质,降解程度越高,基线鼓包越大,对化合物的峰面积积分影响越大。C_4 样品是中度降解原油,用 GC-MS 分析它的饱和烃组分时会形成一个大的基线鼓包,伽马蜡烷在样品中含量低,受鼓包影响大,GC-MS 的峰面积积分结果与 GC×GC-TOFMS 的积分结果较其他样品偏差更大。

　　③ 两台仪器上均能检出,高含量的化合物由于进样量过大,出现峰拖尾现象,使得峰面积积分不准确,从而影响相关参数的计算结果,如 Pr/Ph。

　　④ 在 GC-MS 上不能检出,但在 GC×GC-TOFMS 上能检出。如与金刚烷类化合物有关的地球化学参数 MDI,MAI,DMAI。

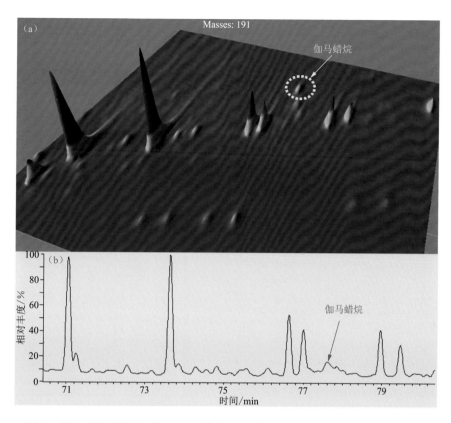

图 4　GC×GC-TOFMS 和 GC-MS 分析 LG7 饱和烃组分在 m/z 191 下的谱图
(a) m/z 191 下的全二维 3D 图；(b) m/z 191 下 GC/MS 的色谱图；
从图(a)中可以看出，伽马蜡烷的出峰很明显，没有其他干扰物质

金刚烷类化合物是原油中一种特殊的化合物，性质极为稳定，具有很强的抗热降解和生物降解能力，是油气地球化学工作者常用的参数。从图 5(c)可以看出，在用 GC-MS 分析 6 个样品时，由于金刚烷类化合物含量很低而未能检出。但用 GC×GC-TOFMS 检测，无共馏峰干扰，基线真实，可以直接检测出金刚烷类化合物。图 5 展示了用 GC-MS(图 5c)和 GC×GC-TOFMS(图 5a,b)的分析结果。

综上全二维气相色谱-飞行时间质谱在分离能力及对痕量物质的分析能力上要强于常规的一维色谱-质谱。对于表 4 中列出的偏差范围在 5% 以内的地球化学参数，两种仪器的分析结果都是可信的；超出 5% 的地球化学参数，用全二维气相色谱-飞行时间质谱分析得到的结果更为准确。

3　小　结

全二维气相色谱-飞行时间质谱具有较高的峰容量和分辨率，可以有效消除"共馏峰"的干扰，有效减小柱流失的影响，是复杂样品或者痕量化合物检测的有效工具。全二维气相色谱-飞行时间质谱分析石油地质样品比常规的气相色谱-质谱分析结果更为准确，是开展油气地球化学研究更为有效的工具之一。

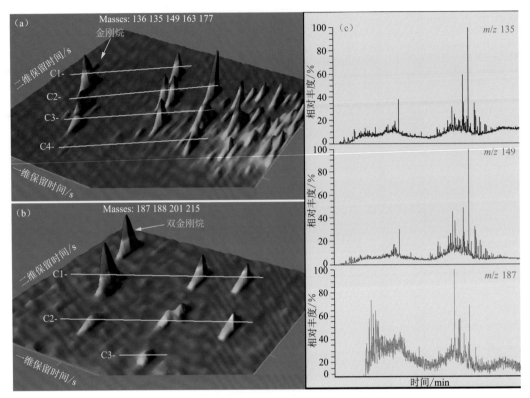

图 5 GC×GC 和 GC/MS 分析 LG7 饱和烃组分中金刚烷类物质的谱图
(a) m/z 136,135,149,163 和 177 下的全二维 3D 图;(b) m/z 187,188,201 和 215 下的全二维 3D 图;
(c) GC-MS 在 m/z 135,149 和 187 下的色谱图;
从图中的标记可以看出,取代基数量不同的金刚烷类化合物呈瓦片状排列

论文 4　2015 年发表于《Analyst》

Discovery and identification of a series of alkyl decalin isomers in petroleum geological samples

Huitong Wang, Shuichang Zhang, Na Weng, Bin Zhang, Guangyou Zhu, Lingyan Liu

Abstract　The comprehensive two-dimensional gas chromatography/time-of-flight mass spectrometry (GC×GC-TOFMS) has been used to characterize a crude oil and a source rock extract sample. During the process, a series of pairwise components between monocyclic alkanes and mono-aromatics have been discovered. After tentative assignments of decahydronaphthalene isomers, a series of alkyl decalin isomers have been synthesized and used for identification and validation of these petroleum compounds. From both the MS and chromatography information, these pairwise compounds were identified as 2-alkyl-decahydronaphthalenes and 1-alkyl-decahydronaphthalenes. The polarity of 1-alkyl-decahydronaphthalenes was stronger. Their long chain alkyl substituent groups may be due to bacterial transformation or different oil cracking events. This systematic profiling of alkyl-decahydronaphthalene isomers provides further understanding and recognition of these potential petroleum biomarkers.

1　Introduction

In geologists' research on fossils, geochemists utilize special biomarkers to trace the ancient environment and discuss the organic composition of stratigraphic origin. By tracing the isomerization of some biomarkers, the extent of specific thermal evolution could be determined. According to the characteristic structures of these biomarkers, geologists and geochemists can determine the correlation between different oil-source rocks, and explore the hydrocarbon migration with geochromatographic action theory. For example, steranes and hopanes are essential in calculating the ratio of various sources of mixed oil, tracing the secondary migration of oil and gas, and determining the extent of biodegradation of crude oil.

However, the analysis of petroleum samples usually involves extensive efforts due to the sample complexity. It is believed that even a part of an unresolved complex mixture consists of thousands of components, presenting as a big "hump" in the gas chromatograph (GC) spectrum, which remains as a challenging problem in traditional one-dimensional (1-D) GC. With higher separation and identification capabilities, the comprehensive two dimensional gas chromatography coupled time-of-flight mass spectrometry (GC×GC-TOFMS) has been applied to different types of oil samples. The comprehensive two dimensional GC utilizes two capillary columns filled with different stationary phases such as non-polar and polar column combination that provides a close to orthogonal separation capability. After eluting from the first column, the analytes are collected and refocused by a cryogenic modulator system periodically. The effluent is then sent to the second column for further separation, creating narrow modulated peaks with better sensitivity. Both the flame ionization detector (FID) and TOF were used for identification and quantification. In recent years, the comprehensive GC×GC-TOFMS has been widely used in the analysis of condensates, crude oil, source rock extracts, etc.

In our GC×GC-TOFMS analysis of source rocks and crude oil, we discovered a series of components located between monocyclic alkanes and mono-aromatics. The tentative assignments of these compounds were decalins with different alkyl substituent groups. The compounds found in heavy gas oil with similar mass spectra once attracted attention of Ávila, B. et al., but these compounds were not studied in further detail. Decalins and their short and branched alkyl substituents were also observed by Tran. T. C. et al. in biodegraded oil samples, while no further qualitative proof has been provided. The present study provides the first systematic profiling of alkyl-decahydronaphthalene isomers for further understanding and recognition. To further validate these isomers, we synthesized a series of trans-2-alkyl-decahydronaphthalene isomers for validation. These series of alkyl-decahydronaphthalenes can be potential petroleum biomarkers.

2 Experimental

2.1 Samples and chemicals

Two petroleum samples were studied. The first sample was the condensate from well ND1 in Baxian Sag of the Jizhong Depression, located in the west of the Bohai Bay Basin, China. The other was a source rock sample from the Xiahuayuan Town of Hebei Province, China. For GC×GC/TOF-MS analysis, the whole crude oil sample was injected into the instrument without any pretreatment or dilution. The well-developed protocol of hydrocarbon extraction has been applied on the source rock sample, including (a) 200 g rock samples were extracted with 250 mL chloroform and dried; (b) dissolving 20 mg extracts in appropriate amount of n-hexane (re-distillated); (c) packing the extract solution in a glass column with 2 g activated silica gel (100 mesh-200 mesh, activated at 200 ℃ for 4 h); (d) washing the silica gel 3 times with 15 mL dichloromethane; (e) collecting the hydrocarbon fraction to 1.5 mL by using a nitrogen evaporator; (f) transferring the hydrocarbon fraction into a GC sample bottle before analysis.

1-Bromobutane, triphenylphosphane, methylbenzene and all other chemicals used for synthesis were purchased from J & K Scientific Ltd, China (http://www.jk-scientific.com/).

2.2 Synthesis of 1-alkyl decalins and 2-alkyl decalins

1-Butyldecahydronaphthalene,-pentyldecahydronaphthalene,-hexyldecahydronaphthalene, and 1-heptyldecahydronaphthalene were synthesized following the steps shown in Scheme 1.

2-Butyldecahydronaphthalene, 2-pentyldecahydronaphthalene, and 2-hexyldecahydronaphthalene were produced according to steps shown in Scheme 2.

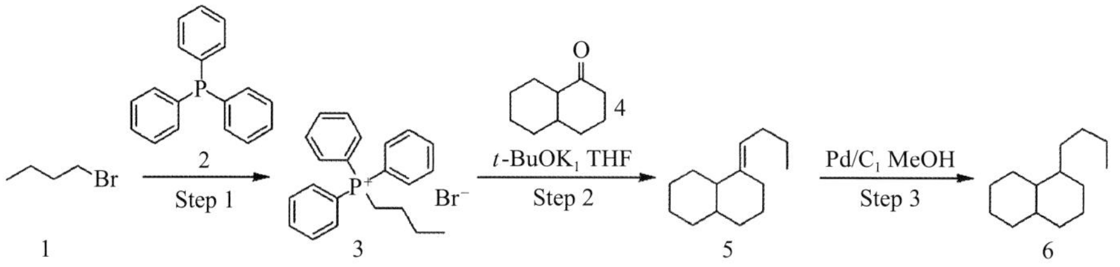

Scheme 1 Synthesis route of 1-alkyl decalins.

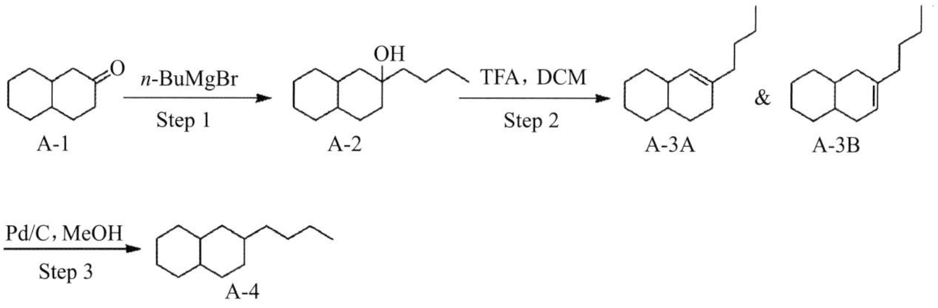

Scheme 2 Synthesis route of 2-alkyl decalins.

The details of all synthesis are included in the ESI. The structures of synthesized compounds were supported by ^{13}C-NMR DEPT 135 analysis.

2.3 Analysis of GC×GC/TOF-MS

The crude oil and source rock extract samples were analyzed by GC×GC/TOF-MS. The Synthesized Standard 1 (STN1), which has three synthesized 1-alkyl decalins, 1-butyldecahydronaphthalene, 1-pentyldecahydronaphthalene, and 1-hexyldecahydronaphthalene, was mixed and analyzed using the same parameters as the petroleum sample. 1-Heptyldecahydronaphthalene ($C_{17}H_{32}$) was excluded due to the mixed impurity. Similarly, the Synthesized Standard 2 (STN2) comprising four synthesized 2-alkyl decalins was analyzed in the same batch as STN2.

The GC×GC/TOF-MS analysis of both petroleum and standards was performed on the Pegasus 4D system (GC×GC/TOF-MS, LECO Corporation, St. Joseph, MI). The instrument was equipped with an Agilent 7890 gas chromatograph (Agilent Technologies, Palo Alto, CA, USA) and a liquid nitrogen-cooled pulse jet modulator. The detailed instrument settings are described below. A 50 m HP-Petro capillary column (100% dimethylpolysiloxane, 0.2 mm I.D. and 0.5 μm film thickness) was used as the first-dimension (1-D) column and a 3 m DB17-HT capillary column (50%-phenyl-methylpolysiloxane, 0.1 mm I.D. and 0.1 μm film thickness) was used as the second-dimension (2-D) column. Both columns were purchased from Agilent Technologies, China. The initial temperature was 35 ℃ and held for 0.2 min. Then it was increased to 210 ℃ at a rate of 1.5 ℃/min. After holding for 0.2 min, the second round of temperature increase was performed at a rate of 2.5 ℃/min until 300 ℃ and kept constant for 20 min. The second oven and the modulator oven were operated with the same temperature gradient but with a temperature offset of 5 ℃ and 45 ℃. Inlet temperature was set at 300 ℃, with a split ratio of 700 : 1, and the sample injection volume was 0.5 μL. He was used as the carrier gas, with a flow rate of 1.5 mL/min. The modulation time was 10 s, of which 2.5 s was the hot pulse time. The temperatures of the transfer line and the ion source were set at 300 ℃ and 240 ℃, respectively. The detector voltage was 1 600 V, and the mass range was 40 − 520 amu with an acquisition rate of 100 spectra per second (Hz). The delay time of the condensate oil was 0 min, and the delay time of the source rock was 11 min. The ChromaTOF software version 4.0 was used to process the raw data from the instrument. The compound assignments were performed by searching the mass spectra in the US National Institutes of Standards and Technology (NIST) MS database (NIST MS Search 2.0, NIST/EPA/NIH Mass Spectral Library; NIST 2002), and further confirmed with synthesized standard compounds.

3 Results and discussion

3.1 Alkyl decalins in petroleum samples

The comprehensive two-dimensional GC spectra with the nonpolar/polar column system used in this study have the features as follows: (1) compounds elution on the first column is according to the boiling point of each compound, which means compounds of lower boiling points will elute earlier than the ones with higher boiling points; (2) on the second column, compounds elution is according to the polarity, which means the compounds of lower polarity will be eluted first. 4 955 compounds were found by GC×GC/TOF-MS in the condensate sample. A series of n-alkanes (C_{11}−C_{38}, red), n-alkyl cyclopentane (green), n-alkyl cyclohexane (pink), n-alkyl benzene (blue), etc. were identified and selectively presented in Fig. 1a using extracted ion chromatograms (EICs) of m/z 68, 82, 92, and 137, respectively. Among them, a group of pairwise peaks was discovered. These peaks eluted between n-C_{11} and n-C_{38} on 1-D. On the 2-D chromatograph, they appeared from 2.4 to 4.8 s, in the range of polycyclic alkanes. They were further highlighted with EIC of m/z 137 in an expanded area (Fig. 1b). Similarly, a total of 5 594 compounds were found in the rock extract sample, and the corresponding EICs and expanded m/z 137 areas showing the pairwise peaks of interest are presented in Fig. 1. In the following content, the crude oil was mainly discussed.

The pairwise peaks were labeled from peak 1 (PK1) to peak 11 (PK11) in the crude oil sample. Details of PK1 to PK11 (GC temperature program changed after PK7) are listed in Table 1. Each pair of peaks could be distinguished by 'a' and 'b' according to their sequence of 2-D eluting times (ex. PK1a eluted earlier than PK1b on the 2nd column), which are also labeled in Fig. 1b. Mass spectra of the first six pairs are presented in Fig. 2, and the rest are provided in Fig. 1. The mass range of molecular ions was chosen from 152 to 292. The ChromaTOF software deconvoluted spectra were compared with those in the NIST library. The search results indicated that the first four eluted peaks were 2-methyldecahydronaphthalene (PK1a), 1-methyldecahydronaphthalene (PK1b), 2-ethyldecahydronaphthalene (PK2a), and 1-ethyldecahydronaphthalene (PK2b) with similarities all greater than 900. The difference of the molecular weights of each adjacent pair was always 14, suggesting a methylene unit (—CH_2—) insertion.

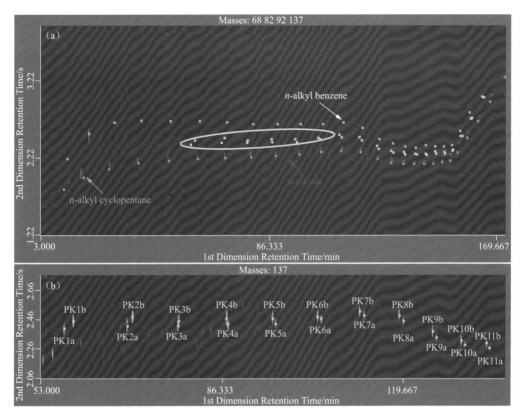

Fig. 1 (a) GC×GC-TOFMS EIC (m/z 68,82,92,137) showing petrochemical compounds in an oil sample. (b) Representative GC×GC-TOFMSEIC (m/z 137) showing the group of pairwise compounds.

Table 1 GC×GC/TOF-MS details of PK1 to PK8 from the crude oil sample.

Peak no.	1st Retention time(RT1)/min	2nd Retention time (RT2)/s	ΔRT1/min (PKb−PKa)	ΔRT2/s (PKb−PKa)	Molecular ion (m/z)	Fragment ions (m/z)
PK1a	57.27	2.40	1.60	0.05	152	41,55,67,81,95,109,123,137,152
PK1b	58.87	2.45			152	41,55,67,81,95,109,123,137,152
PK2a	68.73	2.42	0.93	0.06	166	41,55,67,81,95,109,123,137,166
PK2b	69.67	2.48			166	41,55,67,81,95,109,123,137,166
PK3a	78.07	2.42	0.13	0.03	180	41,55,67,81,95,109,123,137,180
PK3b	78.20	2.45			180	41,55,67,81,95,109,123,137,180
PK4a	87.27	2.43	−0.27	0.04	194	41,55,67,81,95,109,123,137,194
PK4b	87.00	2.47			194	41,55,67,81,95,109,123,137,194
PK5a	96.07	2.43	−0.53	0.04	208	41,55,67,81,95,109,123,137,208
PK5b	95.53	2.47			208	41,55,67,81,95,109,123,137,208

continue

Peak no.	1st Retention time(RT1)/min	2nd Retention time (RT2)/s	ΔRT1/min (PKb−PKa)	ΔRT2/s (PKb−PKa)	Molecular ion (m/z)	Fragment ions (m/z)
PK6a	104.33	2.47	−0.67	0.01	222	41,55,67,81,95,109,123,137,222
PK6b	103.67	2.48			222	41,55,67,81,95,109,123,137,222
PK7a	112.20	2.49	−0.80	0.03	236	41,55,67,81,95,109,123,137,236
PK7b	111.40	2.52			236	41,55,67,81,95,109,123,137,236
PK8a	119.53	2.45	−0.80	0.04	250	41,55,67,81,95,109,123,137,250
PK8b	118.73	2.49			250	41,55,67,81,95,109,123,137,250
PK9a	125.67	2.34	−0.67	0.04	264	41,55,67,81,95,109,123,137,264
PK9b	125.00	2.38			264	41,55,67,81,95,109,123,137,264
PK10a	131.00	2.29	−0.67	0.03	278	41,55,67,81,95,109,123,137,278
PK10b	130.33	2.32			278	41,55,67,81,95,109,123,137,278
PK11a	135.53	2.27	−0.53	0.03	292	41,55,67,81,95,109,123,137,292
PK11b	135.00	2.30			292	41,55,67,81,95,109,123,137,292

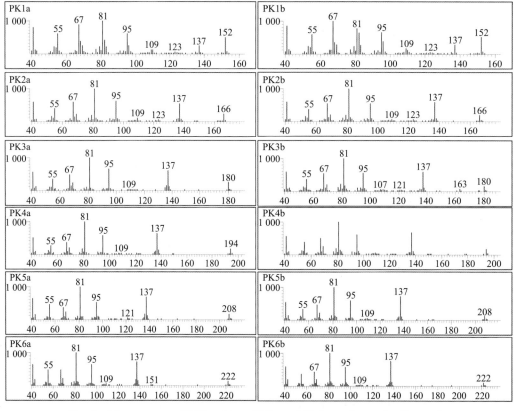

Fig. 2 Mass spectra of PK1a, 1b to PK6a and 6b from the crude oil sample.

The whole series had the same characteristic ions of m/z 137, indicating a decalin cation. Besides, there were no other ions with significant abundance between molecular ions and the ion with m/z 137. This suggested that the single substituent on the decalin ring has high probability, as the loss of whole alkyl groups was the main product. Besides, the substituents were more likely long-chain alkyl groups rather than branched chain groups; otherwise there should be a strong fragment ion in between. The number of C in the long alkyl chain could be 1–26. Other fragmental ions of m/z 41, 55, 67, 81, 95, and 109 were commonly observed in the whole series. A representative mass spectrum of PK6 shown in Fig. 3 provided the proposed MS fragmentation details.

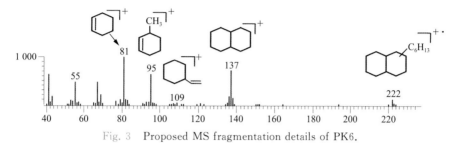

Fig. 3 Proposed MS fragmentation details of PK6.

There are two types of alkyl decalins, cis-decalin and transdecalin. Generally, the trans-decalins are more stable. In addition, they have two isomers each. Meanwhile, the alkyl group can be substituted on C-1, C-2 or C-9. However, 9-alkyl decaline is hard to be formed because of the steric effect. Comparing 1- and 2-decalines with the same substituted alkyl group, 1-alkyl decaline has larger polarity, while 2-alkyl decalinehas higher stability. Since the compounds are relatively stable in geological samples, these pairwise compounds could be trans-1-, and trans-2-alkyl decalins. From the boiling point variation, the alkyl groups increase from C1 to Cn. Furthermore, chromatography information and synthesized standards were used to discover and validate their structure details.

3.2 1-Alkyl decalins and 2-alkyl decalins in synthesized samples

The expanded GC×GC-TOFMS spectrum of STN1 and STN2 is shown in Fig. 4b and c, respectively. The two spectra were obtained in adjacent runs; hence the retention index difference from column aging could be ignored. The GC×GC retention indexes of each compound are summarized in Table 2, together with their fragment ions. The NIST search result suggested the peak at 87.07 min, 2.48 s (RT1, RT2) from Fig. 3B and the peak at 87.33 min, 2.46 s (RT1, RT2) were butyldecahydronaphthalene, which conversely verify the synthesized products.

The isomers of each alkyl decalin were sharing the same MS information (Fig. 4). For example, 1-butyldecahydronaphthalene had the same MS information as 2-butyldecahydronaphthalene, including both the molecular ion (m/z 194) and fragment ions (m/z 137, 41, 55, 67, 81, 95 and 109). 1-Pentyldecahydronaphthalene and 2-pentyldecahydronaphthalene had identical mass spectra. 1-Hexyldecahydronaphthalene was almost the same as 2-hexyl-decahydronaphthalene in their mass details.

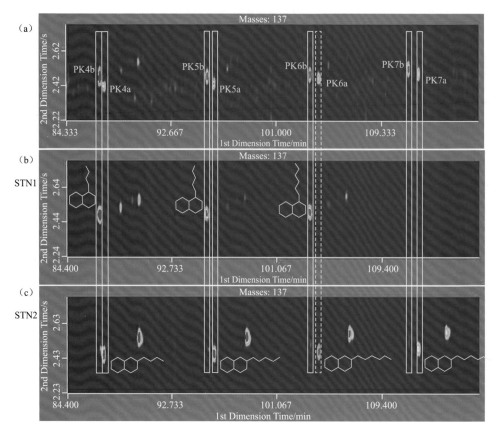

Fig. 4 (a) Expanded GC×GC/TOF-MS area of the PK4 − PK7 area of the crude oil sample; (b) expanded GC×GC/TOF-MS area of STN1; (c) expanded GC×GC/TOF-MS area of STN2.

Table 2　GC×GC/TOF-MS details of peaks from STN1 and STN2.

Sample info	Name	1st Retention time (RT1)/min	2nd Retention time(RT2)/s	ΔRTI/min (PKb−PKa)	Molecular ion(m/z)	Fragment ions (m/z)
STN2	2-Butyldecahydronaphthalene	87.333	2.46	−0.2666	194	41,55,67,81,95, 109,137,194
STN1	1-Butyldecahydronaphthalene	87.067	2.48		194	41,55,67,81,95, 109,137,194
STN2	2-Amyldecahydronaphthalene	96.133	2.46	−0.5333	208	41,55,67,81,95, 109,137,208
STN1	1-Amyldecahydronaphthalene	95.600	2.48		208	41,55,67,81,95, 109,137,208
STN2	2-Hexyldecahydronaphthalene	104.400	2.47	−0.667	222	41,55,67,81,95, 109,137,222
STN1	1-Hexyldecahydronaphthalene	103.733	2.50		222	41,55,67,81,95, 109,137,222
STN2	2-Heptyldecahydronaphthalene	112.133	2.49		236	41,55,67,81,95, 109,137,236

Although MS was not able to distinguish isomers based on m/z ratios, chromatograms provided different retention indexes for each isomer. Compared to the corresponding 2-alkyl decalins, C-1 substituted ones had shorter eluting time on RT1 (Fig. 4b), but longer eluting time on RT2 (Table 2). For example, 1-butyldecahydronaphthalene came out earlier at 87.07 min (RT1) but stayed longer on the 2nd dimension (RT2 = 2.48 s), compared to 2-butyldecahydronaphthalene located at 87.33 min, 2.46 s (RT1, RT2, Table 2). Based on the non-polar/polar GC system combination, it can be inferred that 1-butyldecahydronaphthalene had lower boiling points but higher polarity than 2-butyldecahydronaphthalene.

3.3 Comparison of synthesized samples and the petroleum sample

MS spectra of synthesized compounds were compared with PK1 – PK11 in the petroleum sample. The MS signatures of both synthesized butyldecahydronaphthalene isomers were almost the same with those of PK4a and PK4b from the oil sample. They had the same molecular ion (m/z 194), and characteristic fragment ions (m/z 137, 55, 67, 81, and 95) as shown in Tables 1 & 2.

Similarly, the MS spectra of two pentyldecahydronaphthalene isomers were analogous to those of PK5a and PK5b (Fig. 3). Hexyl-decahydronaphthalene isomers and 2-heptyldecahydronaphthalene were almost identical to PK6 (a & b) and PK7 (a & b) in terms of their mass spectra. Hence, we can conclude that PK4 (a & b), PK5 (a & b), PK6 (a & b), and PK7 (a & b) were the isomers of butyldecahydronaphthalene, pentyldecahydronaphthalene, hexyl-decahydronaphthalene, and heptyldecahydronaphthalene respectively.

The information of the retention index was further investigated. As the STN1 and STN2 were analyzed in the same batch, their retention indexes were comparable. Their chromatogram areas were aligned as shown in Fig. 4b & c. Then the expanded chromatogram area of PK4 – 7 from the oil sample (Fig. 4a) was compared with those of two synthesized standards. Each isomer of PK4 – 7 was neatly arranged with the synthesized 1-or 2-alkyl decalines. For example, PK4a and PK4b were closely aligned with 2-butyldecahydronaphthalene and 1-butyldecahydronaphthalene respectively. PK5a and PK5b were strictly aligned with 2-pentyldecahydronaphthalene and 1-pentyldecahydronaphthalene respectively. The same trend can be found in PK6a and PK6b. The PK7a aligned tightly with 2-heptyldecahydronaphthalene in STN2.

From both the MS and chromatography information, we can conclude that the series of pairwise compounds were isomers of 1-and 2-alkyl decalines. The peak "a" of each pair was C-2 substituted, while the peak "b" was C-1 substituted.

3.4 Impact of alkyl substituents

It is well known that with the growth of the alkyl substituents, boiling points of alkanes increase. It was observed that the boiling point difference (ΔRT1) between the isomers within a pair changed as the substituent chain grew (Table 1). For example, ΔRT1 was 1.60 min when comparing 2-methyldecahydronaphthalene (PK1a) with 1-methyl-decahydronaphthalene (PK1b) while the number decreased to 0.93 when comparing 2-ethyldecahydronaphthalene (PK2a) with 1-ethyldecahydronaphthalene (PK2b). The trend even started to reverse from 4 carbon hydrocarbyl substituted decalins (PK4), the boiling point of 2-alkyl decalins was higher than that of 1-alkyl decalins ($\Delta RT1<0$). This trend was observed from both the petroleum sample and synthesized standards (Table 2). This "inversion" phenomenon in the twodimensional chromatogram was because of the electronic effect of substituents. From PK1 to PK7, the increasing rate of RT1 of 2-alkyl decalins was larger than that of 1-alkyl decalins. When the carbon number in hydrocarbyl substituents increased further than 7, the impact on retention time changes arising from the substituent position could be negligible, which is supported by the consistent $\Delta RT1$ of -0.80 min. On the other hand, 1-alkyl decalins always had longer eluting time on column 2 ($\Delta RT2>0$), due to their stronger polarities than 2-alkyl decalins. It was also observed that $\Delta RT2$ kept decreasing, which means that the substituent position effect decreases on the polarity as the substituents grow.

These alkyl decalins have been discovered in both source rock and crude oil samples. In degradable oil, decalins with short alkyl groups have also been found. Though they always remained in the unresolved complex mixture (UCM) fraction, they started to raise people's interest. The presence of a large number of long-chain substituent groups may suggest that transformation of bacteria plays a significant role in these compounds' formation processes. The different length of side chains may correlate to various degrees of oil cracking.

4 Conclusion

In this study, GC×GC/TOF-MS was used to analyze petroleum samples. A series of pairwise peaks in both condensate and source rock extract samples were observed, which have not been resolved in the common GC/MS. A tentative analysis demonstrated that they were alkyl decalins with different substituted positions. As the MS spectra could not clearly differentiate isomers, a series of alkyl decalins isomers were synthesized. The GC×GC/TOF-MS results of synthesized standards were analyzed together with both petroleum samples. With the analogous MS spectra and close retention index, we can conclude that the pairwise compounds were 1-alkyl-decahydronaphthalenes and 2-alkyl-decahydronaphthalenes, and the former has higher polarity. Their long chain substituent groups may come from bacterial transformation or different oil cracking results. These alkyl decalin isomers could be used as potential petroleum biomarkers representing the source and age related geographical information. More studies are needed to explore their specific geochemical significance.

论文 5　2013 年发表于《Organic Geochemistry》

Use of comprehensive two-dimensional gas chromatography for the characterization of ultra-deep condensate from the Bohai Bay Basin, China

Guangyou Zhu, Huitong Wang, Na Weng, Haiping Huang, Hongbin Liang, Shunping Ma

Abstract　A gas condensate from well ND1 in the Jizhong Depression of the Bohai Bay Basin, China is characterized by two-dimensional gas chromatography with flame ionization detector (GC×GC-FID) and time-of-flight mass spectrometry (GC×GC-TOFMS). This condensate is sourced from the fourth member of the Shahejie Formation (Es_4) but reservoired in the Mesoproterozoic Wumishan Formation carbonate at a depth of 5 641 − 6 027 m and the reservoir temperature is 190 − 201 ℃. It is the deepest and the highest temperature discovery in the basin to date. The API gravity of the condensate is 51° and the sulfur content is < 0.04%. A total of 4 955 compounds were detected and quantified. Saturated hydrocarbons, aromatic hydrocarbons and non-hydrocarbon account for 94.8%, 5.1% and 0.02% of the condensate mass, respectively. Some long chain alkylated cyclic alkanes, decahydronaphthalenes and diamondoids are tentatively identified in this condensate. The $C_6 - C_9$ light hydrocarbon parameters show that the gas condensate was generated at relatively high maturity but its generation temperature derived from the dimethylpentane isomer ratio seems far lower than the current reservoir temperature. Some light hydrocarbon parameters indicate evaporative fractionation may also be involved due to multiple-charging and mixing. The diamondoid concentrations and gas oil ratio (GOR) suggest that the ND1 condensate results from 53.3% − 55% cracking. Since significant liquids remain, the exploration potential of ultra-deep buried hill fields in the Bohai Bay Basin remains high.

1　Introduction

It is generally accepted that C_{15}^+ hydrocarbons are unstable under catagenetic conditions and progressively degrade to methane and pyrobitumen between 150 ℃ and 200 ℃. Gretener and Curtis stated source rocks would pass entirely through the oil window in little more than 10 Ma at 140 ℃. Some kinetic models have invoked major oil accumulations to be unstable at temperatures exceeding 175 − 200 ℃ and with oil able to be preserved to temperatures as high as 200 ℃ in cases where sedimentation is very rapid and recent. In contrast to the traditionally accepted temperature range for petroleum generation-maturation reactions for kerogen, the evidence is that crude oil is much more thermally stable. This is based on field observations, laboratory simulations and theoretical considerations.

Reservoir pressure is one of the reasons for the thermal stability of hydrocarbons, causing some delay in thermal cracking. Price and Wenger documented that increasing fluid pressures retarded all aspects of hydrocarbon generation and maturation in aqueous pyrolysis experiments. Other studies have arrived at similar conclusions based on re-evaluation of kinetic parameters for oils.

Comprehensive two-dimensional gas chromatography (GC×GC) coupled to a time-of-flight mass spectrometer (GC×GC-TOFMS) or a flame ionization detector (GC×GC-FID) is a new emerging gas chromatographic technology. It features two columns with different polarity stationary phases that are connected in series via a modulator. It is a very powerful analytical tool for high resolution separation and analysis of highly complex organic mixtures, as well as a quantitative analysis of their compositions based on internal and external standards. The technique has been used to study the composition of refined products, environmental samples, crude oils, source rock extracts and pyrolysates. However, previous reports focus mainly on crude oil fingerprinting and characterization of the unresolved complex mixture (UCM) in biodegraded oils. There are rare studies reporting GC×GC analysis of condensate samples. Although gas condensates typically consist almost entirely of $C_5 - C_{10}$ range hydrocarbons, in deep accumulations, due to high temperature, relatively high molecular weight hydrocarbons may still be present and unsuitable for direct GC analysis.

For this study, we employed a LECO Pegasus 4D analytical system (GC×GC-TOFMS and GC×GC-FID) to characterize the detailed molecular composition of a condensate from well ND1 in the Jizhong Depression of the Bohai Bay Basin, northern China. The reservoir has been buried to a depth of 5 641 – 6 027 m and has a temperature of 190 – 201 ℃. In addition to full molecular characterization of the condensate, this paper addresses the issue of whether the formation of ND1 condensate was purely the result of thermal generation or migrational fractionation (through phase separation induced by pressure reduction or a later gas influx into a deeper oil accumulation). With exploration targets pushing to ever increasing depth, it is essential to improve our understanding of the evolution of such light fluids by determining their thermal maturity and stability.

2 Geological settings

The Bohai Bay Basin, with an area of 200 000 km^2, is a rift basin of the Cenozoic age and an important petroliferous basin in eastern China. The tectonic environment of Bohai Bay Basin changed from extension to compression during the Neogene as a result of the collision of the Pacific Plate with Eurasia. The Jizhong Depression is one of the subunits situated at the west of the Bohai Bay Basin, bounded by Taihang Uplift in the southwest and Cangxian Uplift in the east (Fig. 1a). It is one of the most productive petroleum provinces in the basin and relies to a large extent on syn-rift sediments for source and reservoir rocks. The Shahejie Formation deposited during the Eocene contains the main exploration targets (Fig. 1b). However, abundant oil has also been found in pre-Tertiary buried hill reservoirs.

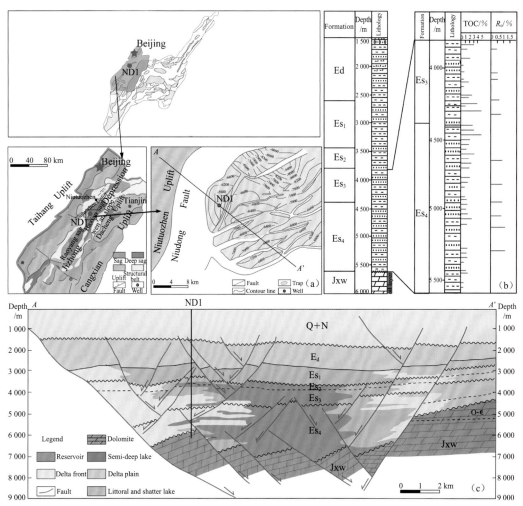

Fig. 1 Integrated geological map of well ND1, Bohai Bay Basin. (a) Basin and well location with geological outline; (b) lithological column of the Baxian Sag with insert figure showing organic carbon content and vitrinite reflectance profiles; (c) cross section showing oil and gas accumulation of well ND1. N+Q: Neogene+Quaternary; Ed: the Eogene Dongying Formation; Es_1: the first member of the Eogene Shahejie Formation; Es_2: the second member of the Eogene Shahejie Formation; Es_3: the third member of the Eogene Shahejie Formation; Es_4: the fourth member of the Eogene Shahejie Formation; O: Ordovician; $\in$: Cambrian; Jxw: the Wumishan Formation of the Mesoproterozoic Jixian Group.

Well ND1, located in the Baxian Sag of the Jizhong Depression, is one of the largest discovery wells in the buried hill exploration province (Fig. 1c). The Baxian Sag has an area of 2 400 km² with a maximum thickness of sedimentary strata of 10 000 m. The fourth member of the Shahejie Formation (Es_4) in the Baxian Sag, which consists mainly of semi-deep lacustrine mudstones and carbonaceous mudstones, is the main source rock for the ND1 condensate. This member is nearly 1 000 m thick and has total organic carbon (TOC) content ranging from 0.4% to 5.0% with an average value of 1.48%. The organic matter, dominated by Type II kerogen, is highly matured with vitrinite reflectance up to 1.5%, within the condensate to wet gas generation stage. The Niudong structure in the Baxian Sag is a faulted buried hill resulting from the periodic movement of the Niudong Fault. Well ND1 is located in the structural high surrounded by this suite of source rocks. The condensate is reservoired in the Wumishan Formation of the Mesoproterozoic Jixian Group (Jxw) dolomite buried hill, which experienced a long period of weathering and leaching. Dissolution pores, cavities and fractures are well developed with effective porosity in the range of 2.2%–19.5% (6.3% on average) and an average permeability value of 213 mD. This well has penetrated a minimum hydrocarbon column height of 427 m since it did not reach gas water contact.

Well ND1 is the most productive well with liquid hydrocarbon rate of 642.9 m^3/d and gas rate of 56.3×10^4 m^3/d. The produced fluid has gas oil ratio of 875 m^3/m^3 (3 685 scf/bbl), API gravity of 50.1° and viscosity of 1.2 mPa·s at 20 ℃. PVT calculation suggests this high API gravity liquid at surface conditions was a gas at reservoir conditions. It contains 0.03%–0.04% sulfur, 13%–17% wax and trace amounts of resin and asphaltene. The associated gas is wet with 81.8%–86.5% methane and >10% of C_2^+ heavy components. The reservoir is slightly over pressured with pressure 1.01 times higher than the hydrostatic pressure.

Two other condensates were investigated for comparison purpose in the present study. One is from recently dilled well ND101. This condensate is reservoired in the same strata as the sample from well ND1 at a depth of 5 584.2–5 930.0 m. Its API gravity is 52.6° with 0.44% NSO compound content. Another condensate is from well TZ 83 at a depth of 5 666.1–5 681 m in the Tarim Basin NW China. This condensate has API gravity of 41.1° at 20 ℃ and viscosity of 5.59 mPa·s at 50 ℃. It is a wax rich condensate with gas oil ratio of 60 300 m^3/m^3.

3 Experimental methods

For GC×GC analysis, the whole condensate sample was dissolved in a solution of 5% DCM in *n*-hexane. Two LECO Pegasus 4D GC×GC systems were used in this study coupled with a TOFMS and a FID, respectively. They were equipped with an Agilent 6 890 N GC (TOFMS) and an Agilent 7 890 A GC (FID system) and configured with a split/splitless auto-injector and a dual stage cryogenic modulator. Two capillary GC columns were fitted in the GC. The first dimension chromatographic separation was performed by the non-polar HP-5MS column (50 m×0.2 mm I.D. and 0.5 μm film thickness). The second column connected to the TOFMS instrument was a DB-17HT column (3 m long×0.1 mm I.D. and 0.1 μm film thickness). Cryofocusing by liquid nitrogen and a quad jet dual stage modulator (Zoex, Houston, TX, USA) was applied.

For GC×GC-TOFMS and GC×GC-FID analysis, split injection of 0.5 μL sample solution (split ratio 1∶700) was applied at an injection temperature of 300 ℃. Temperature programming was performed at an initial temperature of 35 ℃ at the primary GC oven. Then the temperature increased at a rate of 1.5 ℃/min up to 210 ℃, followed by an increase at the rate of 2 ℃/min to the final temperature of 300 ℃ and kept constant for 20 min. The secondary oven was programmed 5 ℃ above of the primary GC oven gradient. The modulation period was 10 s with a 2.5 s hot pulse duration and a 45 ℃ modulator temperature offset to the primary oven temperature. The carrier gas (helium) flow rate was 1.5 mL/min. The TOFMS instrument was operated in the electron ionization mode with a range of m/z 40 – 520. Ion source temperature was 240 ℃, the detector voltage was set at 1 600 V, the applied electron energy was 70 eV and the acquisition rate was 100 spectra/s. The FID signal acquisition rate was 200 spectra/s at detector temperature of 320 ℃. The flow rate of carrier gas, hydrogen and air were set up at 50, 40 and 450 mL/min, respectively. Instrument control and data processing were done using the ChromaTOF (LECO) software and Microsoft Excel. The condensate was analyzed in total ion chromatogram (TIC) and extracted ion chromatogram (EIC) modes. The deconvoluted spectra were compared with the National Institute of Standards and Technology (NIST) software library for compound identification.

On GC×GC, the x-axis represents retention time on the first dimension column where components are primarily separated by boiling point. The y-axis represents retention time on the second dimension column where separation is based on polarity. Most non-polar components are situated at the bottom of the chromatographic plane and most polar ones are situated on the top. Therefore, the simplified 2D elution order for the compound classes is the following: n-alkanes, iso-alkanes, cyclic alkanes, mono-aromatic hydrocarbons, polycyclic aromatic hydrocarbons, and heterocyclic non-hydrocarbons.

Homologue components are collected in specific retention time area of the contour plot based on 2D separations (Fig. 2a). The 3D chromatogram represented as GC×GC contour plot illustrates that the n-alkane series is dominant in the studied gas condensate (Fig. 2b). Assignment of each compound is tedious and difficult because of the complexity of the composition, i.e. different structures, isomers and heteroatom components.

4 Results and discussion

4.1 Comprehensive GC GC characterization

4.1.1 Bulk composition of ND1 condensate

Using the ChromaTOF software, a total of 4 955 compounds were detected by GC×GC-FID in the sample. These compounds were classified into eleven groups and the proportion of each group was quantified (Fig. 2c).

Molecular geochemical analyses of the condensate reveal a wide range of compounds, including n-alkanes, iso-alkanes, cyclic alkanes, aromatic hydrocarbons, and trace amounts of complex multi-ring heterocyclic non-hydrocarbons. Saturated hydrocarbons, aromatic hydrocarbons and non-hydrocarbons are 94.8%, 5.1% and 0.02%, respectively.

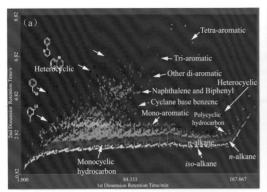
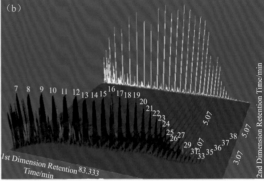
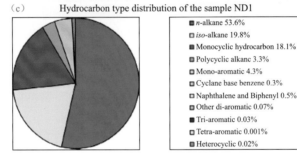

Fig. 2　GC×GC-FID analysis results for ND1 condensate oil. (a) color contour chromatogram; (b) 3D contour plot, the numbers stand for carbon numbers of *n*-alkanes and peak intensity is scaled from east intense to most intense in different colors; (c) quantitative results in groups of compound.

The GC×GC chromatogram of ND1 sample consists dominantly of n-alkanes, ranging from n-C_5 to n-C_{35} with a peak at n-C_{15}. Lower carbon number homologues predominate, and there is no odd/even carbon number predominance. *n*-alkanes are the most abundant compounds and accounts for 53.6% of the condensate. Branched (*iso*-) alkanes including pristane (Pr) and phytane (Ph) are the second richest components and accounts for 19.8% of the condensate. Among the cyclic alkane class, monocyclic alkane series is the richest one, which is 18.1% of condensate. In aromatic hydrocarbon classes, mono-aromatic hydrocarbons including alkylbenzenes and alkyltoluenes are the most abundant compound class while polycyclic aromatic hydrocarbons are also present. Trace amounts (0.02%) of non-hydrocarbon compounds are detected in the sample, including chlorotoluene, benzaldehyde, methylquinoline and phenylpyridine. Chlorotoluene and benzaldehyde have not been reported in a condensate before.

4.1.2　Long chain alkylated cyclic alkanes

Fig. 3 shows a surface plot GC×GC extracted ion chromatogram for sample ND1. In addition to n-alkanes, some long chain alkylated cyclic alkanes such as long chain alkylated cyclopentanes, cyclohexanes, methylcyclohexanes and long chain alkylated benzenes were detected. The alkylated cyclohexanes and cyclopentanes have carbon numbers of up to C_{37} on the side chain, while those in alkylated benzene and methylcyclohexanes are up to C_{27} and C_{34}, respectively.

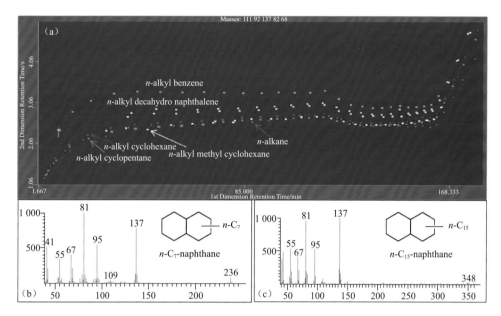

Fig. 3 GC×GC-TOFMS extracted ion chromatograms illustrating elution order and separation results for long chain alkylated compounds in the well ND1 condensate sample. (a) Color contour chromatograms of extracted ions of m/z 68,82,92,111 and 137. Each circle stands for a compound, different color circles stand for different classes of compound; m/z 82 is the base peak for n-alkylcyclohexanes; m/z 92 is the base peak for n-alkylbenzenes; m/z 111 is the base peak for n-alkylcyclohexanes; m/z 137 is the base peak for n-alkylnaphthanes; A pseudohomologous family of alkylated naphthanes was tentatively identified. (b and c) are the mass spectra of two compounds obtained by TOFMS and their inferred structures.

A homologous series with a doublet eluted between n-C_{11} and n-C_{38} in the first dimension gas chromatography and in the range of polycyclic alkanes in the second dimension gas chromatography. Molecular weights of the compounds are range from 152 to 502 and the molecular mass of each member differs by 14 atomic mass units, indicating an extra methylene unit ($-CH_2-$) inserted in the chain. The mass spectra of the compounds have a base peak of m/z 137 and characteristic fragment peaks at m/z 41, 55, 67, 81, 95, and 109. The deconvoluted spectra were compared with the ChromaTOF software library for correct matching. The library searching similarity to the decahydronaphthalene (decalin) series is greater than 900. The doublet is tentatively assigned to alkylated cis-+ trans-decahydronaphthalene. The carbon numbers of the side chain are in the range from C_1 to C_{26}. Fig. 3b and c shows the mass spectra for n-C_7-decalin and n-C_{15}-decalin, respectively. Using conventional gas chromatography-mass spectrometry, only short chain alkylated decalins can be detected ($<C_6$). The origin of homologous series of long chain decalin is not well defined, however, their occurrence in the ND1 condensate is worthy of further investigation.

4.1.3 Diamondoid hydrocarbons

The diamondoids are thermodynamically the most stable compounds in petroleum, and occur naturally in condensates, organic rich rocks and coals. Diamondoids are so named because their chemical structures consist of rigid fused cyclohexane rings, resembling diamonds. These compounds include adamantanes with one cage, diamantanes with two cages and triamantanes with three cages. Because of their structure, they resist thermal decomposition and have been used to assess thermal destruction of oil. However, they are difficult to separate and accurately quantify by conventional geochemical methods due to their low concentrations in oil. GC×GC-TOFMS improves the resolution and separation efficiency of these compounds. It not only separates the compounds that coelute in conventional GC-MS but also allows the identification of compounds that were not previously recognized. Ions were monitored at m/z 135,136,140,149,163,177,191 for adamantanes, m/z 187, 188, 201 and 215 for diamantanes, m/z 239, 240 and 253 for triamantanes. Higher diamondoids may also occur in petroleum that has undergone thermal cracking, but these have not been detected in the present study. In 3D view, homologues of diamondoids are arranged in a unique pattern on the retention plane and classes of similar compounds cluster together in specific regions of the chromatographic plane (Fig. 4). Twenty-five (25) alkylated adamantanes, sixteen (16) alkylated diamantanes and three (3) alkylated triamantanes are identified from ND1 condensate sample. In addition to the commonly detected 17 compounds, 5 C_3-alkyladamantanes, 2 C_4-alkyladamantanes and 1,2,3,5,7-pentamethyladamantane are new identifications among the adamantane series. Of the diamantanes, 4 C_2-diamantanes and 3 C_3-diamantanes are also newly recognized besides the commonly detected 9 compounds (Table 1).

A new homologue series marked with a white asterisk in Fig. 4a has not been previously reported in the literature. They are arranged regularly in the first dimension gas chromatography with mass difference of 14 atomic mass units (—CH_2—) between adjacent compounds and their retention time on the second dimensional gas chromatography is similar to 2-methyladamantane and 2-ethyladamantane. The mass spectra show the base peak of m/z 135 and characteristic fragmental peaks of m/z 67, 79, 93, 107, which are identical to that of methyladamantane. Within a given cage type, methyl (or ethyl) groups can occur at positions on the cage that are either sites of tertiary (a carbon directly bonded to three other carbons) or secondary (a carbon atom that is singly bonded to two other carbon atoms) carbons. This compound series has similar second dimensional retention time to 2-methyladmantanes and 2-ethyladmantanes, but differs from 1-methyladmantanes and 1-ethyladmantanes. Fig. 4b shows mass spectra of n-C_2 and n-C_9 alkylated adamantanes. Based on retention time, mass spectrum and NIST library searching, this series is inferred as secondary alkyl chain substituted adamantanes. The presence of long alkyl chain diamondoids is surprising given that most long alkyl chain substituted compounds should not survive through thermal cracking. In order to verify their genuine

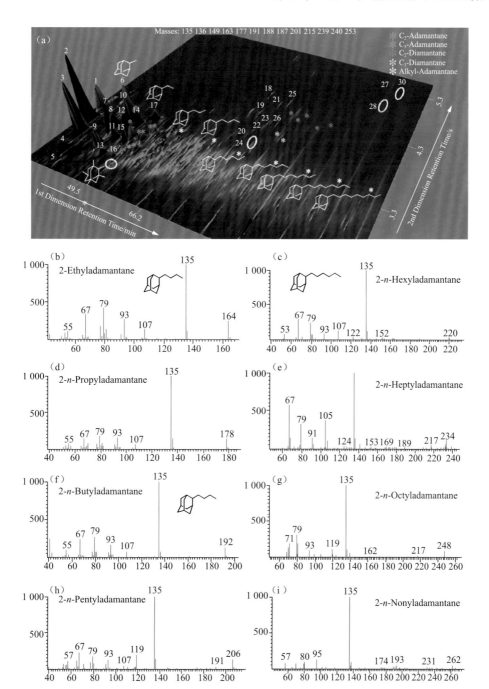

Fig. 4 Diamondoid hydrocarbon distributions in the ND1 condensate oil.
(a) GC×GC-TOFMS 3D plot for extracted ion chromatograms of m/z 135, 136, 149, 163, 177, 191, 188, 187, 201, 215, 239, 240 and 253. (b)—(i) Mass spectra of 2-n-C_2- to 2-n-C_9-adamantanes obtained by TOFMS.

occurrence, condensates from wells ND101 and TZ83 were subjected to the same GC× GC-TOFMS analysis. In contrast to the sample from ND1, these condensates show a doublet C_n-adamantane peaks, implying two isomers occurring at the same molecular

Table 1 Diamondoid compounds identified in the condensate from well ND1 by GC×GC-TOFMS analysis. See Fig. 4 for peak distribution.

Peak #	Name	1st D time /min	2nd D time /s	Base peak	M. weight	M. formula	Abbreviation
1	Adamantane	34.50	3.37	136	136	$C_{10}H_{16}$	A
2	1-Methyladamantane	36.50	3.07	135	150	$C_{11}H_{18}$	1-MA
3	1,3-Dimethyladamantane	38.00	2.89	149	164	$C_{12}H_{20}$	1,3-DMA
4	1,3,5-Trimethyladamantane	39.33	2.59	163	178	$C_{13}H_{22}$	1,3,5-TMA
5	1,3,5,7-Tetramethyladamantane	40.17	2.42	177	192	$C_{14}H_{24}$	1,3,5,7-TeTMA
6	2-Methyladamantane	42.17	3.44	135	150	$C_{11}H_{18}$	2-MA
7	1,4-Dimethyladamantane(cis)	43.50	3.08	149	164	$C_{12}H_{20}$	1,4-DMA (Z)
8	1,4-Dimethyladamantane(trans)	43.83	3.11	149	164	$C_{12}H_{20}$	1,4-DMA (E)
9	1,3,6-Trimethyladamantane	44.83	2.80	163	178	$C_{13}H_{22}$	1,3,6-TMA
10	1,2-Dimethyladamantane	46.17	3.26	149	164	$C_{12}H_{20}$	1,2-DMA
11	1,3,4-Trimethyladamantane(cis)	47.17	2.95	163	178	$C_{13}H_{22}$	1,3,4-TMA (Z)
12	1,3,4-Trimethyladamantane(trans)	47.50	3.12	163	178	$C_{13}H_{22}$	1,3,4-TMA (E)
13	1,2,5,7-Tetramethyl-adamantane	48.17	2.70	177	192	$C_{14}H_{24}$	1,2,5,7-TeTMA
14	1-Ethyladamantane	48.50	3.26	135	164	$C_{12}H_{20}$	1-EA
15	1-Ethyl-3-methyladamantane	49.83	2.98	149	178	$C_{13}H_{22}$	1-M-3-EA
16	1-Ethyl-3,5-dimethyladamantane	50.83	2.74	163	192	$C_{14}H_{24}$	1-E-3,5-DMA
17	2-Ethyladamantane	51.00	3.43	135	164	$C_{12}H_{20}$	2-EA
18	Diamantane	71.00	4.50	188	188	$C_{14}H_{20}$	Diamantane (D)
19	4-Methyldiadamantane	72.33	4.06	187	202	$C_{15}H_{22}$	4-MD
20	4,9-Dimethyldiamantane	73.50	3.68	201	216	$C_{16}H_{24}$	4,9-DMD
21	1-Methyldiamantane	75.17	4.34	187	202	$C_{15}H_{22}$	1-MD
22	1,2-+2,4-Dimethyldiamantane	75.67	3.88	201	216	$C_{16}H_{24}$	1,2-+2,4-DMD
23	4,8-Dimethyldiamantane	76.00	3.96	201	216	$C_{16}H_{24}$	4,8-DMD
24	1,4,9-Trimethyldiamantane	76.33	3.57	215	230	$C_{17}H_{26}$	1,4,9-TMD
25	3-Methyldiadamantane	77.00	4.45	187	202	$C_{15}H_{22}$	3-MD
26	3,4-Dimethyldiamantane	78.17	4.01	201	216	$C_{16}H_{24}$	3,4-DMD
27	Triamantane	99.00	5.39	240	240	$C_{18}H_{24}$	Triamantane (T)
28	9-Methyltriamantane	99.50	4.91	239	254	$C_{19}H_{26}$	9-MT
29	5-Methyltriamantane	101.83	5.22	239	254	$C_{19}H_{26}$	5-MT

mass. Retention time comparison on the second dimensional axis shows that compounds labeled as 1−7 are 2-alkyladamantanes with side chain carbon number up to n-C_8 for well ND101 and to n-C_6 for TZ83 condensate. Another series with slightly lower boiling point is inferred to be 1-alkyladmantanes (Fig. 5). As 1-methyladamantane and 2-methyladamantane are verified by internal standards in our laboratory, similar polarity on the second dimensional retention time indicates that they are the same homologue series. In general, diamondoids become more concentrated as the extent of oil cracking increases. Their occurrence in the ND1 and other condensates needs further investigation.

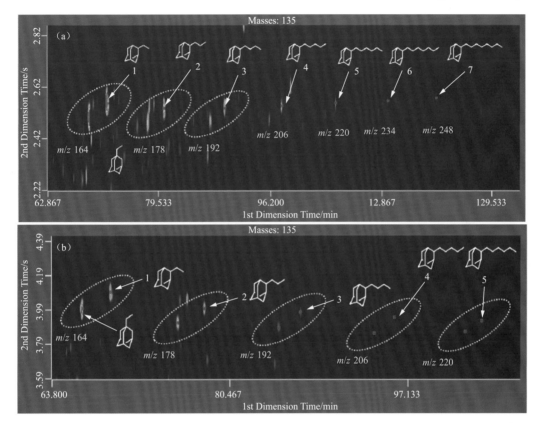

Fig. 5 GC×GC-TOFMS contour plot of long chain alkylated adamantanes in gas condensate samples. (a) well ND101; (b) well TZ83.

4.1.4 Biomarkers

All polycyclic biomarkers were below detection limit.

4.2 Genetic origin of ND1 condensate

Primary controls on gas to oil ratio distribution in a basin include the type and thermal maturity of the source rock, the charge history, and the regional distribution of carrier, reservoir and seal beds. The pressure, volume and temperature conditions in the reservoir influence GOR during hydrocarbon accumulation, whereas several secondary alteration processes influence hydrocarbon composition after trapping. There are two types of gas condensate in a reservoir. One is directly derived from thermally generated products due to high maturity of the source rock and/or crude oil (i.e., thermal condensate). Alternatively, it could be a product of secondary alteration processes in the reservoir or during migration. Evaporative fractionation is one of such alteration processes which can be induced by gas influx into an oil field, stripping its soluble components (gas washing) and condensing liquid in a shallower trap.

The $C_6 - C_9$ hydrocarbons in the ND1 condensate are dominated by n-hexane and n-heptane, indicating highly a paraffinic nature with enriched light ends. A thermal origin for a condensate is believed to reflect the source rock maturity at time of its generation and expulsion. Low relative amounts of branched alkanes, cyclic alkanes and aromatic hydrocarbons indicate a high maturity. The heptane and isoheptane values, which are convertible to equivalent vitrinite reflectance of the potential source rocks using the correlation plot of Thompson, are indicative of late oil window to wet gas stage (Table 2). Thompson introduced several indexes to recognize evaporative fractionation (and other processes) by inspecting the paraffinicity (*n*-heptane/methylcyclohexane) and aromaticity (toluene/*n*-heptane) of both residual oils and derived condensates. The ND1 gas condensate has high n-C_7/methylcyclohexane ratio of 2.7, which seems conform to evaporatively fractionated vapor phase. Other maturity related C_7 parameters, including the 2,4-/2,3-dimethylpentane (DMP) ratio and the derived calculated temperature (C_{temp}) in ℃, indicate a generation temperature of 129 ℃ (Table 2). 2,3-DMP is a primary cracking product of natural precursors while 2,4-DMP has no obvious natural precursor. Although the reliability of such calculation remains controversial, the derived value is much lower than reservoir temperature of 200 ℃. The studied condensate represents a complicated origin and worth for further investigation.

Table 2 Low molecular weight parameters for the ND1 oil.

Parameter	Ratio
i-C_5/n-C_5	0.94
Benzene/n-C_6	0.33
Toluene/n-C_7	0.45
(n-C_6+n-C_7)/(cyclohexane+methylcyclohexane)	3.35
Heptane value	44.68
Isoheptane value	6.88
n-C_7/methylcyclohexane	2.70
Cyclohexane/methylcyclopentane	1.21
n-C_7/2-methylhexane	3.41
n-C_7/methylcyclopentane	10.10
Methylcyclohexane/toluene	1.77
3-Methylpentane/n-C_6	0.33
DMP	1.07
2-MH/3-MH	1.00

continue

Parameter	Ratio
2,4-DMP/2,3-DMP	0.48
C_{temp}/℃	129
K2: $P_3/(P_2+N_2)$	0.15
N_2/P_3	0.70

The heptane value is $100 \times n\text{-}C_7 / \sum$ cyclohexane through to methylcyclohexane. The isoheptane value is (2-MH+3-MH)/(c-1,3-+ t-1,3-+ t-1,2-DMCP). Calculated temperature (C_{temp}, ℃) is $140 + (15 \times (\ln[2,4\text{-DMP}/2,3\text{-DMP}]))$. P2: 2-MH+3-MH; P3: 3,3-DMP+2,3-DMP+2,4-DMP+2,2-DMP; N2: 1,1-DMCP+c-1,3-DMCP+t-1,3-DMCP. MH, methylhexane; DMP, dimethylpentane; DMCP, dimethylcyclopentane; ECP, ethylcyclopentane.

Source rock thermal evolution and hydrocarbon generation is primarily controlled by temperature. As the upper portion the Es_4 is the most possible candidate for condensate formation in the Baxian Sag of the Jizhong Depression, its maturation history is exemplified here. The temperature history profile of well ND1 was modeled based on borehole temperature, vitrinite reflectance and thermal history reconstruction (Fig. 6). The current thermal gradient in the Jizhong Depression is 32.4 ℃/km but it could be as high as 53 ℃/km in the early Paleogene. The hydrocarbon generation is modeled using the chemical kinetic model, taking into account of Type Ⅰ/Ⅲ organic matter (lacustrine, mixed origin). Depth and temperature of the condensate generation are estimated by projecting

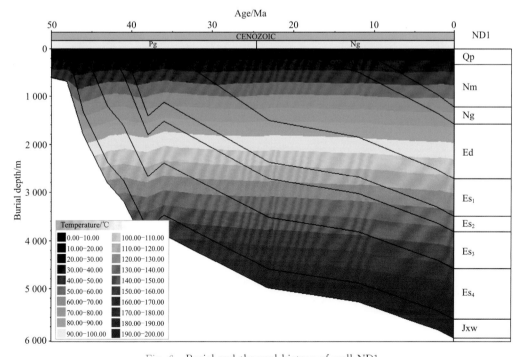

Fig. 6 Burial and thermal history of well ND1.

its equivalent maturity to the curve of Ro increase with depth representative for the same part of the basin. The default values of the kinetic parameters in the BasinMod 1D software were used.

No detailed thermal and maturity history modeling of the source rocks in the Es_4 intervals has been performed in the present study, however, the evolution of a representative geologic crosssection in the Baxian Sag enable us to determine when the source rocks passed through the oil generation window (Fig. 7). Most of the Es_4 has been buried down to depths of more than 2 km with maximum depths over 3 000 m during the deposition of Es_2. Small portion of source rock reached the oil window and limited oil is generated at

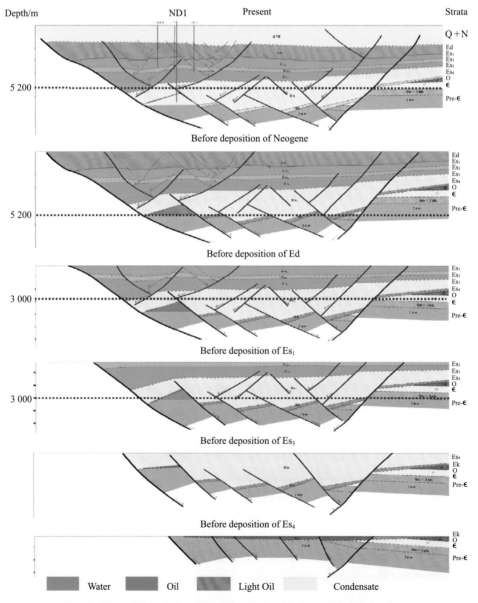

Fig. 7 Thermal evolution of geologic cross-section in the Baxian Sag.

that time. The deposition in the Es_1 proceeded leading to burial depths of the Es_4 source rock up to 3 200 m and the temperature of 100 ℃. Large quantity of oil was generated during this time period. The peak oil generation occurs during the deposition of the Dongying Formation. During the Neogene, the Es_4 source rock was deeply buried to more than 5 200 m resulting in temperatures of up to 200 ℃. Present maturities in some parts of the depocenter reached the overmature state.

Since the oil has experienced mainly lateral migration toward buried hill, the reservoir captures a wide range of charge from varying thermal maturity over geologic time. The initial charged oil could be generated at much lower temperature. Continuous charge results in a mixture of hydrocarbons generated and expelled at varying degrees of thermal maturity, while the following in-reservoir oil cracking increases GOR substantially. The discrepancy of condensate maturity and/or generation temperature may partially be caused by a multiple-charging mechanism and secondary alteration processes including evaporative fractionation.

4.3 Petroleum stability within reservoir

The current reservoir temperature is as high as 201 ℃ in well ND1, which is beyond the temperature limit of crude oil stability. The condensate has a high maturity now and has been cracked. Based on the kinetics model developed by Waples (2000) at the most typical geological heating rates, GOR of 875 m^3/m^3 in ND1 condensate corresponds to 53.3% cracking. It is obvious that more than one factor controls GOR, distorting its ability to reflect inreservoir cracking of oil. However, this simple calculation also agrees with empirical data estimated from diamondoids. Dahl et al. (1999) proposed a diamondoid biomarker method based on their concentrations to determine thermal maturity for any oil and condensate samples at any maturity level. When oil cracking occurs diamondoid concentrations become increasingly elevated with the extent of oil cracking. However, different source rock systems have a different diamondoid baseline (a nearly constant concentration in oil throughout the entire oil generation window). The diamondoid baseline in the present study was referred to the representative oil from the Shahejie Formation in the Bohai Bay Basin and the studied condensate was estimated to be about 55% cracked.

There are several reasons to explain why ND1 condensate remains in the early stages of cracking. Firstly, the basin rapidly subsided since the Neogene and the reservoir was heated at 200 ℃ not longer than a few million years. Limited effective heating time will restrict the degree of thermal cracking. Secondly, the Niudong buried hill was draped completely by Es_4 fine grained sediments and can be regarded as a closed system. Thermal cracking reactions taking place in a closed system (lack of product escape) retards organic matter metamorphism. Thirdly, multiple charges or continuous charge suppresses hydrocarbon thermal destruction. Oil generation starts when Es_4 source rocks were buried to 3 000 m. It reaches to high maturity stage when burial depth reaches 5 200 m. Main oil generation and charge time last from Es_1 to Nm (40 − 12 Ma) (Fig. 7). The early charged oil is much heavier than late charged oil. Meanwhile, thermal cracking in the carbonate reservoir might be much slower than in clastic source rock environment partly due to the lack of clay catalysis effect in reservoir. Lastly, no sulfate developed in the Precambrian reservoirs and the Es_4 source rocks in the Baxian Sag, which prevents TSR caused catalytic conditions for rapid cracking of crude oil.

Thermogenic condensate in nature can be subdivided into two subcategories, one is derived from high maturity kerogen cracking and the other is derived from cracking of oil in high temperature reservoirs. Using diamondoids is probably the only available technology to established oil to gas cracking and to correlate the cracked portion of the oil with a specific source rock. High diamondoid concentrations present in the studied condensates are a reliable indicator for the oil cracking process. Different alkylated chain length on diamondoids may suggest different degree of oil cracking. The occurrence of longer alkyl chain diamondoids in ND1 condensate may suggest a relatively low degree of oil cracking, while relatively shorter alkyl chain diamondoids present in the TZ83 condensate may indicate a higher degree of oil cracking. Understanding which of these processes is active in a basin is important because different processes will result in different vertical and lateral distributions of hydrocarbons, hydrocarbon GOR and condensate API gravity. Although more investigation is required for discerning the source and the geological processes responsible for formation of condensates in the study area, diamondoid compound identification will facilitate deep drilling for light oil and condensate exploration in similar areas.

5 Conclusions

The condensate from well ND1 is reservoired in the highest temperature reservoir found in eastern China. This well is also the deepest commercial condensate production well in the Bohai Bay Basin. The comprehensive GC×GC-FID and GC×GC-TOFMS analysis of ND1 condensate shows that this approach can yield important information about molecular compositions and structural identifications of compounds in a condensate. A total of 4 955 compounds were detected and quantified. Some long chain substituted cyclic alkanes, decahydronaphthalenes and diamondoids are for the first time reported in a gas condensate.

The genetic origin of the study condensate is complicated. It is not a typical high maturity condensate. Evaporative fractionation, coupled with multiple charging and mixing, is the reason for the complexity.

The occurrence of longer alkyl chain diamondoids in the ND1 condensate may suggest a relatively lower degree of oil cracking. Gas/oil ratio, coupled with the concentrations of diamondoids, suggests that slightly $>50\%$ of the original liquid was destroyed by inreservoir thermal cracking.

论文 6　2012 年发表于《石油勘探与开发》

凝析油全二维气相色谱分析

王汇彤，张水昌，翁　娜，李　伟，秦胜飞，马文玲

摘　要　用全二维气相色谱对四川盆地 22 个凝析油样品进行族组分和化合物定性、定量分析。凝析油中各化合物得到很好的分离，有效消除了常规色谱分析时的共馏峰干扰，在 $n\text{-}C_3 \sim n\text{-}C_8$ 色谱段有效定性出 67 个化合物，比常规的色谱轻烃分析结果多 13 个化合物；在全二维气相色谱图上可把凝析油中的各类化合物根据其结构特征分成 12 类组分。与氢火焰离子化检测器联用，过去在色谱上无法定量分析的异构烷烃、烷基环戊烷、烷基环己烷、其他单环烷烃、非甾萜类的多环烷烃、烷基苯、多环芳烃等系列化合物被同时检测，解决了凝析油族组分难以定量的问题。12 类组分、180 多种主要化合物的定量分析数据可为凝析油的成熟度判识、油源对比等油气地球化学研究提供有效数据。

关键词　全二维气相色谱　氢火焰离子化检测器　飞行时间质谱　凝析油　族组分定量

0　引　言

凝析油中轻烃含量很高，包含丰富的地球化学信息。由于其挥发性强，常规石油样品的分析方法无法满足凝析油样品分析的需要。用气相色谱-氢火焰离子化检测器（GC-FID）分析凝析油样品，可以得到正构烷烃等含量较高的少数化合物的定量结果，而大部分化合物由于相互干扰而难以进行定量分析。全二维气相色谱（GC×GC）是一种分离复杂混合物的新技术；与飞行时间质谱联用（GC×GC-TOFMS）能采集到样品中所有物质的质谱信息，为化合物的定性分析提供有效依据；与氢火焰离子化检测器联用（GC×GC-FID）能为化合物的定量分析提供有效手段。本文将 GC×GC-TOFMS 与 GC×GC-FID 相结合，对凝析油中含量较高的 $n\text{-}C_3 \sim n\text{-}C_8$ 的 67 个化合物进行定性分析，对 12 类组分、180 多种主要化合物进行定量分析，探讨全二维气相色谱在石油地质样品分析中的应用。

1　全二维气相色谱实验

本研究所用原油样品取自四川盆地二叠系茅口组和三叠系须家河组、雷口坡组。须家河组油样主要来自煤系烃源岩；雷口坡组和茅口组油样主要来自海相烃源岩。

全二维气相色谱-飞行时间质谱仪由配有氢火焰离子化检测器（FID）的 Agilent 7890A 气相色谱仪和美国 LECO 公司的 Pegasus 4D 飞行时间质谱仪组成；Agilent 7890A 气相色谱仪配有双喷口冷热调制器；工作站为 Chroma TOF 软件。

GC×GC-TOFMS 实验采用的一维色谱柱为 Petro 柱（50 m×0.2 mm×0.5 μm），升温程序为 35 ℃保持 0.2 min，以 1.5 ℃/min 的速率升到 230 ℃保持 0.2 min，再以 4 ℃/min 的速率升到 300 ℃保持 15 min；二维色谱柱是 DB-17HT 柱（3 m×0.1 mm×0.1 μm），采用与一维色谱柱炉箱相同的升温速率，温度比一维色谱柱炉箱高 10 ℃。调制器温度比一维色谱柱炉箱高 35 ℃。进样口温度为 280 ℃，分流进样模式，分流比 700∶1，进样量 0.5 μL。以氦气为载气，流速设定为 1.5 mL/min。调制周期 10 s，其中 2.5 s 热吹时间。质谱传输线和离子源温度分别为 280 ℃和 240 ℃，检测器电压 1 475 V，质量扫描范

围 40～520 amu,采集速率 100 谱图/s,溶剂延迟时间 0。

GC×GC-FID 实验采用与 GC×GC-TOFMS 实验相同的色谱实验条件。FID 检测器中载气、氢气、空气的流速分别为 50,40,450 mL/min。检测器温度 320 ℃,采集频率 200 谱图/s,溶剂延迟时间 0。

用 GC×GC-FID 仪器对广安 125 井样品进行了 7 次重复实验,并对烷烃、环烷烃、单环芳烃、多环芳烃等组分进行定量分析。7 次族组分定量分析结果的相对标准偏差(RSD)小于 3%,满足复杂体系的分析要求。在样品中加入 5α-雄甾烷和 $C_{24}D_{50}$ 标样,定量分析同一个化合物,得到的结果偏差仅为 0.269%,证明所有碳氢化合物在 FID 检测器上的响应因子一致,这与其他学者的研究结果吻合。

2 实验结果与讨论

2.1 凝析油的全二维谱图特征

全二维气相色谱谱图有 2 个特征:族分离特性和"瓦片效应"。根据全二维色谱谱图的这种性质和 TOFMS 提供的化合物质谱信息,利用 Chroma TOF 软件中的分类功能对样品中的化合物进行族组分划分。图 1a 是潼南 101 井凝析油样品在 GC×GC-FID 下的全二维点阵图。根据极性,可识别出 12 类组分:异构烷烃、正构烷烃、烷基环戊烷、烷基环己烷、环庚烷、其他单环烷烃、多环烷烃、单环芳烃、环烷基取代苯、萘和联苯系列、其他双环芳烃及三环芳烃。

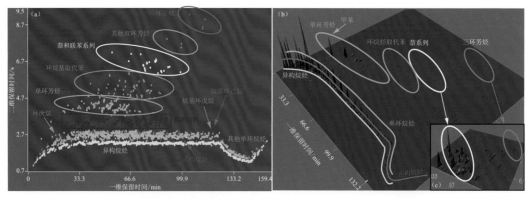

图 1 潼南 101 井凝析油样品在 GC×GC-FID 下的全二维谱图
(a) 全二维点阵图;(b) 全二维 3D 图;(c) 图 1b 中萘系列和三环芳烃的局部全二维 3D 图

潼南 101 井的凝析油样品中,信噪比在 100 以上的化合物共检测出 1 950 个。检测出的正构烷烃范围为 $n\text{-}C_3 \sim n\text{-}C_{31}$,芳烃为单环芳烃到三环芳烃;相对分子质量高的四环、五环芳烃未检出;常见的生物标志化合物如甾烷、藿烷和三环萜烷等未检出,但是存在较高含量的二环倍半萜类和金刚烷类化合物。GC×GC-FID 谱图上清晰可见成对出现的烷基环己烷和烷基环戊烷,同时可见低含量的环庚烷,这些组分在常规色谱上无法识别。图 1b 是潼南 101 井凝析油样品在 GC×GC-FID 下的全二维 3D 图,图中立体峰高度代表各个化合物的响应强度,由该图清晰可见凝析油样品中化合物的分布和含量高低情况。

2.2 $n\text{-}C_3 \sim n\text{-}C_8$ 色谱段化合物

凝析油中 C_8 之前的化合物含量最高,组分最丰富,变化最大,包含了丰富的地球化学信息,一直受到石油地质工作者的关注。用常规色谱、标样共注和保留指数等定性分析轻烃组分的工作已有报道,但由于共馏峰的存在,未能准确识别部分化合物。用全二维气相色谱分析轻组分时,化合物在一维色谱上的分布情况与常规色谱一致,而二维色谱可区分共馏化合物。图 2 是西 35-1 井凝析油样品 n-C_3～n-C_8 色谱段化合物在 GC×GC-FID 下的全二维点阵图,该区域共检测出 67 个化合物,而 GC 检测仅识别出 54 个化合物。用 GC×GC-TOFMS 对这些化合物进行定性识别,用 GC×GC-FID 进行定量分析。由于直接进样分析没有组分的损失,且碳氢化合物在氢火焰上的响应因子近似相等,因此用面积归一法得到的各个化合物面积占化合物总面积的百分比即为该化合物占总化合物的质量分数,结果见表 1。表中 67 个化合物的相似度均在 800 以上,其中有 58 个化合物的相似度在 900 以上。在这 67 个化合物中,有 9 对化合物是一维保留时间相同、二维保留时间不同的共馏峰。如 1,1,2-三甲基环戊烷与 2,3-二甲基己烷在一维色谱柱上的出峰时间均是 18.2 min,是共馏峰。这 2 个化合物一个是环烷烃,一个是异构烷烃,极性的差异使得它们在二维色谱上能很好地被分开。除此之外,环庚烷与 1 反-乙基-2-甲基环戊烷也是共馏峰,其含量相差 10 倍,用普通色谱只能检测到 1 反-乙基-2-甲基环戊烷,而忽略了环庚烷,但用全二维气相色谱两个组分都能检测到。

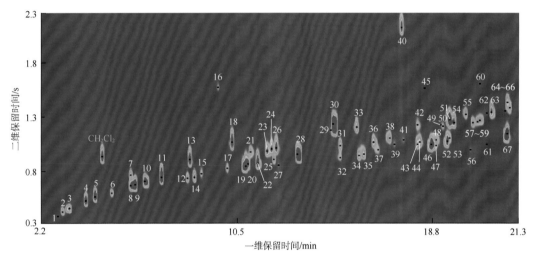

图 2　西 35-1 井凝析油样品 n-C_3～n-C_8 色谱段化合物在 GC×GC-FID 下的全二维点阵图

2.3　环烷烃类化合物

环烷烃类化合物在凝析油样品中占很大比重,是凝析油的重要组分,其中长侧链取代环己烷和环戊烷是 2 类特征化合物,可以作为判别成熟度和沉积环境等的指标,用来指示生物输入和有机物质组成的变化。以往对长侧链取代环己烷和环戊烷的研究主要集中在 C_8 之前的碳数段,很少关注 C_9 之后的碳数段,原因在于 C_9 之后的碳数段这 2 类化合物含量低,用普通气相色谱-质谱联用仪检测时容易受烷烃类化合物的干扰而不易被检测到。用全二维气相色谱检测样品中的环烷烃化合物,2 根色谱柱的正交分离可以消除高含量的烷烃类化合物的影响;不用借助质谱,仅用色谱就能检测该类化合物。

表 1 $n\text{-}C_3 \sim n\text{-}C_8$ 化合物定性定量峰表

峰号	化合物名称	相似度	质量分数/%	峰号	化合物名称	相似度	质量分数/%
1	丙烷	884	0.015	35	2,4-二甲基己烷	975	0.485
2	异丁烷	969	0.021	36	1反,2顺,4-三甲基环戊烷	940	0.275
3	丁烷	976	0.045	37	3,3-二甲基己烷	952	0.177
4	2-甲基丁烷	970	0.096	38	1反,2顺,3-三甲基环戊烷	926	0.228
5	戊烷	968	0.097	39	2,3,4-三甲基戊烷	955	0.012
6	2,2-二甲基丁烷	974	0.016	40	甲苯③	944	6.759
7	环戊烷①	931	0.017	41	2,3,3-三甲基戊烷③	960	0.029
8	2,3-二甲基丁烷①	951	0.033	42	1,1,2-三甲基环戊烷④	901	0.117
9	2-甲基戊烷	959	0.145	43	2,3-二甲基己烷④	964	0.282
10	3-甲基戊烷	971	0.098	44	2-甲基-3-乙基戊烷	958	0.137
11	己烷	969	0.264	45	1-甲基-2-亚甲基环戊烷	840	0.011
12	2,2-二甲基戊烷	977	0.035	46	2-甲基庚烷	976	2.382
13	甲基环戊烷	960	0.303	47	4-甲基庚烷⑤	962	1.074
14	2,4-二甲基戊烷	955	0.056	48	3,4-二甲基己烷⑤	941	0.181
15	2,2,3-三甲基丁烷	966	0.015	49	1顺,2反,4-三甲基环戊烷	844	0.029
16	苯	949	0.184	50	1顺,2顺,4-三甲基环戊烷	885	0.023
17	3,3-二甲基戊烷	976	0.035	51	1顺,3-二甲基环己烷⑥	933	4.100
18	环己烷	972	1.023	52	3-甲基庚烷	979	2.263
19	2-甲基己烷	934	0.109	53	3-乙基己烷⑥	904	0.229
20	2,3-二甲基戊烷	968	0.761	54	1反,4-二甲基环己烷	963	1.781
21	1,1-二甲基环戊烷	926	0.121	55	1,1-二甲基环己烷	939	0.606
22	3-甲基己烷	982	0.749	56	2,2,5-三甲基己烷	954	0.125
23	1顺,3-二甲基环戊烷	955	0.183	57	1反-乙基-3-甲基环戊烷	928	0.178
24	1反,3-二甲基环戊烷	944	0.350	58	1顺-乙基-3-甲基环戊烷	900	0.192
25	3-乙基戊烷	971	0.032	59	1反-乙基-2-甲基环戊烷⑦	931	0.184
26	1反,2-二甲基环戊烷	938	0.498	60	环庚烷⑦	894	0.017
27	2,2,4-三甲基戊烷	942	0.004	61	2,2,4-三甲基己烷⑧	879	0.004
28	庚烷	944	2.768	62	1-乙基-1-甲基环戊烷⑧	861	0.054
29	1顺,2-二甲基环戊烷	881	0.003	63	1反,2-二甲基环己烷	975	1.605
30	甲基环己烷	959	11.419	64	1顺,2反,3-三甲基环戊烷	880	0.009
31	1,1,3-三甲基环戊烷②	966	0.164	65	1顺,4-二甲基环己烷	939	1.211
32	2,2-二甲基己烷②	940	0.138	66	1反,3-二甲基环己烷⑨	971	0.023
33	乙基环戊烷	935	0.476	67	辛烷⑨	954	5.870
34	2,5-二甲基己烷	967	0.503				

注：一维保留时间相同、二维保留时间不同的化合物用相同序号标于上标，按照出峰时间依次标序，序号①～⑨表示共有 9 对共馏峰。

图 3 是潼南 101 井凝析油样品的 GC×GC-FID 全二维点阵图。该图展示了一维保留

时间 79.3～124.6 min、二维保留时间 2.80～3.04 s 内的化合物信息,包含 C_8-烷基环己烷(cyc6)～C_{13}-烷基环己烷、C_9-烷基环戊烷(cyc5)～C_{14}-烷基环戊烷共 12 个主要化合物。从图中可以看出,随着保留时间的增加,同碳数的烷基环戊烷和烷基环己烷在相对出峰位置上有所变化。C_9-烷基环己烷与 C_{10}-烷基环戊烷、C_{10}-烷基环己烷和 C_{11}-烷基环戊烷在一维色谱上是共馏峰,即使借助质谱也无法区分。在全二维气相色谱图上,烷基环己烷与烷基环戊烷的极性差异使得两者可以被分开,在色谱下就可以定性、定量识别这 2 类化合物。

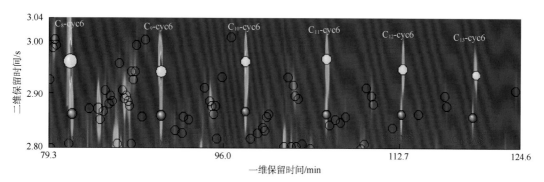

图 3 潼南 101 井凝析油样品的 GC×GC-FID 全二维点阵图
每一个气泡代表一个化合物,黄色的气泡代表烷基环己烷类化合物,红色的气泡代表烷基环戊烷类化合物

在检测的 22 个凝析油样品中均发现了金刚烷类化合物,其中广安 125 井凝析油中金刚烷类化合物含量最高,在 FID 谱图上清晰可见。图 4 中标记出 15 个金刚烷类化合物,不同取代基的金刚烷类化合物以瓦片状排列,从其出峰位置可以看出,金刚烷和 C_1-、C_2-金刚烷的极性与单环芳烃接近,C_3-金刚烷、C_4-金刚烷的极性与多环烷烃接近。因此对多环烷烃和单环芳烃组分进行定量分析时要注意金刚烷类化合物的影响。目前判定凝析油成熟度的方法有限,常用的地球化学参数对凝析油样品都不适用。GC×GC-FID 可以很好地解决特征环烷烃类化合物的定量问题,为找到更多的凝析油成熟度判定参数提供了有效支持。

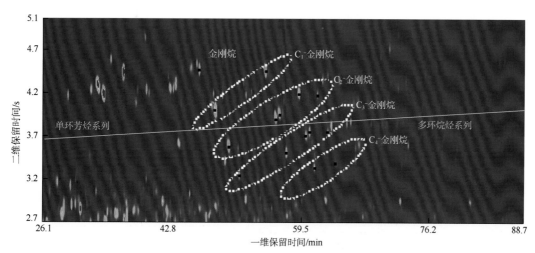

图 4 广安 125 井凝析油样品的 GC×GC-FID 全二维点阵图
黑点为金刚烷类化合物

2.4 GC×GC-FID 的凝析油族组分定量

对 22 个凝析油样品进行族组分划分,12 类族组分的定量分析结果列于表 2。表中出峰数量是指样品中信噪比在 100 以上的峰的数量总和。样品饱和烃含量是正构烷烃、异构烷烃、烷基环己烷、烷基环戊烷、环庚烷、其他单环烷烃和多环烷烃含量的总和,芳烃含量是单环芳烃、环烷烃取代苯、萘系列、其他双环芳烃和三环芳烃含量的总和。表 2 中西 35-1 井和西 20 井样品的饱和烃、芳烃含量差别不大,但化合物组成和族组分定量分析结果有很大区别:西 35-1 井样品的化合物数量少,信噪比在 100 以上的仅有 770 个峰,且以轻组分为主,在 n-C$_{11}$ 之前出峰的化合物质量占总质量的 98.4%,芳烃组分以单环芳烃为主,萘系列化合物含量低,其他双环芳烃和三环芳烃未检出;而西 20 井样品的化合物数量和种类多,信噪比在 100 以上的有 2 296 个峰,在 n-C$_{11}$ 之前出峰的化合物质量仅占总质量的 65.9%,芳烃组分从单环芳烃到三环芳烃均能检测到,有较高含量的萘及联苯化合物。

表 2 22 个凝析油样品族组分定量结果

井 名	出峰数量/个	正构烷烃/%	异构烷烃/%	烷基环戊烷/%	烷基环己烷/%	环庚烷/%	其他单环烷烃/%	多环烷烃/%	单环芳烃(苯系列)/%	环烷基取代苯/%	萘(及联苯)系列/%	其他双环芳烃系列/%	三环芳烃系列(菲等)/%
潼南 101	1 950	26.86	40.54	1.91	8.82	0.003	11.12	1.31	8.49	0.66	0.27	0.015	0.014
潼南 104	1 465	24.97	48.15	1.52	7.50	0.003	9.58	0.90	6.88	0.38	0.11	0.003	0.002
潼南 105	1 126	21.32	48.38	1.38	8.50	0.004	11.36	0.91	7.87	0.24	0.04	0.001	0.001
女 103	1 443	22.97	51.14	1.53	7.43	0.006	9.69	0.84	6.18	0.18	0.04		0.001
合川 124	1 126	19.67	58.18	0.97	6.37	0.004	8.76	0.79	5.07	0.16	0.03		
合川 001-12	1 549	22.20	54.44	1.15	6.25	0.005	9.65	0.73	5.37	0.18	0.03		
广安 002-40	715	19.80	24.15	4.03	20.93	0.013	18.02	0.95	11.87	0.18	0.06		
广安 002-H1	801	22.08	24.01	3.32	18.66	0.012	18.60	1.17	11.82	0.25	0.09		
广安 002-X27	752	20.85	24.69	3.44	19.87	0.013	17.75	1.01	12.17	0.18	0.06		
广安 125	2 041	19.73	36.34	1.08	7.98	0.010	16.46	5.17	11.73	1.25	0.23	0.007	0.004
西 35-1	770	15.16	22.38	0.89	16.51	0.017	19.58	1.63	23.55	0.25	0.04		
西 20	2 296	23.69	27.23	1.65	12.69	0.007	13.33	2.35	15.44	2.34	1.14	0.091	0.036
角 33	2 001	24.14	25.94	1.21	14.39	0.010	14.82	2.37	15.24	0.61	1.19	0.055	0.023
金 2	2 367	17.11	21.39	2.15	15.54	0.008	14.59	3.03	20.58	3.47	1.71	0.289	0.142
遂 37	1 625	26.17	33.64	1.72	12.32	0.003	12.26	1.58	11.23	0.75	0.31	0.011	0.004
遂 56	1 825	25.99	42.67	1.76	9.32	0.003	9.94	1.09	8.40	0.58	0.23	0.011	0.010
丹 11	1 645	24.76	42.24	0.66	4.22	0.003	17.33	2.04	8.64	0.09	0.02		
包 16	916	14.25	19.59	5.27	29.41	0.015	21.98	1.50	7.70	0.21	0.08		
岳 101	2 014	26.90	44.13	1.52	7.27	0.001	11.04	1.22	7.24	0.46	0.20	0.013	0.009
音 27	1 400	25.23	44.60	1.01	6.86	0.007	9.79	1.53	10.42	0.37	0.19		0.001

续表

井　名	出峰数量/个	正构烷烃/%	异构烷烃/%	烷基环戊烷/%	烷基环己烷/%	环庚烷/%	其他单环烷烃/%	多环烷烃/%	单环芳烃（苯系列）/%	环烷基取代苯/%	萘（及联苯）系列/%	其他双环芳烃系列/%	三环芳烃系列（菲等）/%
界6	1 090	20.10	32.50	2.68	18.67	0.010	17.51	1.89	6.20	0.25	0.18		
中23	1 431	25.59	27.05	1.29	5.66	0.003	13.02	1.92	24.57	0.67	0.21	0.001	0.004

广安地区 4 个凝析油样品饱和烃和芳烃总量相近，且正构烷烃、环庚烷、其他单环烷烃和单环芳烃含量相差不大。但在化合物组成、出峰数量和其他族组分定量数据上，广安 125 样品与其他 3 个样品有一定区别，据此将这 4 个样品分为 2 组：广安 002-40、广安 002-H1、广安 002-X27 为第 1 组，广安 125 为第 2 组。第 1 组的 3 个样品采自须六段，化合物均以轻组分为主，在 $n\text{-}C_{11}$ 之前出峰的化合物质量占总质量的 94% 以上，烷基环己烷和烷基环戊烷含量高，异构烷烃、多环烷烃、环烷烃取代苯和萘系列含量低，其他双环芳烃和三环芳烃未检出；第 2 组样品采自须四段，化合物种类和数量多、分布广，在 $n\text{-}C_{11}$ 之前出峰的化合物质量仅占总质量的 61.9%，能检测到三环芳烃和芴等其他双环芳烃系列化合物。此样品的金刚烷含量在 22 个样品中最高，约占总质量的 1.86%。这 2 组数据的差异反映出这 2 组样品可能来自不同的气源。前人的研究结果也验证了这一点。通过以上样品的对比可以看出，全二维气相色谱提供的族组分定量结果能丰富样品的地球化学信息，为凝析油样品的研究提供了更多的有效数据。

3　结　论

用全二维气相色谱分析凝析油，无需对样品进行前处理，简化了方法，避免了轻组分损失，全二维气相色谱图可以清晰展现样品的组分分布。结合氢火焰离子化检测器分析，不仅可以获得常规色谱、色谱-质谱无法得到的轻组分共馏化合物、长侧链取代环烷烃类化合物和高含量金刚烷类化合物的直接定量结果，还可以得到凝析油的族组分定量信息。全二维气相色谱对凝析油的定量分析解决了轻质油组分定量难的问题，为研究凝析油的次生蚀变、成熟度变化、沉积环境以及油源对比等提供了新的方法。一些在传统色谱分析中无法定量的化合物或族组分有可能成为今后石油地质研究的新参数。

论文 7　2014 年发表于《中国科学:地球科学》

生物标志化合物甾烷、藿烷的定量分析新方法

王汇彤,张水昌,翁　娜,张　斌,魏小芳,于　菣,魏彩云

摘　要　通过全二维气相色谱-飞行时间质谱和全二维气相色谱-氢火焰离子化检测器检测,选择合适的分析条件,建立了石油地质样品饱和烃中生物标志化合物甾烷、藿烷类化合物的定量分析新方法。新方法利用全二维气相色谱对甾烷、藿烷的有效分离,实现了甾烷、藿烷的色谱方法定量分析。12 个标准物质在 GC×GC-FID 上定量分析结果与原始数据的相对偏差小于 5‰,7 次重复对甾烷、藿烷类生物标志化合物定量分析结果的相对标准偏差小于 5‰。该方法与传统的色谱-质谱定量分析方法相比具有标样易于选用、分离度高、无共馏化合物和无相同特征离子干扰等优点。

关键词　全二维气相色谱　氢火焰离子化检测器　飞行时间质谱　饱和烃　生物标志化合物　定量分析　甾烷　藿烷

原油或者烃源岩的可溶有机质中甾烷和藿烷类生物标志化合物的鉴定和定量分析在油气勘探中具有重要意义,无论是在混源油识别、混源比例计算,还是油气二次运移示踪、生物降解原油程度判识等方面的研究中都发挥了重要作用。因为生物标志化合物的含量较低,受其他化合物共溢出的干扰,无法实现对某个化合物分子的直接定量分析。长期以来,各类生物标志化合物的定量结果都是利用色谱-质谱分析,得到化合物的特征离子碎片峰,通过内标法计算得到的。显然,特征离子碎片的响应受控于裂解机理。在此原理下,分析化学中严格的绝对定量应该采用与目标化合物结构相同、具有相同特征离子的内标化合物。但是受标样条件的限制,国内外对甾烷类和藿烷类化合物的定量都是采用单一的标样,要么是 5α-雄甾烷,要么是氘代的甾烷。众所周知,甾烷的分子结构格架是四元环,其离子电离的特征离子是 m/z 217,无论是裂解机理还是裂解碎片离子都与其他种类的生物标志化合物(如萜烷等)有很大不同。所以目前色谱-质谱分析中采用雄甾烷或氘代甾烷内标方法对藿烷等其他生物标志化合物的定量结果准确性值得商榷。即使相应的标样足够多,柱流失对低含量生物标志化合物的掩盖或者干扰,具有相同特征离子的化合物共馏的相互干扰等现象也无法避免。色谱-质谱方法之所以被油气地球化学工作者广泛使用,是受分析条件的限制不得已而为之。

全二维气相色谱仪的问世被称为气相色谱界的一次革命。该仪器的柱系统由分离机理不同而又相互独立的两根色谱柱通过调制器连接组成,调制器利用冷热喷头将一维色谱分离出的组分切割成若干份进入二维柱再次分离,一维柱不能分开的组分在二维柱上根据极性的大小在二维方向展开。与普通一维气相色谱相比,全二维气相色谱具有峰容量大、分辨率高、检测灵敏等特性,非常适合石油等复杂样品的分析。本世纪初,前人提出用全二维气相色谱分析石油组成有很好的发展前景;Frysinger 等在 2001 年首次用全二维气相色谱联用氢火焰离子化检测器(GC×GCFID)分析了石油样品,但由于当时还没有能满足全二维气相色谱分析容量的质谱,很多化合物的定性无法解决,尤其是低含量的生物标志化合物图谱信息无法识别。飞行时间质谱仪(TOFMS)的问世以及

全二维气相色谱-飞行时间质谱(GC×GC-TOFMS)仪器的商品化,为石油样品分析提供了有效的工具。一些学者首先利用 GC×GC-TOFMS 为化合物定性,然后利用 GC×GCFID 为化合物定量,如 Ventura 等将 GC×GCTOFMS 和 GC×GC-FID 技术应用到石油样品中化合物的指纹识别。本文在对石油地质样品 GC×GC-TOFMS 的图谱认识基础上,首次利用 GC×GC-FID 对石油地质样品中饱和烃组分中的甾烷和藿烷类生物标志化合物进行了定量分析研究,建立了石油地质样品中甾烷和藿烷类生物标志化合物定量分析的新方法。

1 实验方法

1.1 仪器

7890A 气相色谱仪和双喷口冷热调制器组成的全二维气相色谱仪(GC×GC)(美国 LECO 公司产品),氢火焰离子化检测器(FID)(美国 Agilent 公司产品),Pegasus 4D 飞行时间质谱仪(TOFMS)(美国 LECO 公司产品)。工作站软件 Chroma TOF,支持 GC×GC 与 TOFMS 和 FID 的联用。

Trace 气相色谱/DSQⅡ质谱仪(GC-MS),美国 Thermo 产品。

1.2 样品

本文共选取 7 个样品进行分析:1 个原油样品采自玉门油田的柳 43 井(O43),其余 6 个是采自渤海湾盆地南堡凹陷地区的泥岩抽提物样品,具体样品信息列于表 1。

表 1 样品信息

井 号	O43	NP4-15	NP4-15	NP5-96	NP509	NP509	NP511
井段/m	4 321~4 364	3 096.64	3 098.65	3 902.5	3 344.95	3 345.9	3 564.52
层 位	K_1g	Ed_3	Ed_3	Es_1	Ed_3	Ed_3	Ed_3
样品类型	黑色原油	深灰色泥岩	深灰色泥岩	深灰色泥岩	灰色泥岩	灰色泥岩	灰色泥岩

饱和烃制备:取原油或者岩石抽提后样品 30 mg 左右,用正己烷溶解后吸附于 0.5 g 硅胶(100~200 目,200 ℃活化 4 h)中,正己烷挥发后转入硅胶层析柱,用正己烷淋洗。当柱子下端溶液流出时,分 3 次共加 10 mL 正己烷(重蒸后的分析纯),收集饱和烃馏分,加入标样,用氮吹仪吹至约 1.5 mL 转至进样瓶。

1.3 标样

选择 5α-雄甾烷、全氘代二十四烷 $C_{24}D_{50}$、氘代蒽,D_{16}-金刚烷作为内标化合物。溶剂均为 CH_2Cl_2(重蒸后的分析纯)。

1.4 实验条件

1.4.1 GC×GC-TOFMS 分析条件

一维色谱柱是 50 m×0.2 mm×0.5 μm 的 Petro 柱,升温程序为 80 ℃保持 0.2 min,以 4 ℃/min 的速率升到 290 ℃保持 0.2 min,再以 0.2 ℃/min 的速率升到 305 ℃保持 1 min。二维色谱柱是 2.5 m×0.1 mm×0.1 μm 的 DB-17HT 柱,采用与一维色谱柱炉箱相同的升温速率,温度比一维色谱柱炉箱高 5 ℃。调制器温度比一维色谱高 35 ℃。进样口温度为 300 ℃,分流进样模式,分流比 15∶1,进样量 1 μL。以 He 气为载气,流速设定为

1.8 mL/min。调制周期 10 s,其中 2.5 s 热吹时间。飞行时间质谱的传输线和离子源温度分别为 300 ℃ 和 240 ℃,检测器电压 1 500 V,质量扫描范围 40~520 amu,采集速率 100 谱图/s,溶剂延迟时间 10 min。

1.4.2 GC×GC-FID 分析条件

采用与 GC×GC-TOFMS 相同的色谱实验条件。FID 中载气、氢气和空气的流速分别为 50,40 和 450 mL/min。检测器温度 310 ℃,采集频率 200 谱图/s,溶剂延迟时间 10 min。

1.4.3 GC-MS 分析条件

柱子为 60 m×0.25 mm×0.25 μm 的 HP-5MS 柱,氦气为载气,流速为 1 mL/min。升温程序为 70 ℃ 保持 5 min,以 4 ℃/min 的速率升到 220 ℃,再以 2 ℃/min 的速率升到 320 ℃ 保持 15 min。进样口温度为 300 ℃,不分流进样模式,进样量 1.0 μL。质谱检测器电压 1 700 V,选择离子方式扫描,溶剂延迟时间为 8 min。

1.5 定量方法

采用内标法定量。首先利用 GC×GC-TOFMS 和 GC×GC-FID 分别分析制备的饱和烃样品,得到外观基本一致的色谱-质谱全离子流图谱和色谱-氢火焰离子化检测器采集的色谱图,然后根据色谱-质谱全离子流图谱上各生物标志化合物的保留时间和质谱信息,定性识别出甾烷和萜烷类的各个化合物。对比质谱检测的全离子流图和一维、二维保留时间,在色谱图上找到相应的各生物标志化合物,由标样峰的峰面积和各化合物的峰面积积分结果,计算得到各生物标志化合物的含量。需要说明的是本文样品中甾烷、萜烷的含量是原油或者抽提物中的量。

2 讨论与结果

2.1 分析条件的选择

在最短的时间内达到满意的分离度是选择色谱条件的主要依据。由于生物标志化合物的相对分子质量相对较高,在一维方向出峰相对较晚,所以全二维的色谱条件选择尽量满足后半段生物标志化合物的最大分离度。升温程序前段以 4 ℃/min 的速率升温,后段以 0.2 ℃/min 升温,以保证甾烷类化合物的较好分离。由于升萜烷系列在二维方向上出峰时间较晚,因此采用 10 s 的调制周期以满足分析要求。此条件下热吹时间选用 2.5 s,载气流速选择 1.8 mL/min 能保证目标化合物的分离。为了使 GC×GC-FID 分析得到的结果与用 GC×GC-TOFMS 分析的结果具有可比性,GC×GC-FID 选用与 GC×GC-TOFMS 相同的色谱分析条件。质谱条件上,传输线选择 300 ℃ 的加热温度,以保证高沸点的化合物在由色谱传入质谱时,不会因为传输杆中冷部位的影响而使分辨率降低。

图 1 是 NP511 井饱和烃在 GC×GC-TOFMS 下的全离子流图谱,图 2 是 NP511 井饱和烃在 GC×GCFID 下的分析图谱。由图 1 和图 2 也可看出:用于化合物定性分析的 GC×GC-TOFMS 全离子流图和用于定量分析的 GC×GC-FID 分析图谱接近一致,生物标志化合物甾烷类、萜烷类、三环萜类等与正构烷烃、异构烷烃等完全分开,相互之间没有干扰,保证了生物标志化合物的准确定量。

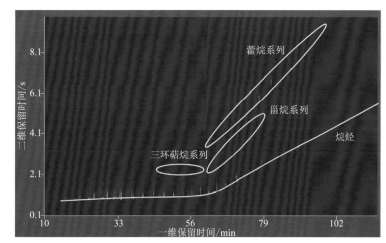

图 1　NP511 井饱和烃在 GC×GC-TOFMS 下的全离子流图谱

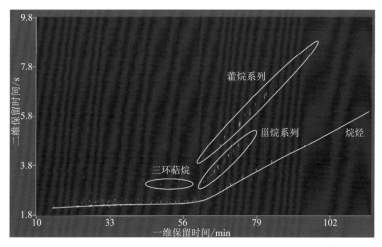

图 2　NP511 井饱和烃在 GC×GC-FID 下的全离子流图谱

2.2　GC×GC-FID 定量结果的可靠性

气相色谱-氢火焰离子化检测器(GC-FID)和气相色谱-质谱(GC-MS)分析方法是石油样品中化合物定量的常用手段。用 GC-FID 定量分析时,化合物经过分离到达检测器后,会在极化电压的作用下产生电流信号,经放大后得到相应的色谱图谱。由于任何到达检测器的化合物均能得到响应,且所有碳氢化合物在检测器中的响应因子均一致,因此该方法可以得到碳氢化合物较为准确的定量结果。由于 FID 检测器只是质量型检测器,无法分辨色谱信号峰是由单一化合物产生的,还是由多个化合物重叠产生的。因此在样品实际分析中,GC-FID 只能得到分离度较高的化合物的定量结果,大量共馏峰的存在使得多数化合物,尤其是生物标志化合物无法用 GC-FID 的方法进行定量分析。

GC-MS 主要用于化合物的定性分析,当其用于化合物的定量分析时,需要结构相同的化合物作为标样才能得到较为准确的定量结果,但在实际工作中很难得到满足,生物标志化合物的定量分析常常使用甾烷标样作为甾烷、藿烷和萜烷等的共同内标。

GC×GC-FID 与传统 GC-FID 的分析原理一样,都是将色谱分离出来的化合物送入 FID 检测器检测,区别在于色谱分离阶段,化合物在全二维色谱上比在传统色谱上多一次二维色谱柱的分离。在二维色谱柱上,化合物被调制器分割成若干个碎片,依次进入检测器检测,图谱显示的信号是计算机将同一化合物的若干信号拟合起来的结果。因此有人担心拟合信号是否能代表化合物的真实响应。为此,实验选择了12个沸点、极性均不同的纯物质配制成标准溶液,分别在GC-FID,GC×GC-FID和GC-MS仪器上进行分析,计算得到这些化合物在总离子流谱图下的定量结果,并将这三台仪器的定量结果与原始标样含量作对比,列于表2。

表2 GC-FID,GC×GC-FID 和 GC-MS 分析所配标准溶液的定量结果与原始含量比对表

化合物名称	原始含量/g	标准溶液			与原始含量对比		
		GC-FID 定量结果/g	GC×GC-FID 定量结果/g	GC-MS 定量结果/g	GC-FID 定量结果 Δ/%	GC×GC-FID 定量结果 Δ/%	GC-MS 定量结果 Δ/%
金刚烷	0.030 59	0.032 68	0.033 73	0.023 24	3.30	4.88	13.65
1-甲基金刚烷	0.013 63	0.014 32	0.014 43	0.010 36	2.47	2.85	13.63
萘	0.017 24	0.017 93	0.018 73	0.014 00	1.96	4.14	10.37
双金刚烷	0.017 03	0.018 72	0.018 46	0.016 82	4.73	4.03	0.62
芴	0.019 24	0.019 82	0.020 12	0.021 01	1.48	2.24	4.40
氘代蒽	0.011 42	0.010 57	0.010 78	0.011 33	3.87	2.88	0.40
5a-雄甾烷	0.013 82	0.013 17	0.012 65	0.013 83	2.41	4.42	0.04
萤蒽	0.013 63	0.012 77	0.012 57	0.014 83	3.26	4.05	4.22
芘	0.019 94	0.018 71	0.019 28	0.023 60	3.18	1.68	8.41
$C_{24}D_{50}$	0.028 71	0.028 13	0.026 47	0.029 40	1.02	4.06	1.19
苊	0.007 44	0.006 96	0.007 68	0.008 42	3.33	1.59	6.18
胆甾烷	0.029 13	0.028 06	0.026 95	0.035 01	1.87	3.89	9.17

注:表中 Δ 表示相对偏差,其计算公式为 $\Delta=|A_1-A_2|/(A_1+A_2)$,其中 A_1 表示 GC-FID,GC×GC-FID 和 GC-MS 的定量结果,A_2 表示化合物原始含量。

从表2的数据可以看出:GC×GC-FID 和 GC-FID 两台仪器的定量结果与原始含量的相对偏差均小于5%,满足复杂体系的分析要求;而 GC-MS 仪器的定量结果与原始含量相差较大,12个标样的定量结果有6个的偏差大于6%。原因在于这12个纯物质均是碳氢组成的化合物,在 FID 检测器上的响应因子一致,因此在定量计算时可以排除响应因子的干扰。而在 GC-MS 上,12个化合物的响应因子均不相同,需要用不同的内标化合物去标定,才能得到较为准确的定量结果。表2的数据也说明用色谱方法定量结果较为真实可靠,全二维气相色谱的信号拟合不影响化合物的定量。

为了进一步验证全二维气相色谱仪器定量的可靠性,将上面所配的包含12个纯物质的标准溶液与已有的正构烷烃标样溶液混合,可得到一个包含26个纯物质的标准溶液,并在 GC-FID 和 GC×GC-FID 下得到定量结果。由于样品是两种溶液的混合,所以无法得到准确的化合物含量值,仅能根据样品中所有组分全部出峰,采用归一化法得到这些化合物的质量百分比作为定量结果(表3)。从表3可以看出,GC-FID 和 GC×GC-FID 两种检测方法的结果非常一致,相对偏差都在小于4%,说明应用 GC×GC-FID 对有机化合物定量分析的方法是可靠的。

表 3　GC×GC-FID 与 GC-FID 对 26 种已知化合物的分析结果对比

化合物名称	GC×GC-FID 定量结果/%	GC-FID 定量结果/%	Δ/%	化合物名称	GC×GC-FID 定量结果/%	GC-FID 定量结果/%	Δ/%
金刚烷	8.62	8.19	2.56	5α-雄甾烷	3.13	3.06	1.13
1-甲基金刚烷	3.71	3.56	2.06	萤蒽	3.15	2.98	2.77
萘	4.81	4.51	3.22	芘	4.80	4.73	0.73
正十三烷	4.37	4.52	1.69	正二十二烷	3.52	3.68	2.22
正十四烷	4.46	4.62	1.76	$C_{24}D_{50}$	6.06	6.32	2.10
正十五烷	4.38	4.59	2.34	正二十四烷	3.29	3.44	2.23
双金刚烷	4.76	4.53	2.48	正二十七烷	2.53	2.61	1.56
芴	5.18	4.92	2.57	䓛	1.67	1.56	3.41
正十七烷	4.25	4.50	2.86	胆甾烷	8.64	8.66	0.12
姥鲛烷	2.18	2.22	0.91	正三十烷	2.42	2.50	1.63
氘代蒽	2.76	2.62	2.60	正三十二烷	1.20	1.25	2.04
正十八烷	4.15	4.37	2.58	正三十四烷	1.16	1.14	0.87
正二十烷	3.83	4.05	2.79	正三十六烷	0.92	0.87	2.79

注：表中相对偏差的计算公式为 $\Delta=|A_1-A_2|/(A_1+A_2)$，其中 A_1 表示 GC×GC-FID 的分析结果，A_2 表示 GC-FID 的分析结果。

2.3　饱和烃中主要生物标志化合物的定量结果

根据 GC×GC-TOFMS 提供的化合物定性信息，在 GC×GC-FID 谱图上依据相对保留时间对主要的生物标志化合物进行标注（图 3 和 4），并得到 18 个常用藿烷、伽马蜡烷以及 12 个常用甾烷的峰面积积分结果，根据 5α-雄甾烷标准样品的峰面积计算得到各样品中主要生物标志化合物含量，表 4 列出 7 个样品中 NP511 井样品的主要生物标志化合物定量结果。

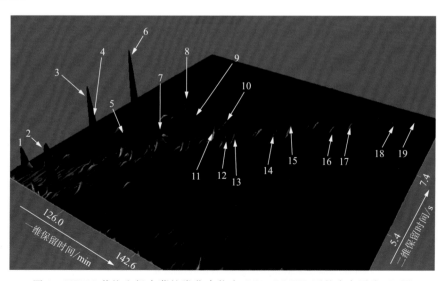

图 3　NP511 井饱和烃中藿烷类化合物在 GC×GC-FID 下的全离子流 3D 图

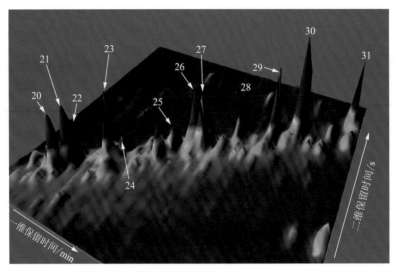

图 4 NP511 井饱和烃中甾烷类化合物在 GC×GC-FID 下的全离子流 3D 图

表 4 NP511 井样品中主要生物标志化合物的 GC×GC-FID 定量结果

序号	化合物	含量/(mg·kg^{-1})	序号	化合物	含量/(mg·kg^{-1})
1	Ts	1 063.7	17	17α,21β-29-C$_{34}$四升藿烷(22R)	488.6
2	Tm	1 720.11	18	17α,21β-29-C$_{35}$五升藿烷(22S)	138.7
3	17α,21β-30-C$_{29}$藿烷	5 430.77	19	17α,21β-29-C$_{35}$五升藿烷(22R)	77.66
4	18α-30-C$_{29}$Ts	3 757.96	20	5α,14α,17α-C$_{27}$甾烷(20S)	2 182.48
5	17β,21β-30-C$_{29}$莫烷	1 646.9	21	5α,14β,17β-C$_{27}$甾烷(20R)	531.39
6	17α,21β-C$_{30}$藿烷	9 260.87	22	5α,14β,17β-C$_{27}$甾烷(20S)	510.18
7	17β,21α-C$_{30}$莫烷	1 448.89	23	5α,14α,17α-C$_{27}$甾烷(20R)	2 677.65
8	17α,21β-29-C$_{31}$升藿烷(22S)	4 551.21	24	13β,17α-C$_{29}$重排甾烷(20R)	1 570.65
9	17α,21β-29-C$_{31}$升藿烷(22R)	3 723.06	25	5α,14α,17α-C$_{28}$甾烷(20S)	389
10	伽马蜡烷	198.54	26	5α,14β,17β-C$_{28}$甾烷(20R)	581.72
11	17α,21α-29-C$_{31}$升莫烷(22S+22R)	1 672.15	27	5α,14β,17β-C$_{28}$甾烷(20S)	246.78
12	17α,21β-29-C$_{32}$二升藿烷(22S)	1 379.29	28	5α,14α,17α-C$_{28}$甾烷(20R)	1 800.22
13	17α,21β-29-C$_{32}$二升藿烷(22R)	1 398.83	29	5α,14α,17α-C$_{29}$甾烷(20S)	982.64
14	17α,21β-29-C$_{33}$三升藿烷(22S)	1 376.85	30	5α,14β,17β-C$_{29}$甾烷(20R+20S)	2 581.82
15	17α,21β-29-C$_{33}$三升藿烷(22R)	1 375.16	31	5α,14α,17α-C$_{29}$甾烷(20R)	2 469.55
16	17α,21β-29-C$_{34}$四升藿烷(22S)	461.72			

2.4 重复性

取 O43 井原油样品的饱和烃组分用新建方法重复分析 7 次,得到 18 个常用藿烷、伽马蜡烷及 9 个常用甾烷的定量结果(表 5),7 次重复实验的 RSD%(相对标准偏差)小于 5%,说明该方法重复性好,满足生物标志化合物定量分析的要求。

第 3 章　全二维气相色谱的石油地质应用

表 5　O43 井样品 7 次重复分析的定量结果

序号	化合物	第 1 次实验结果 /(mg·kg^{-1})	第 2 次实验结果 /(mg·kg^{-1})	第 3 次实验结果 /(mg·kg^{-1})	第 4 次实验结果 /(mg·kg^{-1})	第 5 次实验结果 /(mg·kg^{-1})	第 6 次实验结果 /(mg·kg^{-1})	第 7 次实验结果 /(mg·kg^{-1})	RSD /%
1	Ts	74.87	74.82	72.43	75.11	67.03	74.08	74.35	3.93
2	Tm	243.17	235.55	226.49	240.76	219.6	244.23	245.08	4.17
3	17α,21β-30-C$_{29}$霍烷	793.85	767.73	740.67	91.82	715.55	805.57	810.72	4.60
4	18α-30-C$_{29}$Ts	108.12	104.65	99.79	107.26	94.65	106.89	107.54	4.86
5	17β,21α-30-C$_{29}$莫烷	92.07	95.66	99.12	101.35	98.66	93.27	94.25	3.56
6	17α,21β-C$_{30}$霍烷	1 994.05	1 957.02	1 900.99	2 023.14	1 764.59	1 998.69	2 012.54	4.70
7	17β,21α-C$_{30}$莫烷	277.19	276.76	267.8	281.6	268.29	279.49	244.13	4.76
8	17α,21β-29-C$_{31}$升霍烷(22S)	383.57	380.14	368.54	379.59	332.88	384.65	385.93	5.05
9	17α,21β-29-C$_{31}$升霍烷(22R)	271.41	270.43	271.36	276.49	250.92	263.8	264.2	3.12
10	伽马蜡烷	374.62	365.46	357.88	377.87	343.02	372.25	374.85	3.39
11	17β,21α-29-C$_{31}$升莫烷(22S+22R)	50.74	48.22	48.45	50.5	44.57	50.98	51.67	4.98
12	17α,21β-29-C$_{32}$二升霍烷(22S)	209.71	189.14	184.91	199.05	182.72	195.77	196.15	4.77
13	17α,21β-29-C$_{32}$二升霍烷(22R)	162.28	155.06	151.43	160.31	159.92	154.96	145.77	3.71
14	17α,21β-29-C$_{33}$三升霍烷(22S)	109.36	107.89	107.52	110.97	95.36	109.25	109.38	4.95
15	17α,21β-29-C$_{33}$三升霍烷(22R)	75.35	72.13	69.73	72.96	69.39	71.987	7.04	3.83
16	17α,21β-29-C$_{34}$四升霍烷(22S)	42.34	43.17	44.78	46.49	45.54	44.01	46.71	3.69
17	17α,21β-29-C$_{34}$四升霍烷(22R)	27.57	28.32	27.26	30.61	29.82	29.51	29.16	4.24
18	17α,21β-29-C$_{35}$五升霍烷(22S)	30.11	27.77	27.44	28.98	26.79	30.61	29.88	5.14

续表

序号	化合物	第1次实验结果 /(mg·kg^{-1})	第2次实验结果 /(mg·kg^{-1})	第3次实验结果 /(mg·kg^{-1})	第4次实验结果 /(mg·kg^{-1})	第5次实验结果 /(mg·kg^{-1})	第6次实验结果 /(mg·kg^{-1})	第7次实验结果 /(mg·kg^{-1})	RSD /%
19	$17\alpha,21\beta$-29-C_{35}五升藿烷(22R)	18.54	18.66	18.08	17.16	17.09	18.52	18.35	3.67
20	$5\alpha,14\alpha,17\alpha$-$C_{27}$甾烷(20S)	566.6	553.17	529.19	557.49	516.02	567.7	571.08	3.82
21	$5\alpha,14\beta,17\beta$-C_{27}甾烷(20R)	589.02	572.82	549.15	577.67	531.03	587.02	588.44	3.92
22	$5\alpha,14\beta,17\beta$-C_{27}甾烷(20S)	641.61	639.18	612.01	628.99	567.74	619.46	628.18	4.05
23	$5\alpha,14\alpha,17\alpha$-$C_{27}$甾烷(20R)	808.04	782.34	750.79	791.67	721.78	801.42	804.98	4.14
24	$13\beta,17\alpha$-C_{29}重排甾烷(20R)	357.32	355.02	332.26	351.95	390.49	352.21	353.95	4.85
25	$5\alpha,14\alpha,17\alpha$-$C_{28}$甾烷(20R)	770.57	761.96	736.84	763.96	819.37	754.47	757.55	3.35
26	$5\alpha,14\alpha,17\alpha$-$C_{29}$甾烷(20S)	1 449.46	1 417.97	1 530.5	1 529.59	1 448.84	1 413.19	1 470.05	3.28
27	$5\alpha,14\beta,17\beta$-C_{29}甾烷(20R+20S)	2 522.67	2 477.58	2 411.02	2 498.91	2 296.57	2 576.43	2 601.74	4.18
28	$5\alpha,14\alpha,17\alpha$-$C_{29}$甾烷(20R)	1 224.24	1 196.57	1 161.62	1 231.83	1 296.62	1 233.24	1 239.47	3.37

2.5 标样的选择与比较

为验证不同内标化合物对 GC×GC-FID 化合物定量结果的影响,实验选择 4 种化学结构不同的标准物质,分别是 $C_{24}D_{50}$、D_{16}-金刚烷、5α-雄甾烷和氘代蒽,作为内标化合物加入到分析样品中,计算 4 种标准物质对表 4 中各个化合物的含量并作比较(列出其中对 Tm 的定量结果于表 6)。结果显示,虽然选择的 4 个标样物质有烷烃、多环烷烃和芳烃的结构区别,但是在 7 个样品中由它们分别计算得到的 Tm 化合物的定量结果偏差均在 5% 以内。碳氢化合物对 FID 检测器的响应因子基本一致,任何一种无共馏干扰的烃类都可用作定量分析的标样。说明该方法标样选用方便,易于推广。

表 6 7 个样品中 4 个标准化合物对 Tm 的定量结果

井 号	O43	NP4-15	NP4-15	NP5-96	NP509	NP509	NP511
井深/m	4 321~4 364	3 096.64	3 098.65	3 902.5	3 344.95	3 345.9	3 564.52
层 位	K_1g	Ed_3	Ed_3	Es_1	Ed_3	Ed_3	Ed_3
选择的内标物质/(mg·kg^{-1})							
氘代蒽	53.96	714.29	741.27	390.94	968.92	714.24	984.01
5α-雄甾烷	52.29	668.68	674.48	364.92	1 032.42	668.38	1 063.70
D_{16}-金刚烷	55.99	708.36	668.02	379.90	995.93	712.68	979.86
$C_{24}D_{50}$	51.54	659.25	692.39	410.68	1 062.31	730.76	1 063.83
RSD/%	3.69	4.03	4.77	4.98	4.03	3.78	4.62

2.6 GC×GC-FID 与 GC-MS 方法分析结果的比较

GC-MS 是目前国内外定量生物标志化合物的常用方法,GC×GC-FID 是全新的化合物定量方法,两者对相同样品的定量结果到底存在多大差别一直未讨论过。本文用新建的 GC×GC-FID 方法和传统的 GC-MS 方法分别分析 NP511 井样品饱和烃馏分,根据 5α-雄甾烷计算得到常用生物标志化合物的定量结果并作比较,结果列于图 5。从图 5 上可以看出,藿烷类化合物在两台仪器上得到的定量结果偏差较大,GC-MS 仪器得到的结果普遍偏高,其中 $17\alpha,21\alpha$-C_{30} 莫烷在 GC-MS 上的定量结果是其在 GC×GC-FID 上的近 9 倍。造成这种现象的原因首先与内标化合物的选择有关。实验用于定量的内标化合物 5α-雄甾烷是四元环结构,而藿烷类化合物是五元环结构。两者无论是在化学结构上还是在裂解机理和裂解碎片离子上都有很大不同。用选择离子 m/z 217 下 5α-雄甾烷的峰面积积分结果去计算选择离子 m/z 191 下藿烷类化合物的含量,得到的结果与实际结果偏差较大。其次,与共馏峰以及柱流失物质干扰有关。藿烷类化合物与三环萜烷类化合物以及柱流失物质具有相同的特征碎片离子 m/z 191,GC-MS 在此离子下做藿烷类化合物的峰面积积分时,易受后两者的干扰而得到错误的积分结果。

相比之下,甾烷类化合物在两台仪器上得到的结果偏差较小。原因在于 5α-雄甾烷与甾烷类化合物一样分子结构架均是四元环,它们有相同的特征离子 m/z 217。因此在特征离子 m/z 217 下用 5α-雄甾烷计算得到的甾烷类定量结果较为精确,与 FID 检测器的定量结果也具有可比性。当然,也有几个甾烷化合物在两台仪器上的定量结果有较大差别,如 $5\alpha,14\alpha,17\alpha$-$C_{27}$ 甾烷 20R 和 $5\alpha,14\alpha,17\alpha$-$C_{29}$ 甾烷 20R 等。造成这种差别的原因与 GC-

MS 的色谱分离度、共馏峰干扰以及基线噪音干扰等因素有关。从图 4 可以看出,甾烷类化合物在色谱柱上出峰比较密集,且各个化合物间沸点差异小。常规 GC-MS 的色谱柱分辨率低、柱容量小,含量较高的化合物不易被分开,含量较低的化合物易受基线噪音影响,从而影响定量结果。

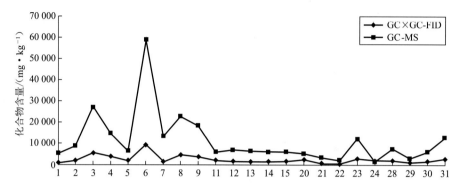

图 5 用 GC×GC-FID 和 GC-MS 分别分析 NP511 井样品得到的常规生物标志化合物的定量结果比对图
横坐标是化合物序号,该序号与表 4 中的化合物序号一致

从图 5 还可以看出,虽然两台仪器得到的化合物含量相差很大,但同系列化合物之间的含量比值具有一定的规律。在正常原油或岩石抽提物样品中,对不需要准确定量的一些常规的地球化学参数,在排除共馏峰和基线噪音干扰等情况下,GC-MS 和 GC×GC-FID 两台仪器的认识应该是一致的。但在非正常原油或混源油中,GC×GC-FID 的定量结果能够提供更精确的数据以帮助混源油比例的计算。在生物降解原油中,随着降解程度的增加,某些地球化学参数变化不明显,但是单个生物标志化合物的绝对含量变化巨大。Later 等(2012)指出,一些生物标志化合物的绝对含量变化更能反映生物降解的程度。因此,GC×GC-FID 得到的定量结果具有更重要的意义。

3 结 论

利用全二维气相色谱的高分辨率和化合物在二维空间上的展布,使得饱和烃中的甾烷、藿烷类生物标志化合物与其共馏的化合物完全分开,实现了氢火焰离子化检测器对生物标志化合物的绝对定量检测。通过人工配制标准溶液,得到相同样品在不同仪器上的定量结果。GC×GC-FID 的结果与原始浓度以及 GC-FID 的结果对比,验证了该定量方法的可靠性。7 次重复性实验的误差小于 5%,说明本方法重复性好。与目前广泛使用的 GC-MS 方法相比,该方法的分析结果更加真实可信,能够为混源油混合比例的计算、油源对比、生物降解程度判别等提供客观、有效的数据。同时该方法具有标样选用范围广(任何一种无共馏干扰的烃类都可用作定量分析的标样)、分离度高、无共馏化合物、无相同特征离子干扰等优点,能够得到客观、准确的生物标志化合物定量分析方法,为石油地球化学工作者开展研究提供了科学、有效的技术新手段。

论文 8 2012 年发表于《中国科学: 化学》

稠油中饱和烃复杂混合物成分解析及其意义

王汇彤,张水昌,翁 娜,魏小芳,朱光有,于 菡,毕丽娜,马文玲

摘 要 利用两根极性不同的毛细柱,在全二维气相色谱上分析辽河油田遭受严重生物降解形成的稠油饱和烃组分,可以将传统色谱分析时形成的"基线鼓包"即不可分辨的复杂混合物(unresolved complex mixtures)分开。根据饱和烃全二维气相色谱谱图的族分离特点和"瓦片效应",结合飞行时间质谱提供的质谱信息初步解析不可分辨的复杂混合物主要成分。发现常规色谱分析时形成的所谓"基线鼓包"是由成千上万、含量相对较低的不同取代基的环状化合物组成,这些化合物在一维色谱上以相对分子质量递增的顺序排列,在二维色谱上以极性的差异或者环的多少排列。C_{24} 之前的第一组不可分辨的复杂混合物主要由环己烷为基本单元的单环、双环和三环烷烃类化合物组成,信噪比在 100 以上的化合物数量约为饱和烃总数量的 75%,质量分数是饱和烃总量的 80% 以上,是饱和烃的主要组成部分。C_{24} 之后出现的第二组不可分辨的复杂混合物主要由四个环或者五个环为基本单元的化合物组成,信噪比在 100 以上的化合物数量约为饱和烃总数量的 17%,质量分数是饱和烃总量的 0.5%。对稠油中这些不可分辨的复杂混合物的解析有助于对其成因机理的认识和高效开采方案的制定。

关键词 全二维气相色谱 飞行时间质谱 稠油 生物降解原油 饱和烃 不可分辨的复杂混合物

1 引 言

稠油因赋存状态、流动性差等因素导致其直接开采效率极低,被称为储层中的难动用资源。原生稠油所占稠油的比例十分有限,绝大部分稠油为次生稠油,即正常原油经历热液蚀变或遭受生物降解形成。这类原油在世界各地十分普遍,在我国的塔里木、渤海湾、准噶尔、松辽等盆地生物降解油超过原油产量的 20%。受分析条件的限制,长期以来人们对稠油成分中的化合物认识水平远不及对正常原油的认识水平。常规的气相色谱-质谱(GC-MS)是分析石油样品的有效手段。但用它分析稠油时,其峰容量和分辨率的限制会导致色谱基线抬高而形成一个大鼓包,被称为"基线鼓包"。由于长期以来对鼓包中的化合物无法识别,因此被称为"不可分辨的复杂混合物"(unresolved complex mixture, UCM)。对于饱和烃组分,浓度高的化合物会在"基线鼓包"上面出峰,但有时看上去是一个峰,实际可能是多个化合物的共馏峰;浓度低的化合物会被抬起的"基线鼓包"掩盖而无法被检出。一般"基线鼓包"中不可分辨的化合物含量占饱和烃总量的 30%~80% 以上。因此,有效解析"基线鼓包"中的化合物分子组成不仅是次生蚀变原油研究中的基础科学问题,而且对稠油的成因机理认识、开发方案的进步、石油炼制方案的制定等有重要指导意义。

剖析"基线鼓包"中化合物的分子组成一直是分析工作者的追求。国外有学者尝试用化学方法如分子筛、尿素络合和四氧化钌氧化法来处理 UCM,再结合 GC-MS 分析化合物组成,但这些方法还不能分析 UCM 组成的 10%。还有些学者采用多种仪器结合的方法分析,毛细管气相色谱法、傅里叶变换核磁共振谱法(FT-NMR)、傅里叶变换红外光谱法(FT-IR)、紫外光谱法(UV)、薄层色谱法(TLC)、气相色谱-电子轰击离子源-质谱法(GC-

EI-MS)、气相色谱-化学电离源-质谱法(GC-CI-MS)、气相色谱-双质谱法(GC-MSMS)等技术都被采用过,得到的也是部分化合物的信息。

全二维气相色谱(GC×GC)是一种全新的分析复杂化合物的技术。Frysinger 等利用全二维气相色谱分析 UCM 组分,发现在 GC-MS 上呈现"基线鼓包"的物质在全二维气相色谱上可以很好地被分开。但由于当时没有找到和全二维气相色谱搭配的质谱,Frysinger 等只能依靠 GC-MS 的分析结果和全二维气相色谱谱图的特点来推测一些常规的化合物。全二维气相色谱-飞行时间质谱(GC×GC-TOFMS)的问世为 UCM 的分析提供了有效手段。Ventura 等用 GC×GC-TOFMS 分析石油沉积物抽提物中的 UCM 组成,发现在抽提物中环烷烃类化合物种类较多,从单环到六环烷烃均存在,这些化合物有异构化的趋势,作者认为可能与有机质的热液温度改变有关。2010 年 Tran 等用 GC×GC-TOFMS 分析生物降解油中的 UCM 组成,发现烷基十氢化萘系列对于 UCM 组分是一个重要的特征。他在文章中指出十氢化萘系列化合物在 UCM 组分中是规律排列的,但是只定性出 $C_1 \sim C_3$ 取代的十氢化萘,对于其他化合物并未给出结构信息。

与国外对 UCM 组成研究的进展相比,国内在这方面的研究尚未见报道。本文采用 GC×GC-TOFMS 对辽河油田的严重生物降解原油饱和烃组分进行分析,利用 GC×GC 的谱图特征和 TOFMS 的质谱图将饱和烃组分的 UCM 进行定性。从整体上认识 UCM 的组成,并尝试定性其中单个化合物,为今后 UCM 的研究提供借鉴和参考。

2 实验部分

2.1 仪器与试剂

全二维气相色谱-飞行时间质谱仪(GC×GC-TOFMS)(美国 LECO 公司),GC×GC 系统由配有氢火焰离子化检测器(FID)的 Agilent7890 气相色谱仪和双喷口热调制器组成;飞行时间质谱仪为美国 LECO 公司的 PegasusⅣ,系统为 Chroma TOF 软件。氮吹仪(美国)。

正己烷(分析纯)购自北京化工厂,使用前进一步提纯;细硅胶(100~200 目,200 ℃下活化 4 h)。

$C_{24}D_{50}$ 标准样品,溶剂为 CH_2Cl_2;5α-雄甾烷标准样品,溶剂为 CH_2Cl_2。

2.2 样品

本文样品取自辽河油田东北部凹陷的茨榆坨矿区,属于经历严重生物降解的原油。样品处理方法如下:① 取原油样品 60 mg 左右,用适量正己烷(重蒸过的分析纯)溶解。② 取细硅胶 6 g 转入玻璃柱中,震荡压实。少许正己烷淋洗后,将原油样品溶液分几次全部转入玻璃柱中。③ 当柱子下端溶液流出时,分批加 10 mL 正己烷,收集饱和烃馏分;过柱子同时用紫外灯照射,通过观察荧光来控制加入溶剂的量。用氮吹仪吹至约 1 mL 转至进样瓶。

2.3 GC×GC-TOFMS 实验条件

一维色谱柱为 Petro 的 50 m×0.2 mm×0.5 μm,升温程序为 40 ℃保持 0.2 min,以 2 ℃/min 的速率升至 300 ℃保持 0.2 min,再以 0.5 ℃/min 的速率升至 320 ℃保持 5 min。二维色谱柱是 DB-17HT 的 2.5 m×0.1 mm×0.1 μm,采用与一维色谱柱炉箱相同的升温速率,温度比一维色谱柱炉箱高 5 ℃。调制器温度比一维色谱柱炉箱高 35 ℃。进样口温度为 300 ℃,不分流进样模式,进样量 1 mL。以 He 气为载气,流速设定为

1.5 mL/min。调制周期 8 s,其中 2 s 热吹时间。质谱传输线和离子源温度分别为 280 ℃ 和 240 ℃,检测器电压 1 500 V,质量扫描范围 40～520 amu,采集速率 100 谱图/s,溶剂延迟时间 11 min。

3 结果与讨论

3.1 实验方法的建立

用 GC×GC-TOFMS 分析正常原油饱和烃组分的方法已建立。但用该方法分析稠油样品时,由于其饱和烃中异构体数量相对于正常原油繁多且极性差异小,部分化合物分离效果不理想,因此需将原有分析方法加以改进。在升温程序上,高温区域采用梯度升温程序,使得更多的高沸点化合物可以被检测出。减少一、二维色谱柱温箱温差,增加调制器与一维色谱柱温箱温差,使化合物在二维色谱上分离效果更好。

3.2 饱和烃组分的全二维谱图特征

图 1 是严重生物降解原油饱和烃组分在 GC×GC-TOFMS 下的全二维轮廓图和 3D 图。从图上可以看出,与正常原油的饱和烃相比,该样品中烷烃类化合物含量极低,常规的甾烷类化合物部分降解,常规的藿烷类化合物轻微降解,相对含量最高的是二环倍半萜类和藿烷类化合物。其他大量的未知低含量化合物聚集在一起,形成不可分辨的基线鼓包物质。从图 1(b) 的一维投影图上看,该饱和烃中存在两组 UCM。一组由 C_{12}～C_{24} 碳数范围内化合物组成的,记为 UCM-A;另一组由 C_{26}～C_{34} 碳数范围内化合物组成,记为 UCM-B。

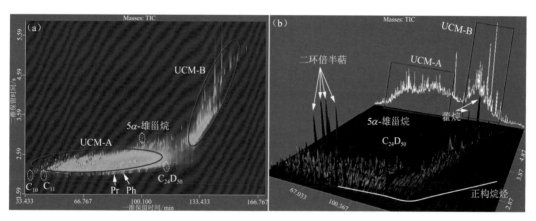

图 1 茨 20-138 的饱和烃组分在 GC×GC-TOFMS 下的全二维谱图
(a) 全二维轮廓图;(b) 全二维 3D 图,图中白线谱图为全二维 3D 图在一维色谱上的投影图

3.3 UCM-A 的化合物组成

UCM-A 是指一维出峰时间在 19～119 min 之间,二维出峰时间在 2～3 s 之间的所有化合物形成的基线鼓包物质。从图 1 可以看出:该区域化合物数量众多,信噪比在 100 以上的化合物在出峰数量上占饱和烃化合物出峰总数量的 75%,在质量上占饱和烃质量的 80% 以上,是饱和烃的主要组成部分。

根据全二维气相色谱的谱图特点可以看出,组成 UCM-A 的化合物极性大于烷烃等无环类异戊二烯物质,小于甾烷等四环烷烃物质,沸点介于 C_9 和 C_{24} 之间。将这些化合物细分成以下三大类(图 2):极性大于正构烷烃和无环类异戊二烯,小于十氢化萘和二环倍半萜的化合物为第一类,在图 2 上以白色表示;极性大于第一类化合物,小于金刚烷的化合物为第二类,在图 2 上以红色表示;极性大于第二类化合物,小于 5α-雄甾烷和孕甾烷、升孕甾烷的化合物为第三类,在图 2 上以黄色表示。全二维气相色谱谱图有"族分离"的特点,即同族化合物在二维色谱上保留时间相似。因此推断第一类化合物与 C_3-环己烷结构相似,主要为单环烷烃类化合物;第二类化合物与十氢化萘结构相似,主要为双环环烷烃类化合物。第三类化合物与金刚烷和三环萜烷结构相似,主要为三环环烷烃类化合物。详细解析如下。

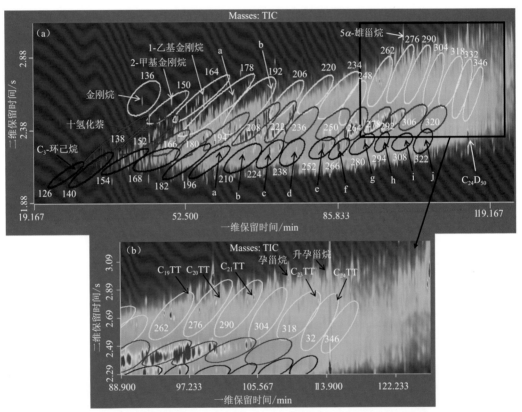

图 2　苏 20-138 的饱和烃组分中 UCM-A 在 GC×GC-TOFMS 下的总离子流全二维谱图
(a) 全二维点阵图;(b) 局部放大图,图中 $C_{19}TT$ 表示 C_{19}-三环萜烷

(1) 单环烷烃化合物。

根据飞行时间质谱(TOFMS)提供的质谱信息,将组成 UCM-A 的化合物按照相对分子质量的不同进行划分,如图 2 所示。第一类化合物按照相对分子质量递增的顺序被分为 15 小类,其相对分子质量范围 126~322,按照相差一个"—CH_2—"的规律排列。根据相对分子质量计算该类化合物的通式为 C_nH_{2n},与单环烷烃化合物通式一致。在这类化合物中,碎片离子以 m/z 55,69,83,97,111,125,139,153,167,181,195 等为主,这与单环的饱和脂环烃开环后先发生氢重排,随后发生 rd 反应,产生 $[M—C_nH_{2n+1}]^+$($n=1,2,3,\cdots$)偶电子离子系列,即 m/z 55,69,83,97,$\cdots$一致。因此组成 UCM-A 的第一类化合物主要是单环烷烃类化合物这个推测是正确的。

在单环烷烃化合物中发现有 10 个化合物的分布存在一定规律(如图 2a 中标记的 a~

j)。质谱信息表明它们具有相同的特征离子和碎片离子,且相邻化合物的相对分子质量相差 14,说明它们是取代基碳数不同的同系列化合物。图 3 列出了其中 c 和 g 两个化合物的质谱图。从图 3(a)上看,该化合物的分子离子是 224,推断出它的分子式是 $C_{16}H_{32}$。从特征离子 m/z 69 和主要碎片离子 m/z 83,97,111,125,$\cdots$上看,该化合物的基本骨架是环己烷结构。通过 m/z 97,111,125 三个碎片离子的丰度相差不大,推测环己烷上应该有三个甲基的取代基,取代基位置无法从质谱图上确定。如图 3(a)中标记,该化合物的断裂方式是先去掉一个"—CH_3",再依次断裂一个"—CH_2",由此推断在环己烷上还有一个长侧链的取代基。综上分析:c 化合物是三甲基-C_7-环己烷,同理 g 化合物是三甲基-C_{11}-环己烷,该系列化合物是三甲基-C_n-环己烷。图 3 中标记了其可能的化学结构。

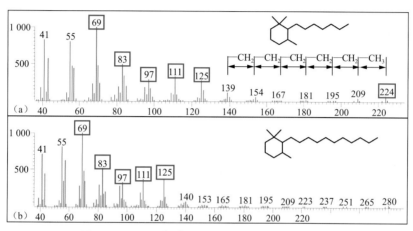

图 3 (a)和(b)分别是 c 和 g 两个化合物的质谱图

(2) 双环烷烃化合物。

仿照单环烷烃的分析方法,将第二类化合物按照相对分子质量递增的特点分为 14 小类,如图 2 所示。其相对分子质量范围 138~320,计算得到的化合物通式是 C_nH_{2n-2}。第二类化合物的相对分子质量均比第一类少 2,说明少了两个 H 原子。这可能是两个单环烷烃在共用两个相邻的碳原子时掉下两个 H,或者是一个长链的单环烷烃在发生成环反应时掉下两个 H。由此验证第二类化合物主要是由双环环烷烃组成的推断成立。

从图 2 上看,第二类化合物中含量最高的区域是 194,208,222,该处的化合物主要是二环倍半萜系列。除此之外,264,278,292 区域也存在相对含量较高的化合物,图 4 标记出几个化合物并列出其质谱信息。图 4(b)是 b 化合物的质谱图,从分子离子上看,该化合物的分子式是 $C_{20}H_{38}$。从 m/z 151 的碎片上看,它断开一个 C_9H_{19} 取代基,生成一个甲基取代的十氢化萘离子,甲基取代的位置无法从质谱信息中获得。从 m/z 179 碎片丰度高、m/z 165,193,207,$\cdots$碎片丰度低上看,C_9H_{20} 是有一个甲基取代的长侧链,甲基的取代位置应该是在贴近双环烷烃的一端。在十氢化萘环上,"α 位"取代的十氢化萘稳定性差,温度高时易受影响。"β 位"取代的十氢化萘 空间阻力小,一旦生成比较稳定。由于长侧链取代基是大基团,因此在"β 位"上取代更为稳定。综上,b 化合物的推测结构如图 4(b)中所示,它所发生的主要离子碎裂反应如公式(1)所示。a,d 所标记的化合物与 b 有相同的特征离子,分子离子分别是 264 和 292,因此 a,d 化合物与 b 化合物的结构相似,只是长侧链取代基碳数和取代基位置有差异。

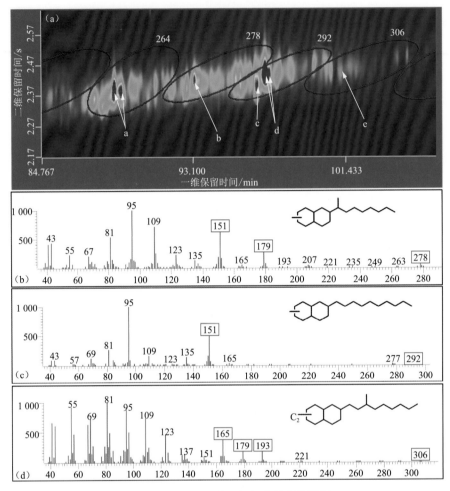

图 4　茨 20-138 的饱和烃组分局部的谱图
(a) 全二维点阵图;(b),(c),(d)分别是(a)中标记的 b,c,e 三个化合物的质谱图

图 4(c)是 c 化合物的质谱图,从它的相对分子质量是 m/z 292 推测出它的分子式是 $C_{21}H_{40}$。m/z 151 的碎片表示它也有一个甲基取代的十氢化萘离子,在 m/z 292 和 m/z 151 之间的碎片离子丰度低且按相差—CH_2 排列,说明化合物有一个 C_{10} 的直链取代。综上,该化合物的推测结构如图 4(c)中所示。

图 4(d)是 e 化合物的质谱图,该化合物的分子式是 $C_{22}H_{42}$。从 m/z 165 的碎片上看,它断开一个 $C_{10}H_{21}$ 取代基,生成一个 C_2 取代的十氢化萘离子,取代基位置无法得知。根据 m/z 179,193 碎片离子推测 $C_{10}H_{21}$ 上的甲基取代位置。该化合物的推测结构式列于图 4(d),主要离子的碎裂反应如公式(2)所示。

$$\left[\begin{array}{c}\vcenter{\hbox{\includegraphics[scale=0.5]{structure}}}\end{array}\right]^{+\cdot} \longrightarrow \underset{(m/z\ 193)}{C_2\text{—}\bigcirc\!\bigcirc\!\text{CH}_2^+} \longrightarrow \underset{(m/z\ 179)}{C_2\text{—}\bigcirc\!\bigcirc\!\overset{+}{\text{CH}}_2}$$

$$\searrow \underset{(m/z\ 165)}{C_2\text{—}\bigcirc\!\bigcirc^{+}} \qquad\qquad\qquad (2)$$

(3) 三环烷烃化合物。

同理,仿照单环烷烃的分析方法将第三类化合物按照相邻化合物相对分子质量相差"—CH_2"的规律划分成 16 小类,如图 2 所示。相对分子质量范围为 136~346,推测出的化合物通式是 C_nH_{2n-4}。第三类化合物的相对分子质量又均比第二类少 2,说明又少了两个 H 原子。这可能是双环烷烃和一个单环烷烃共用两个相邻碳原子时掉下的两个 H,或者发生成环反应时掉下的两个 H。由此证明第三类化合物主要是由三环环烷烃组成的推断成立。

如图 2 所示,第三类化合物中相对含量较高的分别是出峰较早的金刚烷类化合物和出峰较晚的三环萜烷类化合物(图 2b)。在这两类化合物之间,利用质谱解析的方法对两个未知化合物进行鉴定。图 2 中标记的 k 和 l 两个化合物的质谱图列于图 5(a)和(b)。

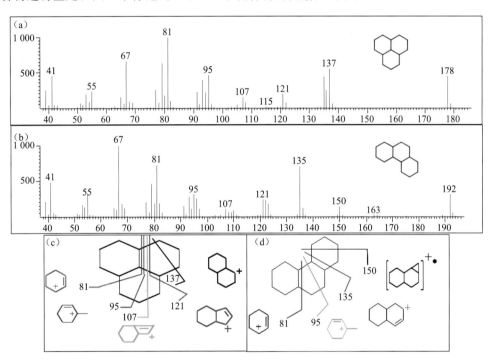

图 5 两个未知化合物的信息

(a),(b)分别是图 2 中标记的 k,l 化合物的质谱图;(c),(d)分别展示 k,l 两个化合物的碎裂情况

从图 5(a)上看,该化合物不属于三环萜烷和金刚烷类化合物,因为它没有 m/z 123 和 m/z 135,149,163,177 的特征离子。从该未知化合物的相对分子质量是 178 推断它的分子式是 $C_{13}H_{22}$,m/z 137 碎片的存在意味着可能有 ⌬⁺ 的结构。m/z 178 和 137 之间相差 41,是 C_3H_5 的结构,不是一个 C_3 的饱和侧链取代,可能是一个 C_3 的环状取代。综上推测该化合物的结构如图 5(a)中标注。该化合物在发生一系列碎裂时能生成 m/z 137,121,107,95 和 81 的碎片(图 5c),因此推测结构可靠性很高。

从图 5(b)上看,该化合物的特征离子是 m/z 135,与金刚烷类化合物相似。但是它没有金刚烷 m/z 79,93 的特征碎片,因此排除它是金刚烷类化合物的可能性。该化合物的相对分子质量是 192,推测的分子式是 $C_{14}H_{24}$。从质谱图上看,它与 k 化合物的质谱图有相似之处,它们都有 m/z 81,95,107,121 的碎片离子,说明它们在这一部分的碎裂方式一致,推测 l 化合物也有双环烷烃的结构。双环烷烃生成的 m/z 135 的碎片可能是 结构。m/z 192 与 m/z 135 之间相差一个 C_4H_9,与 m/z 150 之间相差一个 C_3H_6,说明双环烷烃上的 C_4 取代不是一个饱和的长侧链,很有可能是一个环状取代,取代位置是 m/z 135 碎片中的双键位置。综上推测 l 化合物的结构如图 5(b)中所示,图 5(d)展现了其发生一系列碎裂反应产生碎片的情况。

3.4 UCM-B 的化合物组成

UCM-B 是指一维出峰时间在 123~160 min 之间,二维出峰时间在 2.5~5.5 s 之间的所有化合物形成的基线鼓包物质。该区域信噪比在 100 以上的有近千个化合物,它们虽然在数量上占到饱和烃出峰总量的 17%,但在质量上仅占 0.5%,说明多数都是痕量化合物。根据质谱信息将 UCM-B 中相对含量较高的化合物简单分成五大类(图 6a),分别是烷烃、三环萜烷、甾烷、单芳甾以及藿烷。在 UCM-B 中,甾烷附近化合物的出峰情况最为复杂,除了常规甾烷,还有一些特征离子是 m/z 123,231 和 232 的化合物(图 6b)。特征离子是 m/z 123 的化合物相对分子质量范围是 372~428,且以"—CH_2"的顺序递增,说明它们是同一系列的化合物。图 6(d)是其中一个化合物的质谱图,该图与 Schmitter 等在 1982 年测得的 $C_{27\text{-}8}$,14-断藿烷-Ⅰ的质谱图相似度很高。此外该化合物在二维色谱上的出峰时间与甾烷相近,说明它也是四环烷烃的结构。因此推断 m/z 123 这个系列的化合物是断藿烷系列。同理,特征离子是 m/z 231 和 232 的化合物相对分子质量范围是 386~414,也是一个系列的化合物。从质谱图上看(图 6e)与 Каюкова 等在 1981 年发表的 4α-甲基-$5\alpha(H)$,$14\beta(H)$,$17\beta(H)$-胆甾烷(20S)相似,因此推断这系列化合物是胆甾烷系列。

以上五类化合物中未知化合物的定性都属于推理定性,单靠质谱信息无法推断取代基的位置,还需要借助其他手段。UCM-B 中除了上面鉴定出的五大类化合物外,还有许多特征离子是 m/z 189,193,…的化合物,这些化合物的特征离子丰度不高且多数没有得到准确的分子离子,给化合物定性带来很大难度。

4 结 论

运用全二维气相色谱-飞行时间质谱对辽河油田东北部凹陷的茨榆坨油田的严重生物降解原油的饱和烃组分进行分析。通过饱和烃的全二维气相色谱图特征和质谱解析的手段,认识到严重生物降解原油饱和烃组分中不可分辨的复杂混合物是由不同取代基的环状类化合物组成;C_{24} 之前的第一组不可分辨的复杂混合物主要是由环己烷为基本单元的单环、双环和三环烷的烃类化合物组成;C_{24} 之后出现的第二组不可分辨的复杂混合物主要由四个环或者五个环类的化合物组成。不同的环状化合物又包含若干类不同的化合物,不同的化合物在稠油中所占的比重不同。这些初步认识有助于稠油开发方案的制定、炼制技术的进步,对生物降解原油的定量表征,对探讨经历过生物降解和热蚀变的非正常原油成熟度变化、沉积环境和油源对比等地球化学问题提供有力支持。

第 3 章 全二维气相色谱的石油地质应用

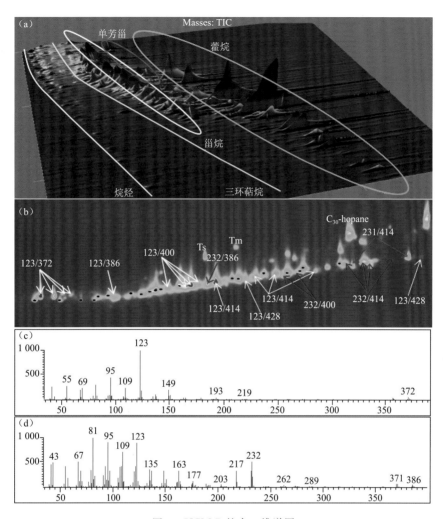

图 6 UCM-B 的全二维谱图

(a) 是在总离子流下的全二维点阵图；(b) 是 m/z 123,217,231,232 下的全二维点阵图，图中标记的 123/372 表示该化合物的特征离子是 m/z 123,分子离子是 m/z 372;黑点标记的是 m/z 217 的甾烷化合物；(c),(d)是特征离子分别是 m/z 123,232 的化合物的质谱图

论文9　2015年发表于《Journal of Chromatography A》

Insight into unresolved complex mixtures of aromatic hydrocarbons in heavy oil via two-dimensional gas chromatography coupled with time-of-flight mass spectrometry analysis

Na Weng, Shan Wan, Huitong Wang, Shuichang Zhang,
Guangyou Zhu, Jingfu Liu, Di Cai, Yunxu Yang

Abstract　The aromatic hydrocarbon fractions of five crude oils representing a natural sequence of increasing degree of biodegradation from the Liaohe Basin, NE, China, were analyzed using conventional gas chromatography-mass spectrometry (GC-MS) and comprehensive two-dimensional gas chromatography (GC×GC). Because of the limited peak capability and low resolution, compounds in the aromatic fraction of a heavily biodegraded crude oil that were analyzed by GC-MS appeared as unresolved complex mixtures (UCMs) or GC "humps". They could be separated based on their polarity by GC×GC. UCMs are composed mainly of aromatic biomarkers and aromatic hydrocarbons with branched alkanes or cycloalkanes substituents. The quantitative results achieved by GC×GC-FID were shown that monoaromatic hydrocarbons account for the largest number and mass of UCMs in the aromatic hydrocarbon fraction of heavily biodegraded crude oil, at 45% by mass. The number and mass of diaromatic hydrocarbons ranks second at 33% by mass, followed by the aromatic biomarker compounds, triaromatic, tetraaromatic, and pentaaromatic hydrocarbons, that account for 10%, 6%, 1.5%, and 0.01% of all aromatic compounds by mass, respectively. In the heavily biodegraded oil, compounds with monocyclic cycloalkane substituents account for the largest proportion of mono-and diaromatic hydrocarbons, respectively. The C_4-substituted compounds account for the largest proportion of naphthalenes and the C_3-substituted compounds account for the largest proportion of phenanthrenes, which is very different from non-biodegraded, slightly biodegraded, and moderately biodegraded crude oil. It is inferred that compounds of monoaromatic, diaromatic and triaromatic hydrocarbons are affected by biodegradation, that compounds with C_1-, C_2-substituents are affected by the increase in degree of biodegradation, and that their relative content decreased, whereas compounds with C_3-substituents or more were affected slightly or unaffected, and their relative content also increased. The varying regularity of relative content of substituted compounds may be used to reflect the degree of degradation of heavy oil. Moreover, biomarkers for the aromatic hydrocarbons of heavily biodegraded crude oil are mainly aromatic steranes, aromatic secohopanes, aromatic pentacyclotriterpanes, and benzohopanes. According to resultant data, aromatic secohopanes could be used as a specific marker because of their relatively high concentration. This aromatic compound analysis of a series of biodegraded crude oil is useful for future research on the quantitative characterization of the degree of biodegradation of heavy oil, unconventional oil maturity evaluation, oil source correlation, depositional environment, and any other geochemical problems.

Keywords　Comprehensive two-dimensional chromatography　Time-of-flight mass spectrometry　Heavy oil　Biodegraded crude oil　UCM

1 Introduction

Heavy oil has attracted increased attention because of its large reserves, intensive resources, and advancing oil development techniques. Based on genesis, heavy oil can be divided into original and secondary heavy oil. Original heavy oil accounts for only a limited proportion of the total heavy oil, whereas most heavy oil is secondary heavy oil. Most secondary heavy oil has been subjected to biodegradation. Hunt estimated that approximately 10% of all the world's petroleum reserves have been lost to biodegradation and an additional 10% altered. Compounds with different chemical structures in crude oil can be biodegraded selectively, thus the heavy oil composition changes significantly. This variation makes genesis discrimination, maturity evaluation or even the prediction of heavy oil resources more difficult. The physical properties of the heavy oil are easy to know. Nevertheless, given the current analytical conditions, the chemical composition of the heavy oil when compared to the crude oil is still not well identified and requires further investigation. Understanding the chemical composition of heavy oil is of great importance for the qualitative and quantitative determination of the chemical composition of heavy oil, oil source correlation analysis, evaluation of the degree of biodegradation, geochemical problems such as genesis discrimination, heavy oil developing and refining, and solving environment problems such as oil spills.

Gas chromatography (GC) is the most conventionally used tool for normal crude oil sample analysis. Because of the limited peak capability and low resolution ratio in GC analysis of heavy oil samples, compounds with the same boiling point and different polarities are co-eluted, which results in a rise of the chromatogram baseline (the so-called "humps"). Since it is difficult to identify compounds in the "humps", they are termed unresolved complex mixtures (UCMs). The high-concentration compounds produce discernible peaks in UCMs, but even these peaks are likely to consist of multiple overlapping compounds. The low-concentration compounds can hardly be detected because they are covered by the rising baseline. As the "humps" become more obvious with increased degree of biodegradation, fewer compounds in the UCMs can be identified. Although tandem quadrupole mass spectrometry(MS) can assist with compound identification in the UCMs to some extent, this technique cannot provide pure mass spectra of the target compounds, and offers strong evidence for unknown compound identification because of the interference of co-eluting peaks and the baseline.

To explore the chemical composition of the UCMs, a number of analytical methods such as molecular sieve, urea complexation, and chemical oxidation have been used to separate the UCM petroleum compounds into distinct chemical fractions or groups, which were further analyzed with GC-MS. Some researchers have used combined apparatus to identify UCM constituents. However, these methods can only identify some of the UCMs, where as some complex compound mixtures cannot be resolved and the nature of the compounds that form these UCMs remains unclear.

Comprehensive two-dimensional chromatography (GC×GC) originated in the 1990s and is a new technology for complex compound analysis. The GC×GC column system consists of two independent columns with a distinct separation mechanism, and a modulator that connects two columns in series. Because of the powerful orthogonal separation capacity of GC×GC, compounds with similar boiling points that cannot be separated in the first column can be separated in the second column, according to different polarities. With the development of GC×GC coupled with time-of-flight mass spectrometry (GC×GC-TOFMS), the mass spectra of UCM compounds can be acquired. The data processing system has a deconvolution function, which can provide an effective basis for compound identification. GC×GC-TOFMS has been applied to UCM analysis by an increasing number of researchers. Currently, many previous studies reported the constituents of UCMs in aromatic hydrocarbons focus mainly on toxic contaminants in marine organisms. In geological sample, more reports have been paid to the analysis of UCMs of saturated hydrocarbons, known aromatic compounds, or heterocyclic compounds such as sulfur and nitrogen containing compounds. Other compounds found in aromatic UCMs, on the other hand, are still lacking sufficient attention.

The purpose of this work is to identify feature compounds in the UCMs. A statistical method based on GC×GC-TOFMS and GC×GC-FID has therefore been developed to analyze for aromatic hydrocarbons in the UCMs of a series of crude oil with different biodegradation degrees. It is expected that this developed method can provide a reference for further research on aromatic hydrocarbon UCMs.

2 Experimental

2.1 Equipment and materials

The comprehensive two-dimensional GC×GC-TOFMS system (LECO Corporation, San Jose, CA, USA) is composed of an Agilent 7890A gas chromatography with flame ionization detector and a Pegasus 4D time-of-flight mass spectrometry (LECO Corporation, San Jose, CA, USA). The Agilent 7890A GC is equipped with a liquid nitrogen-cooled pulse jet modulator. All data were processed using the Chroma TOF software (LECO Corporation, San Jose, CA, USA). DSQ II single stage quadrupole GC-MS (Thermo Scientific Cor-poration, Waltham, Massachusetts, USA) and GC columns (Agilent Technologies, Santa Clara, CA, USA) were used. n-hexane (analytical reagent grade, redistilled), dichloromethane (DCM, analytical reagent grade, redistilled) and silica gel (100 – 200 mesh, activated at 200 ℃ for 4 h) were used for the sample retreatment. D_{10}-anthracene was selected as internal standard.

2.2 Sample preparation

The sample suite comprises five progressively biodegraded crude oils were from the Ciyutuo oilfield in the north eastern Depression of the Liaohe Basin, northeast China, and the information of samples is shown in Table 1. Oil biodegradation level was based from scale of Wenger et al.

Table 1 Oils analyzed in this study.

No.	Sample	Biodegradation level
1	CI 50-85	None
2	CI 54-82	Slight
3	CI 11	Moderate
4	CI 20-138	Heavy
5	CI 8-340	Heavy

The protocol of aromatic hydrocarbons extraction is as follows. (1) Approximately 60 mg crude oil was dissolved in an appropriate amount of n-hexane. (2) Silica gel packed in a glass column was washed using n-hexane, and the oil sample was loaded onto the glass column. (3) As soon as the fluid eluted out the column, 10 mL n-hexane was added into the glass column to wash the column three times, and the saturated fraction was collected. Then, 20 mL DCM was used to wash the silica gel six times to collect the aromatic fraction. Ultraviolet light was used to observe the column and control solvent addition. The collected aromatic fraction was condensed to 1 mL using a nitrogen evaporator, and then transferred into a GC sample bottle for analysis.

2.3 GC×GC analysis

A DB-Petro column (50 m × 0.2 mm × 0.5 μm) was used as the first dimension (1D) chromatographic column for GC×GC-TOFMS analysis. The temperature program was set to start from 60 ℃ (hold for 0.2 min), heat to 300 ℃ at 2 ℃/min (hold for 0.2 min), and heat to 320 ℃ at 0.5 ℃/min (hold for 5 min). The second dimension (2D) separation was performed by a DB-17HT column (3 m × 0.1 mm × 0.1 μm). The 2D and modulator ovens were operated with the same temperature gradient but with a temperature offset of 5 ℃ and 30 ℃ higher than the 1D oven, respectively. The samples (1 μL) were injected into a heated (300 ℃) splitless injector. Helium was used as the carrier gas, with a constant flow rate of 1.5 mL/min. The modulation period was 8 s, with a 2 s hot-pulse duration. The MS transfer line and ion-source temperature were 280 ℃ and 240 ℃, respectively. The detector was operated at 1 450 V. The acquisition rate was 100 spectra/s with a collected mass range of 40−520 amu, and the acquisition delay was 13 min. The GC×GC-FID system had the same separation conditions and columns as the GC×GC-TOFMS system. The flow rates of carrier gas, hydrogen and air in the FID detector were 23, 30, 400 mL/min. The detector temperature was 320 ℃, the acquisition rate was 200 spectra/s, and the acquisition delay was 13 min.

2.4 GC-MS analysis

A HP-5MS column (60 m × 0.25 mm × 0.25 μm) was chosen as the chromatographic column for GC-MS analysis. The temperature program was set to start from 60 ℃ (hold for 0.2 min), heat to 300 ℃ at 2 ℃/min (hold for 0.2 min), and heat to 320 ℃ at 0.5 ℃/min (hold for 5 min). The samples (1 L) were injected into a heated (300 ℃) splitless injector. Helium was used as the carrier gas, with a constant flow rate of 1 mL/min. The MS transfer line and ion-source temperature were 280 ℃ and 250 ℃, respectively. The detector was operated at 1 500 V with a collected mass range of 40 – 520 amu, and acquisition delay of 8 min.

2.5 Quantitative method

The aromatic hydrocarbon samples were prepared according to Section 2.2 and analyzed by using both GC×GC-TOFMS and GC×GC-FID. The total ion current (TIC) chromatograms of GC×GC-TOFMS and GC×GC-FID of samples were also obtained. The compounds present in the aromatic hydrocarbon of samples were identified by GC×GC-TOFMS and quantitatively by GC×GC-FID. The data were processed by the Chroma TOF software. After setting appropriate parameters, Chroma TOF software was able to search the baseline and the peaks, calculate the peak area, search the database (NIST 11) and form the peak table automatically. The peak table contains the number of compounds, their retention time and peak area etc. According to the chemical properties of the different compounds, the software can classify them into certain groups, generating the corresponding number and peak area in the peak table. In the data processing results of GC×GC-FID, because the response factor of the FID detector toward all the hydrocarbons was close to 1, the quantification of compounds was achieved by means of peak area normalization. This means the ratio between the area of one certain compound and the total area of all the aromatic hydrocarbons equals the weight percentage of this compound.

3 Results and discussion

3.1 Comprehensive two-dimensional chromatogram characteristics of aromatic hydrocarbon

Fig. 1a shows the GC-MS (total ion current, TIC) chromatogram of the aromatic hydrocarbon fraction of heavily biodegraded crude oil. The aromatic composition of heavily biodegraded crude oil is complex. Because of the limited chromatographic column capacity in GC-MS, compounds with the same boiling point accumulated, and this resulted in a rise of the chromatogram baseline (so-called "humps"), and an increase in UCMs formed. Since compounds are co-eluted based on their boiling point in GC-MS, it can be seen from Fig. 1a that the boiling points of the compounds in the UCM are higher than naphthalene, and that the UCM contains compounds such as phenanthrene, pyrene, and perylene, which almost covers the entire chromatogram. Fig. 1b is the GC×GC-TOFMS (TIC) color contour chromatogram of the aromatic hydrocarbon fraction in heavily biodegraded crude oil, in which each black dot represents a compound. The UCMs obtained on a conventional GC-MS can be separated further by GC×GC according to different polarities. Data processing by Chroma TOF shows that nearly 9 000 aromatic hydrocarbons showed a signal-to-noise ratio (SNR) greater than 100, and the UCM components accounted for more than 98% of this in number and over 95% of this in mass. Given the extremely complex composition of the UCM, we divided it into six groups (UCM-1, UCM-2, UCM-3, UCM-4, UCM-5, and UCM-6) with different polarities to study their compositions further.

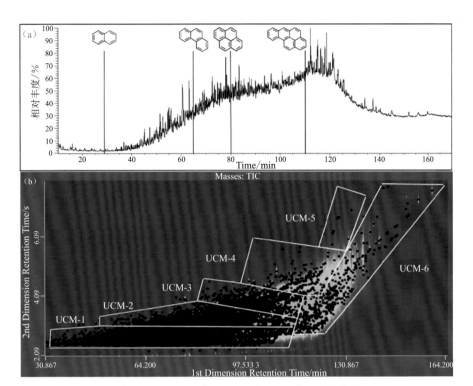

Fig. 1 Chromatograms of aromatic fraction from Well CI 20-138:
(a) TIC chromatogram by GC-MS; (b) TIC chromatogram by GC×GC-TOFMS.

3.2 Composition of UCM-1

As shown in Fig. 2, UCM-1 refers to compounds with first dimension retention times between 25 and 117 min and second dimension retention times between 2.3 and 3.0 s. A number of compounds exist in this area, i.e., the total number of components with an SNR greater than 100 is nearly 4 000, of which the UCM components account for more than 44% of all the aromatic compounds, and more than 45% in mass.

The GC×GC chromatogram characterization showed that the polarity of the UCM-1 compounds is higher than that of the benzene compounds, and lower than that of the naphthalene compounds, suggesting that the basic chemical structure of the compounds in UCM-1 is monoaromatic hydrocarbon. These compounds can be divided into three classes and 43 subclasses based on the different molecular weights, as well as the information provided by TOFMS and the GC×GC chromatogram shown in Fig. 2, in which the first, second, and third UCM-1 classes are marked in red, white, and orange, respectively. Because of the characteristic of "family separation" of the GC×GC chromatogram, the family compounds(with a similar chemical structure) have a similar retention time in the two-dimensional chromatogram. Whereas the first class compounds of UCM-1 are suggested to have a similar chemical structure to trimethyl-benzene (mainly benzene compounds), the second class compounds of UCM-1 are similar to indane in chemical struc-ture, and are mainly a monoaromatic hydrocarbon with monocyclic cycloalkane substituent. The polarity of the third class compounds is higher than that of the second class compounds, which is lower than that of the diaromatic hydrocarbons. Thus, we presume that the third class compounds of UCM-1 consist mainly of monoaromatic hydrocarbons with polycyclic substituents. Based on mass spectra data provided by TOFMS and statistical methods, the characteristics of compounds in UCM-1 are listed in Table 2.

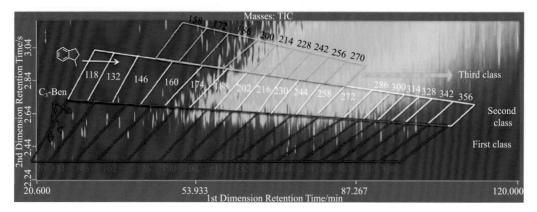

Fig. 2 GC×GC-TOFMS color contour chromatograms of UCM-1 from Well CI 20-138. According to the molecular weights (listed in this figure), compounds in UCM-1 are divided into three classes and 43 subclasses. First class compounds are marked in red (on the bottom), second class compounds are marked in white (in the middle), and third class compounds are marked in orange (at the top). (For interpretation of the references to color in this figure citation, the reader is referred to the web version of this article.)

The data in Table 2 demonstrate that compounds of the first class account for a lower percentage in the UCM-1, and only 4% of all aromatic compounds in mass. Based on a further analysis of the first class compounds, it is found that compounds with characteristic ions with m/z of 92, 105, 120, and 133 are distributed irregularly. It is inferred that benzene, toluene, and xylene compounds with long straight-chain substituents in this heavily degraded crude oil have been fully biodegraded, which leaves only the benzene compounds with complex branched-chain substituents, and the concentration of these compounds is relatively low.

Table 2　Characteristics of compounds of UCM-1.

	First class	Second class	Third class
Group number percentage	14%	20%	10%
Group weight percentage	4%	24%	17%
Molecular weight range	120–330	118–356	158–270

continue

	First class	Second class	Third class
Chemical formula	$C_nH_{2n-6}, n \geq 9$	$C_nH_{2n-8}, n \geq 9$	$C_nH_{2n-10}, n \geq 12$
Degree of unsaturation	4	5	6
Characteristic ions	91, 92, 105, 119, 120, 133, ...	131, 145, 159, 173, 187, 201, ...	129, 143, 157, 171, 185, 199, ...
Possible fragmentation ion or fragmentation method	(structures with m/z 91, 92, 105, 106, 119, 120, 133)	(structures with m/z 131, 145, 159, 173, 187, 201, 145, 159, 173, 187)	(structures with m/z 129, 143, 157, 171, 185, 199)
Possible chemical structure	Alkyl-benzenes	Benzene compounds with monocyclic cycloalkane substituents	Benzene compounds with two monocyclic cycloalkane substituents, or with cyclopentene substituents

The second class compounds account for 20% of all aromatic compounds in number, and 24% in mass. This result suggests that benzene compounds with monocyclic cycloalkane substituents account for the majority of UCM-1, which is also the main form of the benzene compounds.

Several compounds of relatively high content in the third class were marked as "characteristic/molecular ions" in Fig. 3a, and compounds with the same mark have similar mass spectra. We chose two of these compounds to interpret their chemical structure according to their mass spectra. Fig. 3b is the mass spectrum of compound 1. The characteristic ion of the compound is m/z 129, the main fragment ions are m/z 63, 77, 91, 102, 115, and 143 respectively, and the molecular ion is m/z 158. The molecular formula of compound 1 is inferred to be $C_{12}H_{14}$ from the molecular ion. The difference of 15 between fragment ions with m/z 158 and 143 indicates that a "—CH_3" is lost, and that between fragment ions with m/z 143 and 129 is 14, which indicates that a "—CH_2" is lost.

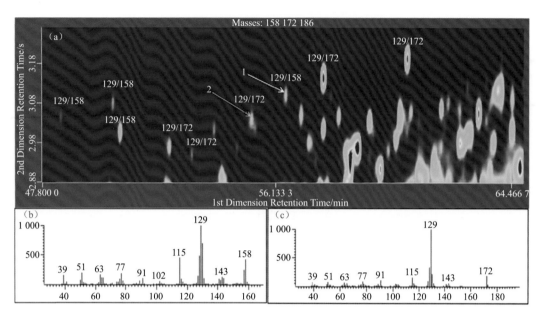

Fig. 3 (a) GC×GC-TOFMS color contour chromatogram for m/z 158, 172, 186, in which compounds marked with 129/158 mean that the characteristic ion of the compounds is 129, and the molecular ion is 158; (b) and (c) mass spectra of compounds 1 and 2, respectively. (For interpretation of the references to color in this figure citation, the reader is referred to the web version of this article.)

As the characteristic ion is m/z 129, and not 143, the basic chemical structure of compound 1 could not be 1,2,3-trimethyl-indene, 1,1,3-trimethyl-indene or 1,1,2-trimethyl-indene, and may be monoaromatic with two monocyclic cycloalkanes. These two monocyclic cycloalkanes should consist of two cyclopentanes, or a cyclopentane and a cyclohexane, rather than two cyclohexanes according to the molecular weight. If it consists of a cyclopentane and a cyclohexane, then the chemical structure should be like compound A (shown in Table 3). If it consists of two cyclopentanes, then the chemical structure should be like compound B, C, or D (shown in Table 3). ChemDraw Ultra is a professional chemical software with which the boiling point can be predicted according to the compound structure. The boiling points of the four compounds above were calculated using this software, and the four compounds were arranged sequentially according to the ascending order of the boiling points, i.e., compound D < compound A < compound B = compound C. Because of limitations of this software, the predicted boiling points of the compounds with similar chemical structure but different substituent positions are the same. More disperse and more intensive compound spatial distributions have a higher and lower boiling point, respectively. For example, the boiling point of phenanthrene is lower than that of anthracene. Therefore, the four compounds were arranged according to ascending boiling point, i.e., compound D < compound A < compound B < compound C. The four compounds marked as "129/158" in Fig. 3a were arranged in ascending order according to the first dimension retention time, i.e., compound D < compound A <

compound B < compound C.

Table 3 The chemical structures of compounds in Section 3.2 compound.

Compound	Chemical structure	Compound	Chemical structure
A		F	
B		G	
C		H	
D		I	
E			

Fig. 3c shows the mass spectrum of compound 2. According to the analytical results of compounds 1, the structures of the five compounds marked as "129/172" are probably compound E < compound F < compound G < compound H < compound I (their chemical structure as shown in Table 3).

3.3 Composition of UCM-2

UCM-2 refers to compounds in "humps" with a first dimension retention time of 37 – 111 min and a second dimension retention time of 3 – 4 s. In this zone, nearly 2 000 compounds exist with an SNR above 100. Their peak number accounts for 26% of the total aromatic hydrocarbons and their mass accounts for 33% of the total. Based on the TOFMS mass spectra, the compounds in UCM-2 are divided into three classes (as shown in Fig. 4a), i.e., area with black, red, and pink lines for the first, second, and third classes, respectively. According to the analytical method used for compounds in UCM-1, the characteristics of compounds in UCM-2 are listed in Table 4.

It can be seen from Table 4 that the main structure of compounds of UCM-2 is diaromatic hydrocarbons, with branched-chain or cycloalkane substituents, and compounds of the second class account for the majority of UCM-2. Two compounds exist with high concentration in the second class (shown in Fig. 4a) and their mass spectra are shown in Fig. 4b and c. The mass spectrum in Fig. 4b obtained a 97.7% match with the acenaphthene mass spectrum from the National Institute of Standards and Technology (NIST) spectra database. Because acenaphthylene consists of three rings and five double bonds, the compound in Fig. 4a (compound (1)) is confirmed to be acenaphthylene, with the structure presented in Fig. 4b. Fig. 4c shows the mass spectrum of compound (2), the fragment ions of the compound are m/z 76, 89, 115, and 141, which is similar to dimethylnaphthalene. This indicates that a naphthalene ring exists in the chemical structure of compound (2). The difference of 11 between m/z 154 and 165 indicates that cyclic substituents rather than chain substituents may exist on the naphthalene ring. From MS analysis it can be speculated that the fragment structure is likely to be [structure] •+ (m/z 154) and [structure] (m/z 165). Possible structures for compound (2) are listed in Fig. 4c.

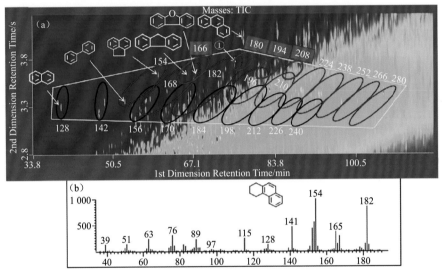

Fig. 4 Chromatogram of UCM-2 (a) GC×GC-TOFMS color contour chromatogram. Peak range of UCM-2 compounds is marked by yellow lines. According to the molecular weights (listed in this figure), compounds in UCM-2 are divided into three classes and 23 subclasses which are marked in black, red, and pink, respectively; (b) and (c) mass spectra of compounds (1) and (2), respectively. (For interpretation of the references to color in this figure legend, the reader is referred to the web version of this article.)

Table 4 Characteristics of compounds of UCM-2.

	First class	Second class	Third class
Group number percentage	8%	17.5%	0.5%
Group weight percentage	13%	19%	1%
Molecular weight range	128–240	154–280	166–208
Chemical formula	$C_nH_{2n-12}, n \geqslant 10$	$C_nH_{2n-14}, n \geqslant 12$	$C_nH_{2n-16}, n \geqslant 13$
Degree of unsaturation	7	8	9
Characteristic ions	115,128,142,155,156, 169,170,184,…	154,168,182,196, 210,…	166,180,194,208
Basic chemical structure	[naphthalene]	[biphenyl / fluorene structures]	[dibenzofuran structure]

It can be seen from Fig. 4 that substituted naphthalenes are distributed widely in these compounds, up to C_8-substituted naphthalenes. From the statistics, the peak number and mass of naphthalene compounds account for one third of the UCM-2, with the highest proportion of C_4-substituted naphthalene compounds, and the relatively highest content compound is endalin. Fig. 5 shows that the variation in the ratio between C_n-substituented naphthalenes and naphthalenes in the aromatic hydrocarbons of five crude oil samples with different degrees of biodegradation. The figure shows that in the non-biodegraded crude oil, the content of C_2-substituted naphthalenes was the highest of all the naphthalene compounds present. In the slightly biodegraded crude oil, the contents of C_2- and C_3-substituted naphthalenes were basically the same, while the contents of C_3-substituted naphthalenes was slightly higher. In the moderately biodegraded crude oil, the content of C_4-substituted naphthalenes was slightly higher than that for C_3-substituted naphthalenes and C_4-substituted naphthalenes far exceed the others in the heavily biodegraded crude oil. If the degree of biodegradation goes a step further, the C_5-substituted naphthalenes have the highest relative content among all the naphthalenes present. Therefore, the content of naphthalene compounds with different substituents was affected by an increase in the degree of biodegradation. Naphthalene compounds with low number of carbon substituents decreased with increase in the degree of biodegradation, whereas the naphthalene compounds with relatively higher number of carbon substituents were only slightly affected or unaffected. Naphthalene compounds with low number of carbon substituents are affected by the increase in degree of degradation, and their relative content decreased, whereas the naphthalene compounds with high carbon number

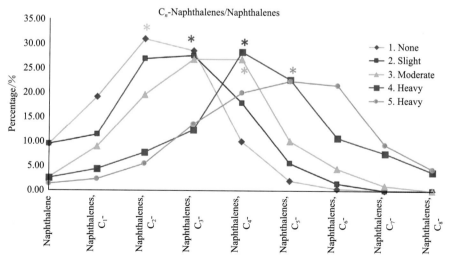

Fig. 5 The change in the relative content of napthalenes in the aromatic fraction of crude oils with different degrees of biodegradation. The data, obtained by GC×GC-FID, reflects the respective ratios between the contents of naphthalenes with different substituents and the total content of all naphthalene compounds.

substituents are slightly affected or unaffected, and their relative content increased as well. The varying regularity of relative content of substituted naphthalene compounds may be used to reflect the degree of degradation of heavy oil.

3.4 Composition of UCM-3, 4, 5

UCM-3 refers to compounds in "humps" with a first dimension retention time of between 73 and 111 min, a second dimension retention time between 3.4 and 4.6 s (as shown in Fig. 6). Approximately 900 compounds exist with an SNR over 100. The peak numbers account for 10% of the total peak numbers of the aromatic hydrocarbons and the mass accounts for 6% of the total. UCM-3 compounds can be divided into four classes, and characteristics of the compounds in UCM-3 are listed in Table 5.

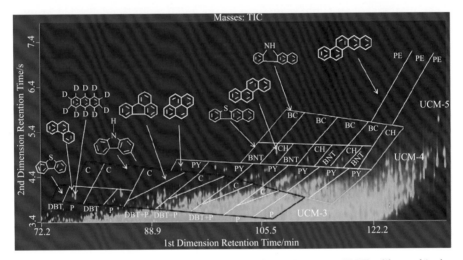

Fig. 6 GC×GC-TOFMS color contour chromatogram of UCM-3, 4, 5 (DBT: dibenzothiophene; P: phenanthrene; C: carbazole; PY: pyrene; BNT: benzo-naphtho-thiophene; CH: chrysene; BC: benzo-carbazole; PE: perylene.) (For interpretation of the references to color in this figure citation, the reader is referred to the web version of this article.)

Table 5 **Characteristics of compounds of UCM-3.**

	First class	Second class	Third class	Fourth class
Group number percentage	9.0%	0.6%	0.4%	0.05%
Group weight percentage	5.4%	0.6%	0.1%	0.01%
Molecular weight range	178 – 248	184 – 240	167 – 237	216
Chemical formula	$C_n H_{2n-18}, n \geq 14$	$C_n H_{2n-16} S, n \geq 12$	$C_n H_{2n-15} N, n \geq 12$	$C_n H_{2n-22}, n \geq 16$
Degree of unsaturation	10	9	9	12
Characteristic ions	178, 192, 204, 206, 220, 230, 234, 248	184, 198, 212, 226, 240	167, 181, 195, 209, 223, 237	108, 189, 216
Basic chemical structure				

It can be seen from Table 5 that compounds of the first class account for the majority of UCM-3. The second and third class compounds of UCM-3 which are tentative identified by the professional references are dibenzothiophenes and carbazoles, respectively. They are difficult to detect in aromatic fractions using conventional GC-MS, because of their relatively low content and the interference of co-eluting peaks. The interference of co-eluting peaks can be eliminated using GC×GC-TOFMS. Data processing results reveal that a large number of isomers exist in the first, second, and third classes, and the highest proportion is C_3-substituted compounds in the first class. In contrast, of all the non-biodegraded and slightly biodegraded crude oil, the highest proportion is C_1-substituted phenanthrenes (as shown in Fig. 7). It is inferred that compounds of phenanthrenes are affected by biodegradation.

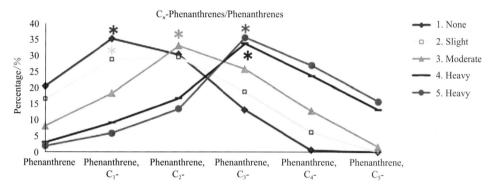

Fig. 7 The change of relative content of phenanthrenes in aromatic fraction of crude oils with different biodegradation degrees. The data, obtained by GC×GC-FID, reflects the respective ratios between the contents of phenanthrenes with different substituents and the total content of all phenanthrene compounds.

UCM-4 refers to compounds in the "humps" with a first dimension retention time of between 92 and 126 min, a second dimension retention time between 4 and 5.8 s (as shown in Fig. 6). Approximately 300 compounds exist with an SNR over 100, the peaknumbers account for 4% of the total peak numbers of aromatic hydrocarbons, and the mass accounts for 1.5% of the total. Compounds of UCM-4 can be divided into four classes, the first class compounds are mainly pyrene compounds according the relative retention time and mass spectra, and the substituent can be up to C_5. Among those compounds, C_3-substituted pyrenes account for the highest content and the largest number. The second class is mainly chrysene compounds, whose degree of unsaturation is one higher than the pyrene compounds. The polarity and boiling point of chrysene compounds is higher than that of pyrene compounds, which is also confirmed by the relative retention time reflected on the GC×GC chromatogram. The third class compounds of UCM-4 are benzonaphthothiophene compounds, the polarity of which is close to the chrysene compounds. The fourth class is that of the benzocarbazole compounds. The polarity of these compounds is higher than the tetraaromatic hydrocarbons, and lower than the pentaaromatic hydrocarbons. These compounds are difficult to obtain by conventional analysis methods.

UCM-5 refers to compounds in the "humps" with a first dimension retention time of between 119 and 132 min and a second dimension retention time of between 5.8 and 7 s (as shown in Fig. 6). Only seven compounds exist with an SNR over 100 in this area, and the mass accounts for 0.01% of the total. Compounds in this area are confirmed to be pentaaromatic compounds according the relative retention time and mass spectra.

3.5 Composition of UCM-6

More than 1 000 compounds exist with an SNR over 100 in UCM-6, with the peak numbers accounting for 14% of the total peak numbers of the aromatic hydrocarbons, and the mass accounting for 10% of the total. Because of the large span of the boiling point and polarity of the compounds in the region, they are difficult to identify. UCM-6 compounds can be divided into five classes (as shown in Fig. 8). The main component of UCM-6 is the first class, which has the most complex components and the highest content. These compounds were analyzed in detail as follows:

1) First class

Aromatic steranes, aromatic terpanes and their corresponding steranes and terpanes have the same sources, but their transformation pathways are different. So the chemical structures of the aromatic steranes and aromatic terpanes are slightly similar to the steranes and terpanes. In addition, the peak position of these aromatic compounds can be estimated by the peak position of the steranes and terpanes in GC×GC chromatogram.

These compounds can be divided into six subclasses (as shown in Fig. 8), and the same color dots in Fig. 8 represent the family compounds.

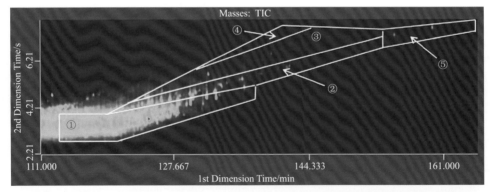

Fig. 8 GC×GC-TOFMS color contour chromatogram of UCM-6. Compounds in UCM-6 are divided into five classes, which are marked with number 1 to 5. (For interpretation of the references to color in this figure citation, the reader is referred to the web version of this article.)

(1) Compounds with characteristic ions m/z 253 are marked as yellow dots in Fig. 9a. The mass spectra of this kind of compound are simple, with a unique characteristic ion m/z 253. Besides the characteristic ion and fragment ion (m/z, 143), other fragment

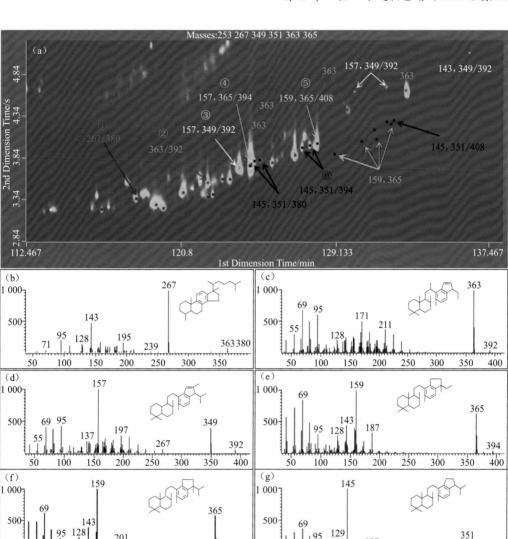

Fig. 9 GC×GC chromatogram: (a) GC×GC-TOFMS color contour chromatogram of fragment ion m/z ratios of 253, 267, 349, 351, 363, and 365, in which compound marked with 157, 349/392 means that the characteristic ions of the compound are 157 and 349, and the molecular ion is 392. Compounds with characteristic ions m/z 253, 267, 363, 349, 365, 351 are marked as yellow, red, pink, white, brown, black dots, respectively; (b)-(g) mass spectra of compounds (1)-(6) in (a). All compound structures are reasonably inferred, and are not accurate structures. (For interpretation of the references to color in this figure citation, the reader is referred to the web version of this article.)

ions occur in low abundance. Therefore, the first class of compounds is the C-ring mono-aromatic steranes, which is inferred from the peak position of normal oil samples, and the possible fragment ion is listed in Table 6 (No. 1).

Table 6　The possible chemical structures for the marked compounds in UCM-6.

No.	Possible structure	No.	Possible structure
1	(m/z 253)	5	(m/z 365)　(m/z 159)　(m/z 187)
2	(m/z 267)	6	(m/z 351)　(m/z 145)
3	(m/z 363)　(m/z 171)	7	(m/z 195)　(m/z 207)　(m/z 221)　(m/z 165)
4	(m/z 349)　(m/z 157)		

(2) Compounds with characteristic ions m/z of 267 are marked as red dots in Fig. 9a. These compounds have a similar retention time to the monoaromatic steranes, which indicates that they have a similar boiling point, polarity, and chemical structure. One of these compounds was chosen and its mass spectrum is shown in Fig. 9b. Compound (1) has a molecular ion m/z of 380 and a characteristic ion m/z 267. Its characteristic ion is of high abundance, and the other fragment and molecular ions occur in low abundance. The mass spectrum characterization of compound (1) is similar to monoaromatic sterane. The molecular ion of this compound is m/z 380, with a molecular formula of C_nH_{2n-12} ($n=28$). The degree of unsaturation of compound (1) is calculated to be 7, which suggests that its structure may include a benzene ring and three cycloalkanes. As seen from the mass spectrum, compound (1) and monoaromatic sterane, whose characteristic ion has a m/z of 253, has a difference of 14 (possibly a "—CH_2") in the characteristic ion, indicating that one methyl substituent may exist in the monoaromatic sterane structure. Fragment ion m/z of 143 shows that the substituted methyl is on the A ring (the left one). Therefore, it is suggested that compound (1) is a methyl-C-ring-monoaromatic sterane, and its possible fragment ion is listed in Table 6 (No. 2). The position of the methyl substituent marked in red is uncertain. Compounds with characteristic ions with m/z of 267 are methyl-C-ring-monoaromatic steranes. The inferred structure of compound (1) in Fig. 9a is shown in Fig. 9b.

(3) Compounds with characteristic ions m/z of 157, 171, and 363 are marked as pink dots in Fig. 9a. The second dimension retention time of these compounds is close to the monoaromatic steranes, which suggests that the compounds have a similar chemical structure to the monoaromatic steranes, and the unknown compounds' chemical structure contains a benzene ring and three cycloalkanes. Since monoaromatic sterane is the sterane aromatization product, it can be inferred that the unknown compounds are the aromatization product of some tetrahydrocarbons. Secohopane compounds, which are tetrahydrocarbons, are discovered in the saturated hydrocarbon of this heavily degraded crude oil, and have a similar second dimension retention time to steranes. We can therefore speculate that the unknown compounds may be aromatic secohopanes. Fig. 9c is the mass spectrum of compound (2) in Fig. 9a. The characteristic ions are m/z 171 and 363, and the molecular ion peak is m/z 392. The general chemical formula of the compound is $C_n H_{2n-14}$ ($n=29$), and the degree of unsaturation is calculated as $(2 \times n + 2 - 2n + 14)/2 = 8$. Considering that the degree of unsaturation of an aromatic secohopane is 7, and one less than the degree of unsaturation of the above compound, it can be speculated that a double bond probably exists in the unknown compound. Fragment ions m/z of 171 indicates that the double bond may be on the D ring (the right one). The compound is therefore speculated to be monoaromatic secohopene, and its possible fragment ions are listed in Table 6 (No. 3). Therefore, compounds with characteristic ion m/z 157, 171, and 363 may be monoaromatic secohopenes. The inferred structure of compound (2) in Fig. 9a is shown in Fig. 9c.

(4) Compounds with characteristic ion m/z ratios of 143, 157, and 349 are marked as white dots in Fig. 9a. Fig. 9d is the mass spectrum of compound (3). The molecular ion is exactly the same as compound (2), which suggests that they have the same chemical formula and degree of unsaturation. The characteristic ions are m/z 143, 157, and 349, which indicates that a difference of methyl "—CH_2" exists from compound (2). The other fragment ions are similar to compound (2), which means that they have a similar fragmentation pattern. Therefore, the compound (3) is speculated to be monoaromatic secohopene, and its possible fragment ions are listed in Table 6 (No. 4). The inferred structure of compound (3) in Fig. 9a is shown in Fig. 9d. Therefore, compounds with characteristic ion m/z 143, 157, and 349 are probably monoaromatic secohopenes, which contain one less methyl than the compounds with characteristic ion m/z 157, 171, and 363.

(5) Compounds with characteristic ion m/z 159 and m/z 365 are marked as brown dots in Fig. 9a. Fig. 8f and e are mass spectra of compounds (4) and (5). From the molecular ions m/z of 394 and 408, the general molecular formula of the compounds is $C_n H_{2n-12}$, and the degree of unsaturation is 7, which is the same as the aromatic secohopane. The mass spectrum of compound (5) is the same as that of 8, 14-monoaromatic secohopane, identified by Hussler. Thus, the compound is speculated to be C_{29}-8, 14-monoaromatic secohopane with characteristic fragment ions of m/z 365, m/z 159, and

m/z 187, its possible fragment ions are listed in Table 6 (No. 5). Compound (5) has similar fragment ions to compound (4) but a molecular ion at m/z 408 (Fig. 9f). It is inferred that compound (5) is C_{30}-8, 14-monoaromatic secohopane. The inferred structure of compounds (4) and (5) in Fig. 9a is shown in Fig. 9e and 9f. Therefore, compounds with characteristic ions m/z 159 and 365 are probably 8,14-monoaromatic secohopanes.

(6) Compounds with characteristic ions m/z 145 and m/z 351 are marked as black dots in Fig. 9a. Fig. 9g is the mass spectrum of compound (6) in Fig. 9a. The molecular ion of compound (6) is similar to that of compound (4), which indicates that they have the same molecular formula and degree of unsaturation. The characteristic ion of compound (6) (m/z 351) is 14 less than that of compound (4) (m/z 365), and shows a difference of a methyl "—CH_2". Other fragment ions are similar to those of compound (4), which suggests that the compound is probably aromatic secohopane. Its possible fragment ions are listed in Table 6 (No. 6). The inferred structure of compound (6) in Fig. 9a is shown in Fig. 9g. Therefore, compounds with characteristic ions m/z 145 and 351 are probably monoaromatic secohopanes, with one less methyl group than the compounds with characteristic ions m/z 159 and 365.

In conclusion, the basic chemical structure of the first class compounds in UCM-6 is composed of an aromatic ring and three monocyclic cycloalkanes. The main compound type is monoaromatic sterane (m/z 253), methyl monoaromatic sterane (m/z 267), monoaromatic secohopane (m/z 145, 159, 351, 365), and monoaromatic secohopene (m/z 143, 157, 171, 349, 363). The relative content of monoaromatic secohopanes (m/z 159, 365) is the highest, which indicates that the monoaromatic secohopanes may be characteristic compounds of the aromatic fraction in heavily degraded crude oil.

2) Second class

As seen from the GC×GC chromatogram (Fig. 8), the second dimension retention time of the second class compounds is higher than that of the first class compounds, which indicates that the polarity of the second class compounds is higher. As the compound structure of the first class is tetracyclic, in which one of the rings is an aromatic ring, the compound structure of the second class may be tetracyclic, in which two or three of the rings are aromatic rings, or a pentacyclic structure with an aromatic ring. By interpreting the mass spectra of the second class compounds, it is found that the characteristic ions of these compounds are m/z 231, 245, and 259 (as shown in Fig. 10). The first two characteristic ions are the characteristic ions of triaromatic steranes and methyl triaromatic steranes (the mass spectra are shown in Fig. 10b and c). These structures are tetracyclic, in which three of the rings are aromatic, and these compounds have a similar second dimension retention time in heavily biodegraded oil than in normal oil. Therefore, compounds with m/z 231 and 245 are speculated to be triaromatic steranes and methyl triaromatic steranes. The characteristic ion with m/z of 259 is 14 more than the characteristic ion with m/z of 245, which indicates a difference of a methyl "—CH_2", and therefore the compound is speculated to be C_2-substituted triaromatic sterane (the mass spectra are shown in Fig. 10d). Therefore, the second compounds of UCM-6 are composed mainly of triaromatic steranes, methyl triaromatic steranes, and C_2-triaromatic steranes.

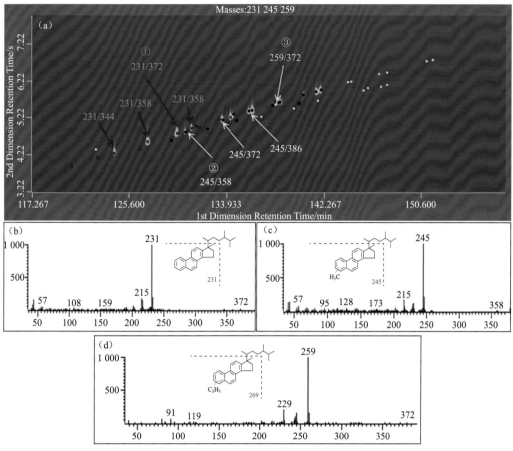

Fig. 10 GC×GC chromatogram: (a) GC×GC color contour chromatogram for m/z 231,245,and 259,in which compounds marked with 231/358 mean that the characteristic ion of the compounds is 231,and the molecular ion is 358. Red,black,and yellow dots represent compounds with a m/z of 231,245,and 259,respectively; (b) to (d) mass spectra of compounds (1)-(3) in (a). All compound structures are reasonably inferred,and are not accurate structures. (For interpretation of the references to color in this figure legend,the reader is referred to the web version of this article.)

3) Third class

The polarity of the third class compounds is higher than that of the second class compounds. As seen from the second dimension retention time of the third class compounds and the distribution of homologous compounds that have the same characteristic ion(as shown in Fig. 11a),the distribution of these compounds is similar to the pentacyclic triterpane distribution of the saturated hydrocarbons. Thus, it is inferred that the third class compounds are aromatic pentacyclotriterpanes. Fig. 11a shows the distribution of compounds with characteristic ions m/z 195 and 209. The mass spectra of the two selected compounds are shown in Fig. 11b and c. It can be seen from Fig. 11b that the molecular weight of compound (1) is 360, its molecular formula is inferred as $C_n H_{2n-18}$ ($n =$ 27), and the degree of unsaturation is 10. The degree of unsaturation of a mono- and diaromatic pentacyclotriterpane is eight and ten, respectively. This suggests that the basic chemical structure of the third compounds is diaromatic pentacyclotriterpane. It can be inferred from the characteristic ions m/z 195 that aromatization is likely to occur on rings A and B, and the possible fragments are listed in listed in Table 6 (No. 7). The characteristic ion of compound (2) is m/z 209 (Fig 11c), and either its molecular or fragment ions are 14 more than compound (1), which indicates a difference of a methyl. According to the fragment ions m/z 209 and 221, the most likely methyl-substituted location is at the junction of rings C and D, and the inferred structure of compound (2) is listed in Fig. 11c. The third class compounds of the UCM-6 are therefore speculated to be A, B-ring-diaromatic-pentacyclotriterpanes.

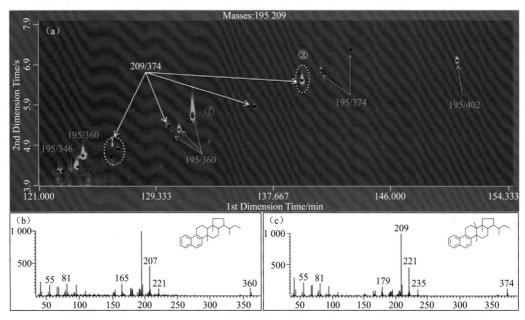

Fig. 11 GC×GC chromatogram. (a) GC×GC-TOFMS color contour chromatogram for m/z 195 and 209, in which compounds marked with 195/346 mean that the characteristic ion is 195, and the molecular ion is 246; (b) and (c) mass spectra of compounds (1) and (2), respectively. (For interpretation of the references to color in this figure citation, the reader is referred to the web version of this article.)

4) Fourth class

The fourth class of UCM-6 consists of two compounds (Fig. 12) with mass spectra as shown in Fig. 12b and c. The molecular formulae of the two compounds are $C_{24}H_{26}$ and $C_{25}H_{24}$, which are calculated according to the molecular weight. The degree of unsaturation is calculated to be 12 and 14, respectively. The possible structure of compound (1) is inferred to be heptacyclic with five double bonds (like tetradecahydrocoronene), hexacyclic with six double bonds (like dodecahydrobenzo[pqr]picene), or pentacyclic with seven double bonds (like octahydropicene). As can be seen from Fig. 8, the location of the fourth compounds in the GC×GC chromatogram is adjacent to the third class compounds. It is inferred that the structure of compound (1) is similar to the third class. The mass spectrum suggests that compound (1) has a triaromatic-pentacyclotriterpane structure. Because of the abundance of fragment ions m/z 271, 299, and 314, its possible structure is listed in Fig. 12b. Similarly, compound (2) is speculated to be tetraaromatic-pentacyclotriterpane, with a possible structure listed in Fig. 12c.

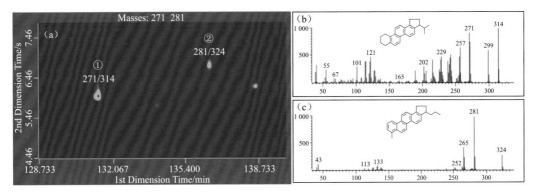

Fig. 12   GC×GC-TOFMS color contour chromatogram for m/z 271 and 281, in which compounds marked with 271/314 mean that the characteristic ion is 271, and the molecular ion is 314; (b) and (c) mass spectra of compounds (1) and (2), respectively. (For interpretation of the references to color in this figure citation, the reader is referred to the web version of this article.)

5) Fifth class

The fifth class of UCM-6 consists of three compounds (Fig. 13a), and its characteristic ion is m/z 191. The three compounds were arranged according to the molecular weight difference of 14, which indicates that these three components are of the same series. Fig. 13b is the mass spectrum of compound (1). The characteristic ion is m/z 191, the feature fragment ions are m/z 211 and 226, and the molecular ion is m/z 432. The chemical formula is $C_n H_{2n-16}$ ($n=32$), which is obtained from the molecular weight of the compounds, and the degree of unsaturation is 9. The mass spectrum of the compound is the same as that of C_{32}-benzohopane, as identified by Hussler. Thus, the compound is speculated to be C_{32}-benzohopane, and these compounds are C_{32}-benzohopane to C_{34}-benzohopane. The inferred structure of compound (1) is shown in Fig. 13b.

In conclusion, the compositions of compounds of UCM-6 have been speculated qualitatively from the GC×GC chromatogram and mass spectra. The UCM-6 compounds are composed mainly of various biomarkers, including aromatic steranes (including mono- and triaromatic steranes), aromatic secohopanes, and aromatic pentacyclotriterpane and benzohopanes. Aromatic secohopanes are of the highest relative content, followed by the triaromatic steranes. The monomer content of the benzohopanes is relatively high, but only three compounds were identified because of limitations in the analytical method. UCM-6 contains some other compounds with low content, low characteristic ion abundance, and unobvious molecular ions. The characterization of these compounds is difficult because of their low quality and irregularity, and is not very meaningful.

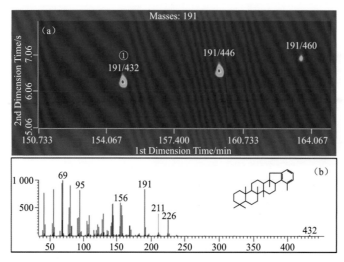

Fig. 13　GC×GC-TOFMS color contour chromatogram with m/z of 191, in which compounds marked with 191/432 mean that the characteristic ion is 191, and the molecular ion is 432; (b) mass spectrum of compound (1). (For interpretation of the references to color in this figure citation, the reader is referred to the web version of this article.)

4　Conclusions

The aromatic hydrocarbon fraction in a series of crude oils with different degrees of biodegradation from the Ciyutuo oilfield in the north eastern depression of Liaohe Basin was investigated by GC×GC-TOFMS and GC×GC-FID. Also, the constituents of the UCMs of the aromatic hydrocarbons in heavily biodegraded crude oil were analyzed in detail. The chromatogram for GC×GC provides a direct and clear image of the constituents of the aromatic hydrocarbon UCMs. It was found that based on their polarity, the UCMs of the aromatic hydrocarbons can be classified into 6 major groups including mono-, di-, tri-, tetra- and pen-tacyclic aromatic hydrocarbons, and aromatic biomarkers. Each group could be further subdivided according to their polarity and molecular weight, and finally 116 subdivisions were obtained. Mass spectroscopy was used to analyze the UCMs and 108 compounds were successfully tentative identified. Among them, the analytical and identification procedure for 24 compounds was introduced in detail. Then, using GC×GC-FID the quantification of the UCM was achieved and showed that the monoaromatic hydrocarbons account for the largest in terms of both weight percentage and peak numbers, followed by diaromatic hydrocarbons, aromatic biomarkers, tri-, tetra- and pentaaromatic hydrocarbons.

In each major group relatively low amounts or even no long side-chain substituted compounds exist because of the biodegradation effect; instead, compounds are mainly in the form of isomers or substituted cycloalkanes. For example, monocyclic substituted benzenes account for the majority of the monoaromatic hydrocarbons and the amount of monocyclic substituted naphthalenes is overwhelming in the diaromatic hydrocarbons. The changes in the relative contents of naphthalenes and phenanthrenes with different substitutes were discovered to be influenced by the degree of biodegradation and became an obvious indication that they were capable of indicating the degree of biodegradation in crude oil. In the heavily biodegraded oil, more types of aromatic biomarkers were detected than in normal crude oil. The aromatic biomarkers primarily include aromatic steranes and aromatic pentacyclotriterpanes, and the contents of aromatic secohopanes and benzohopanes were relatively high. The result of this study indicated that the influence of the degree of biodegradation in crude oil on the constitution of aromatic hydrocarbons was also investigated. Compounds with low polarity and low molecular weights decompose more quickly resulting in their relatively low content in the hydrocarbons. On the contrary, the speed of decomposition for compounds with high polarity and high molecular weight is significantly decreased and show a relatively high content in the aromatic hydrocarbons. These results provide useful information for future research on the quantitative characterization of the degree of biodegradation in heavy oil, the maturity stage of unconventional oil, oil source correlation, depositional environment and any other geochemical problems.

论文 10　2015 年发表于《Marine and Petroleum Geology》

Geochemistry, origin and accumulation of continental condensate in the ultra-deep-buried Cretaceous sandstone reservoir, Kuqa Depression, Tarim Basin, China

Guangyou Zhu, Huitong Wang, Na Weng, Haijun Yang,
Ke Zhang, Fengrong Liao, Yuan Neng

Abstract A continental condensate field, with the deepest burial depth (7 084 m) so far in China, was recently discovered from the Cretaceous sandstone reservoir in the Bozi area of the Kuqa Depression, Tarim Basin. In this paper, we report general features of this field, including geochemistry of the condensates, their origin, and migration and accumulation. General geochemistry of the condensate reflects that it is an over pressured condensate reservoir. Analytical results by newly-developed high-resolution GC × GC-TOFMS method indicate that the condensate is rich in n-alkanes from n-C_3 to n-C_{34} and in diamondoid hydrocarbons with a high content of adamantanes. The condensate has a high abundance of aromatic hydrocarbons. The gas $\delta^{13}C_2$ values in these are $-23.3‰$, which is characteristic of coal-derived gases. Compared with the contents of conventional condensate oils, the BZ1 well has a high content of aromatic hydrocarbons, suggesting that the condensate is in a stage of highly mature cracking. Analyses of oil biomarkers, carbon isotopes, gas compositions, etc., revealed that the condensate is dominantly sourced from the highly mature coals of the Jurassic Qiakemake Formation in the western Kuqa Depression. This set of coal source sequences has generated large quantities of condensate since 3 Ma, when the study area was subjected to a rapid subsidence with sediments over 2 000 m and consequently attained at the condensate-generating stage. Such condensates then migrated vertically into the Cretaceous traps along faults. Therefore, the Bozi condensate field is a typical case of large-scale oil/gas field characterized by late-stage accumulation. Multiple elements favor the formation of this large-scale condensate field. The timing of trap formation accords with that of the condensate generation (since 3 Ma). The traps and source rocks are stacked vertically. Faults connect the source rock sequences and the reservoirs. The hydrocarbon charge is strong and efficient. Thick Paleogene gypsum beds act as an excellent cap rock, favoring the accumulation and preservation of the condensate.

Keywords　Deep-buried reservoir　Coal-derived hydrocarbons　Diamondoid　GC × GC-TOFMS Kuqa Depression　Tarim Basin

1 Introduction

The Tarim Basin is a typical superimposed petroliferous basin in northwest China with the northern Kuqa Depression being an exploration highlight for decades, e.g., the well-known Kela 2 (KL2) giant gas field. This area has recently been an exploration success in the Cretaceous sandstone reservoirs at a record burial depth of more than 7 000 m in the Bozi area of the western Kuqa Depression, with a daily oil production of 29.5 m³ and a daily gas production of 245 040 m³. This is the deepest well (BZ1 well) ever drilled in continental sediments with commercial oil and gas production, and thus is of significance from both a scientific and exploration point of view, which, up to date have been scarce. The stability of crude oil is now the focus of research around the world. The main conclusion drawn is that crude oil begins to crack and generate gas at temperatures higher than 160 ℃ or at burial depths of more than 6 000 m, under which conditions the liquid oil gradually disappears. However, exploration for oil and gas in deep strata has demonstrated that some oil pools exist in the oil phase at even higher temperatures, suggesting that the stability of crude oil is much greater than anticipated. For example, a high hydrocarbon flow was obtained in a buried hill reservoir at depths of 5 641 – 6 027 m and temperatures of 190 – 201 ℃ in Well ND1 in the Jizhong Depression, Bohai Bay Basin, China. A number of condensate gas reservoirs have been discovered in the Ordovician strata of the Tarim Basin. These strata have a low thermal gradient at a depth of about 5 500 – 6 500 m(a reservoir temperature of about 140 – 160 ℃). The BZ1 well in the Tarim Basin is the deepest continental industrial oil and gas well ever found in China and one of deeply buried condensates around the world. The area has a low thermal gradient, with an observed reservoir temperature of 130.6 ℃.

Therefore, in order to extend our understanding of the genesis and phase of the reservoir fluid and to provide references for the exploration of deeply-buried oil and gas, we examine general features of these accumulations in this paper, mainly focusing on their geochemistry of the accumulations, in order to provide information about their origin, migration and accumulation.

2 Geological setting

The Tarim Basin is the largest petroliferous basin in China, and has many deep-buried oil and gas reservoirs with complex migration and accumulation histories. The Kuqa Depression is located in the northern margin of the basin, with Mesozoic and Cenozoic deposits in dominance (Fig. 1a). The depression is structurally NEE-trending, with an area of approximately 3.7×10^4 km^2 (Fig.1a). It can be divided into three structural zones (Kelasu, Yiqikelike, and Qiulitage), three sags (Wushi (the west, beyond Fig. 1a), Baicheng, and Yangxia), one monoclinal zone northern), and one slope zone (southern) (Fig. 1a). Of these zones and sags, the Kelasu structural zone is the area where the BZ1 Well is located (Fig. 1a – d).

The Kuqa Depression is rich in natural gas resources in general and is an important base for the giant Chinese project of "transporting gas from west to east". In this depression, a salt bed of the Paleogene Kumugeliemu Group (E_{1-2}km in Fig. 1e) has significant impacts on the hydrocarbon accumulation. The gypsum salts and gypsum mudstones form a complex salt structural deformation under compression (Fig. 1b). The detachment of the gypsum salt led to the development of thrust belts below the salt (Fig. 1b). Controlled by the separation of the Kelasu, Keshen, and Baicheng faults, the studied Kelasu structural zones were divided into three substructural zones including Kela zone,

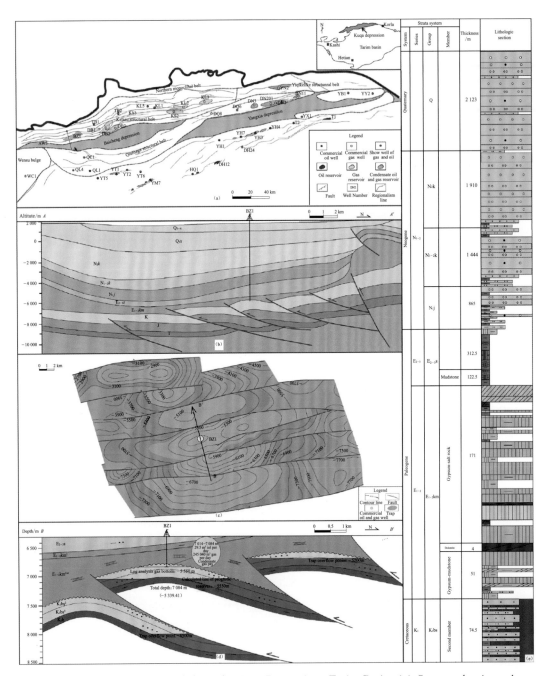

Fig. 1 The BZ1 condensate field in the Kuqa Depression, Tarim Basin. (a) Structural units and well locations in the Kuqa Depression; (b) geological section crossing the BZ1 condensate field along south-north trending, see Fig. 1a for the position of the section; (c) structural configuration of the BZ1 condensate reservoir; (d) cross section showing the BZ1 condensate reservoir, see Fig. 1a for the position of the section; (e) generalized lithological column of the BZ1 condensate field (in right).

the northern Keshen zone, and the southern Keshen zone. Five areas with the development of traps are present from west to east: Awate (AW), Bozi (BZ), Dabei (DB), Keshen (KS), and Kela (KL) (Fig. 1a), providing favorable conditions for extensive hydrocarbon accumulations under the salt bed. The Cretaceous sandstones under the gypsum salt (K_1bs in Fig. 1b) contain abundant natural gas resources, and a number of large to giant gas fields with 100 billion cubic meters in proven reserve have been found, including Kela 2 (KL2), Dina 2 (DN2), Dabei (DB2), and Keshen (KS2) fields.

The newly-discovered Bozi condensate field is located in the western margin of the Kelasu structural zones (the west of Fig. 1a), with the BZ1 well being the discovery well (Fig. 1). The well is located in the Bozi 1 structure of the Keshen block of the Kelasu structural zone, 40 km west of the well-known Dabei gas field. This well completed drilling in May 2012, attaining a bottom-hole depth of 7 084 m. The Cretaceous Bashijiqike Formation (K_1bs) at 7 014 – 7 084 m was tested-with oil and gas producing from a 5 mm nozzle at an oil pressure of 67.229 MPa. The daily oil and gas production are 29.5 m and 245 040 m^3, respectively. The measured reservoir temperature in the well is 130.6 ℃ and the pressure is 125.505 MPa, the formation pressure coefficient is 1.83, indicating an ultra-high-pressure condensate gas pool. It is a condensate layer based on PVT analysis of the reservoir fluid.

The sandstones of braided-river delta-front subaqueous channels and mouth-bars are the reservoir rocks for these condensates. Reservoir pores are poorly developed in general because of deep burial and tectonic compression. The porosity is 1.5%–9.5% with an average of 6.42%, and the permeability is 0.01 – 0.1 mD with an average of 0.08 mD. Thus, these reservoirs are characterized by an extra-low porosity and middlee-low permeability. Nevertheless, some residual primary pores and fractures are developed to accumulating oil and gas. The Bozi 1 (BZ1) gas reservoir is located beside the Baicheng hydrocarbon-generating center (Fig. 1a), thus having a sufficient supply of oil and gas source. The thick gypsum salt, gypsum mudstone, and mudstone that developed in the Paleogene Kumugeliemu Group (Fig. 1b) provide good regional seals overlying the sandstones of the Cretaceous Bashijiqike Formation. Therefore, an excellent combination of source, reservoir, and seal was formed. The faults act as pathways for the vertical migration of oil and gas, leading to the accumulation of oil and gas in faulted anticline trap below the gypsum salt (Fig. 1b). In terms of structural configuration, the faulted anticline (trap) area is 83.0 km^2 and the trap height is 350 m (Fig. 1c). Combined with an effective reservoir thickness of 66.7 m, the reserves of oil and gas in the Bozi 1 structure can reach over 100 billion cubic meters. This is the most deeply buried condensate field ever discovered in China.

3 Data and methodology

Oil and gas samples were collected from the reservoir interval of Wells BZ1, QL1, YT101, YT5, YN, DN1, DB3, YM, RP, XK and H7-2. The BZ1 well condensatewas subjected to the new comprehensive twodimensional gas chromatography with flame ionization detector (GC×GC-FID) and time-of-flight mass spectrometry (GC×GC-TOFMS). GC×GC-TOFMS has been used for advanced biomarker and diamondoid evaluation. Sound statements of the advantages provided by this technique in organic geochemistry have been recently reported. Other Oil and gas samples were completed conventional geochemical analyses of oil and gas were, involving physical properties, chemical components, isotopes and biomarkers.

The BZ1 well condensate was directly analyzed by comprehensive GC×GC-TOFMS and GC×GC-FID without any pretreatment. D16-adamentane (the solvent is dichloromethane) was added in the condensate sample as an internal standard and the quantitative results were gained by using internal standard method. The GC×GC-TOFMS system was a Pegasus IV model made by the LECO Corporation. The GC×GC system was composed of an Agilent 7890A GC coupled to a hydrogen flame ionization detector (FID) and a liquid-nitrogen-cooled pulse jet modulator (LECO Corporation). The first column on GC×GC was DB-Petro (100% dimethylpolysiloxane phase, 50 m×0.2 mm×0.5 μm) and the second column was DB-17HT ((50%-phenyl)-methylpolysiloxane phase, 3 m× 0.1 mm×0.1 μm). The temperature program used for the first column was: 0.2 min at a temperature of 35 ℃; increased to 210 ℃ at a rate of 1.5 ℃/min and held for 0.2 min; and increased to 300 ℃ at the rate of 2 ℃/min and held for 20 min. The temperature program used for the second column was the same as that used for the first one but the temperatures were 5 ℃ higher. The modulator temperature was 45 ℃ higher than for the first column. The inlet temperature was set at 300 ℃, the inlet mode was split injection, the split ratio was 700∶1 with the sample volume of 0.5 μL. Helium was the carrier gas with a flow rate of 1.5 mL/min. The modulation cycle was 10 s, including 2.5 s for heating blow. The TOFMS was performed under the following conditions: ionizing voltage −70 eV, monitoring voltage 1 600 V, rate of data acquisition 100 spectra/s, mass scanning range 40−520 μm, and dwell time 9 min. The temperatures of the transfer line and the ion source were 300 ℃ and 240 ℃, respectively. The GC×GC-FID system had the same separation conditions and columns as the GC×GC-TOFMS system. The flow rates of the carrier gas, hydrogen and air for FID were 50, 40, and 450 mL/min, respectively. The detector temperature was 320 ℃ with the acquisition rate of 200 spectra/s. All the data were processed with the Chroma TOF software (Version 4.50). Normalized peak area was used for quantification of compound groups and D16-adamantane was used as an internal standard for quantification of diamondoids.

Gas Chromatographice-Mass Spectrometry (GC-MS): a TRACE GC ULTRA/DSQⅡ instrument, equipped with an HP-5MS silica column (60 m, 0.25 mm i.d., film thickness is 0.25 μm), was used. The initial GC oven temperature is 100 ℃, held for 5 min, and then programmed to 220 ℃ at the rate of 4 ℃/min; finally it was programmed to 320 ℃ at the rate of 2 ℃/min and held isothermally for 20 min. The biomarker parameters were calculated by peak area of each component.

Compound specific stable carbon isotope ($\delta^{13}C$) ratios were determined on a Finnigan Mat Delta S mass spectrometer interfaced with a HP 5890Ⅱ gas chromatograph. Gas components were separated on the gas chromatograph in a stream of helium, converted into CO_2 in a combustion interface and then introduced into the mass spectrometer. Individual hydrocarbon gas components ($C_1 - C_5$) were initially separated using a fused silica capillary column (PLOT Q 30 m×0.32 mm). The GC oven temperature was ramped from 35 to 80 ℃ at 8 ℃/min, then to 260 ℃ at 5 ℃/min, and the oven maintained its final temperature for 10 min. Gas samples were analyzed in triplicate. Stable carbon isotopic values are reported in the customary d notation in per mil (‰) relative to PDB (VPDB). Measurement precision is estimated to be ±0.3‰ for $\delta^{13}C$.

The chemical composition of gas samples was determined using an Agilent 6890N gas chromatography (GC) equipped with a flame ionization detector and a thermal conductivity detector. Individual hydrocarbon gas components from methane to pentanes ($C_1 - C_5$) were separated using a capillary column (PLOT Al_2O_3 50 m×0.53 mm). The GC oven temperature was initially set at 30 ℃ for 10 min, and then ramped to the maximum temperature of 180 ℃ at a rate of 10 ℃/min where it was held for 20 – 30 min.

PVT analysis: One high-pressure condensate sample from the BZ1 well was analyzed for PVT properties. On the basis of analyzing fluid compositions of fluids in well in Bozi area, oil and gas reservoirs can be classified according to the methods of fluid PVT phase diagram, triangle diagram of fluid tricomponent, and block diagram, thus qualitatively evaluating gas invasion strength. Fluid PVT phase curves: the PVT phase curves of condensate reservoir are in parabolic shape and the fluid phase point is in the right of the critical point; the fluid PVT phase point of volatile reservoir is between the critical point and critical condensate pressure point; and the fluid PVT phase point of oil reservoir is in the left of the critical condensate pressure point (Fig. 2).

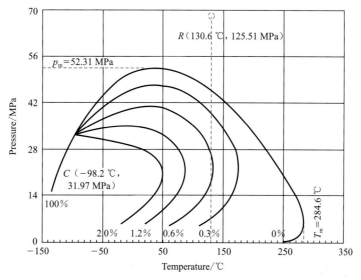

Fig. 2 Reservoir fluid phase in the BZ1 condensate reservoir.

4 Results and discussion

4.1 Physical properties and PVT

According to the analysis of the condensate from the BZ1 well, the oil retrograded from the gas is characterized by low density, low viscosity, and low wax content in general (Table 1). The API gravities are 51.10°–46.60° at 20 ℃ and viscosities are 0.701–1.026 mPa·s at 50 ℃. Condensates have low pour point ($<$4 ℃), low sulfur content ($<$1.13%), low asphaltene contents ($<$0.02%) and low wax content (3.4%–5.7%).

Table 1 **General oil properties of the BZ1 condensate.**

Depth/m	API gravity/(°)		Viscosity/(mPa·s)	Wax /%	Sulfur /%	Asphaltene /%	Pour point /℃
	20 ℃	50 ℃	50 ℃				
7 014–7 084	51.10	56.51	0.701	3.40	0.00	0.000	2
	46.60	51.86	1.026	4.20	0.00	0.000	4
	50.07	55.45	0.772	5.70	1.13	0.020	−4

As for the BZ1 well gas, it is generally characterized by high abundance of methane (89.13%) and low abundance of nonhydrocarbon gases (e.g., 0.88% N_2 and 0.36% CO_2), showing a high-quality characteristic. The GOR ratio is 8 620 m^3/m^3.

As for the phase characteristic of the reservoir fluid (Fig. 2), a low critical pressure of 31.97 MPa, low critical temperature of −98.2 ℃, low dew-point pressure of 45.92 MPa, and large difference between the formation pressure and the dew-point pressure (79.59 MPa) were observed. These indicate that the fluids are predominantly light in nature, with a few heavy hydrocarbon components. The critical condensate pressure and temperature are 52.31 MPa and 284.6 ℃, respectively, while the formation temperature (130.6 ℃) is greater than the critical temperature. During the gas depletion analysis under constant volume, the maximum retrograde pressure for the condensate is 14 MPa, the maximum retrograde fluid amount is 1.17%, and the condensate content on the surface is low at 85.183 g/m^3. According to these basic data, we can infer that the BZ1 reservoir fluid has the typical characteristics of condensate with low abundance of liquid hydrocarbons (Fig. 2).

4.2 Gas geochemistry

Table 2 presents the general geochemistry (e.g., chemical component and carbon isotopic value) of natural gas in the studied BZ1 well, and the values of the gases in the neighboring DB, KS, and KL2 fields were also given for comparison. It is showed that the natural gas in the Bozi gas field has a high content of heavy hydrocarbons and is clearly wet, with a dryness coefficient of approximately 90%. The gas dryness ratio ($C_1/\sum C_{1-4}$) in the other three gas fields are all more than 97%, and their C_4^+ contents of less than 0.5 suggest that the maturities of these gases are significantly different. The

Bozi gas is relatively less mature.

Table 2 Natural gas compositions and carbon isotopes in the BZ1 condensate and neighboring gas fields.

Well	Interval /m	Formation	Natural gas composition/%						Dryness ratio ($C_1/\sum C_{1-4}$)	Carbon isotopes/‰			
			C_1	C_2	C_3	C_4^+	N_2	CO_2		C_1	C_2	C_3	C_4
BZ 1	7 014 – 7 084	K_1bs	89.4	6.80	1.56	1.00	0.54	0.68	90.52	−35.3	−23.3	−21.3	−20.9
			89.90	6.82	1.56	0.99	0.37	0.41	90.56	—	—	—	—
			88.60	6.73	1.55	1.42	1.50	0.17	90.13	—	—	—	—
			88.60	6.76	1.57	1.69	1.12	0.19	89.84	—	—	—	—
DB 2	5 658 – 5 669	K_1bs	95.26	2.25	0.38	0.53	1.19	0.39	96.79	−30.8	−21.5	−19.8	—
DB 301	6 930 – 7 012	K_1bs	96.2	1.66	0.168	0.17	1.26	0.507	97.97	−29.9	−22.3	—	—
KS 2	6 573 – 6 697	K_1bs	97.5	0.514	0.043	0.104	1.100	0.834	99.33	−28.3	−17.7	−15.7	—
KS 5	6 703 – 6 742	K_1bs	97.4	0.267	0.011	0.016	0.973	1.36	99.70	−26.5	−17.8	−19.2	−28.2
KL 2	3 888 – 3 895	K_1bs	97.84	0.12	0	0	1.355	0.685	99.88	−27.8	−19.0	—	—

In terms of carbon isotope composition, the carbon isotopes of the BZ1 gas are relatively light: methane −35.3‰ and ethane −23.3‰ (Table 3). These values are lighter than those in the high-maturity KL2 gas field, implying a lower maturity.

Table 3 Quantification of diamondoids in the BZ1 and ND1 condensate.

No.	Compound	BZ1 mg/kg	ND1[a] mg/kg
1	Adamantane	652.6	492.7
2	1-Methyladamantane	1 703.8	1094.7
3	1,3-Dimethyladamantane	1 154.9	668.8
4	1,3,5-Trimethyladamantane	427.8	223.8
5	1,3,5,7-Tetramethyladamantane	158	217.0
6	2-Methyladamantane	779.8	728.7
7	1,4-Dimethyladamantane (cis)	944	767.6
8	1,4-Dimethyladamantane (trans)	726.7	608.2
9	1,3,6-Trimethyladamantane	424.8	244.2
10	1,2-Dimethyladamantane	673.5	432.4
11	1,3,4-Trimethyladamantane (cis)	420.8	258.2
12	1,3,4-Trimethyladamantane (trans)	428.2	327.2
13	1,2,5,7-Tetramethyladamantane	291.6	156.3
14	1-Ethyladamantane	208.5	135.4
15	1-Ethyl-3-methyladamantane	258.6	160.5
16	1-Ethyl-3,5-dimethyladamantane	157.7	67.2

continue

No.	Compound	BZ1 mg/kg	ND1[a] mg/kg
17	2-Ethyladamantane	336	248.5
18	1,2,3,5,7-Pentamethyladamantane	78	—
19	Diamantane	61.2	64.7
20	4-Methyldiamantane	88.6	52.4
21	4,9-Dimethyldiamantane	10.3	17.1
22	1-Methyldiamantane	23.8	30.9
23	1,4-+2,4-Dimethyldiamantane	13.9	20.5
24	4,8-Dimethyldiamantane	38.6	35.7
25	1,4,9-Trimethyldiamantane	29.8	12.4
26	3-Methyldiamantane	60.3	39.2
27	3,4-Dimethyldiamantane	29.9	46.7

[a] ND1: after Zhu et al.

$\delta^{13}C$ of ethane ($\delta^{13}C_2$) is an important parameter for distinguishing oil-type gas and coal-derived gas. According to the standards introduced by Dai et al. which has been widely and effectively used in China, $\delta^{13}C_2 > -28‰$ indicates a coal-derived gas. The gas $\delta^{13}C_2$ values in these four gas fields range from $-17.7‰$ to $-23.3‰$ (Table 3), a characteristic of coal-derived gases.

In summary, from the natural gas chemical components and their carbon isotopic composition, the gas in the deeply buried Bozi condensate field is wetter and isotopic lighter than those in the shallower-buried DB, KS, and KL2 gas fields, suggesting that the Bozi natural gas is less mature and the gases of the DB, KS, and KL2 gas fields are thus sourced from deeper strata as their gas sources are similar (with similar $\delta^{13}C_2$ values).

4.3 Oil geochemistry

4.3.1 Carbon isotope and GC×GC

The carbon isotopes of the BZ1 oil are relatively heavy. The whole-oil $\delta^{13}C$ is $-25.81‰$, the saturated hydrocarbon $\delta^{13}C$ is $-26.99‰$, and the aromatic hydrocarbon $\delta^{13}C$ is $-25.28‰$ (Fig. 3). These values are generally identical to those of the Jurassic coal-derived oils (e.g., YT and YN), but are significantly different from those of the lacustrine (e.g., YM) and marine origin (e.g., RP) oil fields (Fig. 3). Thus, the BZ1 oil here may also be Jurassic coal-derived. According to the rule of isotopic fractionation during organic matter maturation in the study area, the oil carbon isotope should be close to or slightly lighter than the kerogen carbon isotope. In the study area, the kerogen carbon isotope of the Jurassic coal/coaly mudstone is approximately $-24‰$, which is slightly heavier than the carbon isotopes of the BZ1 condensate. This suggests that the condensate is mainly sourced from Jurassic coal/coaly mudstone.

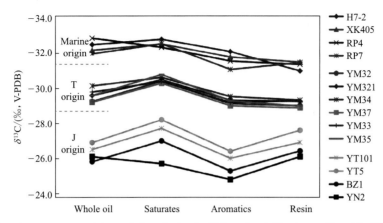

Fig. 3 Carbon isotope compositions of the oils in the BZ1 and neighboring wells with different origins.

In terms of biomarkers, it is implied that the oil from the BZ1 well is typical of continental origin. Representative evidences include a Pr/Ph value of 2.04, abundant rearranged compounds (e.g., rearranged steranes and hopanes), high abundance of C_{24} tetracyclic terpane relative to C_{26} tricyclic terpane, distribution of tricyclic terpanes of $C_{19} > C_{20} > C_{21} > C_{23}$, and little detection of gammacerane, high relative abundances of C_{29} steranes compared to C_{27} and C_{28}. Oil-oil correlation shows that these characteristics of the BZ1 oil are generally similar to those of the Quele (QL) and Yangta (YT) oils (Fig. 4), which have been suggested to sourced from the coals and coaly mudstones of the Jurassic Qiakemake Formation. As such, the BZ 1 well oil is inferred to be derived from the Jurassic Qiakemake Formation, further supporting the understanding obtained from carbon isotopes of oil above.

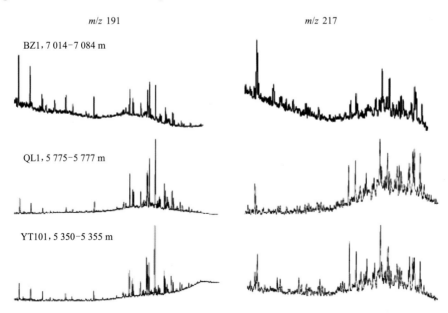

Fig. 4 Oil biomarkers in the BZ1, QL1, and YT101 condensate fields.

As for the oil maturity, a Ts/Tm value of 1.72 is indicative of a moderate to high maturity. This can also be implied from the analysis based on the relationship of heptane vs. isoheptane in light hydrocarbons (Fig. 5). It is showed that the oil is less mature than the condensate from eastern Dabei (DB) filed (Fig. 5). This difference is similar to that between the gases at these two fields (wet gas vs. dry gas). In Figure 5, a paraffin index of 1, the i-heptane index = (2-methylhexane + 3-methylhexane)/(cis-1,3-cyclopentane + trans-1,3-cyclopentane); a paraffin index of 2, the n-heptane index = $100 \times n$-heptane/(cyclohexane + 2-methylhexane + 1,1-dimethylcyclopentane + 2,3-dimethylpentane + 3-methylhexane + cis-1,3-dimethylcyclohexane + trans-1,3-dimethylcyclohexane + trans-1,2-dimethylcyclohexane + 3-ethylpentane + 2,2,4-trimethylpentane + n-heptane + methylcyclohexane).

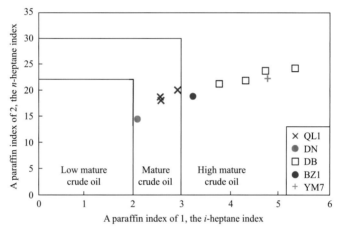

Fig. 5 Relationship between heptane and isoheptane in light hydrocarbons of the oils in the BZ1 and neighboring wells.

4.3.2 GC×GC-TOFMS and diamondoids

A total of 3 177 compounds above a signal-to-noise ratio of 100 were detected in the condensate (oil) sample from the BZ1 well. According to the mass spectra, these compounds can be classified into 10 groups. The proportion of each groupwas quantified (Fig. 6), with the contents of saturated hydrocarbons, aromatic hydrocarbons, and nonhydrocarbons being 2.19%, 17.66%, and 0.15%, respectively. N-alkanes range from C_7 to C_{34}, and the aromatic hydrocarbons range from monocyclic to tricyclic.

Compared with Chinese conventional condensate oils, the content of aromatic hydrocarbons in the BZ1 well is relatively high (Fig. 6). Of the various types of aromatic hydrocarbons, the contents of monocyclic aromatic hydrocarbons are the highest, with the contents of benzene, methylbenzene, and dimethylbenzene accounting for 1.55%, 3.51%, and 3.63% of the all compounds, respectively. The contents of naphthalene and phenanthrene are 0.67% and 0.29%, respectively. In addition, the abundance of monocyclic cycloalkanes in the total cycloalkanes is relatively high (15.27%) compared with that of Chinese conventional condensates, of which the methylcyclohexanes account for 2.16%.

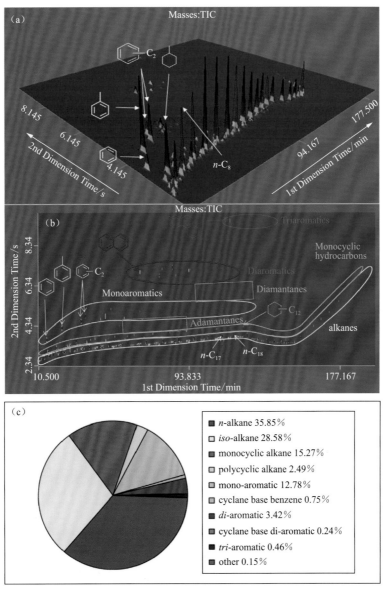

Fig. 6　GC×GC-FID analysis of the BZ1 condensate (a: 3D plot chromatogram; b: color contour chromatogram) and quantification of group components in the BZ1 condensate determined by GC×GC-FID (c).

In addition, long-side-chain-substituted alkyl cyclohexane, benzene, and decalin compounds were detected in the condensate from the BZ1 well. Long-side-chain-substituted decalins range from decalin to tridecyl decalin. The alkyl benzenes range from benzene to n-nonadecylbenzene, and the alkyl cyclohexanes range from cyclohexane to 27 alkyl cyclohexane.

Diamondoids have widely been regarded as an indicator of oil cracking during highly mature stage. Because compounds with the same boiling point can be effectively separated by GC×GC-FID based on their different polarities, GC×GC-FID is a good means to detect adamantanes in the samples from the BZ1 well. Analytical results show that the condensate from the BZ1 well contain abundant diamondoids, among which the adamantanes have a high abundance with various types (Table 3). Besides the commonly-detected 18 adamantanes, we identify 74 more diamondoids (Fig. 7 and 8).

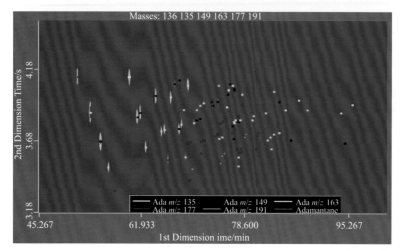

Fig. 7 GC×GC-TOFMS color contour chromatogram of extracted ion of m/z 135, 136, 149, 163, 177, and 191 of the BZ1 condensate. The distribution of adamantanes is labeled in the Figure. Ada, adamantane.

As shown in Fig. 8a, 11 compounds were identified in m/z 135 and m/z 136 chromatograms. Based on the TOFMS mass chromatograms, 10 compounds can be ordered into five families based on their molecular weights (ranging from 150 to 206), with a mass difference of 14 atomic mass units (—CH_2—) between each family. The mass spectra of these 10 compounds have a base peak in m/z 135 chromatogram and characteristic fragment peaks in m/z 79, 93, and 107 chromatograms. Their retention times on the second dimension are similar to those of the methyladmantanes (no. 2 and 6 compounds labeled in Fig. 8a) and ethyladmantanes (no. 14 and 17 compounds labeled in Fig. 8a). According to the characteristics detected by GC×GC-FID, the compounds in the sample can be classified into each group. These five compounds were inferred to be alkyladamantanes (C_n-adamantanes) with a single substituent, and the substituted positions are "1" ( ) and "2" ( ), respectively, provide a molecule with carbon numbers to let the reader know C-1 and C-2. Based on the substituted positions in the methyladmantanes and ethyladmantanes, we infer that the retention time both on the first and second dimensions of 1-C_n-adamantane is shorter than that of the 2-C_n-adamantane.

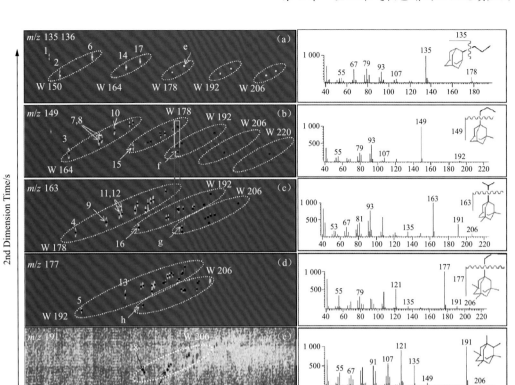

Fig. 8 GC×GC-TOFMS color contour chromatogram of the extracted ion of m/z 135,136,149, 163,177,and 191 in the BZ1 condensate. Adamantanes with different substituent distributions are labeled in the figure and compounds 1－18 correspond to the compounds in Table 2. Mass spectra of compounds e,f,g,h,and j are provided. W,molecular weight.

As shown in Fig. 8b, 21 compounds were identified in m/z 149 chromatogram. Based on the TOFMS mass chromatograms, the compounds (ranging from 164 to 220) were ordered into five families by molecular weight, with a mass difference of 14 atomic mass units (—CH_2—) between neighboring compounds. According to the mass spectra, these compounds are similar to the adamantanes with two substituents (3, 7, 8, 10, and 15 in Table 3); they have the same base peak in m/z 149 chromatogram and characteristic fragment peaks in m/z 79, 93, 107, and 121 chromatograms, and their retention times on the second dimension are similar. Therefore, we infer that these 21 compounds are adamantanes with two substituents. The mass spectrum of compound f in Fig. 8b shows that the molecular weight difference between ion 192 and ion 149 is 43 according to the mass spectrum and with no other fragment ions, implying that the ion 149 is generated from the loss of a propyl group. Furthermore, considering the retention time, compound f is inferred to be 1-propyl-3-methyl-adamantane. In Fig. 8b, compound f in ion chromatograms of m/z 149 has a similar onedimensional retention time as the other two compounds (As shown in the yellow (in the web version) box of Fig. 8b, compound f and other two compounds are the same in one dimension reservation time but different in two dimension reservation times), indicating that these three compounds coelute on one-dimensional chromatography. In m/z 192 chromatogram, these three compounds can be separated from each other, but compound f still coeluted with the other two compounds in Fig. 8b. Therefore, compound f cannot be distinguished by MS alone. Many other compounds in Fig. 8 show the same characteristics as compound f.

Using the methods described above, the 29 compounds labeled in Fig. 8c were presumed to be adamantanes with three substituents, with molecular weights ranging from 178 to 206. The mass spectrum of compound g suggests that the substituted alkyl group at C-1 is more likely to be isopropyl than n-propyl because of the presence of the fragment ion m/z 191 in the mass spectrum. The inferred structural formula of compound g is shown in Fig. 8c. Similarly, the 25 compounds labeled (As shown in the red (in the web version) box of Fig. 8b and c, compound f and other two compounds are the same in one dimension reservation time but different in two dimension reservation times) in Fig. 8d are all adamantanes with four substituents, and the six compounds labeled in Fig. 8f are all adamantanes with five substituents.

The content of diamantanes is not high in the sample. Apart from the nine identified diamantanes, two other diamantane compounds were found (Fig. 9a). The mass spectrum of one of these two compounds is shown in Fig. 9b. From this mass spectrum, the compound is similar to the known dimethyl (21, 23, 24, and 27 in Table 3), with a similar retention time. As such, these two compounds are inferred to be dimethylcongressanes, but the substituent positions are unknown.

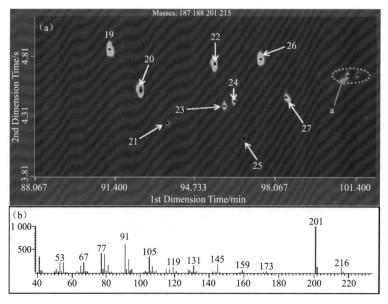

Fig. 9 GC×GC-TOFMS color contour chromatogram of the extracted ion chromatograms of m/z 187, 188, 201 and 215 in the BZ1 condensate. Replace by diamantanes with different substituent distributions are labeled in the figure and compounds 19−27 correspond to the compounds in Table 3.

In summary, the high abundance and diversity of the diamondoids identified in the BZ1 condensate here suggest that the condensate is highly mature during the cracking stage.

4.4 Source of the Bozi condensate

Base on the above geochemical results and discussion, it is implied that the Bozi condensate is most likely sourced from the Jurassic coal sequences in the study area. The Jurassic of the Kuqa Depression mainly developed coaly sequences of shallow lake-swamp facies as hydrocarbon source rock, whose effective thickness is about 250 m and organic carbon content exceeds 4%. The organic maceral compositions of these coal sequences are predominantly vitrinite, semivitrinite, and semi-inertinite, favorable for generating gas.

From the present-day R_o contour line of the Jurassic Qiakemake Formation source rock in the Kuqa Depression, most of the R_o is higher than 1.0%, with the highest level in the Dabei-Kelasu area, at more than 1.5%. This agrees with the dry gases discovered here. In contrast, the Qiakemake Formation source rock near the BZ1 well has a R_o value of about 1.3%. This is in the highmaturity stage, and mainly generate condensate. This is consistent with the discovery of condensate here. Thus, the differences of hydrocarbon source rock maturity agree with the different maturity of condensates and gases in these fields as discussed above. This also implies that the hydrocarbon charge and accumulation is characterized by vertical migration close to the source.

In summary, based on the above analytical results of diamondoids, biomarkers, and thermal evolution of source rock, it is suggested that the Bozi condensate is a product at the high-evolution stage of Jurassic Qiakemake Formation source rock. The thermal maturity of the source rocks and the generated source rock hydrocarbon products correlate significantly well in the Kuqa Depression. These preliminarily results indicate that the Bozi reservoir hydrocarbon may be condensate, especially when in comparison with the characteristics of neighboring Dina 2 (DN2) condensate and Kela 2 (KL2) dry gas fields in the Kuqa Depression and other condensates worldwide. Dry gas is predominant in the Keshen area (KS), Kela 2 (KL2), where the source rock maturity is highly to post mature (Fig. 1a). It changes to condensate reservoirs and volatile-oil reservoirs as the thermal evolution of source rock decreases to highly mature, suggesting that the charge of oil and gas is largely their vertical migration close to the source (Fig. 1b and d).

4.5 Accumulation of the Bozi condensate

Based on the above discussion, the Bozi condensate is derived mainly from Jurassic coal sequences. These coal-derived hydrocarbons then migrated and accumulated under the control of tectonic event. Major tectonic event in the Bozi area generally occurred after the deposition of the Neogene Kangcun Formation, reached peak activity during the deposition of the Neogene Kuqa Formation, and continued activating up to the present. As a consequence, major traps were formed during this period. The structural evolution in this area can be divided into three stages (Fig. 10).

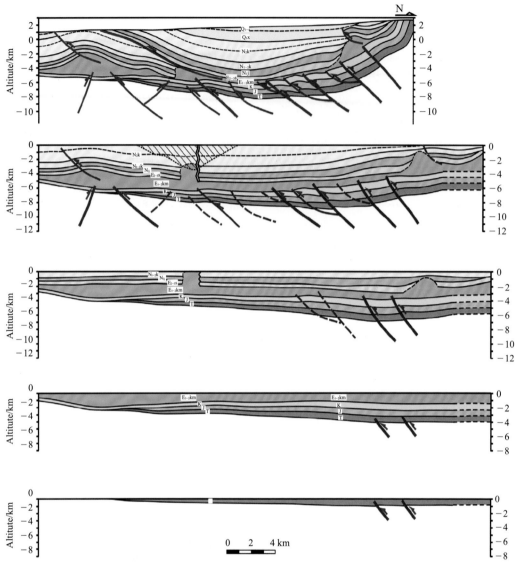

Fig. 10 Cross section showing the structural evolution of the western Kuqa Depression in south-north trending.

The first stage is before the deposition of the Neogene Kangcun Formation, when the entire Kuqa Depression was subjected to a stable deposition. In the Paleogene Kumugeliemu Group, a set of thick gypsum salt rocks was deposited in the Kelasu structural belt. The Suweiyi Formation was stably deposited. In contrast, weak tectonic movements occurred in the Jidike Formation, with a development of slight diapirism. However, little structural deformation is evident and few faults are developed in general.

The second stage starts from the deposition of the Kangcun Formation and continues to the deposition of the Kuqa Formation. During this period, the Kuqa Depression was slightly compressed in an approximately south-north direction. The reverse faults formed during the Mesozoic were reactivated, thereby producing two second-order reverse faults with relatively small throws in the Kelasu area. As a consequence, the Kelasu structural belt was roughly formed. Thick sediments from the piedmont areas deposited rapidly in the southern Baicheng sag of the Kelasu area. This results in the uplift of gypsum rocks in some sites, thus accelerating the diapirism.

The third stage starts from the deposition of the Kuqa Formation and continues to the present. This period is the main stage when the Kelasu structural belt was formed. During the deposition of the middle and lower parts of the Kuqa Formation, the compression strengthened rapidly. The two previously-formed reverse faults below the salt strata in the Kelasu structural belt were strongly activated and several "forward spreading" reverse faults were formed in their footwall and spread southwards. Thus, the thrust fault structure was formed below the salt beds in the Kelasu structure. From the depositional period of the upper Kuqa Formation to the present, the south-north compression continued. The structural deformation was complex, the faults continued to activate, the diapirism in the salt rock became stronger and stronger, and thrust faults and folds were formed in the strata above the salt, which were covered by the Quaternary strata.

In summary, the western Kuqa Depression where the Bozi condensate field of this study is located has subsided rapidly since the period of the Pliocene Kuqa Formation to the Quaternary in response to tectonic movement, i.e., after 3 Ma. The thickness of the Neogene sediment is greater than 2 000 m (Fig. 11). This rapid subsidence has caused the Jurassic source rock to be mature rapidly and to enter the stage of condensate generation.

During this period, a number of thrust faults have developed in the deep layers below the salt layer in the Kelasu structural belt. These faults connect with the deep-buried Jurassic source rocks and act as the main paths for the vertical migration of the oil and gas (Fig. 10). A strong gas charge inhibited the compaction of the sandstone reservoirs during the deep burial. Strong tectonics caused an imbricated thrust structure below the salt layer, providing trap conditions for the late gas charge and accumulation. The fault penetrating into the gypsum salt gradually disappeared in the salt because of the plastic flow in the gypsum salt rock.

Thus, the formation of the Bozi condensate field is much related to the Paleogene gypsum salt rock. It has played important roles in the evolution of the Kelasu structure, in reservoir protection, and in hydrocarbon preservation. The gypsum salt rocks are prone to deformation, thus contributing to the release of structural stress under compression, and facilitating the preservation of the large traps below the salt. Gypsum salt rocks have a low density and high thermal conductivity, which inhibit reservoir compaction below the salt and preserve the reservoir porosity and permeability. Both pressure sealing and physical sealing occur in the gypsum salt rock, forming an excellent regional seal, thus controlling the accumulation and preservation of the oil and gas. Under the control of this set of gypsum salt rocks, the Bozi condensate reservoirs in the study area were finally formed, typical of a late accumulation (i.e., after 3 Ma).

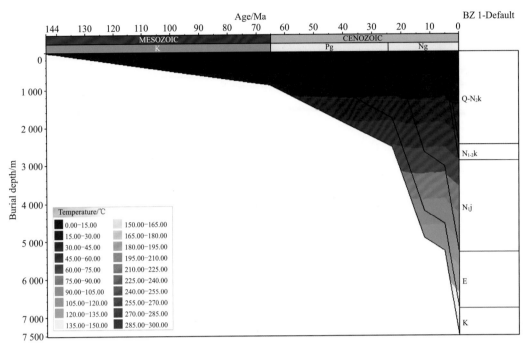

Fig. 11　Burial and thermal history of the BZ 1 well.

4.6　Conclusions

(1) The condensate in the BZ 1 well of the Tarim Basin has abundant n-alkanes, ranging from n-C_3 to n-C_{34}, is rich in diamondoid hydrocarbons, including nine identified diamantanes and two other new dimethylcongressane, and has high levels of aromatic hydrocarbons. Compared with the content of conventional condensate oils, the samples from the BZ 1 well have high levels of aromatic hydrocarbons. These data suggest that the BZ 1 condensate is in the stage of highly mature cracking.

(2) The condensate is dominantly sourced from highly mature coals and/or coaly mudstones of the Jurassic Qiakemake Formation in the western Kuqa Depression. Affected by tectonic activity since 3 Ma, the study area has subsided rapidly, greatly increasing the organic maturity of the source rock and allowing the generation of great quantities of condensates, which migrated vertically into the traps along faults. Therefore, the Bozi condensate gas field is characterized by typical late accumulation.

(3) The formation of traps accompanies the massive generation of condensates in terms of time. The traps and source rocks were stacked vertically, with faults connecting them as migrating pathways. A strong and efficient charge formed the large gas field. Thick Paleogene gypsum salt rocks form an excellent ultimate cap rock and control the accumulation and preservation of the oil and gas.

(4) The studied BZ 1 condensate here is the most deeply buried condensate reservoir with continental origin ever found in China. The discovery can extend our understanding of the oil and gas resources in deep strata in the Tarim Basin, and can be referred in other condensate studies worldwide. Thus, this study has both scientific and practical implications.

论文 11　2017 年发表于《Marine and Petroleum Geology》

Non-cracked oil in ultra-deep high-temperature reservoirs in the Tarim Basin, China

Guangyou Zhu, Alexei V. Milkov, Feiran Chen, Na Weng,
Zhiyao Zhang, Haijun Yang, Keyu Liu, Yongfeng Zhu

Abstract　Thermal stability of liquid petroleum in the subsurface is closely linked to reservoir temperature. Most oil accumulations occur at temperatures <120 ℃. Oil cracks into gas at temperatures >150 – 160 ℃ leading to the dominance of gas condensate and free gas accumulations in ultra-deep high-temperature reservoirs. The recently drilled Fuyuan-1 exploration well (northern Tarim basin) produced high-quality noncracked single-phase (black) oil from a carbonate reservoir located at maximum depth 7 711 m and temperature 172 ℃. This is the deepest oil discovery in China to date and among the deepest in the world. The oil density (0.825 g/cm^3 at 20 ℃ or API gravity 40°), relatively low gas/oil ratio (135 m^3/m^3 or 758 scf/bbl), low variety and abundance of adamantanes as well as lack of thiaadamantanes and dibenzothiophenes indicate that the oil was expelled from a source rock at moderate thermal maturity and has not been cracked. The molecular and isotopic composition of oil-associated gases are consistent with this interpretation. 1D modeling suggests that the oil remained uncracked because the residence time at temperatures >150 – 160 ℃ was insufficient (<5 my). We conclude that there is potential for finding unaltered liquid petroleum in other high-temperature reservoirs with relatively low geothermal gradient and recent burial in the Tarim basin and around the world.

Keywords　Ultra-deep reservoirs　Oil　Adamantane　Tarim basin

1　Introduction

The vast majority (~85%) of oil reserves occur at temperatures between 60 ℃ and 120 ℃. This temperature interval is optimal for petroleum accumulation because of good reservoir quality, proximity to oil-mature source rocks and low risk of biodegradation, gas-flushing and cracking. Oil usually starts cracking into gas at temperatures >150 – 160 ℃ and little reservoired oil occur at greater temperatures and depths. Although there are theoretical kinetic models suggesting that oil can be preserved as a separate phase at temperatures above 170 ℃ and as high as slightly over 200 ℃, little data on such accumulations exist. Deep hot petroleum accumulations in the Central Graben in the North Sea (e.g., Frankin and Elgin fields, Isaksen, 2004) are often cited as examples of preserved oil at reservoir temperatures 174 – 203 ℃. However, these accumulations have relatively high gas-oil ratios (typical GOR>3 500 scf/bbl), often classified as condensates and experienced significant (35% – 91%) oil cracking. Therefore, petroleum explorers commonly assume that ultra-deep (>6 000 m) reservoirs with temperatures

>150 – 160 ℃ contain largely gas and gas condensate accumulations and provide rare opportunities for finding single-phase (black) oils.

Large petroleum reservoirs at >7 000 m depth in China, such as the Yuanba field in the Sichuan basin and the Keshen field in the Tarim basin, contain only dry gas (Dai, 2016). However, recent exploration in the Tarim basin revealed that gas with oil rims can occur in ultra-deep (>6 000 m) reservoirs at high temperatures exceeding 160 ℃. The Fuyuan-1 (FY1) exploration well drilled on the Corresponding author. south slope of the Tabei uplift in the Tarimbasin (Fig.1) resulted in significant oil and gas flows from the Ordovician carbonate reservoir at depth of 7 711 m and temperature >170 ℃. This reservoir contains single phase (black) oil with co-generated dissolved gas and no evidence of significant oil cracking. The discovered Fuyuan field is the deepest oil field in China and is among the deepest oil fields in the world. This discovery is a significant breakthrough in our understanding of petroleum geology, as it extends the empirical depth and temperature limits for the existence and thermal stability of black oil, provides support for kinetic models of oil cracking, creates new oil exploration opportunities in ultra-deep high-temperature reservoirs.

2 Geological setting and discovery of the Fuyuan field

The Tarim basin is the largest petroleum-bearing basin in China covering an area of 560 000 km^2 (Fig. 1). It is one of the most complex cratonic basins in the world. This old basin is characterized by a large thickness of sediments, multiple cycles of uplift and erosion, and a complex distribution of oil and gas accumulations.

The Tabei uplift is a long-term paleo-uplift developed on metamorphic Precambrian basement. It consists of Precambrian (Sinian) to Devonian marine strata, Carboniferous to Permian alternating sedimentary sequences of marine and continental formations, and Triassic to Quaternary continental deposits. Uplift of the Tabei area was initiated during the middle and late Caledonian orogeny. A large-scale nose-shaped structural bulge, plunging to the southwest and extending in a northeasterly direction, was formed by strong regional compression during the early Hercynian orogeny. Long-term exposure and erosion during the uplifting process resulted in the current lack of Silurian-Devonian and Middle-Upper Ordovician sedimentary deposits in most of the bulge area. The Lower Ordovician strata were also partially eroded. The Late Hercynian orogeny caused the Lunnan low uplift and erosion. As a result, sediments overlying the Lower Carboniferous strata are absent in most areas. Ordovician carbonate rocks were locally exposed and eroded. Since the Triassic (Indo-Chinese Epoch), the Tabei area underwent continuous deposition and deep burial. Rapid burial has occurred since 5 my and the current depth of the Ordovician reservoirs ranges from 6 000 m to 9 000 m.

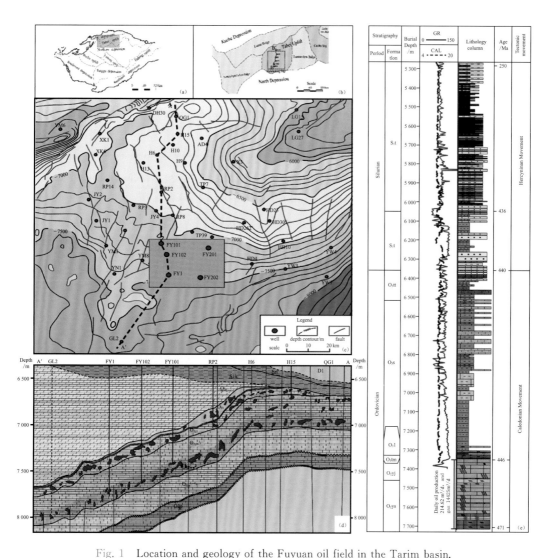

Fig. 1　Location and geology of the Fuyuan oil field in the Tarim basin.
(a) structural elements of the Tabei area; (b) structural map of the top of the Ordovician Yingshan Formation and the location of the Fuyuan block; (c) S-N cross section of reservoir in Tabei area, showing drilled wells, faults and oil reservoirs (red blobs). Location of the section is shown in Fig. 1d; (d) stratigraphy and lithology of rocks penetrated by well FY1. (For interpretation of the references to colour in this figure legend, the reader is referred to the web version of this article.)

The newly discovered Fuyuan oil field is located on the southern slope of the Lunnan low upliftin the Tabei area of the Tarim basin (Fig. 1). The Fuyuan-1 (FY1) discovery well intersected the oil-bearing reservoir in the middle section of the Ordovician Yingshan Formation. The well was spudded on May 22, 2015 and completed on September 15, 2015 with a total drilling depth of 7 711.65 m. The well section between 7 322.17 and 7 711.65 m is classified as a petroleum pay-zone. It tested oil from a completion through a 4 mm choke under 69.08 MPa. The daily production amounts of oil and associated gas were 214.62 m³ (1 350 bbl) and 28 970 m³ (1 023 mscf), respectively. The measured temperature at the bottom of the FY1 well (7 711 m) is 172 ℃. A completion test in the nearby FY102 well was performed for the interval between 7 177.25 m and 7 568.99 m through a 4 mm choke. The test shows daily oil and gas production of 156.87 m^3 (987 bbl) and 27 828 m^3 (983 mscf), respectively. This interval is also classified as a petroleum pay-zone. Significant oil flows were also obtained from the middle section of the Yingshan Formation intersected by wells FY101, FY201 and FY202 (Fig. 1). Preliminary analysis suggests that the distribution of oil and gas in the Fuyuan oil field is controlled by the characteristics of the reservoirs, rather than by local structural lows and highs, similar to that in the adjacent Halahatang oil field.

In the Fuyuan block, carbonate karst reservoirs were developed in the Ordovician rocks, principally within the Yijianfang Formation and in the middle section of the Yingshan Formation. Oil-bearing reservoirs are composed predominantly of heterogeneous fractured-vuggy and fractured limestones formed by the combination of karsts, faulting and fracturing. The matrix porosity and permeability are low, with porosity generally < 8% and averaging around 3.6%. The strata above the oil reservoirs are composed of dense muddy limestones and mudstones of the Ordovician Tumuxiuke, Lianglitage and Sangtamu Formations (Fig. 1).

3 Materials and methods

3.1 Sample preparation

One oil sample was selected from each of the FY1 and FY102 wells. Approximately 20 mg of oil were dissolved in a CH_2Cl_2 (dichloromethane) solution. Silica gel (100−200 mesh, activated at 200 ℃ for 4 h) packed in a glass column was washed using CH_2Cl_2, and the oil sample in a CH_2Cl_2 solution was loaded onto the glass column. The collected fraction (saturated hydrocarbons, aromatic hydrocarbons and some non-hydrocarbons) was condensed to 1 mL using a nitrogen evaporator, and then transferred into a GC sample bottle for analysis. A D16-admantane (ISOTEC brand, 98 atom % D) standard solution, using CH_2Cl_2 as a solvent, was added to the remaining sample fraction and the total solution was analyzed.

3.2 GC×GC-TOFMS analysis

The oil analyses were performed using a LECO GC×GC-TOF-MS. The GC×GC consists of an Agilent 7890A gas chromatograph equipped with a flame ionization detector and two-nozzle temperature modulators. The time-of-flight (TOF) mass spectrometer (MS) is a Pegasus 4D model (LECO Corporation, USA). The size of the first column (1D) is 50 m×0.2 mm×0.5 μm (Petro). The oven temperature was initially set at 35 ℃ for 0.2 min, then increased at a rate of 2 ℃/min until reaching 300 ℃ and held at that temperature for 20 min. The size of the second column (2D) is 3 m× 0.1 mm×0.1 μm (DB-17HT). The heating process used for the 2D column was similar to that for the 1D column but the former had temperature 10 ℃ higher than the latter. The temperature modulator for 2D column was kept at 40 ℃ higher than for the 1D column. The temperature of the injection port was maintained at 300 ℃. 1 μL of sample was injected in the split flow mode at a ratio of 10∶1. Helium was used as the carrier gas with a flow rate of 1.5 mL/min. The modulation period was set at 6 s, 1.5 s of which is the heat blowing time. During the mass spectrometer analysis, the transfer line was maintained at a temperature of 300 ℃ and the ion source was set at 240 ℃. The detector voltage was 1 600 V. Mass spectra were collected from 40 to 520 amu at a rate of 100 spectra/s. The solvent delay time was 10 min. All the data were processed with Chroma TOF software (Version 4.51). After setting appropriate parameters, ChromaTOF software searched the baseline and the peaks, calculated peak areas, searched the database (NIST 11) and automatically generated the peak table listing compounds, their retention time, peak area etc. By adding D_{16}-adamantane into the sample as an internal standard, with CH_2Cl_2 being the solvent, we obtained a quantitative analysis of admantanes. The formula used is:

$$\omega(i) = \frac{m_{standard} \times A_{target}}{A_{standard} \times m_{oil}}$$

where $\omega(i)$ is mass fraction of target compound, mg/kg; $m_{standard}$ is mass of standard sample, mg; A_{targe} is peak area integration of the target compound; $A_{standard}$ is peak area integration of standard sample; m_{oil} is mass of oil samples, kg.

3.3 Gas chromatographic (GC) and gas chromatographice-mass spectrometry (GCeMS) analyses

Gas chromatographic analyses of whole oils was performed using an HP 7890A instrument equipped with HPDB-5 silica column (Model no. J&W 122-5-32, 30 m, 0.25 mm i.d., film thickness=0.25 μm). The initial GC oven temperature was 40 ℃, held for 2 min and then programmed to 310 ℃ at the rate of 6 ℃/min, with a final hold time of 30 min.

A TRACE GC ULTRA/DSQ Ⅱ instrument, equipped with an HP-5MS silica column (60 m, 0.25 mm i.d., film thickness is 0.25 μm), was used for GC-MS biomarker analysis. The initial GC oven temperature is 100 ℃, held for 5min, and then programmed to 220 ℃ at the rate of 4 ℃/min; finally it was programmed to 320 ℃ at the rate of 2 ℃/min and held isothermally for 20 min. The biomarker parameters were calculated by peak area of each component.

3.4 Gas analysis

The chemical composition of gas samples was determined using an Agilent 6890N gas chromatograph (GC) equipped with a flame ionization detector and a thermal conductivity detector. Individual hydrocarbon gas components from methane to pentanes (C_1 – C_5) were separated using a capillary column (PLOT Al_2O_3, 50 m×0.53 mm). The GC oven temperature was initially set at 30 ℃ for 10 min and then ramped to the maximum temperature of 180 ℃ at a rate of 10 ℃/min where it was held for 20 min. Stable carbon isotope ($\delta^{13}C$) ratios were determined on a Finnigan Mat Delta S mass spectrometer interfaced with a HP 5890 Ⅱ gas chromatograph. Gas components were separated on the gas chromatograph in a stream of helium, converted into CO_2 in a combustion interface and then introduced into the mass spectrometer. Individual hydrocarbon gas components ($C_1 - C_5$) were initially separated using a fused silica capillary column (PLOT Q, 30 m × 0.32 mm). The GC oven temperature was ramped from 35 to 80 ℃ at 8 ℃/min and then to 260 ℃ at 5 ℃/min, and the oven was maintained at its final temperature for 10 min. Stable carbon isotopic values were reported in the customary δ notation in per mil (‰) relative to Vienna Pee Dee Belemnite (PDB) standard. Measurement precision was estimated to be ±0.3‰ for $\delta^{13}C$.

4 Results

4.1 Oil is apparently not cracked

Crude oils collected from stock tanks of five wells in the Fuyuan oil field (Table 1) have densities ranging from 0.825 g/cm³ (API gravity 40°) to 0.843 g/cm³ (API gravity 35°) and viscosities of 3.1−5.2 mPa·s, and are thus classified as light and medium oils. The sulfur content is <0.3% and the wax content varies considerably, ranging from 2.8% to 16.0%. Saturated hydrocarbons make up 69%−83.7% of the crude oil composition, having a saturated to aromatic ratio of 3.95−7.12 and a low content of non-hydrocarbon and asphaltenes. The bulk composition suggests that Fuyuan crude oil has high quality.

Table 1 Physical properties and bulk composition of crude oil samples from the Fuyuan oil field.

Well	Depth/m	Density /(g·cm⁻³)		Viscosity /(mPa·s)		Wax /%	Sulfur /%	Saturate /%	Aromatic /%	Resin /%	Asphaltenes /%	Saturate/ Aromatic
		20 ℃	50 ℃	20 ℃	50 ℃							
FY1	7 322−7 711	0.825	0.803	4.231	3.398	16.0	0.051	83.71	11.76	2.31	0.38	7.12
FY101	7 211−7 415	0.843	0.821	3.100	2.546	2.8	0.241					
FY102	7 177−7 569	0.832	0.811	5.155	4.180	8.7	0.083	74.85	13.58	2.25	1.04	5.51
FY201	7 003−7 106	0.827	0.805	3.138	2.526	4.1	0.250	75.73	19.17	2.28	0.15	3.95
FY202	7 347−7 530	0.829	0.807	3.264	2.634	5.2	0.221	69.11	16.38	2.95	1.74	4.22

For the two samples collected from the FY1 and FY102 wells and analyzed by GC×GC-TOFMS, we provide two-dimensional stereograms (Fig. 2) and color contour chromatograms which show the position, elution time and abundance of specific compounds (Fig. 3). The data processed using the Chroma TOF software suggest that there are 3 630 and 3 162 compounds with S/N ratios >100 in these two samples.

Figs. 2 and 3 illustrate a similar distribution of compounds for the two samples. The peaks height and color intensity reflect the relative abundance of compounds. A higher, sharper and dark red peak in the stereogram (Fig. 2) represents a higher content of the corresponding compound. For the saturated hydrocarbons, the nalkanes range from $n\text{-}C_6$ to $n\text{-}C_{32}$. The distributions of n-alkanes in the two samples are very similar. The peak areas for $n\text{-}C_{11}$ to $n\text{-}C_{15}$ are similar for the two samples and they are highest among all nalkanes. Propyl cyclohexane has the highest content among detected monocyclic naphthenic compounds. The detection range of long-chain alkyl cyclohexanes can reach C_{25}-cyclohexane. The dicyclic sesquiterpenoid series compounds have the highest content among the polycyclic cycloalkanes. Some monomeric compounds in dicyclic sesquiterpenoids (e.g., the marked C_{15}-dicyclic sesquiterpenoids in Figs. 2 and 3) have a relatively high content, similar to C_2- and C_3-benzene and higher than the naphthalene series compounds. The principal aromatics are monocyclic aromatic, di- and tri-aromatic compounds (in the order of decreasing content). Tetramantanes and pentamantanes and sulfur compounds such as benzothiophene and dibenzothiophene were not detected.

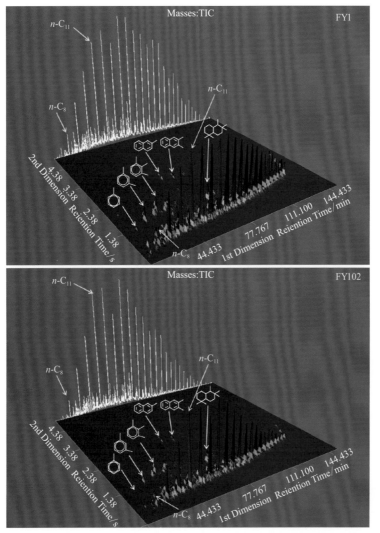

Fig. 2 Two-dimensional stereograms of oil samples from the FY1 and FY102 wells. The peak positions of some typical compounds are indicated by arrows.

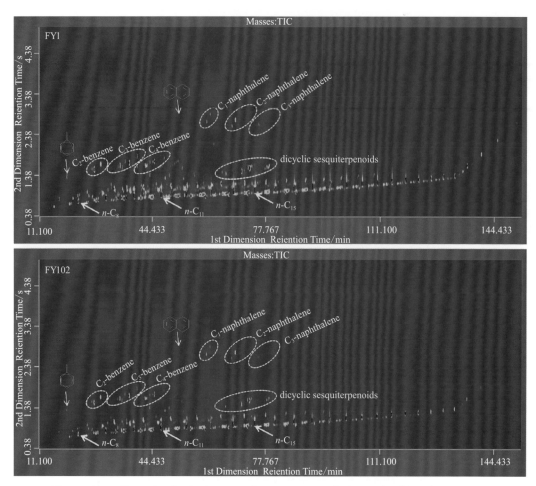

Fig. 3 Two-dimensional color contour chromatograms of oil samples from the FY1 and FY102 wells. The peakpositions of some typical compounds are indicated by yellow arrows. (For interpretation of the references to colour in this figure legend, the reader is referred to the web version of this article.)

The abundance of adamantanes (rigid polymeric cyclic hydrocarbon compounds with a structure similar to diamond) is often high in high-maturity crude oils and gas condensates which often experienced in-reservoir cracking. Fig. 4 presents a two-dimensional comparative diagram of adamantanes found in the two Fuyuan oil samples. Two-dimensional stereograms reflect the relative contents of each compound. Sharper peaks with flatter baselines and darker background colors define compounds of greater contents and vice versa. The distribution and concentrations of adamantane compounds are quite similar for these two samples. Oils from the FY1 and FY102 wells have a low concentration and poor variety of adamantanes. This is in sharp contrast with the highly mature condensate samples from the ZS1C, ZS1, ND1, BZ1 and TZ83 wells in the Tarim basin which are characterized by high content of adamantanes. Fig. 4 reveals very low diversity of adamantanes in the two Fuyuan oil samples with only 10 compounds being detected. Based on qualitative results, two of the 10 adamantanes have one substituent in selected ion detected at m/z 135, five in selected ion at m/z 149, and three in selected ion at m/z 163. Adamantanes with four substituents as well as diadamantanes (diamantanes) were not detected (Fig. 5). Quantitative GC×GC-TOF-MS analysis using D_{16}-adamantane as the internal standard compound confirmed the generally low concentrations of adamantanes (Table 2).

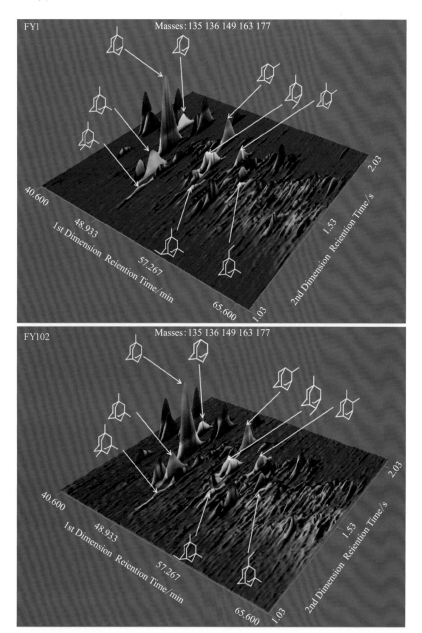

Fig. 4　Two-dimensional stereograms of two oil samples collected from the FY1 and FY102 wells. The selected ions are at m/z 135, 136, 149, 163 and 177. Typical adamantane compounds are indicated by white arrows.

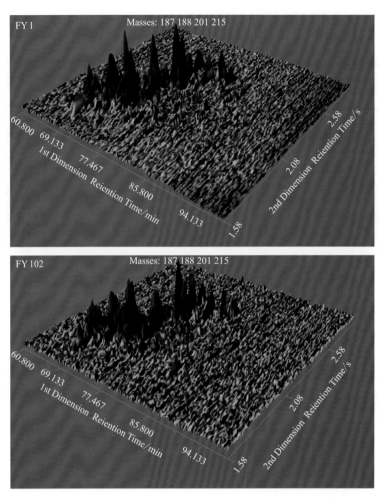

Fig. 5 Two-dimensional bitmaps of oil samples collected from the FY1 and FY102 wells with selected ions at m/z 188, 187, 201, 215 used for the detection of diadamantanes (diamantanes). The enlarged areas show where diadamantanes should elute. Diadamantanes were not detected.

Table 2 Concentration of diamondoids in two crude oil samples obtained from the FY1 and FY102 wells.

No.	Compound	Concentration/(mg · kg^{-1})	
		FY1	FY102
1	Adamantane	4.56	21.31
2	1-Methyladamantane	24.85	87.54
3	1,3-Dimethyladamantane	15.95	56.59
4	1,3,5-Trimethyladamantane	5.56	20.81
5	2-Methyladamantane	8.67	47.46
6	1,4-Dimethyladamantane (cis)	7.74	29.91
7	1,4-Dimethyladamantane (trans)	7.81	29.53
8	1,3,6-Trimethyladamantane	4.70	15.40
9	1,2-Dimethyladamantane	0.04	28.71

continue

No.	Compound	Concentration/(mg · kg^{-1})	
		FY1	FY102
10	1,3,4-Trimethyladamantane (cis)	—	—
11	1,3,4-Trimethyladamantane (trans)	—	—
12	1,2,5,7-Tetramethyladamantane	—	—
13	1-Ethyladamantane	—	—
14	1-Ethyl-3-methyladamantane	3.26	14.17
15	Diamantane	—	—
16	4-Methyldiamantanee	—	—
17	1-Methyldiamantane	—	—
18	3-Methyldiamantane	—	—

Based on their bulk and molecular composition (Tables 1 and 3; Fig. 6), crude oils from the Fuyuan field have maturity consistent with expulsion from source rocks at temperatures ~140 ℃ and vitrinite reflectance 0.8%–1% (middle-upper oil window). This moderate maturity is also consistent with the type and low concentrations of adamantanes. Adamantane-containing compounds may be constantly generated and accumulated during cracking of oil at high temperatures. The presence and abundance of adamantane is often used to measure the degree of cracking of crude oil. Our adamantane data indicate that the Fuyuan crude oil samples have not started to crack. Furthermore, the lack of thiaadamantanes suggests that oils did not undergo thermochemical sulfate reduction (TSR) alteration. Therefore, the geochemical characteristics of these crude oils support a well-preserved unaltered ultra-deep hot oil reservoir.

Table 3

Well	Diasterane/Sterane	Moretane/Hopane	C_{29}(20S)/(20S+20R)	C_{29}($\beta\beta$)/($\alpha\alpha+\beta\beta$)	C_{31}(22S)/(22S + 22R)	Ts/(Ts+Tm)	n-C_{17}/Pr	n-C_{18}/Ph	CPI	OEP
FY1	0.4	0.3	0.5	0.6	0.3	0.9	2.2	2.1	1	1
FY102	0.4	0.4	0.5	0.6	0.2	0.7	1.8	1.8	1.1	0.9

Selected biomarker ratios for crude oil samples from the Fuyuan oil field. Only C_{29} carbon number was used to calculate the diasterane/sterane ratio. Ts/(Ts+Tm) is the ratio of 18α(H)-22,29,30 trisnorneohopane (Ts) to the sum of 18α(H)-22,29,30 trisnorneohopane and 17α(H)-22,29,30 trisnorhopane (TM). Carbon preference index (CPI) and odd-to-even predominance (OEP) are calculated according to these formulas:

$$CPI = \frac{1}{2}\left(\frac{C_{25}+C_{27}+C_{29}+C_{31}+C_{33}}{C_{24}+C_{26}+C_{28}+C_{30}+C_{32}} + \frac{C_{25}+C_{27}+C_{29}+C_{31}+C_{33}}{C_{26}+C_{28}+C_{30}+C_{32}+C_{34}}\right)$$

$$OEP = \frac{C_{21}+6C_{23}+C_{25}}{4(C_{22}+C_{24})}$$

Fuyuan oils have GOR ranging from 135 to 186 m^3/m^3 (758−1 044 scf/bbl) indicating that the reservoirs contain single phase black oil. The oil-dissolved gas in Fuyuan reservoirs is relatively wet, with $C_1/(C_1-C_5)$ ranging from 0.71 to 0.90 (Table 4). Methane constitutes 68%–83% of the total gas and the concentration of ethane is 5%–15%. The CO_2 content is relatively low (<2%), especially for high-temperature carbonate reservoirs. Carbon isotopic composition of hydrocarbon gases (Table 4) show enrichment in 12 °C (e.g., methane has $\delta^{13}C$ values of −49.5‰ and −46.9‰) relative to other gases in the Tarim basin. Molecular and isotopic composition of oil-dissolved gases suggest that the gases were co-generated at the similar thermal maturity levels as the oils (standard thermal stress 120−∼140 °C according to the nomogram of Dzou and Milkov). Previous investigations have shown that methane and ethane in reservoirs with cracked oil in the Tarim basin are relatively depleted in ^{12}C and have $\delta^{13}C$ values ranging from −35‰ to −45‰ and from −28‰ to −36‰ respectively. Natural gases from wells FY102 and FY201 are much more enriched in ^{12}C than gases associated with oil cracking. We interpret that natural oil-dissolved gases in the Fuyuan field likely have the same genesis as the crude oil.

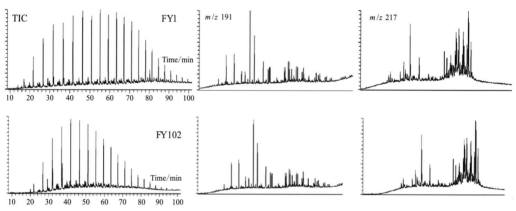

Fig. 6. GC and GC-MS (m/z 191, m/z 217) traces for crude oil samples from the Fuyuan oil field. Labelled peaks: n-alkanes (n-C_{15}, n-C_{20}, n-C_{25}), pristine (Pr), phytane (Ph), terpanes (C_{21}TT, C_{23}TT, C_{24}TeT), 18α(H)-22,29,30 trisnorneohopane (Ts), 17α(H)-22,29,30 trisnorhopane (TM), norhopane (C_{29}H), hopane (C_{30}H), steranes (C_{21}p, C_{22}p, C_{27}RS, C_{28}RS, C_{29}RS) and diasteranes (DS).

Table 4　Molecular and carbon isotopic composition of oil-dissolved gases from the Fuyuan oil field.

Well	Depth/m	C_1/C_1^+	Natural Gas Components/%										Isotope $\delta^{13}C/(‰, PDB)$				
			CH_4	C_2H_6	C_3H_8	i-C_4H_{10}	n-C_4H_{10}	i-C_5H_{12}	n-C_5H_{12}	C_6^+	N_2	CO_2	CH_4	C_2H_6	C_3H_8	i-C_4H_{10}	n-C_4H_{10}
FY1	7 322−7 711	0.75	72.75	14.27	6.74	0.878	1.585	0.289	0.280	0.238	1.747	0.745					
FY101	7 211−7 415	0.82	77.90	8.54	4.57	0.949	1.602	0.395	0.394	0.251	0.285	1.536					
FY102	7 177−7 569	0.71	67.83	15.43	7.92	1.076	2.032	0.351	0.296	0.161	0.956	0.873	−49.5	−38.1	−35.1	−35.9	−33.8
FY201	7 003−7 106	0.90	83.00	5.06	2.400	0.623	0.886	0.261	0.228	0.168	4.910	0.641	−46.9	−37.5	34.7	−35.4	−34.7
FY202	7 347−7 530	0.72	65.65	13.34	7.048	1.192	2.11	0.447	0.511	0.371	7.064	1.822					

4.2 Accumulation and preservation of non-cracked oil in high-pressure reservoirs

The Permian Late Hercynian tectonism is crucial to petroleum generation and accumulation in the Tarim basin. Previous studies on petroleum expulsion, migration and accumulation concluded that most oil and gas pools in the Ordovician formations of the Tabei area formed from 288 to 250 my during the Permian Period. 1D modeling of the burial history suggests that the main pool-forming event occurred from the Middle and the Late Permian to the Triassic periods, during the uplift and deformation associated with the late Hercynian orogeny (Fig. 7). Gas charging and flushing was wide-spread in the Tarim

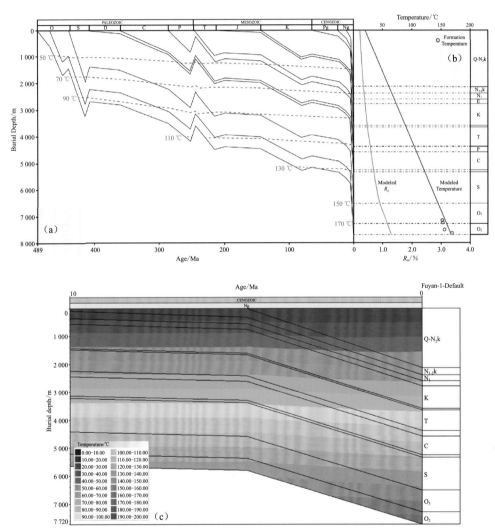

Fig. 7　1D model of burial history and temperatures of the Fuyuan block drilled by the FY1 well (a). Modeling was performed using commercial software Petromod. The model is calibrated to measured bottom-hole temperatures (b), but no vitrinite reflectance (R_o) were available for calibration. The bottom image (c) shows the enlargement of 1D model demonstrating the burial history and temperatures in the last 10 my.

basin during the Late Himalayan tectonism (10 my), but it clearly did not affect the Fuyuan oil field.

Oil cracking into gas is inferred elsewhere in the Tarim basin. For example, Zhao et al. suggest that gas in the Jurassic reservoirs in the YN2 field migrated from deep reservoirs in which oil cracked into gas at temperatures >190 ℃. Most researchers believe that crude oil starts cracking into gas at a temperatures >150 – 160 ℃ leading to a gradual disappearance of liquid petroleum from reservoirs at higher temperatures. The temperature of the Fuyuan oil field reservoir at a depth of 7 711 m exceeds 170 ℃, while the crude oil shows no evidence of significant cracking. This finding extends the empirical limits of black oil thermal stability in the subsurface and supports the theoretical kinetic models of oil cracking.

The Ordovician section of the Tarim basin reached a maximum burial depth during the Quaternary and the geothermal gradient has been decreasing gradually since the Permian Period. The current depth (7 711 m) and temperature (172 ℃) of the Ordovician carbonate reservoir in well FY1 are at their maximum (Fig. 7), and the current geothermal gradient is relatively low (20 – 22 ℃/km). Reservoirs in the Fuyuan fields have reached high oil-cracking temperatures >150 – 160 ℃ only ~4 my ago (Fig. 7). It is apparent that greater burial depth, higher temperatures or longer time at high temperatures are needed for appreciable oil cracking at this location in the Tarim basin. This explains why all current discoveries in the Fuyuan area are oil accumulations.

Assuming that the oil filled the reservoir at the end of the Permian period, the average heating rate for the reservoired oil is very slow (0.25 ℃/my). The kinetic model of Waples predicts that at this slow heating rate the liquid oil should disappear at 170 ℃, which is not consistent with the observation of non-cracked oil in FY1 well. However, the heating rate is modeled to be significantly faster (5.5 ℃/km) over the last 4 my during which the reservoir was located at temperatures >150 ℃. The kinetic model of Waples predicts that the liquid oil should disappear at temperature of about 185 ℃ for such higher heating rate. Thermal simulation experiments of marine normal crude oil and kinetic parameters for oil cracking into gas enabled Zhu et al. to report that minimum 52.8 my are needed to convert 51% of oil to methane at a constant heating temperature of 180 ℃. Moreover, at least 100 my is needed to obtain a similar conversion ratio when crude oil is heated at a rate of 2 ℃/my. Modeling of the geothermal and burial history of the Fuyan study area suggests that massive cracking of crude oil begins at depths between 9 000 m and 9 500 m, corresponding to reservoir temperatures ranging from 210 ℃ to 220 ℃.

Recent studies suggest that TSR in deep carbonate reservoirs can enhance oil cracking. Mixing of pyrolytic gas and crude oil produces phase changes and affects the properties of oil and gas in deep pools. Crude oil cracking induced by TSR alteration is often indicated by high concentrations of thiaadamantanes in the residual reservoirs. Thiaadamantanes were not identified in two crude oil samples collected from the Fuyuan reservoirs. Fuyuan natural gases are relatively enriched in C_2^+ hydrocarbons and do not contain hydrogen sulfide, which also confirms that the Fuyuan reservoirs did not undergo TSR alternation. This lack of TSR may help explain the preservation of deep crude oil at a high temperature.

5 Conclusions

The Fuyuan oil field located in the Tarim basin is the deepest oil reservoir discovered so far in China. The greatest depth of the oil-producing Ordovician Yingshan Formation is 7 711 m. Crude oil from these reservoirs have low variety and concentration of adamantanes and lack dibenzothiophenes. This suggests that in spite of high temperature >170 ℃, the oil has barely been cracked. The preservation of oil in this ultra-deep high-temperature Paleozoic reservoir is due to two factors: (1) low current geothermal gradient of the area, and (2) initial continuously slow burial followed by rapid late burial. More explicitly, it has not been that hot for that long. Moreover, TSR-generated thiaadamantanes and H_2S have not been detected in these crude oils, indicating the carbonate reservoir has not been exposed to TSR alteration.

Discovery of non-cracked oil in the Fuyuan field opens a new frontier of exploration for ultra-deep high-temperature oil reservoirs in the Tarim basin as wells as in similar geological settings in other basins around the world. Other reservoirs with non-cracked oil may occur in areas with low geothermal gradients and initial continuously slow burial followed by late rapid burial.

论文 12　2016 年发表于《Marine and Petroleum Geology》

TSR-altered oil with high-abundance thiaadamantanes of a deep-buried Cambrian gas condensate reservoir in Tarim Basin

Guangyou Zhu，Huitong Wang，Na Weng

Abstract　The first exploratory well, the ZS1C well, with 158 − 545 m^3 daily gas production was discovered in 6 861 − 6 944 m deep strata of the Cambrian gypsolyte layer of the Tarim Basin, China in 2014. The discovery opens a new target for the Cambrian-reservoired oil and gas exploration, and directly leads to large-scale oil and gas exploration of the deep-reservoired Cambrian oil and gas fields in the Basin. Comprehensive two-dimensional gas chromatography/time-of-flight mass spectrometry and a comprehensive two-dimensional gas chromatography-flame ionization detector revealed the presence of abundant adamantane compounds, 2-thiaadamantanes and 2-thiadiamantanes, and a large amount of sulfur-containing compounds in the condensate oil. The formation of organic sulfur-containing compounds, such as 2-thiaadamantanes, is an indication of sulfur incorporation from the gypsum in the stratum into oil and gas in the course of TSR. This reservoir has apparently suffered severe TSR alteration because (1) High content of H_2S, (2) H_2S sulfur isotopes, (3) CO_2 carbon isotopes, and others abundant data to support this findings. Similar sulfur isotopic composition of H_2S, oil condensate and the gypsum in the Cambrian strata indicate that the produced condensate is experienced TSR alteration. Therefore, the deep-accumulated Cambrian oil reservoir has experienced severe TSR alteration, and accumulated natural gas and condensate contains high sulfur content.

Keywords　Adamantane　Thiaadamantane　TSR　Condensate　H_2S　Cambrian system　Tarim Basin

1 Introduction

As a type of rigid polymeric and cyclic hydrocarbon compound with a structure similar to diamond, the abundance of adamantane is often high in high-maturity crude oil and condensate. In the cracking of crude oil at high temperature condition, adamantane-containing compounds may be constantly generated and accumulated. The presence and abundance of adamantane is often used to measure the cracking degree of crude oil. However, thiaadamantane is the compound generated in the process when the carbon atom in the adamantane molecules is replaced by sulfur atom. At present, this compound has been detected in the deep-accumulated crude oil in the Bon Secour Bay in the Mobile Bay gas field, located offshore Alabama in the northern Gulf of Mexico, USA, the Jurassic Smackover (USA) and the Devonian Nisku (Canada) Formations. It is believed that thiaadamantane is the indicator product of the thermochemical sulfate reduction (TSR) between hydrocarbons and sulfates under high temperature in deep-buried strata.

TSR is a reservoir alteration process in which petroleum hydrocarbons are oxidized by inorganic sulfate, ultimately yielding CO_2 and H_2S. TSR is well documented in numerous geological settings from around the world. It is widely believed that the cause for high hydrogen sulfide content in gas and oil reservoir is from TSR. Hydrogen sulfide is generated by the reaction between hydrocarbons and sulfates driven by thermal chemical sulfate reduction. A large amount of excellent research work has identified the trigger mechanism, reaction condition, and dynamic mechanism of TSR. TSR involves the participation of various hydrocarbons, which have different reactivities because of their differences in chemical structures and sulfur contents. The quantity and type of the intermediate products of the reaction are hard to estimate due to the participation of different hydrocarbons. TSR reaction can result in complex changes in the fluid components of oil and gas reservoirs by producing some new sulfur-containing compounds. Therefore, oil and gas reservoirs altered by TSR reaction could contain some unique geochemical features, which could indicate the presence/absence of TSR, the intensity or extent of TSR. In recent years, the development of two-dimensional gas chromatography/time-of-flight mass spectrometry (GC×GC-TOFMS) has provided detection means for fast and accurate identification of new compounds. Stable isotopes of carbon and sulfur have proven to be particularly successful diagnostic tools for monitoring the extent of TSR. These analytical means can help identify whether TSR occurs in the deepburied oil and gas reservoir, its alteration intensity, and the distribution of hydrogen sulfide for risk assessment.

The Tarim Basin is the largest oil-and-gas-bearing basin in China that covers an area of roughly $56×10^4$ km^2. It is a typical superimposed basin with lower Paleozoic carbonate rocks containing rich oil and gas resources. Having been buried underground at a great depth for a long period of time and experiencing multi-cycle superimposition and transformation, the oil and gas distribution is highly complex. The ZS1C well was drilled in the middle of the central uplift of the Tarim Basin (abbreviated as Tazhong area), and the daily gas production is 158 545 m^3 in 6 861 − 6 944 m deep strata of the Cambrian gypsolyte layer. This is the first important discovery of commercial oil and gas accumulations in the Cambrian reservoirs in the Tarim Basin (Fig. 1), leading directly to the exploration of a new stratum under current breakthrough in the ZS1C. Intensive oil and gas exploration work in the deep-buried Cambrian system in the Tarim Basin has been initiated since then, and the depth for oil and gas exploration has been extended to 7 000 − 10 000 m. In the present study, molecular and sulfur and carbon isotopes were analyzed on gases and condensates using GC×GC-TOFMS and GC×GC-FID to identify whether or not and to what extent of TSR alteration occurred in this deep Cambrian reservoir.

2 Materials and methods

2.1 Instruments

The comprehensive GC×GC system for the GC×GC-TOFMS is from LECO Corporation. Studies reporting GC×GC analysis of condensate samples are rare. The GC×GC system was composed of an Agilent 7890 A gc coupled to a hydrogen flame ionization detector (FID) and a liquid-nitrogen-cooled pulse jet modulator. The TOF mass spectrometer is a Pegasus 4D (LECO Corporation). All the data were processed with Chroma TOF software.

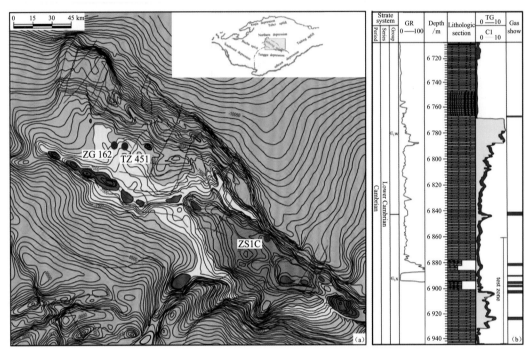

Fig. 1 Structural map of the top Cambrian in the Tazhong area (a) and the lithology column for the Cambrian oil and gas reservoir of ZS1C well (b).

2.2 Analytical method for GC×GC-TOFMS

The one-dimensional chromatographic column was a DB-Petro (50 m×0.2 mm×0.5 μm). The temperature program used was 0.2 min at 35 ℃; increased to 210 ℃ at a rate of 1.5 ℃/min and held for 0.2 min; and increased to 300 ℃ at the rate of 2 ℃/min and held for 20 min. The two-dimensional chromatographic column was a DB-17ht (3 m×0.1 mm×0.1 μm). The temperature program applied was the same as that for the one-dimensional gas chromatography, but the temperatures were 5 ℃ higher. The modulator temperature was 45 ℃ higher than for the one-dimensional gas chromatography. The inlet temperature was 300 ℃, the inlet mode was split injection, the split ratio was 700∶1, and the sample volume was 0.5 μL. Helium was used as the carrier gas, with a flow rate of 1.5 mL/min. The modulation time was 10 s, 2.5 s of which was the hot pulse time. For the mass spectrometry, the temperatures of the transfer line and the ion source were 300 ℃ and 240 ℃, respectively, the detector voltage was 1 600 V, the scan range was 40–520 amu, the acquisition rate was 100 spectra/s, and the delay time of the solvent was 9 min.

2.3 Analytical method for GC×GC-FID

The experimental conditions for chromatography were the same as those for GC×GC-TOFMS. The flow rates of the carrier gas, hydrogen, and the air in the FID detectorwere 50, 40, and 450 mL/min, respectively. The detector temperature was 320 ℃, the acquisition rate was 200 spectra/s, and the delay time of the solvent was 9 min.

2.4 Analytical method for carbon isotopes

Gas samples were collected at the production well site using steel bottles. The carbon isotope analysis was completed using a Thermo Delta V Advantage instrument by CNPC Research Institute of Petroleum Exploration & Development, with an analysis accuracy of ±0.1‰.

2.5 Quantitative method

The group components of the compounds were quantified by peak area normalization. D_{16}-adamantane (using CH_2Cl_2 as a solvent) was added in the condensate sample, and the quantitative results of conventional diamondoids in the condensate were obtained using the internal standard method.

3 Results and discussion

3.1 Physical characteristics of the Cambrian condensate oil

A condensate sample was obtained from the 6 861 – 6 944 m interval of the ZS1C well in the lower Cambrian reservoir. The density of condensate was approximately 0.93 g/cm³ (20 ℃) (Table 1); the viscosity of crude oil was 2.1 mPa·s (50 ℃); the sulfur content 2.06% – 2.68%; and the wax content 0.2% – 4.9%. For the content of resin and asphaltene was quite low, the oil belongs to high-sulfur and low-wax condensate oil.

Table 1 Physical properties of the Cambrian crude oil in Well ZS1C.

Depth /m	Density /(g·cm⁻³)		Viscosity /(mPa·s)		Wax content /%	Sulfur /%	Resin content /%	Asphaltene /%	Solidification point/℃
	20 ℃	50 ℃	20 ℃	50 ℃					
6 861 – 6 944	0.930 0	0.910 8	2.382	2.170	2.80	2.06	0.29	0.18	−30
	0.929 8	0.910 6	2.345	2.136	4.9	2.67	0.19	0.43	−30
	0.927 4	0.908 1	2.401	2.181	0.20	2.68	0.45	0.08	−30

The bulk oil composition in the ZS1C well condensate is dominated by saturated hydrocarbons and aromatic hydrocarbons, which account for over 86% of total hydrocarbons. Saturated hydrocarbon content is 36.0%. The content of aromatic hydrocarbon is 50.54%. Comparatively speaking, the oils from the ZS1C well area contain high aromatic hydrocarbons. With increasing oil maturity level and TSR alteration, oils become enriched in monoaromatic hydrocarbons and the distribution of high molecular weight aromatic hydrocarbons shifts towards more condensed species with a decrease in the number of alkyl carbons. So, the organosulfur compounds are created, and the density of oils becomes heavy with highly aromatic hydrocarbons.

3.2 Chemical composition of crude oil

The composition of the condensate samples was analyzed using GC×GC-TOFMS to obtain a comprehensive two-dimensional color contour chromatogram (Fig. 2). Based on the GC×GC-TOFMS 3D plot (Fig. 3), the hydrocarbon distribution in the condensate ranges from C_6 to C_{30} with the main peak carbon is of C_{11}. Hydrocarbons greater than C_{30} are very minor.

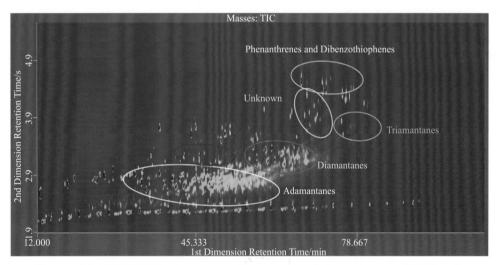

Fig. 2 GC×GC-TOFMS color contour chromatogram of the condensate of ZS1C well.
(For interpretation of the references to color in this figure legend,
the reader is referred to the web version of this article.)

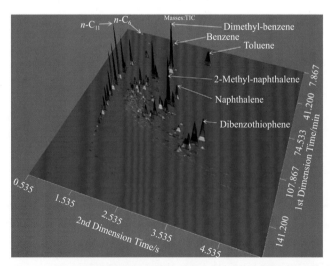

Fig. 3 GC×GC-FID 3D plot for the condensate of ZS1C well.

The condensate contains a high amount of aromatic hydrocarbons, and the amounts of dimethyl-benzene, 2-methyl-naphthalene and dibenzothiophene are even higher than that of some nalkanes. The quantitative analysis of group components based on the GC×

GC-FID of oil samples indicates that the content of saturated hydrocarbons, aromatic hydrocarbons, and resins was 42.2%, 54.9%, and 2.9%, respectively. This result is rarely seen in ordinary condensate (Fig. 3).

3.3 Distribution of adamantane compounds

There are abundant adamantane compounds in the condensate, where adamantanes, diamantanes, and triamantanes can be detected. Fig. 4 shows the GC×GC-TOFMS color contour chromatogram of adamantane-containing compounds. Up to 55 adamantane compounds of different configurations exist with a signal-to-noise ratio of over 100. Twenty-two diamantane compounds of different configurations were observed with a signal-to-noise ratio of over 100. The content of triamantane compounds with different substituents is low and over nine compounds have a signal-to-noise ratio of over 100. The quantitative results of conventional diamondoids were listed in Table 2.

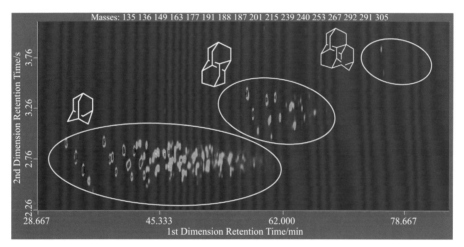

Fig. 4 GC×GC-TOFMS color contour chromatogram for m/z 135, 136, 149, 163, 177, 191, 188, 187, 201, 215, 239, 240, 253, 267, 292, 291, 305 of the condensate of ZS1C well. (For interpretation of the references to color in this figure legend, the reader is referred to the web version of this article.)

Table 2 **Quantification of diamondoids in Well ZS1C.**

No.	Name	Concentration/(mg · kg^{-1})
1	Adamantane	1 585.36
2	1-Methyladamantane	2 394.37
3	1,3-Dimethyladamantane	1 726.95
4	1,3,5-Trimethyladamantane	962.84
5	1,3,5,7-Tetramethyladamantane	165.66
6	2-Methyladamantane	1 795.41

Continue

No.	Name	Concentration/(mg·kg^{-1})
7	1,4-Dimethyladamantan(cis)	1 421.49
8	1,4-Dimethyladamantane(trans)	1 393.12
9	1,3,6-Trimethyladamantane	1 015.44
10	1,2-Dimethyladamantane	1 560.32
11	1,3,4-Trimethyladamantane(cis)	1 280.66
12	1,3,4-Trimethyladamantane(trans)	1 119.50
13	1,2,5,7-Tetramethyladamantane	806.52
14	1-Ethyladamantane	1052.99
15	1-Ethyl-3-methyladamantane	1 058.98
16	1-Ethyl-3,5-dimethyladamantane	1 160.14
17	2-Ethyladamantane	666.80
18	1,2,3,5,7-Pentamethyladamantane	232.55
19	Diamantane	1619.62
20	4-Methyldiamantane	1 125.29
21	4,9-Dimethyldiamantane	783.41
22	1-Methyldiamantane	1 341.10
23	1,4-+2,4-Dimethyldiamantane	740.76
24	4,8-Dimethyldiamantane	876.73
25	Trimethyldiamantane	421.67
26	3-Methyldiamantane	1284.49
27	3,4-Dimethyldiamantane	782.09
28	Triamantane	588.58
29	9-Methyltriamantane	356.12
30	9,15-Dimethyltriamantane	81.89
31	5,9-Dimethyltriamantane	80.12
32	5-Methyltriamantane	214.63
33	Trimethyltriamantane	50.65
34	3,4-Dimethyltriamantane	116.89
35	Methyltriamantane	129.66
36	8-Methyltriamantane	175.56
37	Dimethyltriamantane	114.16
38	16-Methyltriamantane	126.28

3.4 Distribution of sulfur-containing compounds

The largest difference between the ZS1C condensate and other condensates is that it contains abundant sulfur-containing compounds. Either in the total ion current of GC×GC-TOFMS or the chromatogram obtained with GC×GC-FID, high-sulfur-containing compounds are clearly detected (Fig. 3). Fig. 5 displays the distribution of all sulfur-containing compounds in the condensate. It shows the presence of tetrahydrothiophene compounds with a characteristic ion $m/z=101$ (see Fig. 6a for the mass spectrogram of compound a), benzyl mercaptane with a molecular ion peak $m/z=124$ (see Fig. 6b for the mass spectrogram of compound b), alkylbenzothiophene compounds with a characteristic ion $m/z=147$, 161, 175 (see Fig. 6c for the mass spectrogram of compound c), 2-thiaadamantanes with a characteristic ion $m/z=168$, 182, 192 (see Fig. 6d for the mass spectrogram of compound d), diamantane-based sulfides with a characteristic ion $m/z=206$, 220 (see Fig. 6d for the mass spectrogram of compound d), dibenzothiophene compounds with a characteristic ion m/z 184, 198 and 212 (see Fig. 6e for the mass spectrogram of compound e), phenanthrothiophene compounds with a characteristic m/z 208 and 222 (see Fig. 6f and g for the mass spectra of compounds f and g), and benzonaphthothiophenes with a characteristic ion m/z 234 (see Fig. 6h for the mass spectrogram of compound h). Note that the determination of thiadiamantanes is based on their position in the comprehensive two-dimensional chromatogram and mass spectrogram, without verification by references and standard samples.

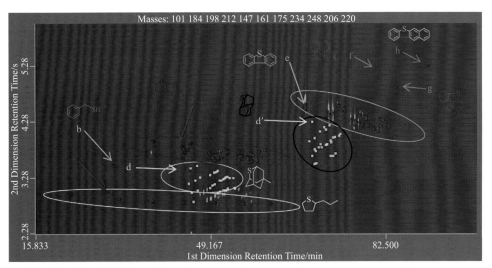

Fig. 5　GC×GC-TOFMS color contour chromatogram for m/z 101, 184, 198, 212, 147, 161, 175, 234, 248, 206, 220 of sulfur-containing compounds in the condensate of ZS1C well.

(For interpretation of the references to color in this figure legend,

the reader is referred to the web version of this article.)

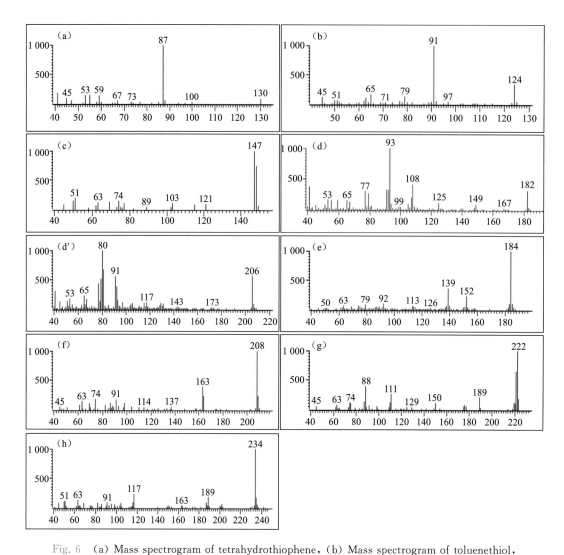

Fig. 6 (a) Mass spectrogram of tetrahydrothiophene, (b) Mass spectrogram of toluenethiol, (c) Mass spectrogram of benzothiophene, (d) Mass spectrogram of methyl-2-thiaadamantanes, (d') Mass spectrogram of 2-thiadiamantane, (e) Mass spectrogram of dibenzothiophene, (f) Mass spectrogram of phenanthrothiophene, (g) Mass spectrogram of methyl-phenanthrothiophene, (h) Mass spectrogram of benzonaphthothiophene.

3.5 Recognition and identification of thiaadamantanes

The abundant thiaadamantanes discovered in the condensate (Fig. 7) include 31 thiaadamantanes (the serial number from Fig. 7 to 10) and 20 thiadiamantanes (Fig. 5). No 2-thiaadamantane was detected. The result only showed the presence of 2-thiaadamantane substituted with $C_1 - C_4$ (Fig. 8 to 10). There are only two 2-thiaadamantanes substituted with C_1 (No. 2 and No. 3 in Fig. 7), namely, 5-methyl-2-thiaadamantane and 1-methyl-2-thiaadamantane, which is in agreement with the literature.

Seven 2-thiaadamantanes substituted with C_2 (Fig. 8) were detected. Apart from No. 4, No. 5 and No. 6 peaks that have mass spectra in the corresponding literature, No. 9 —

12 are the newly discovered 2-thiaadamantanes substituted with C_2. The mass spectra are plotted in Fig. 8. It can be seen that the fragment ions of the four compounds are principally m/z 93, 107 and 125, with all molecular ions being m/z 182, which coincides with the No. 4 – 6 peaks. This suggests that they share a similar way of fragmentation and the four compounds have a consistent twodimensional retention time as No. 4 – 6 compounds. Based on the characteristics of the comprehensive two-dimensional chromatogram, we can infer that No. 9 – 12 compounds have a similar chemical structure with No. 4 – 6 peak compounds. It can be tentatively concluded that they are also the 2-thiaadamantanes substituted with C_2, although the position of substitution is unclear yet.

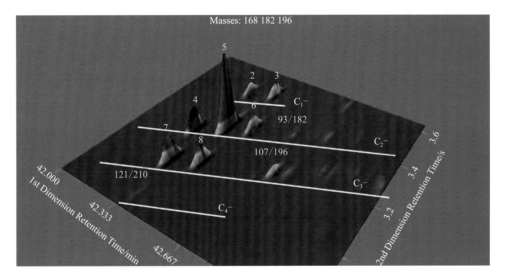

Fig. 7 GC×GC-TOFMS color contour chromatogram for m/z 168, 182, 196 of the 2-thiaadamantane compounds in the condensate of ZS1C well (93/182 means that the characteristic ion of the compound is m/z 93. The molecular ion is m/z 182, and other features are similar). (For interpretation of the references to color in this figure legend, the reader is referred to the web version of this article.)

Thirteen 2-thiaadamantanes substituted with C_3 (Fig. 9) were detected. Apart from the No. 8 peak that has a mass spectrogram that corresponds with literature, all others are the newly discovered 2-thiaadamantanes substituted with C_3. The mass spectra are plotted in Fig. 9. The fragment ions of these compounds are mainly m/z 93, 107 and 125 and all the molecular ions are m/z 196. The same feature can be also seen in the No. 8 compound, and these compounds have the same two-dimensional retention time as the No. 8 compound. Therefore, they are similar in chemical structure to the compound at the No. 8 peak. It is probable that they are also the 2-thiaadamantanes substituted with C_3, but the position of substitution is unclear.

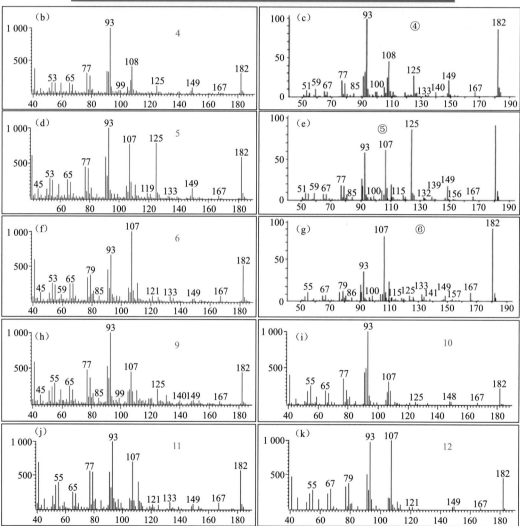

Fig. 8 (a) GC×GC-TOFMS color contour chromatogram for m/z 182 of the condensate of ZS1C well; (b),(d),(f),(h)-(k) are the mass spectra of peaks 4-6 and 9-12, respectively, in the GC×GC chromatogram, (c),(e),(g) are mass spectra from literature.
(For interpretation of the references to color in this figure legend, the reader is referred to the web version of this article.)

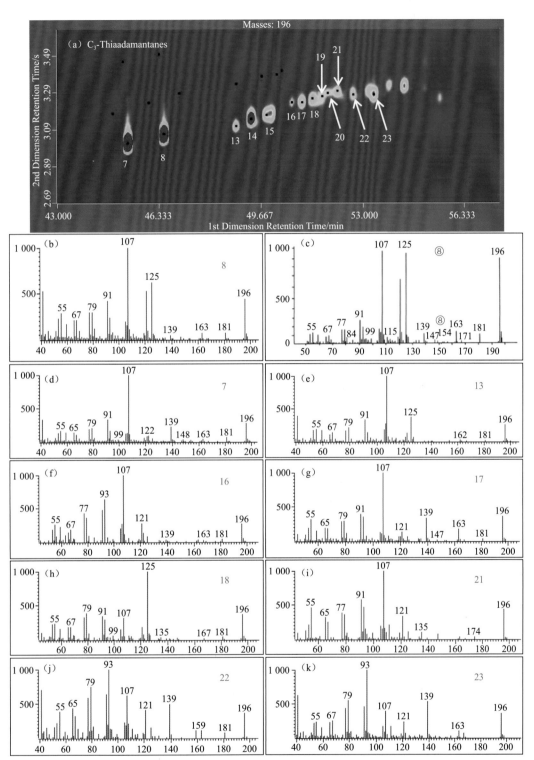

Fig. 9 (a) GC×GC-TOFMS color contour chromatogram for m/z 196 of ZS1C well; (b), (d)–(k) are the mass spectra of peaks 7, 8, 13, 16–18, and 21–23, respectively, in the GC×GC chromatogram, (c) is the mass spectrum from literature. (For interpretation of the references to color in this figure legend, the reader is referred to the web version of this article.)

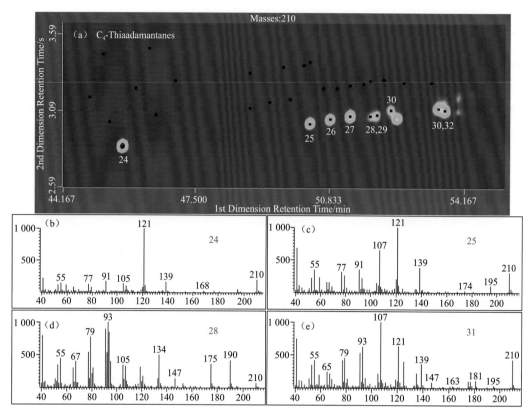

Fig. 10 (a) GC×GC-TOFMS color contour chromatogram for m/z 210 of ZS1C well; (b)–(e) are the mass spectra of peaks 24, 25, 28, and 31, respectively, in the GC×GC chromatogram. (For interpretation of the references to color in this figure legend, the reader is referred to the web version of this article.)

The detection also confirmed the presence of nine 2-thiaadamantanes substituted with C_4 in the condensate. The mass spectra of some of the compounds are presented in Fig.10. Based on Fig. 10, the characteristic ion on the No. 24 mass spectrum peak is m/z 121, with a difference of 14 compared with m/z 107 corresponding to the characteristic ion of the No. 8 peak in Fig. 9. The main fragment ions are m/z 139 and 168, which are the same as the No. 8 compound. The molecular ion peak m/z 210 also has a difference of 14 from m/z 196 of the No. 8 peak, indicating that the fragmentation spectrum of the No. 24 peak is the same as that of the No. 8 peak. There is a difference of "—CH_2" from the No. 8 peak both in the characteristic ion and the molecular ion peaks. We suggest that the No. 24 peak has one more methyl group than the No. 8 peak, and that the position of methyl substitution is on the adamantane ring. We also infer that the No. 24 compound is tetramethyl-2-thiaadamantane. Similarly, the No. 25 – 32 compounds are also the C_4-2-thiaadamantane, and the position of substitution is unclear for now.

3.6 Distribution of other compounds

Most likely because of the high maturity of the condensate, no sterane and terpane compounds were detected at characteristic ion m/z 123, m/z 191, or m/z 217. Compound peaks were still not observed at characteristic peak m/z 213 and m/z 245, indicating that mono-aromatic steranes and tri-aromatic steranes have all disappeared from the condensate.

The dominant aromatic hydrocarbons were multi-substituted series of benzene, naphthalene, and phenanthrene. The content of alkylbenzene series and polycyclic aromatic hydrocarbons was extremely low. Other heteroatom compounds, such as neutral nitrogen-containing and dibenzofuran compounds, were not detected.

3.7 Genesis of the Cambrian oil and gas in the ZS1C well and the generation of organo-sulfur-containing compounds

3.7.1 Chemical and isotopic compositions of natural gas and its genesis

The Cambrian natural gas in the ZS1C well is dry gas and is mainly composed of methane (62.7%–63.0%), CO_2 (24.1%–25.0%) and H_2S (7.22%–8.27%), with the ethane content of less than 0.6% (Table 3). The content of heavier hydrocarbons (C_2^+) was low and the dryness ratio ($C_1/\sum C_{1-4}$) was 0.98. Therefore, the gas belongs to high-H_2S sour gas.

Compared with ordinary oil-type gas in the non-TSR reservoirs of Well TZ451 and Well ZG162 that fall into the oil cracking gas area in the plot of $\delta^{13}C_1$ vs ($\delta^{13}C_2-\delta^{13}C_3$), the $\delta^{13}C_1$ value of the natural gas in Well ZS1C is as heavy as to $-42‰$ (Table 3), indicating the TSR has an obvious effect on the methane carbon isotope. The difference between $\delta^{13}C_2$ and $\delta^{13}C_1$ ($\delta^{13}C_{2-1}$) is less than 10‰, reflecting the high maturity of natural gas. Based on research of Ordovician natural gas in the Tazhong area, the natural gas resulted from high temperature secondary oil cracking.

Table 3 Chemical and carbon isotopic compositions of gases from Well ZS1 and two others non-TSR reservoirs (Well TZ451 and Well ZG162).

Well	Depth /m	Strata	Dry coefficient	Natural gas components/%											Isotope $\delta^{13}C/(‰, PDB)$					
				CH_4	C_2H_6	C_3H_8	$i-C_4H_{10}$	$n-C_4H_{10}$	$i-C_5H_{12}$	$n-C_5H_{12}$	C_6^+	N_2	CO_2	H_2S	CH_4	C_2H_6	C_3H_8	$n-C_4H_{10}$	$i-C_4H_{10}$	CO_2
ZS1C	6 861	$\in_1 x$	0.986	62.70	0.469	0.157	0.054	0.079	0.047	0.030	0.077	4.00	24.2	8.27	−42.24	−34.71	−31.77	−30.02	−31.36	−30.18
			0.985	62.70	0.509	0.160	0.045	0.080	0.039	0.041	0.071	4.02	25.0	8.26						
			0.983	63.00	0.545	0.178	0.060	0.093	0.050	0.043	0.090	3.93	24.1	7.22						
TZ451	6 194	$O_3 l$	0.889	84.12	6.34	2.410	0.510	0.810	0.210	0.190	0.170	5.070	0.170	0	−50.6	−35.9	−30.6	—	−29.7	—
ZG162	6 161	$O_3 y$	0.850	77.08	8.49	2.679	0.660	1.100	0.000	0.000	0.870	7.578	1.540	0	−51.0	−36.5	−31.6	−30.0	−29.7	−13.0

Bulk isotopic compositions of saturated hydrocarbons, aromatic hydrocarbons, resins in ZS1C condensate are $-30.9‰$, $-30.5‰$, and $-29.4‰$, respectively. Oil from Well TD2 oil has the heaviest $\delta^{13}C$ value, which is reservoired in the Cambrian strata. It has been regarded as the standardized end member for the Cambrian sourced oil in all previous studies. The ZS1C condensate from the Lower Cambrian is isotopically heavy and close to oils in Well TD2, indicating that the condensate in Well ZS1C is derived from the Cambrian source rocks.

3.7.2 TSR and generation of organo-sulfur-containing compounds

The formation of high concentration of hydrogen sulfide in the deep-buried oil and gas reservoir is often related to thermochemical sulfate reduction. Based on the characteristics of the Cambrian oil reservoir, the reservoir stratum has a set of dolomite, clay-containing dolomite, argillaceous dolomite, and gypsum layers. The caprock is a set of evaporates, including gypsum-bearing rocks and argillaceous gypsum bearing rocks, providing a material basis for the occurrence of TSR. The content of SO_4^{2-}, Ca^{2+}, and Mg^{2+} ions in the producing formation water from the ZS1C well is 1 970 mg/L, 5 300 mg/L, 4 880 mg/L, respectively, which are favorable formation water chemistry for the occurrence of TSR. The measured temperature of the reservoir is 169.0 ℃, which is consistent with the low geothermal gradient of the Tarim Basin (the geothermal gradient is approximately (2.0 – 2.2 ℃)/100 m). Therefore, hydrocarbons and sulfates may undergo TSR reaction under high temperature, leading to the generation of nonhydrocarbon gases, such as hydrogen sulfide and carbon dioxide.

Thiaadamantanes are known as indicators of TSR or the molecular fingerprint of TSR. The compound 2-thiaadamantane is generated by a reaction in which the C-2 position is substituted by a sulfur atom. The generation of organic sulfur-containing compounds, including thiaadamantanes and tetrahydrothiophene compounds, benzyl mercaptane, long-chain alkylbenzothiophene, dibenzothiophene, benzonaphthothiophene and phenanthrothiophene compounds, in the ZS1C well condensate may be related to the severe TSR. Under high temperature and severe TSR, the crude oil undergoes cracking and the long-chain hydrocarbons evolve into methane. Hydrocarbons react with sulfates to generate a large amount of hydrogen sulfide and carbon dioxide. The sulfur is bonded to the organic matter to form organic sulfur-containing compounds. Therefore, the discovery of a large numbers of sulfur-containing compounds in the condensate of the ZS1C well is an indication of the TSR reaction between hydrocarbons and sulfates under high temperature conditions. During the generation of hydrogen sulfide and carbon dioxide, a lot of organic sulfur-containing compounds and thiaadamantanes are generated and distributed in the oil.

3.7.3 Degree of TSR

The $\delta^{34}S$ of Cambrian gypsum ranges from 32‰ to 37‰, and the $\delta^{34}S$ of hydrogen sulfide in the ZS1C well natural gas was 33.5‰, and the $\delta^{34}S$ of crude oil was 33.0‰. The sulfur isotopes of hydrogen sulfide and crude oil are very similar to those of sulfates in the strata. This indicates that the sources of both are related to sulfates in the strata and that hydrogen sulfide is from TSR. In the TSR, the fractionation of sulfur isotopes is closely related to the extent of TSR, the stronger TSR is associated with the less sulfur isotopic fractionation between the produced H_2S and the reactant of sulfate in the strata. The sulfur isotope values of the hydrogen sulfide in the ZS1C well are almost equivalent to those of the sulfates, suggesting the high intensity of TSR.

The high CO_2 content and its carbon isotope composition indicate strong TSR. The CO_2 accounts for 24.1%–25.0% of the natural gas component, and the $\delta^{13}C_{CO_2}$ is −30.18‰. The deep-buried strata oil reservoir under the severe TSR produces a large quantity of sour gases (CO_2 and H_2S). Almost all of the extremely light carbon isotopes of carbon dioxide originate from the carbon in the organic matter. This natural gas may have the lightest CO_2 carbon isotopes in Chinese natural gas. Chemical and isotopic compositions of natural gases have a lot of differences in the reservoir of ZS1C Well and others reservoirs (Well TZ451 and Well ZG162) in Tazhong region (Table 3). In the non-TSR reservoirs of the Wells TZ451 and ZG162, close to ZS1C well (Fig.1), the carbon isotopes of the natural gases were clearly lighter. The content of heavier hydrocarbons (C_2^+) were high and the dryness ratio ($C_1/\sum C_{1-4}$) was 0.89. So the effects of TSR alteration on gaseous hydrocarbons can be demonstrated clearly in the reservoir of ZS1C well. This suggests that the risk of high content of H_2S and the wide distribution in the deep stratum might be involved with severe TSR.

The co-existence of 2-thiaadamantanes and 2-thiadiamantanes, other organosulfur-containing compounds, the similarity in the sulfur isotopes of crude oil, H_2S and gypsum, the similarity of CO_2 carbon isotopes and condensates and natural gas have provided comprehensive geochemical evidences for the occurrence of the severe TSR in this area. It might be the region with the severest TSR so far in China. The ZS1C well is the first commercial producing oil and gas well which was drilled through the lower Cambrian dolomite layer. As a major breakthrough in the exploration of the Cambrian system over the past two decades, the well has opened up a new exploration stratum (Fig. 11). The depth of oil and gas exploration in the Tarim Basin has extended down to 7 000 – 10 000 m because of the success of this well. Based on the present results, the deep-buried oil reservoir has undergone severe TSR alteration. The deep-buried Cambrian oil and gas reservoirs are dominated by natural gas and condensate with high content of hydrogen sulfide. It is therefore to be cautious of the various safety issues related to hydrogen sulfide prior to or during drilling.

4 Conclusions

With GC×GC-TOFMS and GC×GC-FID, abundant adamantane compounds, 2-thiaadamantanes and 2-thiadiamantanes, and a large number of sulfur-containing compounds have been detected in the condensate oil of the ZS1C well in the Tarim Basin.

The condensate is highly altered by TSR in the deep-buried strata Cambrian oil reservoir. The similarity of the sulfur isotopes among hydrogen sulfide, condensate and gypsum suggests that the TSR occurred in the deep Cambrian oil reservoirs. The carbondioxide had exceptionally light carbon isotopes, which further supports the high intensity of TSR. Newly discovered organo-sulfur-containing compounds, such as tetrahydrothiophene compounds, benzyl mercaptane, alkylbenzothiophene, dibenzothiophene, phenanthrothiophene, and benzonaphthothiophene compounds, were indication of the severe alteration of the hydrocarbons by TSR.

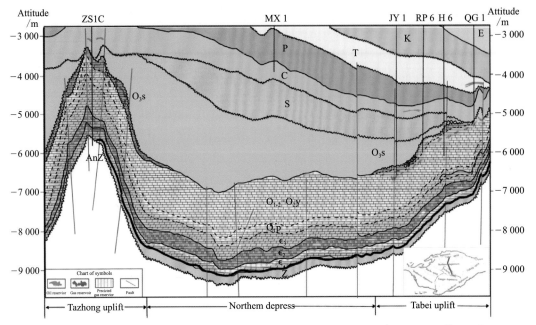

Fig. 11　Cross section of the Tazhong uplift showing oil and gas accumulation at Well ZS1C area.

Therefore, the integrated geochemical TSR indicators as shown in this study can be used to evaluate the intensity of TSR and to predict the nature and phase state of the deep-buried fluid. Our research indicates that the Cambrian deep-buried oil and gas reservoir has undergone intense TSR. It is therefore vital to prevent and guard against various safety hazards caused by hydrogen sulfide.

论文 13　2015 年发表于《Marine and Petroleum Geology》

Origin of diamondoid and sulphur compounds in the Tazhong Ordovician condensate, Tarim Basin, China: Implications for hydrocarbon exploration in deep-buried strata

Guangyou Zhu, Na Weng, Huitong Wang, Haijun Yang, Shuichang Zhang, Jin Su, Fengrong Liao, Bin Zhang, Yungang Ji

Abstract　In this paper, we analyse the Ordovician condensate in the Tazhong area of the Tarim Basin, NW China, using a new high-resolution comprehensive two-dimensional gas chromatography with time-of-flight mass spectrometry (GC×GC-TOFMS). Results show that 4 778 compounds at a signal-to-noise ratio of more than 100 were detected, reflecting the advantage of this method in the analysis of condensate. Abundant adamantane series compounds and some diamantane series compounds, as well as a series of sulphur-containing compounds (including the benzothiophene series, and the dibenzothiophene series) were detected for the first time in the study area. Combined with the results of a study on oil and gas sources and migration/accumulation, we address the origin of these compounds. It is implied that the diamondoids were not formed in-situ due to oil cracking (in reservoir approximately 5 500 m deep at temperature around 139 ℃), whereas the cracking of oil to gas may have occurred in the deeper-buried Cambrian dolomite reservoir at a reservoir temperature of 180 − 220 ℃ (8 500 − 11 000 m depth). The oil-cracked gas (secondary cracking product) might have migrated to and accumulated in the shallower-buried Ordovician reservoir, accompanied by abundant diamondoids. The formation of sulphur-containing compounds may be related to thermochemical sulphate reduction (TSR), which, similar to the oil cracking, took place in the deeper-buried Cambrian reservoir. Under high-temperature TSR, SO_4^{2-} in the formation water is reduced to H_2S and organic sulphur compounds, leading to the enrichment of sulphur in crude oils (e.g., benzothiophene, and dibenzothiophene series). As such, it is implied that the Tazhong area could potentially be explored for deep-buried oil-cracked gas.

Keywords　Deep-buried reservoir　GC×GC-TOFMS　Ordovician　Oil cracking　TSR

1　Introduction

Adamantane, a type of polymerization cyclic hydrocarbon compound similar in structure to diamond, is a product of strong Lewis acid catalyst polymerization of polycyclic hydrocarbons under thermal action. This compound may be generated and enriched during the cracking process, and thus is usually abundant in highly mature crude oil and condensate. Therefore, adamantane is commonly used to measure the degree of cracking of crude oil. Stability and pyrolysis temperature of crude oil are matters of great interest in current researches. It is thought that crude oil begins to crack into gas at a temperature of more than 160 ℃ or at a burial depth of more than 6 000 m, after which liquid oil gradually disappears. With progress in deep oil and gas exploration, some reservoirs that still contain abundant oil reserves have been found at high temperatures or great depths, suggesting that crude oil can be much more stable than expected. For example, a high hydrocarbon flow was obtained in a buried hill reservoir at depths of 5 641 – 6 027 m and temperatures of 190 – 201 ℃ in Well ND1 in the Jizhong Depression, Bohai Bay Basin, China; the crude oil, dominated by gas condensate, contained relatively complete long-chain compounds and adamantane series.

A number of crude oil, cracked gas and condensate gas reservoirs have been discovered in the Ordovician strata of the Tarim Basin. These strata have a low geothermal gradient at a depth of about 5 500 m (a reservoir temperature of about 140 ℃). This type of complex oil and gas geochemical phenomenon is probably related to charging and mixing of oil and gas of multiple origins, or high thermal maturation or secondary alteration, such as TSR, which causes changes in the composition of the reservoir fluid.

The Tazhong area, i.e. the central area of the Tarim Basin, NW China (Fig. 1), contains large amount of hydrocarbon resources, and it is characterized by multiple hydrocarbon phases and associated complex migration and accumulation stories. Recently, a number of crude oil, gas, and condensate have been discovered in the Ordovician carbonate reservoirs of this area, through a discovery Well-Tazhong 83 (TZ83). This suggests the occurrence of complex multi-stage hydrocarbon migration/accumulation and/or reservoir hydrocarbon alteration phenomenon (e.g., oil-cracking and gas washing). However, this complexity is not adequately understood. Especially, the Middle and Lower Ordovician carbonate rocks-the major exploration target are buried in 5 500 – 7 000 m, for which, the reservoir heterogeneity is strong, the fracture-cavity reservoir predominates, the fluid connectivity is poor, and the oil and gas properties are complicated. These characteristics have brought many difficulties for the operations like oil and gas exploration and development and resource estimation. Therefore, the identification of genesis mechanism and distribution rules and the effective prediction of oil and gas phase state will have an important value scientifically for directing the oil and gas exploration of carbonate strata and developing the geologic theory of condensate gas reservoir.

In this paper, we analyse the oil and gas geochemistry (especially of condensate) in the study area by a new high-resolution comprehensive two-dimensional gas chromatography with time-of-flight mass spectrometry (GC×GC-TOFMS). Multiple types of new molecular compounds were detected for the first time in the study area. We discuss the origins of these compounds, especially the diamondoids (including adamantanes and diamantanes) and sulphur compounds, and address their implications for petroleum exploration.

2　Geological setting

The Well TZ83 area in this study is located in the central-east of the Tazhong No. Ⅰ slope break belt in the Tarim Basin (Fig. 1). It borders the Manjiaer sag to the north and the Tazhong 10 structural belt and central fault base belt of the Tazhong low bulge to the southwest. The Tazhong low bulge, located in the middle of the Tarim Basin central uplift, is bordered by the Bachu uplift to the west, the Tadong low bulge to the east, the

Tangguzibasi sag to the south, and the Manjiaer sag to the north. This long-term-developed palaeo-uplift is divided into three second-order structural units from north to south: northern slope zone, central fault zone, and southern slope zone. The Tazhong No. I slope break belt, extending in NW-SE for approximately 260 km, is located on the northern edge of the Tazhong low bulge, immediately adjacent to the Manjiaer sag.

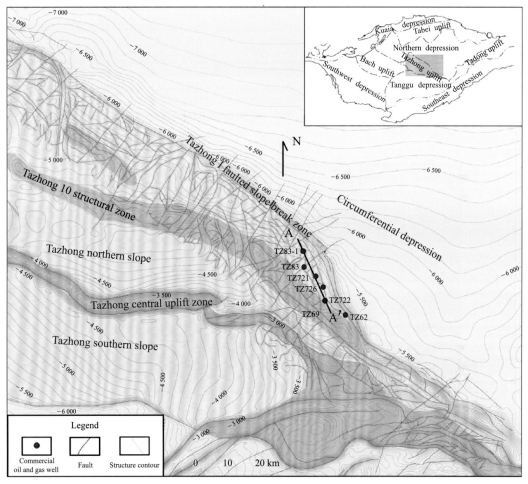

Fig. 1　Structural map of the upper Ordovician in Tazhong area, Tarim Basin.

The studied Ordovician reservoir was discovered when 10.6 m³ oil and 639 177 m³ gas were produced from the Ordovician Yingshan Formation at 5 666.1 − 5 684.7 m in Well TZ83. Then four appraisal wells (TZ722, TZ723, TZ724 and TZ726) were drilled, all of which have demonstrated commercial production (Fig. 2). Test results show that TZ83, TZ726, TZ721 and TZ83-1 are condensate wells, showing reservoir characteristics with GOR ratios of ca. 18 698 − 21 213 m³/m³. However, the adjacent Well TZ722 area is an oil province with a GOR ratio of ca. 180 m³/m³, indicating complex hydrocarbon migration and accumulation histories.

Production test results indicate heterogeneities in the carbonate reservoir and in oil and gas distribution. From the studies of the carbonate reservoir and its pores and fractures, it has been recognized that the oil and gas distribute continuously, but each reservoir is composed of relatively independent pore-fracture networks. Each network contains well-connected channels conducive for fluid flow, whereas between different networks the channels are relatively poorly connected. This is one of the important reasons that the oil and gas occurrence is heterogeneous (Fig. 2). The oil and gas are distributed in a large area and in quasilayered form, reserves have a large scale but a low abundance, there is no unified gas-water interface, reservoirs are mainly low porosity and permeability vuggy reservoir, with strong heterogeneity. The large cavity-type heterogeneous carbonate oil and gas fields consist of a series of composite contiguous small oil and gas reservoirs, so hydrocarbon-water distribution is not completely controlled by local structures, and reservoir boundary is not clear.

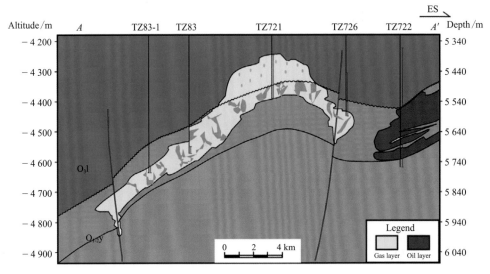

Fig. 2 North-south structural profile of the Tazhong 83 (TZ83) condensate field in Tazhong area, Tarim Basin.

Fluid distribution in the Well TZ83 area is complex. Oil and gas provinces exist side by side at the same level, uncontrolled by the structural height, but they are closely related to the reservoir distribution and the oil and gas charging conditions. The main rocks containing the gas reservoir include the Ordovician Yingshan Formation and Lianglitage Formation limestone. The cap rocks are the Lianglitage Formation limestone and overlying Ordovician Sangtamu Formation mudstone, which are widely distributed in the area, forming a good reservoir-cap assemblage (Fig. 3).

In the Well TZ83 area, there are mainly three categories of reservoir space, including pores, vugs and fractures. Analysis of the physical properties of core samples has shown a porosity of $0.05\%-3.81\%$ (mainly $0.6\%-1.8\%$, or averagely 0.88%) and a permeability of $0.012\,7\times10^{-3}-12.7\times10^{-3}\,\mu m^2$ (largely in $0.01\times10^{-3}-3\times10^{-3}\,\mu m^2$, or averagely $0.144\times10^{-3}\,\mu m^2$). Of the samples, only 9.48% have a porosity greater than

1.8% (the lower limit); 99.14% have a permeability between 0.01×10^{-3} μm^2 and 3×10^{-3} μm^2, but only 0.86% have a permeability more than 3×10^{-3} μm^2. Thus, the reservoirs in this area have extremely low porosity and permeability. The distribution of Yingshan Formation reservoirs is closely related to the lateral distribution of karst, and is controlled by karst vertically. The Yingshan Formation reservoirs are parallel to the direction of the Tazhong No. I slope break belt, 115 – 200 m vertically below the weathered

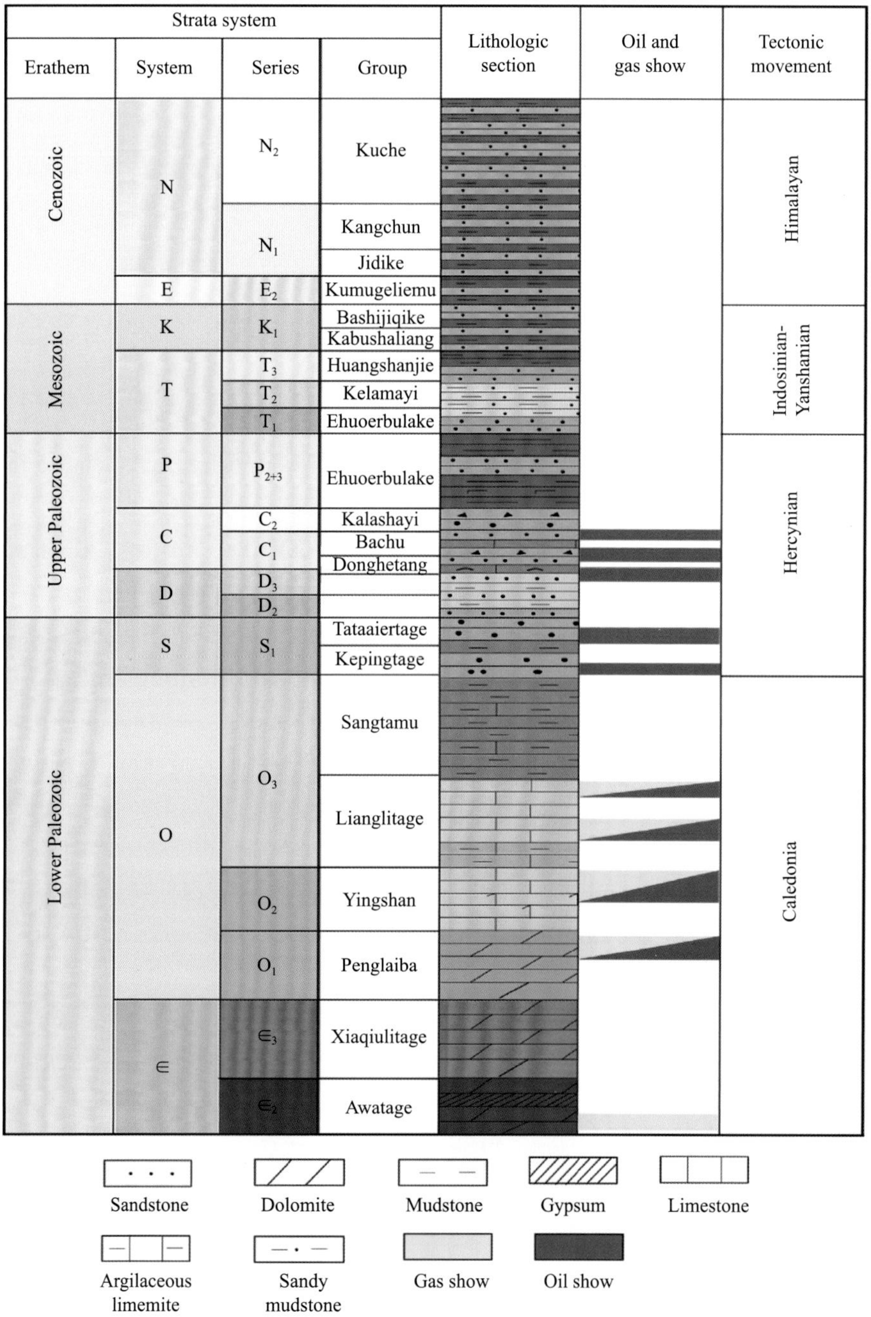

Strata system				Lithologic section	Oil and gas show	Tectonic movement
Erathem	System	Series	Group			
Cenozoic	N	N_2	Kuche			Himalayan
		N_1	Kangchun			
			Jidike			
	E	E_2	Kumugeliemu			
Mesozoic	K	K_1	Bashijiqike			Indosinian-Yanshanian
			Kabushaliang			
	T	T_3	Huangshanjie			
		T_2	Kelamayi			
		T_1	Ehuoerbulake			
Upper Paleozoic	P	P_{2+3}	Ehuoerbulake			Hercynian
	C	C_2	Kalashayi			
		C_1	Bachu			
			Donghetang			
	D	D_3				
		D_2				
Lower Paleozoic	S	S_1	Tataaiertage			Caledonia
			Kepingtage			
	O	O_3	Sangtamu			
			Lianglitage			
		O_2	Yingshan			
		O_1	Penglaiba			
	$\in$	$\in_3$	Xiaqiulitage			
		$\in_2$	Awatage			

Legend: Sandstone, Dolomite, Mudstone, Gypsum, Limestone, Argilaceous limemite, Sandy mudstone, Gas show, Oil show

Fig. 3 Stratigraphic lithology column in Tazhong area, Tarim Basin.

crust, and roughly 160 m thick. The reservoir section occurs in quasi-layers near the Yingshan Formation weathered crust.

There is an oil and gas boundary between Well TZ726 and Well TZ722 in the Well TZ83 area (Fig. 2). West and east of the boundary are gas and oil areas, respectively. Within these two areas, the Ordovician reservoirs have well-developed fractures and karst systems, good reservoir connectivity, similar fluid properties and sources, and good pressure communication. As such, the blocks controlled by Wells TZ83, TZ721 and TZ726 are condensate reservoir areas, while the block controlled by Well TZ722 is an oil reservoir area.

3 Materials and methods

Oil and gas samples were collected from the reservoir interval of Wells TZ83, TZ722, TZ723, TZ724, and TZ726. These samples were subjected to the new comprehensive two-dimensional gas chromatography with flame ionization detector (GC×GC-FID) and time-of-flight mass spectrometry (GC×GC-TOFMS). In addition, conventional geochemical analyses of oil and gas were completed, involving physical properties, chemical components, isotopes and biomarkers.

The GC×GC system for GC×GC-TOFMS was from LECO Corporation. The GC×GC system was composed of an Agilent 7890 GC coupled to a hydrogen flame ionization detector (FID) and a liquid-nitrogen-cooled pulse jet modulator. The time-of-flight mass spectrometer was a Pegasus 4D (LECO Corporation). All the data were processed with the Chroma TOF software.

As for the analysis for GC×GC-TOFMS, the one-dimensional chromatographic column was DB-Petro (50 m×0.2 mm×0.5 μm). The temperature program used was: 0.2 min at a temperature of 35 ℃; increased to 210 ℃ at a rate of 1.5 ℃/min and held for 0.2 min; and increased to 300 ℃ at the rate of 2 ℃/min and held for 20 min. The two-dimensional chromatographic column was DB-17ht (3 m×0.1 mm×0.1 μm). This program was the same as that used for one-dimensional gas chromatography, except that the temperatures were 5 ℃ higher and the modulator temperature was 45 ℃ higher. The inlet temperature was 300 ℃, the inlet mode was split injection, the split ratio was 700∶1, and the sample volume was 0.5 μL. Helium was the carrier gas, with a flow rate of 1.5 mL/min. The modulation time was 10 s, including 2.5 s of hot pulse time. On the mass spectrum, the temperatures of the transfer line and the ion source were 300 ℃ and 240 ℃, respectively, the detector voltage was 1 600 V, the scan range was 40−520 amu, the acquisition rate was 100 spectra/s, and the delay time of the solvent was 9 min.

The experimental conditions for GC×GC-FID were the same as those for GC×GC-TOFMS. The flow rates of the carrier gas, hydrogen, and air were 50, 40, and 450 mL/min. The detector temperature was 320 ℃, the acquisition rate was 200 spectra/s, and the delay time of the solvent was 9 min.

The group compositions of the compounds were quantified by peak area normalization. D_{16}-adamantane (using CH_2Cl_2 as a solvent) was placed in the condensate sample, and the quantitative results of conventional diamonoids in the condensate were obtained using the internal standard method.

Natural gas C-isotope analysis: Carbon isotope compositions of natural gases were measured with an Isochrom II isotope ratio mass spectrometer connected to an Agilent 6890 gas chromatograph through a combustion chamber (GC-IRMS) with a He carrier and temperature program: initial 50 ℃ held for 3 min, increased to 150 ℃ at 15 ℃/min, held 8 min. Carbon isotope compositions are reported in the usual δ notation relative to the PDB standard. The reproducibility of the duplicate C isotope measurement is ±0.5‰.

4 Results

4.1 General geochemistry of reservoir fluid

4.1.1 General geochemistry of condensate and oil

The retrograde oil from condensate in the Well TZ83 area has a density of 0.820 8 – 0.827 6 g/cm³ (at surface 20 ℃), a viscosity of 3.168 – 5.585 mPa·s (at 50 ℃), a sulphur content of 0.18% – 0.40%, a wax content of 10.14% – 24.95% (mostly more than 20%), and low colloid and asphaltene content (Table 1). Thus, the condensate is characterized by low density, low sulphur, and high wax.

The crude oil in the oil reservoir of the Well TZ722 area shows the physical properties that are much different from that in the condensate reservoirs, such as slightly lighter density, lower wax content, and lower compositions of saturates (Tables 1 and 2).

4.1.2 Natural gas properties

The natural gas in the condensate reservoir of the Well TZ83 area is dry with a high non-hydrocarbon content including N_2, CO_2 and H_2S (Table 3). It is quite different from the associated gas from the adjacent oil reservoir, mainly in terms of the levels of ethane and heavier hydrocarbons (Table 3).

4.1.3 PVT phase analysis of formation fluid

PVT phase samples were taken from the Well TZ83 area. Analytical results show that these samples have the characteristics of condensate (Table 4), as showed by two wells below. Samples from Well TZ83 are in two phases (gas and liquid) under formation conditions (trace liquid hydrocarbon). The fluid produced from the well consists of 90.88% C_1+N_2, 8.61% $C_2-C_6+CO_2$, and 0.51% C_7^+, which shows the characteristic of condensate on the triangular phase diagram (Fig. 4). In the phase diagram (Fig. 5), the formation temperature is on the right side of the critical temperature. The reservoir fluid has a condensate oil content of 40.57 g/m³, a large amount of retrograde condensed liquid (0.88%), and a dew pressure difference of zero, indicating that it is a condensate reservoir.

Table 1 Physical properties of Ordovician crude oil in the Well TZ83 area.

Reservoir type	Well	Depth /m	Density /(g·cm⁻³) 20 ℃	Density /(g·cm⁻³) 50 ℃	Viscosity /(mPa·s) 50 ℃	Sulphur content/%	Wax content/%	Colloid content/%	Asphaltene content/%	Water cut/%	Freezing point/℃	Initial boiling point/℃
Condensate gas	TZ83	5 666.10 − 5 681.00	0.824 5	0.803 5	5.585	0.36	22.75	0.00	0.00	1.25	42	—
	TZ83	5 666.10 − 5681.00	0.823 7	0.802 7	3.857	0.34	24.95	0.00	0.00	1.72	38	—
	TZ83	5 666.10 − 5 681.00	0.821 1	0.799 8	3.168	0.40	23.56	0.00	0.00	2.65	42	—
	TZ83	5 666.10 − 5 681.00	0.823 2	0.802 2	3.179	0.38	22.49	0.00	0.00	16.65	44	—
	TZ721	5 355.50 − 5 505.00	0.827 6	0.806 6	3.474	0.27	20.18	0.19	0.25	2.79	32	95
	TZ726	5 386.00 − 5 534.09	0.820 8	0.799 5	2.849	0.18	10.14	0.51	0.09	4.77	10	118
Oil	TZ722	5 356.70 − 5 750.00	0.784 9	0.806 5	1.906	0.08	13.12	1.15	0.30	1.94	2.0	—
	TZ72	5 125.00 − 5 130.00	0.782 6	0.804 5	1.719	0.22	7.72	0.86	0.02	0.02	−2	85
	TZ723	5469.34 − 5 495.55	0.787 5	0.809 1	1.792	0.56	12.6	2.79	0.21	60.04	−30	—
	TZ724	5523.89 − 5 580.00	0.793 2	0.819 7	2.625	0.05	7.06	0.74	0.69	10.06	0.0	—

Table 2 Group composition of Ordovician crude oil in the Well TZ83 area.

Reservoir type	Well	Depth/m	Saturate/%	Aromatic/%	Resin/%	Asphaltene/%	Saturate/aromatic
Condensate gas	TZ83	5 666.00 − 5 684.70	72.21	11.91	4.34	9.75	6.07
	TZ721	5 355.00 − 5 505.00	77.57	6.46	4.19	9.13	12.01
	TZ726	5 386.00 − 5 534.09	83.80	5.59	2.23	5.87	14.99
Oil	TZ722	5 356.70 − 5 750.00	57.80	22.81	4.95	12.17	2.53
	TZ72	5 125.00 − 5 130.00	70.03	12.12	6.06	10.77	5.78
	TZ724	5 529.00 − 5 550.00	68.94	5.68	9.47	13.26	10.14
	TZ724	5 523.89 − 5 580.00	58.53	22.3	5.69	10.71	2.62

第 3 章 全二维气相色谱的石油地质应用

Table 3 Natural gas compositions and carbon isotopes in the Well TZ83 area.

Reservoir type	Well	Depth/m	C_1/C_1^+	Natural gas components/%									Isotope $\delta^{13}C/(‰, PDB)$				
				CH_4	C_2H_6	C_3H_8	$i\text{-}C_4H_{10}$	$n\text{-}C_4H_{10}$	N_2	CO_2	H_2S (mg·kg^{-1})		CH_4	C_2H_6	C_3H_8	$n\text{-}C_4H_{10}$	$i\text{-}C_4H_{10}$	CO_2
Condensate gas	TZ83	5 666.10 – 5 681.00	0.989 5	93.1	0.64	0.10	0.06	0.04	0.97	4.91	32 700	−38.9	−32.2	−28.0	−26.7	−26.2	−2.5
	TZ83	5 666.10 – 5 684.70	0.986 2	91.7	0.68	0.12	0.09	0.05	0.82	6.16	30 400	—	—	—	—	—	—
	TZ83	5 666.10 – 5 684.70	0.989 2	89.9	0.62	0.11	0.06	0.04	3.69	5.39	32 300	—	—	—	—	—	—
	TZ83	5 666.10 – 5 684.70	0.989 8	92.2	0.62	0.10	0.05	0.04	0.89	5.99	34 300	—	—	—	—	—	—
	TZ83-1	5 274.00 – 5 298.00	0.969 4	89.6	1.28	0.41	0.26	0.12	4.20	3.31	32 200	—	—	—	—	—	—
	TZ83-1	5 550.00 – 5 762.77	0.966 3	89.4	1.74	0.64	0.27	0.19	6.69	0.15	9 400	—	—	—	—	—	—
	TZ721	5 355.50 – 5 505.00	0.990 2	94.1	0.63	0.13	0.06	0.03	1.50	3.42	—	−38.8	−36.5	−30.8	−29.1	—	−10.0
	TZ721	5 355.50 – 5 505.00	0.990 0	94.9	0.64	0.13	0.06	0.03	1.14	2.99	—	−37.8	−35.5	−31.8	−30.0	−30.1	−12.0
	TZ721	5 355.50 – 5 505.00	0.989 5	94.7	0.65	0.14	0.07	0.03	0.97	3.31	9.9	—	—	—	—	—	—
	TZ721	5 355.50 – 5 505.00	0.988 5	93.6	0.61	0.18	0.09	0.05	0.86	4.44	95	—	—	—	—	—	—
	TZ721	5 355.50 – 5 505.00	0.989 4	93.7	0.61	0.15	0.08	0.04	0.83	4.48	100	—	—	—	—	—	—
	TZ721	5 355.50 – 5 505.00	0.993 1	95.0	0.62	0.02	0.01	0.00	1.11	3.28	—	—	—	—	—	—	—
	TZ721	5 355.50 – 5 505.00	0.988 1	93.2	0.63	0.16	0.10	0.05	0.80	4.88	32	—	—	—	—	—	—
	TZ721	5 355.50 – 5 505.00	0.992 7	94.5	0.61	0.03	0.02	0.01	0.84	3.93	26	—	—	—	—	—	—
	TZ721	5 355.50 – 5 505.00	0.993 2	94.8	0.61	0.02	0.01	0.00	1.40	3.19	—	—	—	—	—	—	—
	TZ721	5 355.50 – 5 505.00	0.989 7	93.9	0.61	0.13	0.06	0.04	0.74	4.42	8.9	—	—	—	—	—	—
	TZ726	5 355.50 – 5 505.00	0.976 2	92.3	1.76	0.22	0.07	0.08	0.90	4.56	—	—	—	—	—	—	—
Oil	TZ722	5 356.70 – 5 750.00	0.927 9	88.1	3.50	1.28	0.73	0.42	2.45	2.58	4 100	—	—	—	—	—	—
	TZ722	5 356.70 – 5 750.00	0.870 7	80.9	4.03	1.83	1.71	0.83	2.33	4.78	—	—	—	—	—	—	—
	TZ722	5 356.70 – 5 750.00	0.912 4	81.4	3.18	1.58	1.06	0.58	2.50	8.24	2 900	—	—	—	—	—	—
	TZ72	5 125.00 – 5 130.00	0.927 8	88.2	3.16	1.23	0.71	0.39	1.30	3.68	—	—	—	—	—	—	—
	TZ72	5 125.00 – 5 130.00	0.938 9	90.0	3.18	1.04	0.54	0.30	0.93	3.26	750	—	—	—	—	—	—

Table 4 PVT data of Ordovician reservoir fluids in the Well TZ83 area.

Well	TZ83	TZ721
Interval/m	5 666.10 − 5 684.70	5 355.50 − 5 505.00
Daily oil production/m^3	14.20	12.96
Daily gas production/m^3	308 713	254 152
GOR ratio/($m^3 \cdot m^{-3}$)	21 740	19 610
Formation pressure/MPa	61.67	64.46
Dew point pressure/MPa	61.67	56.16
Dew differential pressure/MPa	0	9.30
Critical condensate pressure/MPa	63.44	59.55
Dew point volume factor/($10^{-3}\ m^3 \cdot m^{-3}$)	3.052 1	3.211 5
Dew point pressure deviation factor	1.302	1.253
Formation temperature/ ℃	145.1	143.3
Critical pressure/MPa	29.65	39.36
Critical temperature/℃	−122.1	−56.4
Critical condensate temperature/ ℃	387.3	341.3
Condensate oil content /($g \cdot m^{-3}$)	40.57	46.2
Oil density of oil tank (20 ℃)/($g \cdot cm^{-3}$)	0.812 4	0.821 6
Conclusion	condensate reservoir	condensate reservoir

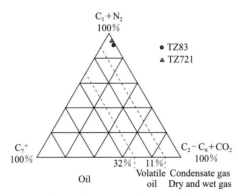

Fig. 4 Ternary geochemical diagram showing the reservoir fluid type.

In contrast, samples from Well TZ721 are in a single gas phase under formation conditions. The fluid produced from this well is composed of 94.74% C_1+N_2, 4.75% $C_2-C_6+CO_2$, and 0.51% C_7^+, representing condensate on the triangular phase diagram. As can be seen on the phase diagram (Fig. 5), the formation temperature is on the right side of the critical temperature. With a condensate oil content of 46.2 g/m^3, a large amount of retrograde condensed liquid (12.51%), and a dew pressure differential of 9.30, the reservoir fluid is condensate (Fig. 5).

In summary, both wells above adequately demonstrate the characteristics of condensate reservoirs in the Well TZ83 area. However, Well TZ83 has a higher GOR ratio and lower condensate oil content than Well TZ721. This may attribute to the presence of the two groups of strike-slip faults to the northwest of the Well TZ83 area, which act as the main gas migration paths (Fig. 2).

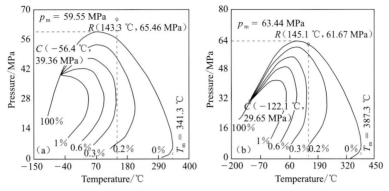

Fig. 5 State of reservoir fluid phase. Well TZ83 (left) and Well TZ721 (right).

4.2 GC×GC-TOFMS geochemistry of condensate

Analytical results of GC×GC-TOFMS on condensate in this study show that a large number (4 778) of compounds at a signal-to-noise ratio of more than 100 were detected, reflecting the advantage of this method in the analysis of condensate. Abundant adamantane series compounds and some diamantane compounds, as well as a series of sulphur-containing compounds (including the benzothiophene series, and the dibenzothiophene series) were detected for the first time in the study area.

4.2.1 Oil compound composition

Based on the information from mass spectrometry provided by TOFMS, and the comprehensive two-dimensional chromatogram characteristics, these compounds were divided into ten groups using Chroma TOF software. The same temperature program was used for the repetitive analysis of oil samples by GC×GC-FID (chromatogram shown in Fig. 6), and the compounds were classified based on the GC×GC-TOFMS classification results. After peak area normalization, groups of quantitative results were obtained for the condensate samples from Well TZ83 (Fig. 6).

The levels of saturated hydrocarbons, aromatic hydrocarbons, and non-hydrocarbons in condensate samples from Well TZ83 were 92.69%, 7.18% and 0.12%, respectively. The alkane (including n-alkane and iso-alkane) content was high, accounting for 72.79% of the total crude oil mass. The detection range was n-C_6 to n-C_{33} with the main carbon peak at n-C_{11}. The naphthene (including monocyclic alkane and polycyclic alkane) content was the next highest, accounting for 19.91% of the total mass. Long-chain alkyl-substituted cyclohexane and cyclopentane series compounds were present in the single naphthene at relatively high levels, accounting for 1.5% and 1.2% of the total mass, respectively. The multi-ring alkanes contain higher levels of adamantane series compounds, accounting for 0.2% of the total mass, which is clearly visible on the FID chromatogram (Fig. 6a). Among the aromatics, the monocyclic aromatic content was the highest, accounting for 5.26% of the total mass. The contents of benzene, toluene, dimethyl benzene series, and trimethyl benzene series compounds were the highest among the monocyclic aromatics; they are 0.10%, 0.45%, 0.98% and 1.00% of the total mass, respectively.

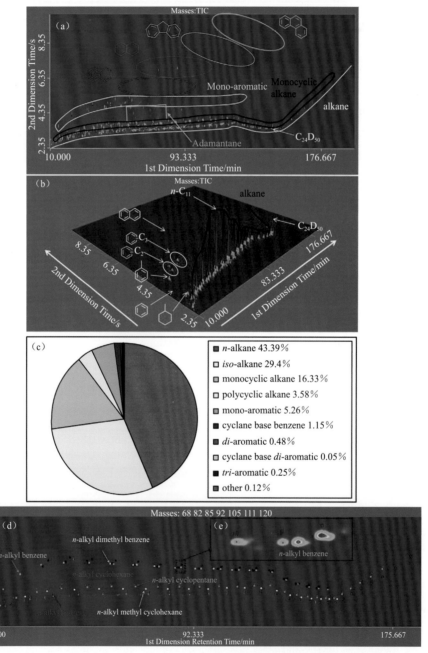

Fig. 6 GC×GC-FID chromatogram of a condensate samples from Well TZ83.
(a) GC×GC-FID colour contour chromatogram, (b) GC×GC-FID 3D plot, (c) quantitative results in groups of compound, (d) colour contour chromatograms of extracted ions of m/z 68, 82, 85, 92, 105, 111 and 120, each spot stands for a compound, different colour spots stand for different groups of long side-chain substituted compounds, (e) local enlargement of figure d. (For interpretation of the references to colour in this figure legend, the reader is referred to the web version of this article.)

A series of long side-chain substituted compounds were present in the TZ83 samples (Fig. 6d), including long side-chain substituted cyclohexane, cyclopentane, methyl cyclohexane, benzene, toluene, and xylene. The detectable compound carbons were C_{33}, C_{26}, C_{24}, and C_{18}. There were abundant long side-chain substituted monocyclic aromatic compounds, of which the long side-chain substituted toluene compounds had three isomers: separate, pair, and adjacent (m, p, o) (Fig. 6e).

4.2.2 Adamantane compounds

A total of 23 adamantane series compounds have been identified, but more abundant adamantane series compounds have been found in condensate samples from Well TZ83 (Fig. 7). On the chromatogram of GC×GC-TOFMS, adamantane series compounds are divided into six groups according to their ion characteristics (as shown in Fig. 7).

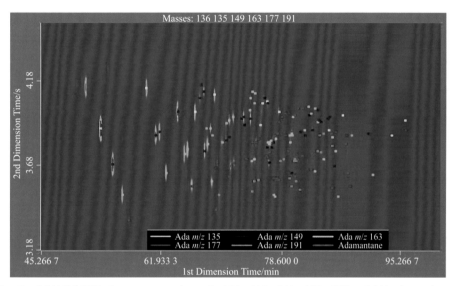

Fig. 7 GC×GC-FID chromatogram for m/z 136, 135, 149, 163, 177 and 191 of acondensate samples from Well TZ83. The distributions of adamantane series compounds are marked.

(1) Compounds with characteristic ions of m/z 135, indicated by yellow (in the web version) dots in Fig. 7:12 compounds of this type were detected, and we conclude from the spectral analysis these contain one substituent group, and the carbon number of the substituent group is from C_1 to C_5.

(2) Compounds with characteristic ions of m/z 149, indicated by brown (in the web version) spots in Fig. 7:32 compounds of this type were detected, and we conclude from the spectral analysis that these contain two substituent groups.

(3) Compounds with characteristic ions of m/z 163, indicated by blue (in the web version) dots in Fig. 7:44 compounds of this type were detected, and we conclude from the spectral analysis that these contain three substituent groups.

(4) Compounds with characteristic ions of m/z 177, indicated by red (in the web version) dots in Fig. 7:51 compounds of this type were detected, and we infer from the spectral analysis that these contain four substituent groups.

(5) Compounds with characteristic ions of m/z 191, indicated by orange (in the web version) dots in Fig. 7. 18 compounds of this type were detected, and from the spectral analysis we conclude that these contain five substituent groups.

(6) Compounds with characteristic ions of m/z 136, indicated by purple (in the web version) dots in Fig. 7. Different substituents of the adamantane series compounds are clearly shown on Fig. 8, together with their ion distributions. The compounds with serial numbers are those identified in the diagram database. The quantitative results were obtained using GC×GC-FID analysis and D_{16}-adamantane as the internal standard compound (Table 5).

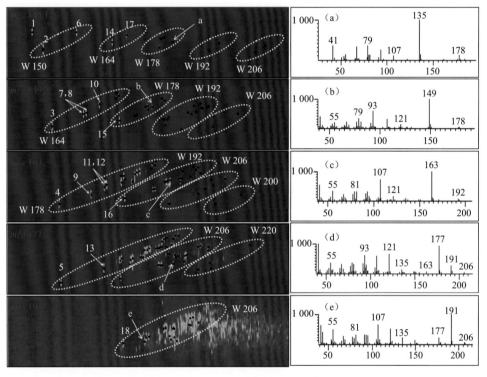

Fig. 8   GC×GC-FID chromatogram for m/z 136,135,149,163,177 and 191 of a condensate sample from Well TZ83. Different substituent adamantane series compound distributions are marked. W represents the molecular weight of compounds. The same series of compounds with the same characteristic ion are arranged with the molecular weight difference of one CH_2.

The labelled compounds (a – e) are on the right-hand side of the mass spectra column.

Table 5   **Quantitative results for diamondoid compounds in condensate samples from Well TZ83.**

No.	Compound	Concentration/(mg·kg^{-1})
1	Adamantane	553.2
2	1-Methyladamantane	1 456
3	1,3-Dimethyladamantane	1 149.9
4	1,3,5-Trimethyladamantane	612.9

continue

No.	Compound	Concentration/(mg·kg^{-1})
5	1,3,5,7-Tetramethyladamantane	1 901.2
6	2-Methyladamantane	395.4
7	1,4-Dimethyladamantane (cis)	620.4
8	1,4-Dimethyladamantane (trans)	458.7
9	1,3,6-Trimethyladamantane	383.1
10	1,2-Dimethyladamantane	706.2
11	1,3,4-Trimethyladamantane (cis)	491.2
12	1,3,4-Trimethyladamantane (trans)	539.6
13	1,2,5,7-Tetramethyladamantane	483.8
14	1-Ethyladamantane	179.8
15	1-Ethyl-3-methyladamantane	278.7
16	1-Ethyl-3,5-dimethyladamantane	214.1
17	2-Ethyladamantane	91.5
18	1,2,3,5,7-Pentamethyladamantane	6.4
19	Diamantane	126.5
20	4-Methyldiamantane	101.3
21	4,9-Dimethyldiamantane	23.4
22	1-Methyldiamantane	67.6
23	1,4＋2,4-Dimethyldiamantane	31.9
24	4,8-Dimethyldiamantane	42.4
25	1,4,9-Trimethyldiamantane	43.7
26	3-Methyldiamantane	55.2
27	3,4-Dimethyldiamantane	47.6

In addition, diamantane compounds were detected in the samples (Fig. 9). Nine compounds, identified in the diagram database, are marked with serial numbers in the figure. These quantitative data were obtained using GC×GC-FID analysis (Table 4). In addition, five new compounds were found. Their mass spectra are similar to those of the diamantane compounds. The mass spectra of two of the compounds are given in Fig. 9b and c. The characteristic ion of this compound is m/z 201, and the main fragment ions are m/z 55, 79, 105, 131, and 159 (Fig. 9b), consistent with dimethyl substituted of adamantane. In addition, the retention times of the compound on the GC×GC chromatogram are also consistent with dimethyl substituted of adamantane, indicating their similarity in polarity. Thus, the three compounds marked with pink (in the web version) circles in Fig. 9a are inferred to be dimethyl substituted of diamantane, but the substitution position is unknown. Similarly, we speculate that the two compounds marked with yellow circles relate to the trimethyl substituted of diamantane, but the substitution position is unknown.

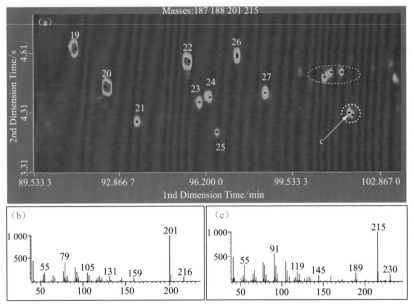

Fig. 9 (a) GC×GC-FID chromatogram for m/z 187, 188, 201 and 215 of a condensate sample from Well TZ83. The distributions of different substituent diamantane compounds are marked. Mass spectra of two compounds are shown in figures (b) and (c).

4.2.3 Sulphur compounds

A series of sulphur-containing compounds were detected in the condensate samples from Well TZ83 (Fig. 10), including the alkyl benzothiophene series (m/z 147, 161, 175) and the dibenzothiophene series (m/z 184, 198, 212). Specifically, they are mainly alkyl benzothiophene series compounds with $C_1 - C_3$ substituent group and alkyl dibenzothiophene series compounds with $C_1 - C_2$ substituent group.

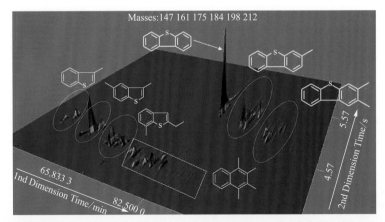

Fig. 10 GC×GC-FID chromatogram for m/z 147, 161, 175, 184, 198, and 212 of condensate samples from Well TZ83. The distributions of different structural series of sulphur compounds are marked.

As shown in Fig. 11, the boiling point of alkyl benzothiophene series compounds is similar to those of benzene, alkyl substituted benzene, and naphthalene series compounds. On conventional one-dimensional chromatography, they often overlap and are difficult to distinguish even using mass spectrometry. Because the alkyl benzene and thiophene ions and fragment ions have the same characteristics as some monocyclic aromatic hydrocarbons, peak fractions are likely to form under the condition of option ions caused by monocyclic aromatics. Thus, in conventional one-dimensional chromatography, small amounts of alkyl benzothiophene series compounds cannot be detected. Even though large amounts can be detected, they give a common fraction peak with an inaccurate peak area

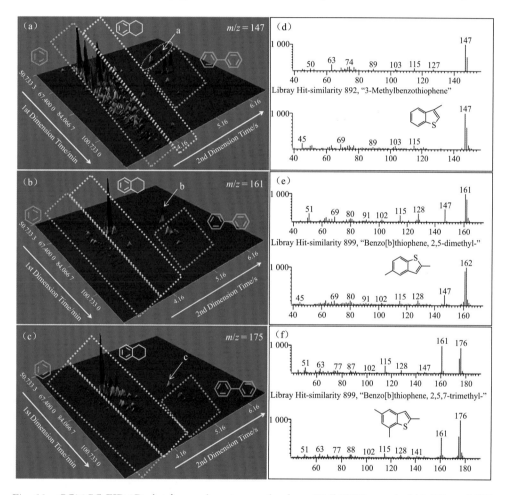

Fig. 11 GC×GC-FID 3D plot for condensate samples from Well TZ83 at m/z 147, 161, and 175. (a) m/z 147 ion; (b) m/z 161 ion; (c) m/z 175. Pink areas are the peak positions of benzene series compounds, yellow areas are the peak positions of naphthene substitute benzene series compounds, and green areas are the peak positions of di-aromatic series compounds. The mass spectra for the compounds marked a, b, and c are shown on the right-hand side of the figure (parts (d), (e), and (f)). (For interpretation of the references to colour in this figure legend, the reader is referred to the web version of this article.)

integral result. GC×GC can therefore be used to separate compounds of different polarity. GC×GC can distinguish alkyl benzothiophene from the naphthalene series and the monocyclic aromatic series because of their polarity differences, which are not affected by the interference of these compounds, resulting in more accurate quantitative results.

Similarly, dibenzothiophene compounds are likely to be affected by a common distillation peak in conventional one-dimensional chromatography, so detection using GC×GC-TOFMS is preferable (Fig. 12). The mass spectra of compounds a, b and c in Fig. 11 (a)–(c) are shown in Fig. 11(d)–(f). The upper part of Fig. 11(d) shows the mass spectrum of a compound detected by TOFMS; while the lower part is the mass spectrum of standard compound in the NIST spectrum library. The spectra of the upper and lower parts have 89.2% similarity. Comparing the upper and lower mass spectra in Fig. 11(d)–(f), the mass spectrum of the collected compounds is almost exactly the same as the library spectrum, suggesting that these compounds are structurally analogous with a similarity of nearly 90.0%, and the compounds (a)–(c) are methyl, dimethyl, and trimethyl substituted of alkyl benzothiophene. However, the substitution position is not certain.

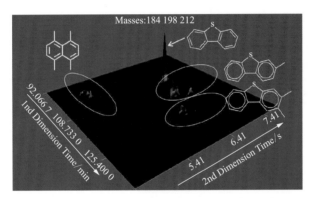

Fig. 12   GC×GC-FID 3D plot for condensate samples from Well TZ83 at m/z 184, 198, and 212. The distribution of dibenzothiophene series compounds is marked.

5   Discussion

5.1   Origin of oil, gas and condensate in the Well TZ83 area

The Well TZ83 area contains condensate reservoirs and oil reservoirs side by side, with complex fluid property. TZ83, TZ83-1, TZ721, TZ726 and other wells are located in the condensate reservoir area, and TZ722 is located in the oil reservoir area (Fig. 2). In terms of physical properties, the condensate and oil reservoirs are markedly different (Tables 1 and 2), implying a complex hydrocarbon accumulation process and alteration process with possibly multiphase hydrocarbon charging and accumulation.

5.1.1   Crude oil

Regarding the crude oil composition, the density of the condensate is significantly higher than that of the normal oil reservoir, which is mainly caused by the high wax content (more than 20%) in the condensate (Table 1). High wax content in marine oil is mainly the result of gas fractionation caused by natural gas charging, during which the gas carries the light components of deeper-buried oil to shallower layers, leaving the heavier components in residue, which results in the high wax component content in condensate. This phenomenon suggests that gas charging and secondary reservoir alteration occurred in this area.

5.1.2 Natural gas

In terms of the composition of natural gas, it is characterized by dry gas in the condensate reservoirs, with a dryness ratio (C_1/C_1^+) of more than 0.96 (Table 3). In contrast, in oil reservoirs of the Well TZ722 area, the natural gas is wet with a dryness ratio of less than 0.94 and an ethane content of more than 3% (Table 3). The gas production rate is high in the condensate reservoirs with a GOR ratio up to 44 897 m^3/m^3, but low in the oil reservoirs with a GOR ratio less than 200 m^3/m^3. In terms of carbon isotope composition, in the condensate field, $\delta^{13}C_1$ and $\delta^{13}C_2$ of natural gas are −38‰ to −39‰, and −32‰ to −36‰, respectively, suggesting a characteristic of highly mature oil-cracked gas in nature (Fig. 13).

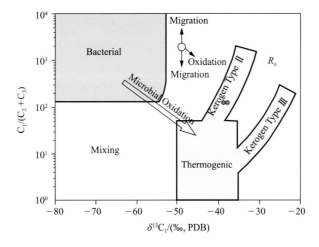

Fig. 13 Plot of $\delta^{13}C_1$ vs. C_1/C_{2+3} for natural gases from the Well TZ83 area

5.2 Generalized formation process of the condensate reservoir

Modelling of stratigraphic and burial evolution indicates that the Tarim basin achieved its maximum burial depth during the Quaternary Period, and the geothermal gradient has been gradually decreasing since the Permian Period (Fig. 14). Particularly in the Tarim Basin, low geothermal gradient and short heating time favours for liquid oil preservation in deep strata. The burial history illustrates that the basin has reached its maximum burial depth within the last 5 Myr (Fig. 14). Recent rapid burial requires a high temperature for crude oil cracking due to time-temperature compensation effect. Actual reservoir temperature in the Well TZ83 area is 139.5 ℃ (at 5 500 m depth), with a temperature gradient of 2.02 ℃/100 m. It is thought that crude oil begins to crack into gas at a temperature of more than 160 ℃ or at a burial depth of more than 6 000 m, after which liquid oil gradually disappears. Thus the Ordovician reservoir does not reach pyrolysis conditions. As a consequence, in-situ cracking of oil to gas (secondary cracking product) in the Ordovician reservoir is not possible, and thus the oil-cracked gas here must have come from depth and migrated into the upper Ordovician reservoir.

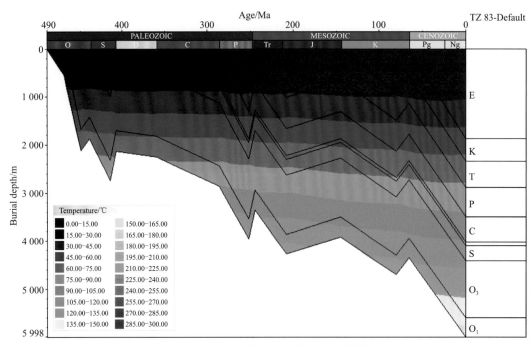

Fig. 14　Burial and thermal history of TZ83.

Two types of reservoir (oil and condensate) are present side by side in the study area. As they have entirely different gas content, it is inferred that their gas charging intensity in late stages was different. In the condensate reservoir, there is an apparent later-stage gas charge. Strong late-stage natural gas charging can alter an early paleo-oil reservoir to form a condensate reservoir. This has been proved in a number of basins, and also verified by experiments. It may be the formation mechanism of the condensate in this study. Marine high-wax oil in the condensate reservoir is mainly the result of gas fractionation caused by natural gas charging into oil reservoirs, leading to high wax content and the heavier oil density. In contrast, in the oil reservoir, the later-stage gas charge is weak, likely due to unfavourable conditions of migration conduits and reservoir properties. Under these circumstances, reservoirs would have undergone little change in the oil properties and remained as oil reservoirs.

This different intensity of late gas charging depends on many factors, of which the fracture conduit system is very important. As outlined above, the late Permian Period natural gas in the Tazhong area originated from gas cracking in deeper-buried oil reservoirs. Strike-slip faults were migration channels for the later-stage gas, and played a decisive role in the formation of condensate reservoirs (Fig. 15). The reservoirs and traps adjacent to these faults were likely to have received late gas charge and further altered to form condensate reservoirs (Well B). In contrast, reservoirs far away from the strike-slip faults would preserve their oil (Well A), because they were located at the end of the gas charge where the charging energy was reduced and so were little affected by gas invasion (Fig. 15).

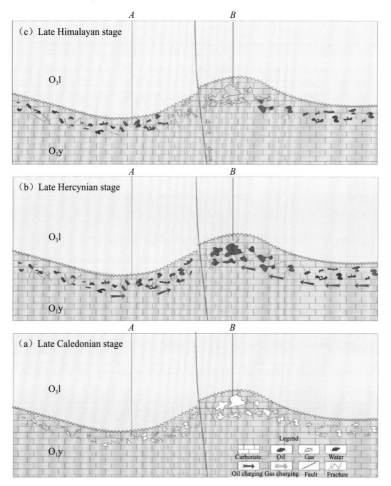

Fig. 15 Oil and gas charging and accumulation model for Ordovician condensate reservoirs in the Tazhong area.

In summary, the Ordovician oil and gas here mainly came from the Cambrian source rocks at the bottom of the sedimentary sequence. Oil and gas could migrate and accumulate in Ordovician reservoirs only via faults cutting through the hydrocarbon source layers. In places far from the faults, oil and gas charging was weak and thus the reservoir fluid is mainly water.

Two N－E trending secondary faults, which are the main faults for hydrocarbon migration, are developed in the Well TZ83 area. During the Himalayan period, natural gas passed through these two faults to reach Ordovician karst reservoirs, causing gas invasion from northwest to southeast. Due to the heterogeneity of the carbonate rocks, the reservoir between TZ726 and TZ722 has poor connectivity, so the area to the east of TZ722 was little affected by gas invasion (Fig. 2). This explains why the area around Wells TZ83 to TZ721 is a gas province, while the Well TZ722 area is an oil province. The GOR ratio decreases from northwest to southeast (21 213 m^3/m^3 in TZ83, 18 698 m^3/m^3 in TZ721, and 180 m^3/m^3 in TZ722). These data show that the late-stage gas invasion occurred along the faults, resulting in a high degree of gas charging in places near the faults and a low degree of gas charging in places distant from the faults (Fig. 15).

5.3 Formation mechanism of adamantane and sulphur compounds

Crude oil may crack at high temperatures, and, during the process of cracking, diamondoid compounds can be continuously generated and accumulated. The Cambrian dolomite reservoir under the salt layers in the study area, which are buried at depths of 8 500 –11 000 m with reservoir temperatures of 180 – 220 ℃, must have undergone oil cracking, forming large amounts of diamondoid compounds.

Sulphur compounds, including benzothiophene series and dibenzothiophene series compounds, are probably related to the thermochemical sulphate reduction (TSR), which may also widespread in the Cambrian reservoirs. Because the cap rock that overlies the Cambrian dolomite reservoir rock consists of gypsum (Fig. 3), the formation water is rich in SO_4^{2-}. The TSR reaction converts the SO_4^{2-} in the formation water into H_2S and organic sulphur compounds at high temperature, resulting in an increase in sulphur-containing compounds such as benzothiophene and the dibenzothiophene series in crude oil. All condensate reservoirs in the study area contain a certain amount of hydrogen sulfide, which has been confirmed to be the result of TSR.

In summary, the abundant diamondoid and sulphur compounds found in the condensate in the Well TZ83 area are related to deep-burial natural gas charging. Which is likely to be the result of oil-cracking, which migrated along deep-rooted faults to the shallower-buried Ordovician reservoirs and mixed with and altered the in-situ oils, resulting in the present-day complex oil-gas-distributed reservoirs. The study area therefore shows promising prospects for exploring deep-burial oil-cracking gas.

6 Conclusions

By using the new high-resolution comprehensive two-dimensional gas chromatography with time-of-flight mass spectrometry (GC×GC-TOFMS), we identified for the first time abundant adamantane series compounds, as well as some diamantane series compounds and a series of sulphur-containing compounds (e.g., the benzothiophene series, and the dibenzothiophene series) in the Ordovician marine condensate in the Tarim Basin.

Because the temperature of the crude oil reservoir is not high enough for pyrolysis, the density of the gas condensate is markedly heavier than that of normal oil, with a high wax content. The condensate gas reservoirs mainly contain dry gas, and carbon isotope values are obviously heavier, suggesting that the gas is high-maturity cracked gas, and so the cracked gas in the reservoirs must have come from deeper reservoir cracked gas, which must have come from deeper-buried Cambrian reservoirs. Marine high wax oil is mainly the result of gas fractionation caused by natural gas charging into oil reservoirs, leading to high wax content and an increasing oil density.

The Cambrian dolomite reservoir in the Tazhong area at burial depths of 8 500 − 11 000 m with reservoir temperatures of 180 − 220 ℃ experienced oil cracking, producing an abundance of diamondoid compounds. The formation of sulphur-containing compounds may be related to sulphate thermal chemical reduction (TSR), which converted SO_4^{2-} in the formation water into H_2S and organic sulphur compounds at high temperature, resulting in the enrichment of sulphur in benzothiophene series and dibenzothiophene series. The hydrocarbons rich in diamondoid and sulphur compounds migrated along deep faults to shallower-buried Ordovician reservoirs, resulting in reservoirs with complex oil and gas occurrence. The Tazhong area therefore has potential for the exploration of oil-cracking gas.

专利 1　专利号:ZL 2012 1 0256743.6,授权公告日:2014 年 06 月 04 日

用全二维气相色谱定量石油样品中金刚烷类化合物的方法

说明书摘要

本发明涉及一种用全二维气相色谱定量石油样品中金刚烷类化合物的方法。该方法包括以下步骤:(1) 制备得到待测样品;(2) 利用全二维气相色谱-飞行时间质谱分析待测样品,得到待测样品中的金刚烷类化合物的出峰信息;(3) 利用全二维气相色谱-氢火焰离子化检测器分析待测样品,得到待测样品的全二维气相色谱-氢火焰离子化检测图谱;(4) 根据步骤(2)得到的金刚烷类化合物的出峰信息,确定金刚烷类化合物和 D_{16}-金刚烷在全二维气相色谱-氢火焰离子化检测图谱上的出峰位置,得到它们的峰面积积分结果;(5) 采用内标法计算并得到金刚烷类化合物的定量结果。该方法适用于所有原油和岩石抽提物样品中的金刚烷类化合物的定量。

权利要求书

1. 一种石油样品中金刚烷类化合物的定量分析方法,其包括以下步骤:

(1) 制备待测样品。

当待测样品为凝析油样品时,取适量凝析油样品于 1.5 mL 自动进样瓶中,加入配制好的 D_{16}-金刚烷标准样品 300 μL,得到待测样品。

当待测样品为其他原油或者岩石抽提物样品时,将 1.8 g 细硅胶在振荡下装入玻璃柱中,取 30 mg 其他原油或岩石抽提物样品,用适量正己烷溶解,加入配制好的 D_{16}-金刚烷标准样品 300 μL,然后全部转入玻璃柱中,用 3 mL 正己烷淋洗,收集洗脱馏分,浓缩至 1.5 mL 自动进样瓶中,得到待测样品。

(2) 全二维气相色谱分析。

① 利用全二维气相色谱-飞行时间质谱分析步骤(1)得到的待测样品,得到待测样品中的金刚烷类化合物的出峰信息,包括金刚烷类化合物的保留时间和定性信息。

全二维气相色谱-飞行时间质谱的分析条件分为全二维气相色谱条件和飞行时间质谱条件,其中:

a. 全二维气相色谱条件为:

一维色谱柱为 DB-1MS 柱,升温程序为 50 ℃保持 0.2 min,以 2 ℃/min 的速率升到 180 ℃保持 0.2 min,再以 8 ℃/min 的速率升到 300 ℃保持 10 min。

二维色谱柱是 DB-17HT 柱,采用与一维色谱相同的升温速率,起始温度和终止温度比一维色谱高 5 ℃;气相色谱进样口温度为 300 ℃,以氢气为载气,流速为 1.5 mL/min;对于采用凝析油样品制备得到的待测样品采用分流进样模式,分流比为 700∶1,进样量为 0.5 μL;对于采用其他原油或岩石抽提物样品制备得到的待测样品采用分流进样模式,分流比为 20∶1,进样量为 1 μL。

调制器采用与一维色谱相同的升温速率,起始温度和终止温度比一维色谱高 40 ℃,调制周期为 10 s,其中热吹时间为 2.5 s。

b. 飞行时间质谱条件为:

质谱传输线和离子源温度分别为 280 ℃和 240 ℃,质谱检测器电压为 1 475 V,质量扫描范围为 40～520 amu,采集速率为 100 谱图/s;对于采用凝析油样品制备得到的待测样品,溶剂延迟时间为 0 min;对于采用其他原油或岩石抽提物样品制备得到的待测样品,溶剂延迟时间为 10 min。

② 利用全二维气相色谱-氢火焰离子化检测器分析步骤(1)制备的待测样品,得到相应的全二维气相

色谱-氢火焰离子化检测谱图。

分析条件为:采用与全二维气相色谱-飞行时间质谱分析相同的全二维气相色谱条件;所述氢火焰离子化检测器中的载气、氢气、空气的流速分别为 50 mL/min,40 mL/min,450 mL/min;氢火焰离子化检测器温度为 320 ℃,采集频率为 200 谱图/s;对于采用凝析油样品制备得到的待测样品,溶剂延迟时间为 0 min;对于采用其他原油或岩石抽提物样品制备得到的待测样品,溶剂延迟时间为 10 min。

(3) 数据处理。

根据步骤(2)得到的金刚烷类化合物的出峰信息,确定金刚烷类化合物和 D_{16}-金刚烷在全二维气相色谱-氢火焰离子化检测谱图上的出峰位置,得到它们的峰面积积分结果。

采用内标法计算并得到金刚烷类化合物的定量结果。

2. 根据权利要求 1 所述的方法,其中,所述 DB-1MS 柱的尺寸为 30 m×0.25 mm×0.25 μm。

3. 根据权利要求 1 所述的方法,其中,所述 DB-17HT 柱的尺寸为 3 m×0.1 mm×0.1 μm。

4. 根据权利要求 1 所述的方法,其中,所述细硅胶的粒径为 100～200 目,并且,所述细硅胶在 200 ℃下进行 4 h 的活化处理。

说明书

用全二维气相色谱定量石油样品中金刚烷类化合物的方法

技术领域

本发明涉及一种用全二维气相色谱定量分析石油样品中金刚烷类化合物的方法,属于石油样品分析技术领域。

背景技术

金刚烷类化合物是原油中一种特殊的环状烃类化合物,具有与金刚石类似的碳骨架结构,其稳定特性决定了它在地质演变过程中具有很强的抗热和抗生物降解能力。在石油成熟演变过程中,金刚烷类化合物始终存在,随着成熟度的增加,它们的相对含量会不断增加。在高成熟原油和凝析油中,甾烷、萜烷等常用生物标志化合物缺失,金刚烷化合物就成为判定成熟度的重要指标。除此之外,金刚烷类化合物还可以用来判别原油裂解程度。Dahl 等(Dahl J E,Moldowan J M,Peters K E,et al. Diamondoid hydrocarbons as indicators of natural oil cracking. Nature,1999,399(6):54-57)研究发现在热模拟和自然地质条件下,原油中的金刚烷既不产生也不裂解,随着其他组分裂解程度的提高,金刚烷的浓度不断增加,因此可以根据金刚烷浓度的变化计算原油的裂解生气量。

金刚烷类化合物在石油地质样品中的含量较低,受共馏峰的干扰和提纯条件的限制无法对其进行色谱定量分析。目前都是采用气相色谱-质谱(GC-MS)方法进行定量分析。用 GC-MS 分析金刚烷类化合物时,需要较多的不同结构的金刚烷标样。在国外,Wei 等(Wei Z B, Molodowan J M, Zhang S C, et al. Diamondoid hydrocarbons as a molecular proxy for thermal maturity and oil cracking: Geochemical models from hydrous pyrolysis. Org Geochem,2007,38:227-249)分别使用 6 种氘代金刚烷化合物对不同结构的金刚烷类化合物定量,得到的结果较为准确。在国内,由于标样的缺乏,只能用一个氘代金刚烷化合物定量所有结构的金刚烷类化合物,得到的结果与实际偏差较大。

全二维气相色谱(GC×GC)是 20 世纪 90 年代发展起来的一种分离复杂混合物的全新手段,它的正交分离系统能够使金刚烷类化合物得到较好的分离。用全二维气相色谱-飞行时间质谱(GC×GC-TOFMS)分析石油样品中金刚烷类化合物的方法已见报道(王汇彤,翁娜,张水昌,等. 全二维气相色谱-飞行时间质谱对饱和烃分析的图谱识别及特征. 质谱学报,2010,31(1):18-27)。但该报道只公开了金刚烷类化合物的定性信息,未涉及此类化合物的定量分析。且报道中所使用的色谱柱成本高,分析方法时间长,分析一个样品至少需要 2 h 以上,在质谱采集期间需要不停灌充液氮,既浪费时间又增加分析成本。

综上所述,国内外至今还没有一种令人满意的金刚烷类化合物的定量分析方法。

发明内容

为解决上述技术问题,本发明的目的是提供一种定量分析石油样品中金刚烷类化合物的方法。该方法是利用全二维气相色谱的分离特点建立的一种有效的定量分析石油样品中金刚烷类化合物,能够填补色谱方法对金刚烷类化合物定量的空白,使得到的金刚烷类化合物定量结果更加真实可靠。

为达到上述目的,本发明提供了一种石油样品中金刚烷类化合物的定量分析方法,其包括以下步骤:

(1) 制备待测样品(样品前处理)。

当待测样品为凝析油样品时,取适量凝析油样品于 1.5 mL 自动进样瓶中,加入配制好的 D_{16}-金刚烷标准样品 300 μL(浓度为 0.063 2 mg/mL,溶剂为 CH_2Cl_2)得到待测样品。

当待测样品为其他原油或者岩石抽提物样品时,将 1.8 g 细硅胶在振荡下装入玻璃柱中,取 30 mg 其他原油或岩石抽提物样品,用适量正己烷(重蒸过的分析纯)溶解,加入配制好的 D_{16}-金刚烷标准样品 300 μL(浓度为 0.063 2 mg/mL,溶剂为 CH_2Cl_2),然后全部转入玻璃柱中,用 3 mL 正己烷淋洗,收集洗脱馏分,浓缩至 1.5 mL 自动进样瓶中,得到待测样品。

(2) 全二维气相色谱分析。

① 利用全二维气相色谱-飞行时间质谱(GC×GC-TOFMS)分析步骤(1)得到的待测样品,得到待测样品中的金刚烷类化合物的出峰信息,包括金刚烷类化合物的保留时间和定性信息。

全二维气相色谱-飞行时间质谱的分析条件分为全二维气相色谱条件和飞行时间质谱条件,其中:

a. 全二维气相色谱条件为:

一维色谱柱为 DB-1MS 柱,升温程序为 50 ℃保持 0.2 min,以 2 ℃/min 的速率升到 180 ℃保持 0.2 min,再以 8 ℃/min 的速率升到 300 ℃保持 10 min;二维色谱柱是 DB-17HT 柱,采用与一维色谱相同的升温速率,起始温度和终止温度比一维色谱高 5 ℃(即升温程序为 55 ℃保持 0.2 min,以 2 ℃/min 的速率升到 185 ℃保持 0.2 min,再以 8 ℃/min 的速率升到 305 ℃保持 10 min);气相色谱进样口温度为 300 ℃,以氢气为载气,流速为 1.5 mL/min;对于采用凝析油样品制备得到的待测样品采用分流进样模式,分流比为 700∶1,进样量为 0.5 μL;对于采用其他原油或岩石抽提物样品制备得到的待测样品采用分流进样模式,分流比为 20∶1,进样量为 1 μL。

调制器采用与一维色谱相同的升温速率,起始温度和终止温度比一维色谱高 40 ℃(即升温程序为 90 ℃保持 0.2 min,以 2 ℃/min 的速率升到 220 ℃保持 0.2 min,再以 8 ℃/min 的速率升到 340 ℃保持 10 min),调制周期为 10 s,其中热吹时间为 2.5 s。

b. 飞行时间质谱条件为:

质谱传输线和离子源温度分别为 280 ℃和 240 ℃,质谱检测器电压为 1 475 V,质量扫描范围为 40～520 amu,采集速率为 100 谱图/s;对于采用凝析油样品制备得到的待测样品,溶剂延迟时间为 0 min;对于采用其他原油或岩石抽提物样品制备得到的待测样品,溶剂延迟时间为 10 min。

② 利用全二维气相色谱-氢火焰离子化检测器(GC×GC-FID)分析步骤(1)制备的待测样品,得到相应的全二维气相色谱-氢火焰离子化检测(GC×GC-FID)谱图。

分析条件为:采用与全二维气相色谱-飞行时间质谱分析相同的全二维气相色谱条件;所述氢火焰离子化检测器中的载气(氮气)、氢气、空气的流速分别为 50 mL/min,40 mL/min,450 mL/min;氢火焰离子化检测器温度为 320 ℃,采集频率为 200 谱图/s;对于采用凝析油样品制备得到的待测样品,溶剂延迟时间为 0 min;对于采用其他原油或岩石抽提物样品制备得到的待测样品,溶剂延迟时间为 10 min。

(3) 数据处理。

根据步骤(2)得到的金刚烷类化合物的出峰信息,确定金刚烷类化合物和 D_{16}-金刚烷在全二维气相色谱-氢火焰离子化检测(GC×GC-FID)谱图上的出峰位置(确定出峰位置就是对照化合物的相对保留时间,根据全二维气相色谱-飞行时间质谱上化合物的保留时间来确定 FID 检测器上化合物的出峰时间),得到它们的峰面积积分结果。

采用内标法计算并得到金刚烷类化合物的定量结果。

在本发明提供的上述方法中,对于数据、谱图等的处理可以按照本领域的常规方式进行。

在本发明提供的上述方法中,优选地,DB-1MS柱的尺寸为 30 m×0.25 mm×0.25 μm。

在本发明提供的上述方法中,优选地,DB-17HT柱的尺寸为 3 m×0.1 mm×0.1 μm。

在本发明提供的上述方法中,优选地,细硅胶的粒径为100～200目,并且该细硅胶在200 ℃下进行 4 h的活化处理。

全二维气相色谱(GC×GC)是20世纪90年代发展起来的分离复杂混合物的一种全新手段,它是把分离机理不同而又相互独立的两支色谱柱通过一个调制器以串联方式连接成二维气相色谱柱系统。与常规的一维气相色谱相比,全二维气相色谱具有分辨率高、峰容量大、灵敏度好和分析速度快等特点。本发明所提供的分析方法通过将全二维气相色谱与飞行时间质谱(GC×GC-TOFMS)搭配,能采集到样品中所有物质的质谱信息,为化合物的定性提供有效依据。同时,通过将全二维气相色谱与氢火焰离子化检测器(GC×GC-FID)搭配进行分析,由于氢火焰离子化检测器几乎对所有挥发性的有机化合物均有响应,且对碳氢化合物响应因子几乎一致,为化合物的定量提供了有效手段。

本发明提供的是一套完整的石油样品中金刚烷类化合物的定量分析方法,包括样品前处理、化合物定性及定量分析。本发明提供的全二维气相色谱分析方法是通过控制两维色谱柱类型及长度,一维、二维色谱及调制器升温程序,载气流速,调制周期,热吹时间,检测器温度等工艺参数,得到在色谱条件下有效分离金刚烷类化合物的方法。在本发明提供的方法中,16个参数是相辅相成的一个整体,更改任意参数都不能达到理想的分离效果。与已公开的全二维气相色谱-飞行时间质谱方法相比,本发明提供的方法所采用的色谱柱成本低,分析时间短,至少节省色谱分析时间49 min以上,节省质谱采集时间45 min以上,节省液氮1/3以上,大大提高了分析效率,节约了分析成本。

用本发明提供的方法定量石油样品中的金刚烷类化合物,只需取原油或者饱和烃组分就能检测,简化了样品前处理过程,避免了低质量数金刚烷化合物在前处理过程中的损失。与传统GC-MS定量方法相比,用一个氘代金刚烷标准样品就能在FID检测器上得到金刚烷类化合物的定量结果,填补了色谱方法对金刚烷类化合物定量的空白。实验结果显示:该方法灵敏度高,检测限低至mg/kg,适合痕量金刚烷化合物的分析。而且,该方法的重复性好,常规单金刚烷、双金刚烷化合物的7次重复试验定量结果RSD(相对标准偏差)小于5%,满足复杂体系的分析要求。

目前在国内石油地质实验领域,全二维气相色谱仪器的普及率很高,但由于建立方法需要投入的成本较高且专业人员较少等问题的限制,至今还没有建立起一种有效地化合物分析方法。

本发明提供的石油样品中金刚烷类化合物的定量分析方法适用于所有原油和岩石抽提物样品中的金刚烷类化合物定量。该方法操作简单、灵敏度高、重复性好,相比于国内现有方法得到的金刚烷类化合物定量结果更加真实、可靠,因此具有良好的应用前景,值得推广。用本发明提供的方法得到的金刚烷类化合物定量数据可以为判定高成熟原油和凝析油的成熟度以及原油的裂解程度提供有效支持。

附图说明

图1a和图1b为克拉205井凝析油样品在GC×GC-FID下的局部全二维3D图,其中图1a上标记出17个单金刚烷系列化合物,图1b上标记出9个双金刚烷系列化合物。

图2为金刚烷系列化合物用飞行时间质谱采集到的质谱图。

具体实施方式

为了对本发明的技术特征、目的和有益效果有更加清楚的理解,现参照说明书附图对本发明的技术方案进行以下详细说明,但不能理解为对本发明的可实施范围的限定。

实施例1

本实施例提供了一种石油样品中金刚烷类化合物的定量分析方法,它是对四川盆地川中中北部区广安125井和塔里木盆地库车地区克拉205井(样品信息见表1)凝析油样品进行定量分析的方法。

表 1 样品信息

井 号	井段/m	层 位	岩 性
广安 125	2 529～2 536,2 544～2 558	须 四	凝析油
克拉 205	3 789～3 952.5	E+K	凝析油

该方法包括以下步骤：
(1) 制备待测样品。
用移液器各取上述凝析油样品 500 μL 分别置于 1.5 mL 样品瓶中，加入配制好的浓度为 0.063 2 mg/mL 的 D_{16}-金刚烷标准样品 300 μL，得到待测样品，迅速密封待用。
(2) 全二维气相色谱分析。
① 利用全二维气相色谱-飞行时间质谱(GC×GC-TOFMS)进行分析，根据飞行时间质谱采集到的化合物质谱图(见图2)与标准物质谱图(例如 NIST 谱库中的标准物质谱图)进行对比，找到匹配度最高的标准物质谱图并进行计算，得到待测样品中金刚烷类化合物的出峰信息，包括待测样品中是否含有金刚烷类化合物，以及所含金刚烷类化合物的种类、数量和保留时间(该步骤可以按照现有的方式进行，例如通过质谱仪自带的 ChromaTOF 软件参考现有方法完成)。
全二维气相色谱-飞行时间质谱仪是美国 LECO 公司的产品，型号是 Pegasus 4D，其分析条件分为全二维气相色谱条件和飞行时间质谱条件。其中：
a. 全二维气相色谱分析条件为：
一维色谱柱为 DB-1MS 柱(30 m×0.25 mm×0.25 μm)，升温程序为 50 ℃保持 0.2 min，以 2 ℃/min 的速率升到 180 ℃保持 0.2 min，再以 8 ℃/min 的速率升到 300 ℃保持 10 min；二维色谱柱是 DB-17HT 柱(1.5 m×0.1 mm×0.1 μm)，采用与一维色谱柱相同的升温速率，升温程序为 55 ℃保持 0.2 min，以 2 ℃/min 的速率升到 185 ℃保持 0.2 min，再以 8 ℃/min 的速率升到 305 ℃保持 10 min；气相色谱进样口温度为 300 ℃，以氦气为载气，流速为 1.5 mL/min，采用分流进样模式，分流比为 700∶1，进样量为 0.5 μL；调制器采用与一维色谱相同的升温速率，升温程序为 90 ℃保持 0.2 min，以 2 ℃/min 的速率升到 220 ℃保持 0.2 min，再以 8 ℃/min 的速率升到 340 ℃保持 10 min，调制周期为 10 s，其中 2.5 s 热吹时间。
b. 飞行时间质谱条件为：
利用飞行时间质谱仪进行测试。该质谱仪是美国 LECO 公司的产品(型号 Pegasus 4D)，质谱传输线和离子源温度分别为 280 ℃ 和 240 ℃，质谱检测器电压为 1 475 V，质量扫描范围为 40～520 amu，采集速率为 100 谱图/s，溶剂延迟时间为 0 min。
数据处理过程可以由全二维气相色谱自带的 ChromaTOF 软件完成，这是本领域公知的。该软件有自动计算基线、峰查找、谱库检索和峰面积积分等功能。在数据处理过程中，可以先编辑好峰查找和谱库检索条件，然后进行数据处理，软件会计算出化合物的保留时间和定性信息。其中得到的定性信息是根据飞行时间质谱采集到的化合物质谱图，与 NIST 谱库中的标准物质谱图进行比对，根据匹配度的高低得到一个检索结果，并据此进行计算，得到化合物的种类、数量和保留时间等信息；对于 NIST 谱库中未收录的化合物，可以通过与专业书籍或已发表文献中标准物质的质谱图进行人工比对，得到该化合物的种类、数量等信息。具体处理方法可以参照现有的方式进行。
② 利用全二维气相色谱-氢火焰离子化检测器(GC×GC-FID)分析，得到相应的全二维气相色谱-氢火焰离子化检测图谱(GC×GC-FID 谱图)。
分析条件为：氢火焰离子化检测器(与 Agilent 7890A 气相色谱仪匹配)是美国 Agilent 公司的产品；采用与全二维气相色谱-飞行时间质谱分析相同的全二维气相色谱条件；FID 检测器中载气(氦气)、氢气、空气的流速分别为 50 mL/min，40 mL/min，450 mL/min；检测器温度为 320 ℃，采集频率为 200 谱图/s，溶剂延迟时间为 0 min。

(3) 数据处理。

根据步骤(2)中 GC×GC-TOFMS 得到的金刚烷类化合物的出峰信息,确定金刚烷类化合物和 D_{16}-金刚烷在全二维气相色谱-氢火焰离子化检测谱图上的出峰位置(如图 1a 和图 1b 所示),得到它们的峰面积积分结果。

由于在 FID 检测器上所有碳氢化合物的响应因子均相等,因此采用内标法计算金刚烷类化合物的定量结果。用该方法能得到凝析油样品中 14 个常用单金刚烷系列化合物的定量结果。另外,经 7 次重复性实验的相对标准偏差 RSD 小于 5%(见表 2),说明该方法重复性很好。

表 2 广安 125 井凝析油样品中金刚烷类化合物 7 次重复性试验的定量分析结果

化合物名称	一维保留时间/min	二维保留时间/s	第一次/(mg·kg^{-1})	第二次/(mg·kg^{-1})	第三次/(mg·kg^{-1})	第四次/(mg·kg^{-1})	第五次/(mg·kg^{-1})	第六次/(mg·kg^{-1})	第七次/(mg·kg^{-1})	RSD/%
金刚烷	46.33	5.06	980.7	943.1	1015.1	954.5	1 017.1	1 012.8	1 052.4	3.89
1-甲基金刚烷	48.33	4.39	2 219.3	2 266.7	2 300.8	2 175.8	2 297.9	2 296.4	2 340.3	2.47
1,3-二甲基金刚烷	50.00	3.78	1 508.5	1 567.7	1 571.6	1 506.0	1 565.8	1 559.9	1 618.9	2.52
1,3,5-三甲基金刚烷	51.33	3.28	485.4	460.2	500.3	474.6	503.5	503.3	473.3	3.53
2-甲基金刚烷	54.67	5.08	1 473.1	1 481.8	1 522.8	1 433.9	1 529.8	1 524.6	1 560.6	2.85
1,4-二甲基金刚烷,顺式	56.00	4.26	1 283.5	1 308.8	1 324.4	1 249.9	1 332.5	1 327.4	1 359.4	2.74
1,4-二甲基金刚烷,反式	56.50	4.32	1 079.1	1 088.2	1 109.7	1 053.9	1 120.8	1 116.9	1 144.6	2.74
1,3,6-三甲基金刚烷	57.50	3.65	834.9	846.5	856.6	813.8	848.8	864.4	878.0	2.44
1,2-二甲基金刚烷	59.00	4.64	1 046.6	1 052.7	1 085.7	1 003.9	1 089.1	1 083.4	1 114.9	3.42
1,3,4-三甲基金刚烷,顺式	60.00	3.92	690.5	708.2	691.7	673.6	714.4	718.4	727.2	2.68
1,3,4-三甲基金刚烷,反式	60.50	3.97	646.1	663.2	673.3	624.8	671.3	666.8	693.5	3.30
1,2,5,7-四甲基金刚烷	61.00	3.40	496.1	498.5	515.2	462.2	520.8	509.2	523.4	4.17
1-乙基金刚烷	61.50	4.59	303.5	338.6	328.8	310.7	321.2	303.8	338.9	4.76
1-乙基-3-甲基金刚烷	62.83	3.97	417.2	429.3	409.3	418.5	435.4	435.8	440.6	2.74

实施例 2

本实施例提供了一种石油样品中金刚烷类化合物的定量分析方法,其是对塔里木地区玛 3 井、轮南 62 井、轮古 38 井、轮南 57 井、哈德 4 井和解放 100 井(样品信息见表 3)的 6 个原油样品进行定量分析的方法,包括以下步骤。

表 3　塔里木地区 6 个样品的信息

样　品	玛 3	轮南 62	轮古 38	轮南 57	哈德 4	解放 100
年　代	O_1	O	O	T	C	T_3
井深/m	1 508～1 518	5 565～5 578	5 619.38～5 740	4 341.8～4 344.0	5 069.64～5 076.32	4 473～4 475.5
外观颜色	棕　色	黑　色	浅黄色	浅黄色	黑　色	深棕色

(1) 制备待测样品。

将约 1.8 g 的细硅胶(100～200 目,200 ℃活化 4 h)在振荡情况下装入玻璃柱中;取样品约 30 mg,用适量正己烷溶解(重蒸过的分析纯),加入浓度为 0.063 2 mg/mL 的 D_{16}-金刚烷标准样品 300 μL,然后全部转入玻璃柱中,用 3 mL 正己烷淋洗,收集洗脱馏分,浓缩至 1.5 mL 自动进样瓶中,得到待测样品,待分析。

(2) 全二维气相色谱分析。

① 利用全二维气相色谱-飞行时间质谱(GC×GC-TOFMS)对待测样品进行分析,根据飞行时间质谱采集到的化合物质谱图(见图 2)与标准物质谱图(例如 NIST 谱库中的标准物质谱图)进行对比,找到匹配度最高的标准物质谱图并进行计算,得到待测样品中金刚烷类化合物的出峰信息,包括待测样品中是否含有金刚烷类化合物,以及所含金刚烷类化合物的种类、数量和保留时间(该步骤可以按照现有的方式进行,例如通过质谱仪自带的 ChromaTOF 软件参考现有方法完成)。

a. 全二维气相色谱分析条件为:

一维色谱柱为 DB-1MS 柱(30 m×0.25 mm×0.25 μm),升温程序为 50 ℃保持 0.2 min,以 2 ℃/min 的速率升到 180 ℃保持 0.2 min,再以 8 ℃/min 的速率升到 300 ℃保持 10 min;二维色谱柱是 DB-17HT 柱(1.5 m×0.1 mm×0.1 μm),采用与一维色谱相同的升温速率,温度比一维色谱高 5 ℃,即升温程序为 55 ℃保持 0.2 min,以 2 ℃/min 的速率升到 185 ℃保持 0.2 min,再以 8 ℃/min 的速率升到 305 ℃保持 10 min;进样口温度为 300 ℃,以氦气为载气,流速为 1.5 mL/min,采用分流进样模式,分流比 20∶1,进样量为 1 μL。

调制器温度比一维色谱高 40 ℃,即升温程序为 90 ℃保持 0.2 min,以 2 ℃/min 的速率升到 220 ℃保持 0.2 min,再以 8 ℃/min 的速率升到 340 ℃保持 10 min,调制周期 10 s,其中 2.5 s 热吹时间。

b. 飞行时间质谱条件为:

质谱传输线和离子源温度分别为 280 ℃和 240 ℃,质谱检测器电压为 1 475 V,质量扫描范围为 40～520 amu,采集速率为 100 谱图/s,溶剂延迟时间为 10 min。

② 利用全二维气相色谱-氢火焰离子化检测器(GC×GC-FID)进行分析,得到相应的全二维气相色谱-氢火焰离子化检测图谱(GC×GC-FID 谱图)。

分析条件为:采用与 GC×GC-TOFMS 相同的全二维气相色谱条件;FID 检测器中载气、氢气、空气的流速分别为 50 mL/min、40 mL/min、450 mL/min;检测器温度为 320 ℃,采集频率为 200 谱图/s,溶剂延迟时间为 10 min。

(3) 数据处理。

根据 GC×GC-TOFMS 得到的金刚烷类化合物的出峰信息,确定金刚烷类化合物和 D_{16}-金刚烷在 GC×GC-FID 谱图上的出峰位置,得到它们的峰面积积分结果。

由于在 FID 检测器上所有碳氢化合物的响应因子均相等,因此采用内标法得到金刚烷类化合物的定量结果。用该方法能得到 6 个样品中 24 个常用单金刚烷、双金刚烷系列化合物的定量结果(见表 4)。解放 100# 样品 7 次重复性实验的 RSD 小于 5%(见表 5),这说明本发明提供的方法的重复性良好,能够满足复杂体系的分析要求。

表 4 金刚烷类化合物的定量分析结果

一维保留时间/min	二维保留时间/s	样品 化合物名称	玛 3 (mg·kg^{-1})	轮南 62 (mg·kg^{-1})	轮古 38 (mg·kg^{-1})	轮南 57 (mg·kg^{-1})	哈德 4 (mg·kg^{-1})	解放 100 (mg·kg^{-1})
22.83	2.90	金刚烷	115.5	178	97.4	13.1	26.5	21.03
24.00	2.70	1-甲基金刚烷	366.6	276.2	156.9	48.1	62.3	60.65
25.17	2.50	1,3-二甲基金刚烷	459.5	428.1	219.5	64.9	89.7	75.59
26.00	2.35	1,3,5-三甲基金刚烷	195.2	170.6	85.2	34.7	62.4	41.14
28.17	3.01	2-甲基金刚烷	264.6	411	202.1	49.8	50.8	37.95
29.17	2.75	1,4-二甲基金刚烷,顺式	376.6	386.4	190.1	83.4	85.8	66.46
29.50	2.76	1,4-二甲基金刚烷,反式	160.3	211.9	107.8	43.5	41.6	28.43
30.17	2.52	1,3,6-三甲基金刚烷	211.8	276.8	98.5	45.2	19.3	17.95
31.17	2.90	1,2-二甲基金刚烷	156	274.3	151.3	56.3	36.3	29.15
31.83	2.65	1,3,4-三甲基金刚烷,顺式	102.5	173.1	102.4	53.1	—	—
32.17	2.68	1,3,4-三甲基金刚烷,反式	134.9	196.2	112.3	63.3	65.4	—
32.50	2.46	1,2,5,7-四甲基金刚烷	52.5	59.1	60.5	51.2		
32.83	2.91	1-乙基金刚烷	48.2	97.6	57.4	29.4	13.7	14.01
33.67	2.70	1-乙基-3-甲基金刚烷	14.1	149.3	48.1	9.9	13.7	
34.83	3.01	2-乙基金刚烷	91.7	154.9	48.1	27.3	32.1	27.98
49.83	3.96	双金刚烷	44.2	161.7	52.1	38.2	13.3	15.45
50.83	3.62	4-甲基双金刚烷	45.2	172.6	63.3	37.7	13.4	7.7
51.67	3.29	4,9-二甲基双金刚烷	12.1	40.5	12.3	11.8	—	3.21
53.00	3.83	1-甲基双金刚烷	15.3	67.4	36.4	20.7	8.8	8.27
53.33	3.47	1,4－＋2,4-二甲基双金刚烷	13.7	64	23.4	14.8	—	4.32
53.50	3.53	4,8-二甲基双金刚烷	20.1	57.7	26.4	17.2		8.1
53.83	3.19	三甲基双金刚烷	12.3	57.4	25.4	—	—	—
54.33	3.92	3-甲基双金刚烷	19.3	87.2	37.2	27.8	10.7	9.59
55.17	3.55	3,4-二甲基双金刚烷	11.2	65.1	26.9	14.2	7.2	7.88

表5　解放100#样品中金刚烷类化合物7次重复性试验的定量分析结果

化合物名称	第一次 /(mg·kg^{-1})	第二次 /(mg·kg^{-1})	第三次 /(mg·kg^{-1})	第四次 /(mg·kg^{-1})	第五次 /(mg·kg^{-1})	第六次 /(mg·kg^{-1})	第七次 /(mg·kg^{-1})	RSD /%
金刚烷	22.18	22.34	22.82	22.76	21.49	21.03	21.74	3.01
1-甲基金刚烷	62.59	60.43	59.34	56.77	62.54	60.65	57.66	3.74
1,3-二甲基金刚烷	86.15	81.80	80.88	78.33	77.99	75.59	80.69	4.21
1,3,5-三甲基金刚烷	40.47	39.35	39.73	40.80	41.28	41.14	41.49	1.99
2-甲基金刚烷	39.18	39.38	39.38	36.19	35.89	37.95	35.07	4.89
1,4-二甲基金刚烷,顺式	65.30	60.94	59.83	64.13	65.30	66.46	66.37	4.12
1,4-二甲基金刚烷,反式	29.96	32.01	31.41	30.46	28.16	28.43	31.05	4.85
1,3,6-三甲基金刚烷	19.12	18.01	17.56	18.49	18.29	17.95	17.47	3.15
1,2-二甲基金刚烷	30.08	29.83	29.34	28.84	28.32	29.15	28.77	2.12
1-乙基金刚烷	14.65	14.06	14.24	15.63	14.29	14.01	15.04	4.08
2-乙基金刚烷	27.31	29.11	28.99	29.32	25.77	27.98	28.07	4.45
双金刚烷	14.97	14.89	14.50	15.64	15.84	15.45	14.83	3.22
4-甲基双金刚烷	7.26	8.07	7.74	7.90	8.07	7.70	7.86	3.56
4,9-二甲基双金刚烷	3.14	3.22	3.06	3.01	3.15	3.21	2.96	3.23
1-甲基双金刚烷	8.87	9.07	8.57	9.12	9.00	8.27	9.59	4.74
1,4-+2,4-二甲基双金刚烷	4.84	4.64	4.84	4.57	4.68	4.32	4.64	3.79
4,8-二甲基双金刚烷	7.93	7.57	7.13	7.75	7.57	8.10	7.23	4.58
3-甲基双金刚烷	10.65	9.63	10.53	10.59	10.66	9.59	10.40	4.60
3,4-二甲基双金刚烷	6.95	7.13	7.15	7.01	7.16	7.88	7.02	4.42

说明书附图

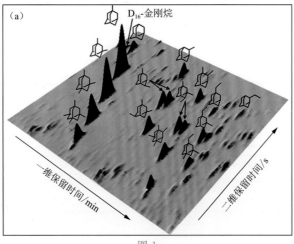

图1

第 3 章　全二维气相色谱的石油地质应用

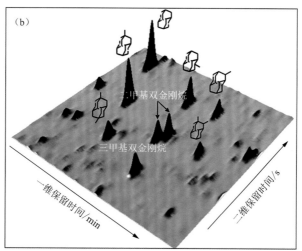

图 1（续）

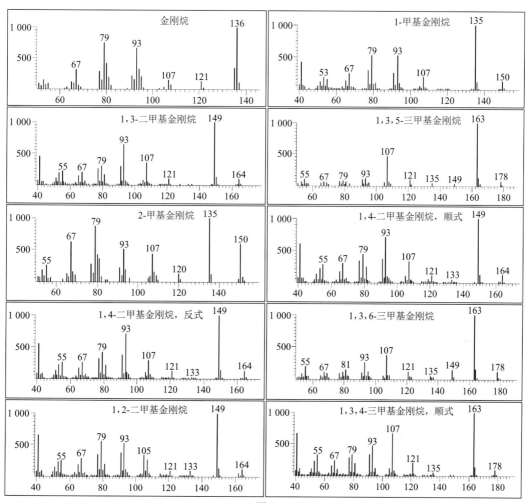

图 2

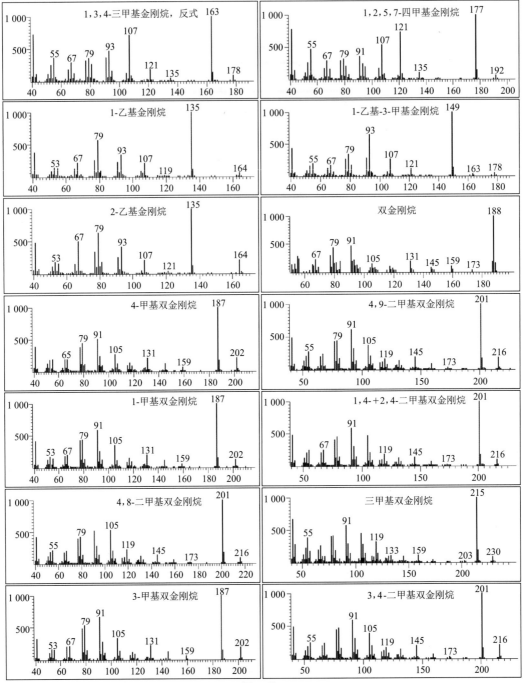

图 2(续)

专利 2　专利号:ZL 2012 1 0256600.5,授权公告日:2014 年 05 月 14 日

石油样品中五环三萜烷类化合物的定量分析方法

说明书摘要

本发明涉及一种石油样品中五环三萜烷类化合物的定量分析方法。该方法包括以下步骤:(1) 制备得到待测样品;(2) 利用全二维气相色谱-氢火焰离子化检测器分析待测样品,得到样品的 GC×GC-FID 谱图;(3) 利用全二维气相色谱-飞行时间质谱分析待测样品,得到待测样品的 GC×GC-TOFMS 谱图;(4) 数据处理:由步骤(3)得到的五环三萜烷类化合物的保留时间和质谱信息,确定五环三萜烷类化合物和标样在 GC×GC-FID 谱图上的出峰位置,根据峰面积计算并得到五环三萜烷类化合物的定量结果。该方法适用于所有原油和岩石抽提物样品中的五环三萜烷类化合物的定量。

权利要求书

1. 一种石油样品中五环三萜烷类化合物的定量分析方法,其包括以下步骤:

(1) 样品前处理。

将 3 g 细硅胶在振荡情况下装入玻璃柱;取 20 mg 原油或岩石抽提物样品,用少量正己烷溶解,加入配制好的 5α-雄甾烷标准样品溶液 200 μL,然后全部转入玻璃柱中,加入 10 mL 正己烷淋洗,收集洗脱馏分,浓缩至 1.5 mL 自动进样瓶中,得到待测样品。

(2) 全二维气相色谱-氢火焰离子化检测器分析。

利用全二维气相色谱-氢火焰离子化检测器分析步骤(1)制备的待测样品,得到待测样品的全二维气相色谱检测谱图。其分析条件分为全二维气相色谱条件和检测器条件,其中:

① 全二维气相色谱条件为:

一维色谱柱为 HP-5MS 柱,升温程序为 80 ℃保持 0.2 min,以 5 ℃/min 的速率升到 280 ℃保持 0.2 min,再以 0.8 ℃/min 的速率升到 305 ℃保持 1 min;二维色谱柱是 DB-17HT 柱,采用与一维色谱相同的升温程序,起始温度和终止温度比一维色谱高 5 ℃;气相色谱进样口温度为 310 ℃,采用分流进样模式,分流比为 15∶1,进样量为 1 μL;以氦气为载气,流速为 1.2 mL/min。

调制器采用与一维色谱相同的升温速率,起始温度和终止温度比一维色谱高 50 ℃,调制周期为 6 s,其中热吹时间为 1.5 s。

② 氢火焰离子化检测器的检测条件为:

载气、氢气、空气的流速分别为 23 mL/min,60 mL/min,400 mL/min;氢火焰离子化检测器的温度为 310 ℃,采集频率为 200 谱图/s,溶剂延迟时间为 1 700 s。

(3) 全二维气相色谱-飞行时间质谱仪分析。

利用全二维气相色谱-飞行时间质谱仪分析步骤(1)制备的待测样品,得到待测样品的全二维气相色谱-质谱检测谱图。其分析条件分为全二维气相色谱条件和飞行时间质谱条件,其中:

全二维气相色谱条件与步骤(2)相同,质谱传输线和离子源温度分别为 280 ℃和 240 ℃,质谱检测器电压为 1 475 V,质量扫描范围为 40~520 amu,采集速率为 100 谱图/s,飞行时间质谱条件中溶剂延迟时间与步骤(2)中溶剂延迟时间相同,为 1 700 s。

(4) 数据处理。

对步骤(2)和步骤(3)得到的待测样品谱图进行数据处理,根据步骤(3)得到的全二维气相色谱-飞行

时间质谱检测谱图得到五环三萜烷类化合物的保留时间和定性信息,确定步骤(2)中五环三萜烷类化合物和 5α-雄甾烷在全二维气相色谱检测谱图上的出峰位置,得到它们的峰面积积分结果,采用内标法计算得到五环三萜烷各个化合物的含量。

2. 根据权利要求 1 所述的方法,其中,所述 HP-5MS 柱的尺寸为 30 m×0.25 mm×0.25 μm。

3. 根据权利要求 1 所述的方法,其中,所述 DB-17HT 柱的尺寸为 1.6 m×0.1 mm×0.1 μm。

4. 根据权利要求 1 所述的方法,其中,所述细硅胶的粒径为 100～200 目,并且所述细硅胶在 200 ℃下进行 4 h 的活化处理。

说明书

石油样品中五环三萜烷类化合物的定量分析方法

技术领域

本发明涉及一种石油样品中五环三萜烷类化合物的定量分析方法,尤其涉及一种石油样品中五环三萜烷类化合物的全二维气相色谱精确定量分析方法,属于石油样品分析技术领域。

背景技术

五环三萜烷是普遍存在于石油样品中的一类五元环类生物标志化合物,此类化合物的分布特征是研究石油的母质类型、沉积环境以及成熟度等方面信息的重要指标,是油源对比和油气二次运移等研究的主要目标化合物。

对五环三萜烷类化合物的定量分析在混源油识别、混源比例计算、油气二次运移示踪、生物降解原油程度判识、石油炼制加工、环境保护等方面的研究中有重要作用。五环三萜烷类化合物在石油样品中含量较低,常规色谱分离时由于共馏峰较多,无法对其进行色谱定量。长期以来,这类化合物的定量分析都是采用色谱-质谱方法。但受标样合成条件的限制,国内外的定量采用的都是单一的标样,要么是 5α-雄甾烷,要么是氘代的甾烷。众所周知,五环三萜烷类化合物的分子结构格架是五元环,其离子电离的特征离子是 m/z 191;而所用标样甾烷的分子结构格架是四元环,其离子电离的特征离子是 m/z 217,无论是裂解机理还是裂解碎片离子都与五环三萜烷类化合物的生物标志化合物有很大不同。依据分析化学的原理,质谱定量至少需要一种结构相似、具有相同特征离子的物质做标样,但目前利用单一的标样对各类生标定量,定量结果的准确性可想而知。即使使用氘代的五环三萜烷标准样品在 GC-MS 上为五环三萜烷类化合物定量,下列几点不足也无法避免:第一,五环三萜烷类化合物与色谱柱的柱流失物质有相同的特征离子,利用 m/z 191 定量这类化合物时,含量低的物质容易被掩盖或者干扰,从而得到错误的定量结果;第二,Wang 等(Wang H T,Weng N,Zhang S C,et al. Comparison of geochemical parameters derived from comprehensive two-dimensional gas chromatography with time-of-flight mass spectrometry and conventional gas, chromatography-mass spectrometry. Science China:Earth Sciences,2011,54(12):1892-1901)实验证实 Ts(18α(H)-22,29,30-三降藿烷)和 Tm(17α(H)-22,29,30-三降藿烷)与 C_{29} 以后的三环萜烷在它们共有的特征离子 m/z 191 下具有共馏现象,用常规 GC-MS 定量时,它们的相互干扰会影响定量结果;第三,Peters 等(Peters K E,Moldowan J M. 生物标志化合物指南:生物标志化合物和同位素在环境与人类历史研究中的应用. 张水昌等译. 北京:石油工业出版社,2011:250-260)指出伽马蜡烷的对称结构有可能产生两个完全相同的 m/z 191 碎片,也有可能是一个,这种不确定性难以保证伽马蜡烷定量的准确性。

全二维气相色谱(GC×GC)是 20 世纪 90 年代发展起来的一种分离复杂混合物的全新手段,它的正交分离系统能够使五环三萜烷类化合物得到较好的分离。用全二维气相色谱-飞行时间质谱(GC×GC-TOFMS)分析石油样品中五环三萜烷类化合物的方法已见报道(王汇彤,翁娜,张水昌,等. 全二维气相色谱-飞行时间质谱对饱和烃分析的图谱识别及特征. 质谱学报,2010,31(1):18-27;李水福,胡守志,曹剑,等. 基于反相柱系统分析的原油烃类化合物全二维色-质谱图特征. 石油实验地质,2011,33(6):645-

651)。但是将已公开的 GC×GC-TOFMS 分析方法直接移植到全二维气相色谱-氢火焰离子化检测器(GC×GC-FID)上应用是不可行的。受一维和二维色谱柱的耐温性限制,应用已公开的全二维气相色谱分析方法无法得到高沸点的 C_{32} 以上的升藿烷系列化合物。延长高温保留时间可能会有更多的升藿烷类化合物馏出,但是会降低色谱柱的使用寿命,且色谱分析时间和质谱采集时间较长。分析一个样品至少需要 2 h 以上,在质谱采集期间需要不停灌充液氮,既浪费时间又增加分析成本。此外,在已发表的方法中没有涉及定量方面的报道。

综上所述,国内外至今还没有一种令人满意的五环三萜类化合物的定量分析方法。

发明内容

为解决上述技术问题,本发明的目的在于提供一种定量分析石油样品中五环三萜烷类化合物的方法,该方法能够实现对于石油样品中五环三萜烷类化合物精确的定量分析。

为达到上述目的,本发明提供了一种石油样品中五环三萜烷类化合物的定量分析方法,即采用全二维气相色谱对石油样品中的五环三萜烷类化合物进行定量分析的方法,能够为油气勘探、石油炼制、环境保护等研究人员提供可靠的定量分析数据。该方法包括以下步骤:

(1) 样品前处理

将 3 g 细硅胶在振荡情况下装入玻璃柱;取 20 mg 原油或岩石抽提物样品,用少量正己烷溶解,加入配制好的 5α-雄甾烷标准样品溶液 200 μL,然后全部转入玻璃柱中,加入 10 mL 正己烷淋洗,收集洗脱馏分,浓缩至 1.5 mL 自动进样瓶中,得到待测样品。

(2) 全二维气相色谱-氢火焰离子化检测器仪分析。

利用全二维气相色谱-氢火焰离子化检测器分析步骤(1)制备的待测样品,得到待测样品的全二维气相色谱检测谱图。其分析条件分为全二维气相色谱条件和检测器条件,其中:

① 全二维气相色谱条件为:

一维色谱柱为 HP-5MS 柱,升温程序为 80 ℃保持 0.2 min,以 5 ℃/min 的速率升到 280 ℃保持 0.2 min,再以 0.8 ℃/min 的速率升到 305 ℃保持 1 min;二维色谱柱是 DB-17HT 柱,采用与一维色谱相同的升温程序,起始温度和终止温度比一维色谱高 5 ℃(即升温程序为 85 ℃保持 0.2 min,以 5 ℃/min 的速率升到 285 ℃保持 0.2 min,再以 0.8 ℃/min 的速率升到 310 ℃保持 1 min);气相色谱进样口温度为 310 ℃,采用分流进样模式,分流比为 15∶1,进样量为 1 μL;以氢气为载气,流速为 1.2 mL/min。

调制器采用与一维色谱相同的升温速率,起始温度和终止温度比一维色谱高 50 ℃(即升温程序为 130 ℃保持 0.2 min,以 5 ℃/min 的速率升到 330 ℃保持 0.2 min,再以 0.8 ℃/min 的速率升到 355 ℃保持 1 min),调制周期为 6 s,其中热吹时间为 1.5 s。

② 检测器条件为:

所述检测器为氢火焰离子化检测器,其中载气、氢气、空气的流速分别为 23 mL/min,60 mL/min,400 mL/min;氢火焰离子化检测器温度 310 ℃,采集速率为 200 谱图/s,溶剂延迟时间为 1 700 s。

(3) 全二维气相色谱-飞行时间质谱仪分析。

利用全二维气相色谱-飞行时间质谱仪分析步骤(1)制备的待测样品,得到待测样品的全二维气相色谱-质谱检测谱图。其分析条件分为全二维气相色谱条件和飞行时间质谱条件,其中:

采用与步骤(2)相同的全二维气相色谱条件,质谱传输线和离子源温度分别为 280 ℃和 240 ℃,质谱检测器电压为 1 475 V,质量扫描范围为 40~520 amu,采集速率为 100 谱图/s,飞行时间质谱条件中溶剂延迟时间与步骤(2)中溶剂延迟时间相同,为 1 700 s。

(4) 数据处理。

对步骤(2)和步骤(3)得到的待测样品谱图进行数据处理,根据步骤(3)得到的全二维气相色谱-质谱检测谱图得到五环三萜烷类化合物的保留时间和定性信息,确定步骤(2)中五环三萜烷类化合物(一般确定常用五环三萜烷类化合物的出峰位置即可)和 5α-雄甾烷在全二维气相色谱-氢火焰离子化(GC×GC-FID)检测谱图上的出峰位置,得到它们的峰面积积分结果,采用内标法计算得到五环三萜烷各个化合物的含量。

在检测分析过程中,相对于现有的方法,通过采用本发明提供的预处理方法,可以减少样品的采样量,并且在较少采样量的情况下仍能够获得良好的检测分析结果。

在数据处理过程中,确定出峰位置就是对照化合物的相对保留时间,根据全二维气相色谱-飞行时间质谱上化合物的保留时间来确定 FID 检测器上化合物的出峰时间。对于峰面积积分结果的计算,可以运用 Chroma TOF 软件的数据处理功能得到,即样品分析后,可以在 Chroma TOF 软件中编辑相应的数据处理方法,然后运行这个数据处理方法得到积分结果。Chroma TOF 软件是二维色谱仪自带的软件,具体处理方法可以参照现有的方式进行。

在本发明提供的上述方法中,优选地,HP-5MS 柱的尺寸为 30 m×0.25 mm×0.25 μm。

在本发明提供的上述方法中,优选地,DB-17HT 柱的尺寸为 1.6 m× 0.1 mm×0.1 μm。

在本发明提供的上述方法中,优选地,细硅胶的粒径为 100～200 目,并且所述细硅胶在 200 ℃下进行 4 h 的活化处理。

本发明提供的是一套完整的石油样品中五环三萜烷类化合物的定量分析方法,包括样品前处理、化合物定性及定量分析。本发明提供的全二维气相色谱-氢火焰离子化检测器分析方法通过控制两维色谱柱类型及长度,一维、二维色谱及调制器升温程序,载气流速,调制周期,热吹时间,检测器温度等工艺参数,得到了在色谱条件下有效分离五环三萜烷类化合物的方法。在本发明提供的方法中,16 个参数是相辅相成的整体,更改任意参数都不能达到理想的分离效果。与已公开的全二维气相色谱-飞行时间质谱方法相比,本发明提供的方法所采用的色谱柱成本低,分析时间短,至少节省色谱分析时间 49 min 以上,节省质谱采集时间 72 min 以上,节省液氮 60% 以上,大大提高了分析效率,节约了分析成本。

与目前常规 GC-MS 定量方法相比,用全二维气相色谱仪分析五环三萜烷类化合物时,正交分离系统能够消除共馏峰的干扰,与氢火焰离子化检测器联用为这些化合物定量,无需昂贵的标准样品,用常规的标准样品就能得到准确的定量结果,且操作简单。实验结果显示:用本发明提供的定量分析方法得到的五环三萜烷类化合物定量结果的重复性好,18 个常用藿烷和伽马蜡烷的 7 次重复实验定量结果 RSD(相对标准偏差)小于 5%,能够满足复杂体系的分析要求。

目前在国内石油地质实验领域,全二维气相色谱仪的普及率很高,但由于建立方法需要投入的成本较高且专业人员较少等问题的限制,至今还没有建立起一种有效的化合物分析方法。本发明提供的定量分析方法适用于所有原油和岩石抽提物样品中五环三萜烷类化合物的定量分析。该方法具有操作简单易学且重复性好的特点,其所得到的五环三萜烷类化合物的定量结果相对于其他方法更加真实、可靠,适于在油气勘探、石油炼制、环境保护等研究和生产领域中应用,为理论研究和生产服务提供服务。该方法值得在行业内推广。

附图说明

图 1a～图 1c 为南堡 511 井灰色泥岩抽提样品的饱和烃组分中五环三萜烷的色谱图。

图 2 为 18 个藿烷化合物和伽马蜡烷用飞行时间质谱采集到的质谱图。

具体实施方式

为了对本发明的技术特征、目的和有益效果有更加清楚的理解,现对本发明的技术方案进行以下详细说明,但不能理解为对本发明的可实施范围的限定。

实施例

本实施例提供了一种石油样品中五环三萜烷类化合物的定量分析方法,采用该方法对渤海湾盆地南堡凹陷地区 6 个泥岩抽提物样品和 1 个柴达木盆地柳 43 井(样品的井号、井深、层位信息列于表 1)原油样品进行分析,包括以下步骤:

(1) 样品前处理。

将约 3 g 的细硅胶(100～200 目,200 ℃活化 4 h)在振荡的情况下装入玻璃柱中;各取 20 mg 左右的样品,用适量正己烷溶解(重蒸过的分析纯),加入 5α-雄甾烷标准样品(浓度为 0.053 3 mg/mL,溶剂为 CH_2Cl_2),然后全部转入玻璃柱中,加入 10 mL 正己烷淋洗,收集洗脱馏分,浓缩至 1.5 mL 自动进样瓶中

第3章 全二维气相色谱的石油地质应用

待用,得到待测样品。

(2) 全二维气相色谱-氢火焰离子化检测器分析。

利用全二维气相色谱-氢火焰离子化检测器(GC×GC-FID)分析步骤(1)制备的待测样品,得到待测样品的全二维气相色谱检测谱图。

全二维气相色谱仪(GC×GC)是美国 LECO 公司产品,它是由美国 Agilent 公司的 7890A 气相色谱仪和 LECO 公司的双喷口冷热调制器组成,氢火焰离子化检测器(FID)是美国 Agilent 公司产品;GC×GC-FID 的分析条件分为全二维气相色谱条件和 FID 检测器条件,其中:

① 全二维气相色谱条件为:

一维色谱柱为 HP-5MS 柱(30 m×0.25 mm×0.25 μm),升温程序为 80 ℃ 保持 0.2 min,以 5 ℃/min 的速率升到 280 ℃ 保持 0.2 min,再以 0.8 ℃/min 的速率升到 305 ℃ 保持 1 min;二维色谱柱是 DB-17HT 柱(1.6 m×0.1 mm×0.1 μm),采用与一维色谱相同的升温程序,起始温度和终止温度比一维色谱高 5 ℃;气相色谱进样口温度为 310 ℃,采用分流进样模式,分流比为 15∶1,进样量为 1 μL;以氦气为载气,流速为 1.2 mL/min。

调制器采用与一维色谱相同的升温速率,起始温度和终止温度比一维色谱高 50 ℃,调制周期为 6 s,其中热吹时间为 1.5 s。

② 检测器条件为:

所述检测器为氢火焰离子化检测器,其中载气、氢气、空气的流速分别为 23 mL/min、60 mL/min、400 mL/min;氢火焰离子化检测器温度 310 ℃,采集速率为 200 谱图/s,溶剂延迟时间为 1 700 s。

(3) 全二维气相色谱-飞行时间质谱仪分析。

利用全二维气相色谱-飞行时间质谱仪(GC×GC-TOFMS)分析步骤(1)制备的待测样品,得到待测样品的全二维气相色谱-质谱检测谱图。

全二维气相色谱-飞行时间质谱仪是美国 LECO 产品,型号是 Pegasus 4D,其分析条件分为全二维气相色谱条件和飞行时间质谱条件,其中:

采用与步骤(2)相同的全二维气相色谱条件,质谱传输线和离子源温度分别为 280 ℃ 和 240 ℃,质谱检测器电压为 1 475 V,质量扫描范围为 40~520 amu,采集速率为 100 谱图/s,飞行时间质谱条件中溶剂延迟时间与步骤(2)中溶剂延迟时间相同,为 1 700 s。

(4) 数据处理。

利用数据处理软件处理步骤(2)和步骤(3)得到的待测样品谱图。

数据处理过程可以是由全二维气相色谱自带的 ChromaTOF 软件完成,这是本领域公知的。该软件适用于 GC×GC-FID 系统和 GC×GC-TOFMS 系统,软件有自动计算基线、峰查找、谱库检索和峰面积积分等功能。在处理 GC×GC-FID 数据过程中,可以先编辑好峰查找的条件,然后进行数据处理,软件会计算出所有化合物的保留时间和峰面积积分结果。在处理 GC×GC-TOFMS 数据过程中,可以先编辑好峰查找和谱库检索的条件,然后进行数据处理,软件会计算出化合物的保留时间和定性信息。其中得到的定性信息是根据飞行时间质谱采集到的化合物质谱图(见图2),与标准物质谱图(例如 NIST 谱库中的标准物质谱图)进行比对,根据匹配度的高低得到的一个检索结果,并据此进行计算,得到化合物的种类、数量等信息。对于 NIST 谱库中未收录的化合物,可以通过与专业书籍(中华人民共和国石油天然气行业标准 生物标志物谱图 SY 5397—91)中标准物质的质谱图进行人工比对,得到该化合物的种类、数量等信息。具体处理方法可以参照现有的方式进行。

根据 GC×GC-TOFMS 得到的五环三萜烷类化合物的出峰信息,确定常用藿烷、伽马蜡烷和 5α-雄甾烷在 GC×GC-FID 谱图上的出峰位置(南堡 511 井的待测样品的谱图如图 1a 所示),并得到它们的峰面积积分结果。

由于在 FID 检测器上所有碳氢化合物的响应因子均相等,因此采用内标法得到 7 个样品中 18 个常用藿烷和伽马蜡烷的定量结果,详见表 1。用上述实验方法重复分析柳 43# 样品 7 次,得到 18 个常用藿烷和伽马蜡烷的定量重复性结果 RSD 小于 5% (见表 2)。

图 1a~图 1c 为南堡 511 号井的灰色泥岩抽提样品的饱和烃组分中五环三萜烷的色谱图,图中标记出 18 个常规藿烷化合物和伽马蜡烷。其中,图 1a 是该样品在 GC×GC-FID 下的全二维 3D 图;图 1b 是伽马蜡烷附近化合物在 GC×GC-FID 下的全二维点阵图,图中每个黑点代表一个化合物;图 1c 是该样品在 GC-MS 选择离子 m/z 191 下的色谱图。从图 1c 可以看出,在 GC-MS 条件下伽马蜡烷由于附近化合物的干扰而未被很好地检出,而在 GC×GC-FID 谱图上,伽马蜡烷可以与附近化合物很好地被分开(如图 1b 所示)。

表 1 7 个样品中 18 个常用藿烷和伽马蜡烷的 GC×GC-FID 定量结果

井 号	一维保留时间/min	二维保留时间/s	柳 43	南堡4-15	南堡4-15	南堡5-96	南堡509	南堡509	南堡511
井深/m			4 321~4 364	3 096.64	3 098.65	3 902.5	3 344.95	3 345.9	3 564.52
层 位			K_1g	Ed_3	Ed_3	Es_1	Ed_3上	Ed_3上	Ed_3
化合物名称(与图 1 中的编号对应)			mg/kg	mg/kg	mg/kg	mg/kg	mg/kg	mg/kg	mg/kg
1:Ts	112.17	3.65	54	714.3	741.3	390.9	940.3	714.2	1 063.7
2:Tm	113.83	3.97	269.6	875.3	1031.4	462.5	1 161.9	904.9	1 720.11
3:17α,21β(H)-30-降藿烷	120.67	4.49	1 006	3 911.3	4 755.3	1 453.2	5 390.2	3 335.5	5 430.77
4:18α(H)-30-降新藿烷	121.00	4.50	—	1 053.3	1 221	1 060.6	1 694.1	2 502.1	3 757.96
5:17β,21α(H)-30-降藿烷	123.17	4.85	77.4	789.6	921.6	—	1 202.9	973	1 646.9
6:17α,21β(H)-藿烷	125.50	5.02	2 158.7	4 968.9	5 966.8	2 656.9	6 805	5 624.1	9 260.87
7:17β,21α(H)-莫烷	127.67	5.29	273.6	796.9	920.9	391.2	1 153.2	880.2	1 448.89
8:17α,21β(H)-29-升藿烷(22S)	131.83	5.61	418.4	2 158.8	2 582.8	1 270.8	2 764	2 267.6	4 551.21
9:17α,21β(H)-29-升藿烷(22R)	132.67	5.69	303.3	1 777.4	2 077.2	1 006.3	2 471.3	2 011.5	3 723.06
伽马蜡烷	133.33	6.53	399	46.1	62.2	54.6	133.6	60.5	198.54
10:17β,21α(H)-29-升藿烷(22S+22R)	134.83	6.03	79.7	955.9	1 062.2	416.7	1 215.8	900.3	1 672.15
11:17α,21β(H)-29-二升藿烷(22S)	137.50	6.00	236.9	956.8	1 101.7	531.7	850.9	667.9	1 379.29
12:17α,21β(H)-29-二升藿烷(22R)	138.67	6.16	166	888.1	995.6	450.9	941.7	730.2	1 398.83
13:17α,21β(H)-29-三升藿烷(22S)	145.00	6.64	90.5	869.3	1 031.7	448.9	578.7	504.9	1 376.85

续表

井 号	一维保留时间/min	二维保留时间/s	柳43	南堡4-15	南堡4-15	南堡5-96	南堡509	南堡509	南堡511
14：$17\alpha,21\beta(H)$-29-三升藿烷(22R)	147.00	6.87	48.3	763.7	907.1	401.4	603.4	538.8	1 375.16
15：$17\alpha,21\beta(H)$-29-四升藿烷(22S)	154.00	7.38	49.9	371.9	448.5	161.2	273.4	176.8	461.72
16：$17\alpha,21\beta(H)$-29-四升藿烷(22R)	156.83	7.67	37.1	333.4	376.2	172.5	313.4	183.7	488.6
17：$17\alpha,21\beta(H)$-29-五升藿烷(22S)	164.00	8.20	30.1	119.3	149.8	144.8	82.1	60.4	138.7
18：$17\alpha,21\beta(H)$-29-五升藿烷(22R)	168.00	8.55	16.6	121.3	138	98.3	89.4	60.3	77.66

表2 柳43#样品分析7次得到的18个常用藿烷和伽马蜡烷的GC×GC-FID定量结果

化合物名称（与图1中的编号对应）	一维保留时间/min	二维保留时间/s	第一次/(mg·kg^{-1})	第二次/(mg·kg^{-1})	第三次/(mg·kg^{-1})	第四次/(mg·kg^{-1})	第五次/(mg·kg^{-1})	第六次/(mg·kg^{-1})	第七次/(mg·kg^{-1})	RSD/%
1：Ts	112.17	3.65	74.87	74.82	72.43	75.11	67.03	74.08	74.35	3.64
2：Tm	113.83	3.97	243.17	235.55	226.49	240.76	219.60	244.23	245.08	3.86
3：$17\alpha,21\beta(H)$-30-降藿烷	120.67	4.49	793.85	767.73	740.60	791.82	715.55	805.57	810.72	4.26
4：$18\alpha(H)$-30-降新藿烷	121.00	4.50	108.12	104.65	99.79	107.26	94.65	106.89	107.54	4.50
5：$17\beta,21\alpha(H)$-30-降藿烷	123.17	4.85	75.00	72.92	69.28	72.81	77.88	72.25	74.33	3.35
6：$17\alpha,21\beta(H)$-藿烷	125.50	5.02	1 994.05	1 957.02	1 900.99	2 023.14	1 764.59	1 998.69	2 012.54	4.35
7：$17\beta,21\alpha(H)$-莫烷	127.67	5.29	249.13	248.02	240.30	253.54	243.37	243.55	244.13	1.68
8：$17\alpha,21\beta(H)$-29-升藿烷(22S)	131.83	5.61	383.57	380.14	368.54	389.59	332.88	384.65	385.93	4.88
9：$17\alpha,21\beta(H)$-29-升藿烷(22R)	132.67	5.69	271.41	270.43	271.36	276.49	250.92	263.80	264.20	2.89
伽马蜡烷	133.33	6.53	374.62	365.46	357.88	377.87	343.02	372.25	374.85	3.14
10：$17\beta,21\alpha(H)$-29-升藿烷(22S+22R)	134.83	6.03	50.74	48.22	48.45	50.50	44.57	50.98	51.67	4.62
11：$17\alpha,21\beta(H)$-29-二升藿烷(22S)	137.50	6.00	194.18	189.14	184.91	199.05	182.72	195.77	196.15	2.99
12：$17\alpha,21\beta(H)$-29-二升藿烷(22R)	138.67	6.16	140.85	132.79	130.70	140.44	133.06	142.43	145.77	3.86

续表

化合物名称（与图1中的编号对应）	一维保留时间/min	二维保留时间/s	第一次/(mg·kg⁻¹)	第二次/(mg·kg⁻¹)	第三次/(mg·kg⁻¹)	第四次/(mg·kg⁻¹)	第五次/(mg·kg⁻¹)	第六次/(mg·kg⁻¹)	第七次/(mg·kg⁻¹)	RSD/%
13：17α,21β(H)-29-三升藿烷(22S)	145.00	6.64	109.36	107.89	107.52	110.97	95.36	109.25	109.38	4.58
14：17α,21β(H)-29-三升藿烷(22R)	147.00	6.87	75.35	72.13	69.73	72.96	69.39	71.98	77.04	3.55
15：17α,21β(H)-29-四升藿烷(22S)	154.00	7.38	42.34	43.17	44.78	46.49	45.54	44.01	46.71	3.42
16：17α,21β(H)-29-四升藿烷(22R)	156.83	7.67	27.57	28.32	27.26	30.61	29.82	29.51	29.16	3.93
17：17α,21β(H)-29-五升藿烷(22S)	164.00	8.20	30.11	27.77	27.44	28.98	26.79	30.61	29.88	4.77
18：17α,21β(H)-29-五升藿烷(22R)	168.00	8.55	18.54	18.66	18.08	17.16	17.09	18.52	18.35	3.39

说明书附图

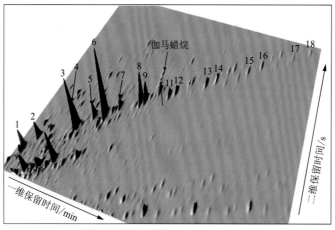

图 1a

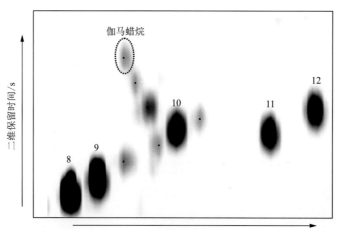

图 1b

第3章 全二维气相色谱的石油地质应用

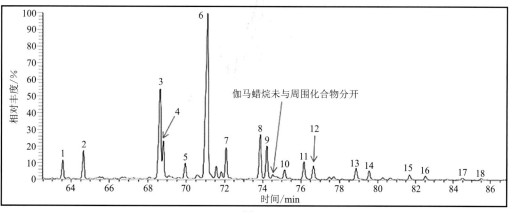

图 1c

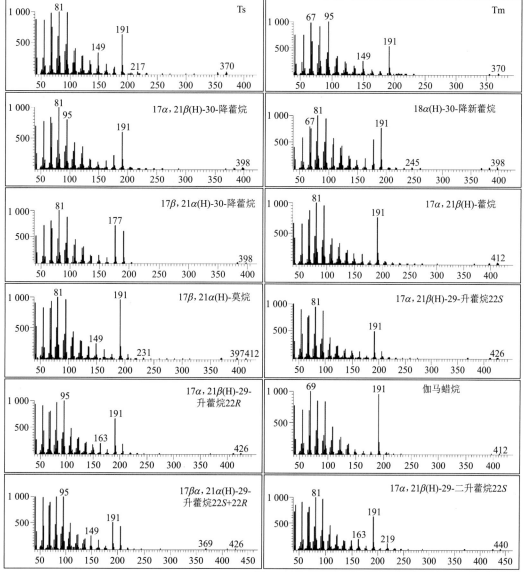

图 2

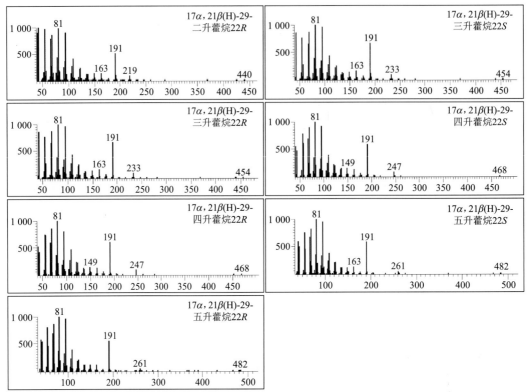

图 2(续)

专利 3　专利号：ZL 2013 1 0508538.9，授权公告日：2015 年 01 月 21 日

一种生排烃模拟实验液态产物的定量分析方法

说明书摘要

　　本发明提供一种生排烃模拟实验液态产物的定量分析方法。该方法包括以下步骤：收集与处理生排烃模拟实验的液态产物，其中包括轻烃收集、重烃收集、加标样、除水、过滤和浓缩；利用全二维气相色谱-氢火焰离子化检测器对生排烃模拟实验的液态产物进行定量分析，其中包括一次全二维分析、自然挥发恒重以及二次全二维分析；计算生排烃模拟实验液态产物全组分定量结果。本发明的定量分析方法避免了轻烃的挥发，实验结果重复性较好且操作简单易学，为生排烃模拟实验的液态产物定量分析提供了可靠的技术方法，使对盆地油气资源量的估算更加客观。

摘要附图

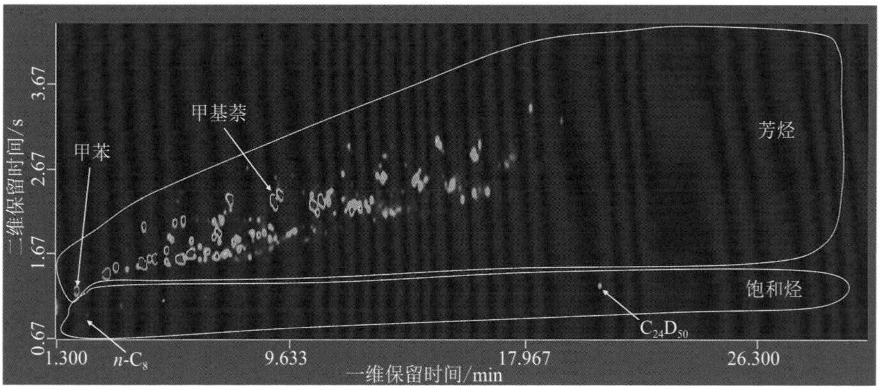

权利要求书

1. 一种生排烃模拟实验液态产物的定量分析方法，包括以下步骤。

　　A. 生排烃模拟实验液态产物的收集与处理。

　　A1. 轻烃收集：将轻烃收集容器连接于生排烃模拟实验装置的釜体，所述轻烃收集容器中盛有 20～30 mL 的二氯甲烷且所述轻烃收集容器放置在液氮冷阱中；将生排烃模拟实验装置的釜体温度保持在 210～220 ℃，打开生排烃模拟实验装置的产物出口阀门，液态产物中的轻烃组分随气态产物进入轻烃收集容器，气态产物进入气体收集容器；待气体计量平衡后，轻烃收集结束，得到含有轻烃的二氯甲烷溶液。

　　A2. 重烃收集：轻烃收集结束后，将生排烃模拟实验装置的釜体温度降至室温，用 20～50 mL 的二氯甲烷清洗生排烃模拟实验装置的管道、釜体内壁及生排烃模拟实验样品各三次，得到含有重烃的二氯甲烷溶液。

　　A3. 加标样：合并含有轻烃的二氯甲烷溶液与含有重烃的二氯甲烷溶液于容量瓶中，得到含有轻烃和重烃的二氯甲烷溶液，然后向其中加入 300～600 μL 浓度为 0.545 4 mg/mL 的氘代正二十四烷烃标样。

　　A4. 除水：再向其中加入二氯甲烷，使加入标样后的含有轻烃和重烃的二氯甲烷溶液中的水升至容量瓶颈部以上，然后用长颈滴管将水去除。

　　A5. 过滤：在塞有脱脂棉的漏斗中过滤除去除水后的含有轻烃和重烃的二氯甲烷溶液中的固体杂质。

A6. 浓缩:将过滤后的含有轻烃和重烃的二氯甲烷溶液在氮吹仪上挥发,浓缩至 1.5 mL,转入气相色谱常规的 2 mL 进样瓶中,得到生排烃模拟实验液态产物定量分析的样品。

B. 生排烃模拟实验液态产物的定量分析。

B1. 利用全二维气相色谱-氢火焰离子化检测器分析步骤 A 得到的样品,根据全二维气相色谱图的峰面积计算出饱和烃的质量和芳烃的质量,分别记为 S_1 和 F_1。

B2. 将步骤 B1 分析后的样品在室温下自然挥发至恒重,得到轻烃挥发后的液态产物及其质量 W_1。

B3. 采用 1~2 mL 二氯甲烷溶解轻烃挥发后的液态产物,然后利用全二维气相色谱-氢火焰离子化检测器分析含有轻烃挥发后的液态产物的二氯甲烷溶液,根据全二维气相色谱图的峰面积计算出饱和烃的质量和芳烃的质量,分别记为 S_2 和 F_2。

C. 生排烃模拟实验液态产物全组分定量结果的计算:

生排烃模拟实验液态产物总质量 $W=W_1+S_1+F_1-S_2-F_2$,生排烃模拟实验液态产物中含有的非烃类物质的质量 $N=W_1-S_2-F_2$。

2. 根据权利要求 1 所述的方法,其中所述全二维气相色谱-氢火焰离子化检测器的分析条件中的全二维气相色谱条件为:一维色谱柱为 Petro 柱,一维色谱的升温程序为 50 ℃ 保持 0.2 min,然后以 10 ℃/min 的速率升到 305 ℃ 保持 5 min;二维色谱柱为 DB-17HT 柱,二维色谱的升温程序为 60 ℃ 保持 0.2 min,然后以 10 ℃/min 的速率升到 315 ℃ 保持 5 min;气相色谱进样口温度为 300 ℃,以氦气作为载气,载气流速为 2 mL/min,采用分流进样模式,分流比为 400∶1,进样量为 1 μL;调制器的升温程序为 85 ℃ 保持 0.2 min,然后以 10 ℃/min 的速率升到 335 ℃ 保持 5 min,调制周期为 6 s,调制周期中的热吹时间为 1.5 s。

3. 根据权利要求 1 所述的方法,其中所述全二维气相色谱-氢火焰离子化检测器的分析条件中的氢火焰离子化检测器条件为:以氦气作为载气,载气、氢气、空气的流速分别为 23 mL/min,60 mL/min,400 mL/min,检测器温度为 310 ℃,采集速率为 200 谱图/s,溶剂延迟时间为 0 min。

4. 根据权利要求 2 所述的方法,其中所述 Petro 柱的尺寸为 10 m×0.2 mm×0.5 μm。

5. 根据权利要求 2 所述的方法,其中所述 DB-17HT 柱的尺寸为 2 m×0.1 mm×0.1 μm。

说明书

一种生排烃模拟实验液态产物的定量分析方法

技术领域

本发明涉及一种生排烃模拟实验液态产物的定量分析方法,属于油气地球化学技术领域。

背景技术

沉积盆地烃源岩成烃潜力与资源评价是石油天然气研究中的核心科学问题之一。生排烃模拟实验是目前广泛应用的烃源岩成烃潜力与区域油气资源评价的重要技术手段。生排烃模拟实验的产物包括气态产物和液态产物。气态产物是指常温常压下为气体的成分,包括甲烷、乙烷、丙烷、丁烷及二氧化碳等。液态产物是指碳数分布从 C_5~C_{40} 的烃类及胶质和沥青质。C_5~C_{15} 的烃类组分称为轻烃,经由液氮冷却收集;C_{15} 以上的组分滞留在生排烃模拟实验系统的管道、内壁中,或者被烃源岩中的黏土矿物吸附,用有机溶剂冲洗、抽提得到。利用各类烃源岩在不同演化阶段的气、液烃产率,结合沉积盆地的具体地质条件(埋藏历史、受热历史和热演化生烃史)进行盆地模拟,可以估算盆地的油、气生成量,在此基础上进一步预测油气资源。近年来在迅速发展的页岩气、致密砂岩油和气等非常规油气资源的勘探和开发实践中,非常规油气原地资源量的估算迫切需要通过生排烃模拟实验来解决富有机质烃源岩中残余油的评价和估算、烃源岩排油效率等关键问题。因此,对生排烃模拟实验的产物进行准确定量在油气勘探中具有重要的实际意义,尤其对以生成轻烃为主的煤系腐殖型母质和高成熟及过成熟烃源岩显得更为重要。

目前,生排烃模拟实验的气体产物收集和定量方法比较成熟,但液态产物全组分的定量一直没有很好的方法,主要是因为对轻烃组分的定量困难所致。轻烃组分经由液氮冷却收集后,往往含有大量的水

(10 mL 左右),而轻烃的量一般在几百毫克左右,如何使油水完全分离、轻烃又无挥发损失是个难题。张文正等(张文正,裴戈.热模拟中轻烃的收集及石油地质意义.天然气地质研究论文集.北京:石油工业出版社,1989:118-125)提出的油水分离称重方法操作难度高,轻烃在油水分离过程及恒重过程中挥发严重,油水完全分离非常困难。轻烃中混入微量的水或者少许的轻烃留在水中都会带来很大的误差,所以实验结果很难重复,得到的轻烃定量结果准确性无法确定。在收集轻烃的容器中加入二氯甲烷等有机溶剂可避免油水分离过程的轻烃逸失,但溶剂的存在无法用称重方法对轻烃定量。郑伦举等(郑伦举,马中良,王强,等.烃源岩有限空间热解生油气潜力定量评价研究.石油实验地质,2011,33(5):452-459)借助常规的气相色谱方法定量轻烃,但由于所收集的轻烃组分中同时含有胶质等成分,饱和烃和芳烃的峰重叠共馏,对标样峰有不同程度的干扰,也无法得到准确的轻烃定量结果。即使有一个轻烃(C_{15} 以下烃类)定量的结果,重烃中由于含有 C_{15} 以下烃类的部分成分,液态烃的量仍然无法确定。因此,在实际的生排烃模拟实验中,往往把收集的轻烃和其他液态产物混在一起进行分析,由于 C_{15} 以下的烃类组分挥发逸失,得到的液态产物结果不完整,从而影响了对资源量的估算。

全二维气相色谱(GC×GC)是 20 世纪 90 年代发展起来的分离复杂混合物的一种全新手段,它的二维正交分离系统能把在普通气相色谱-氢火焰离子化检测器(GC-FID)上无法分离的化合物在第二维色谱柱上分开。

因此,如何利用全二维气相色谱研发出一种生排烃模拟实验液态产物的定量分析方法,仍是本领域亟待解决的问题之一。

发明内容

为解决上述技术问题,本发明的目的在于提供一种生排烃模拟实验液态产物的定量分析方法。本发明的方法是一种利用全二维气相色谱针对不同极性化合物的分离特点而建立的有效定量分析生排烃模拟实验液态产物的方法。

为达到上述目的,本发明提供一种生排烃模拟实验液态产物的定量分析方法,其包括以下几个步骤:

A. 生排烃模拟实验液态产物的收集与处理。

A1. 轻烃收集:将轻烃收集容器连接于生排烃模拟实验装置的釜体,所述轻烃收集容器中盛放有 20~30 mL 的二氯甲烷且所述轻烃收集容器放置在液氮冷阱中;将生排烃模拟实验装置的釜体温度保持在 210~220 ℃,打开生排烃模拟实验装置的产物出口阀门,液态产物中的轻烃组分随气态产物进入轻烃收集容器,气态产物进入气体收集容器,待气体计量平衡后;轻烃收集结束,得到含有轻烃的二氯甲烷溶液(该溶液含水)。

A2. 重烃收集:轻烃收集结束后,将生排烃模拟实验装置的釜体温度降至室温,用 20~50 mL 的二氯甲烷清洗生排烃模拟实验装置的管道、釜体内壁及生排烃模拟实验样品各三次,得到含有重烃的二氯甲烷溶液(该溶液含水)。

A3. 加标样:合并含有轻烃的二氯甲烷溶液与含有重烃的二氯甲烷溶液于容量瓶中,得到含有轻烃和重烃的二氯甲烷溶液,然后向其中加入 300~600 μL 浓度为 0.545 4 mg/mL 的氘代正二十四烷烃($C_{24}D_{50}$)标样(溶剂为二氯甲烷)。

A4. 除水:再向其中加入二氯甲烷,使加入标样后的含有轻烃和重烃的二氯甲烷溶液中的水升至容量瓶颈部以上,然后用长颈滴管将水去除。

A5. 过滤:在塞有脱脂棉的漏斗中过滤除去除水后的含有轻烃和重烃的二氯甲烷溶液中的固体杂质。

A6. 浓缩:将过滤后的含有轻烃和重烃的二氯甲烷溶液在氮吹仪上挥发,浓缩至 1.5 mL 左右,转入气相色谱常规的 2 mL 进样瓶中,得到生排烃模拟实验液态产物定量分析的样品。

B. 生排烃模拟实验液态产物的定量分析。

B1. 利用全二维气相色谱-氢火焰离子化检测器(GC×GC-FID)分析步骤 A 得到的样品,根据全二维气相色谱图的峰面积计算出饱和烃的质量和芳烃的质量,分别记为 S_1 和 F_1。

其中,S_1 和 F_1 的计算公式分别为:

$$S_1 = \frac{A_1 \times M_0}{A_0}, \quad F_1 = \frac{A_2 \times M_0}{A_0}$$

式中，A_1 为饱和烃峰面积积分结果；A_2 为芳烃峰面积积分结果；A_0 为标样峰面积积分结果；M_0 为标样质量。

B2. 将步骤 B1 分析后的样品在室温下自然挥发至恒重（等同于氯仿沥青"A"的恒重条件），得到轻烃挥发后的液态产物（该液态产物不含二氯甲烷，二氯甲烷均已挥发）及其质量 W_1。

B3. 采用 1~2 mL 二氯甲烷溶解轻烃挥发后的液态产物，然后利用全二维气相色谱-氢火焰离子化检测器分析含有轻烃挥发后液态产物的二氯甲烷溶液，根据全二维气相色谱图的峰面积计算出饱和烃的质量和芳烃的质量，分别记为 S_2 和 F_2（S_2 和 F_2 的计算公式与步骤 B_1 中的 S_1 和 F_1 的计算公式一致）。

C. 生排烃模拟实验液态产物全组分定量结果的计算。

生排烃模拟实验液态产物总质量 $W=W_1+S_1+F_1-S_2-F_2$，生排烃模拟实验液态产物中含有的非烃类物质的质量 $N=W_1-S_2-F_2$。

在上述的方法中，优选地，所述全二维气相色谱-氢火焰离子化检测器的分析条件中的全二维气相色谱条件为：一维色谱柱为 Petro 柱，一维色谱的升温程序为 50 ℃保持 0.2 min，然后以 10 ℃/min 的速率升到 305 ℃保持 5 min（共需 30.7 min）；二维色谱柱为 DB-17HT 柱，二维色谱的升温程序为 60 ℃保持 0.2 min，然后以 10 ℃/min 的速率升到 315 ℃保持 5 min（二维色谱的升温程序采用与一维色谱相同的升温速率，不同之处在于起始温度和终止温度比一维色谱高 10 ℃）；气相色谱进样口温度为 300 ℃，以氦气作为载气，载气流速为 2 mL/min，采用分流进样模式，分流比为 400∶1，进样量为 1 μL；调制器的升温程序为 85 ℃保持 0.2 min，然后以 10 ℃/min 的速率升到 335 ℃保持 5 min（调制器的升温程序采用与一维色谱相同的升温速率，不同之处在于起始温度和终止温度比一维色谱高 35 ℃），调制周期为 6 s，调制周期中的热吹时间为 1.5 s。该全二维气相色谱条件为上述方法中的步骤 B1 和 B3 中的全二维气相色谱条件。

在上述的方法中，优选地，所述全二维气相色谱-氢火焰离子化检测器的分析条件中的氢火焰离子化检测器（FID）条件为：以氦气作为载气，载气、氢气、空气的流速分别为 23 mL/min、60 mL/min、400 mL/min，检测器温度为 310 ℃，采集速率为 200 谱图/s，溶剂延迟时间为 0 min。该氢火焰离子化检测器条件为上述方法中的步骤 B1 和 B3 中的氢火焰离子化检测器条件。

在上述的方法中，优选地，所述 Petro 柱的尺寸为 10 m×0.2 mm×0.5 μm。

在上述的方法中，优选地，所述 DB-17HT 柱的尺寸为 2 m×0.1 mm×0.1 μm。

本发明的生排烃模拟实验液态产物的定量方法将生排烃模拟实验收集到的全部液态产物经加标样、除水等处理后，采用全二维气相色谱-氢火焰离子化检测器分析液态产物中饱和烃和芳烃的质量，通过试剂的挥发恒重和二次全二维气相色谱-氢火焰离子化检测器分析得到全部液态烃的质量。全二维气相色谱的正交分离系能够消除共馏峰的干扰和胶质、沥青质的干扰，对标样峰没有影响，可以得到液态烃中饱和烃和芳烃的准确含量。二次全二维气相色谱-氢火焰离子化检测器分析消除了部分烃在轻烃中和重烃中的重复计量，能够计算出全部液态烃（包括胶质、沥青质）的质量，使定量结果更加准确可靠。本发明的定量分析方法避免了轻烃的挥发，实验结果重复性较好且操作简单易学，为生排烃模拟实验的液态产物定量分析提供了可靠的技术方法，使对盆地油气资源量的估算更加客观。

附图说明

图 1 为实施例 1 的步骤 B1 中的全二维点阵谱图。

图 2 为实施例 1 的步骤 B1 中的全二维 3D 谱图。

图 3 为实施例 1 的步骤 B3 中的全二维点阵谱图。

图 4 为实施例 1 的步骤 B3 中的全二维 3D 谱图。

具体实施方式

为了对本发明的技术特征、目的和有益效果有更加清楚的理解，现对本发明的技术方案进行以下详细说明，但不能理解为对本发明的可实施范围的限定。

实施例 1

本实施例提供一种生排烃模拟实验液态产物的定量分析方法。取张家口下花园地区黑色页岩样品进行生排烃模拟实验(该样品信息如表 1 所示),将该样品进行生排烃模拟实验的液态产物按照下述的步骤进行收集、处理、分析与计算,得到定量结果。

表 1 样品中共馏峰的定性分析结果

采样地点	层位	岩性	有机质类型	总有机碳含量/%	镜质体反射率/%	热解烃峰顶温度/℃	含气态烃量/(mg·g^{-1})	含游离烃量/(mg·g^{-1})	热解烃量/(mg·g^{-1})	生烃潜量/(mg·g^{-1})	产率指数	烃指数[mg·(g C)$^{-1}$]	氢指数[mg·(g C)$^{-1}$]	有效碳/%	降解潜率/%	总硫含量/%
张家口下花园地区	青白口系下马岭组	黑色页岩	I	1.55	0.50	434	0.00	1.84	42.56	44.40	0.04	24	564	3.69	49	0.13

本实施例的生排烃模拟实验液态产物的定量分析方法包括以下几个步骤:

A. 生排烃模拟实验液态产物的收集与处理。

A1. 轻烃收集:将轻烃收集容器连接于生排烃模拟实验装置的釜体,所述轻烃收集容器中盛有 20 mL 的二氯甲烷且所述轻烃收集容器放置在液氮冷阱中;将生排烃模拟实验装置的釜体温度保持在 220 ℃,打开生排烃模拟实验装置的产物出口阀门,液态产物中的轻烃组随气态产物进入轻烃收集容器,气态产物进入气体收集容器;待气体计量平衡后,轻烃收集结束,得到含有轻烃的二氯甲烷溶液(该溶液含水)。

A2. 重烃收集:轻烃收集结束后,将生排烃模拟实验装置的釜体温度降至室温,用 20~50 mL 的二氯甲烷清洗生排烃模拟实验装置的管道、釜体内壁及生排烃模拟实验样品各三次,得到含有重烃的二氯甲烷溶液(该溶液含水)。

A3. 加标样:合并含有轻烃的二氯甲烷溶液与含有重烃的二氯甲烷溶液于容量瓶中,得到含有轻烃和重烃的二氯甲烷溶液,然后向其中加入 300 μL 浓度为 0.545 4 mg/mL 的氘代正二十四烷烃($C_{24}D_{50}$)标样。

A4. 除水:再向其中加入二氯甲烷,使加入标样后的含有轻烃和重烃的二氯甲烷溶液中的水升至容量瓶颈部以上,然后用长颈滴管将水去除。

A5. 过滤:在塞有脱脂棉的漏斗中过滤除去除水后含有轻烃和重烃的二氯甲烷溶液中的固体杂质。

A6. 浓缩:将过滤后的含有轻烃和重烃的二氯甲烷溶液在氮吹仪上挥发,浓缩至 1.5 mL 左右,转入气相色谱常规的 2 mL 进样瓶中,得到生排烃模拟实验液态产物定量分析的样品。

B. 生排烃模拟实验液态产物的定量分析。

B1. 利用全二维气相色谱-氢火焰离子化检测器(GC×GC-FID)分析步骤 A 得到的样品,根据全二维气相色谱图的峰面积计算出饱和烃的质量和芳烃的质量,分别记为 S_1 和 F_1,其中,S_1 和 F_1 的计算公式分别为:

$$S_1 = \frac{A_1 \times M_0}{A_0}, \quad F_1 = \frac{A_2 \times M_0}{A_0}$$

式中,A_1 为饱和烃峰面积积分结果;A_2 为芳烃峰面积积分结果;A_0 为标样峰面积积分结果;M_0 为标样质量。

得到的全二维点阵谱图如图 1 所示,全二维 3D 谱图如图 2 所示。图 1 上标记了正辛烷(n-C_8)、甲苯、甲基萘和标样($C_{24}D_{50}$)的出峰位置,以及饱和烃和芳烃在全二维点阵谱图上的分布;图 2 立体直观地反映了饱和烃和芳烃在全二维谱图上的分布情况。

其中,所述全二维气相色谱-氢火焰离子化检测器的分析条件中的全二维气相色谱条件为:一维色谱柱为 Petro 柱(尺寸为 10 m×0.2 mm×0.5 μm),一维色谱的升温程序为 50 ℃保持 0.2 min,然后以 10 ℃/min 的速率升到 305 ℃保持 5 min(共需 30.7 min);二维色谱柱为 DB-17HT 柱(尺寸为 2 m×0.1 mm×0.1 μm),二维色谱的升温程序为 60 ℃保持 0.2 min,然后以 10 ℃/min 的速率升到 315 ℃保持 5 min(二维色谱的升温程序采用与一维色谱相同的升温速率,不同之处在于起始温度和终止温度比一维色谱高 10 ℃);气相色谱进样口温度为 300 ℃,以氦气作为载气,载气流速为 2 mL/min,采用分流进样模式,分流比为 400∶1,进样量为 1 μL;调制器的升温程序为 85 ℃保持 0.2 min,然后以 10 ℃/min 的速率升到 335 ℃保持 5 min(调制器的升温程序采用与一维色谱相同的升温速率,不同之处在于起始温度和终止温度比一维色谱高 35 ℃),调制周期为 6 s,调制周期中的热吹时间为 1.5 s。

所述全二维气相色谱-氢火焰离子化检测器的分析条件中的氢火焰离子化检测器(FID)条件为:以氦气作为载气,载气、氢气、空气的流速分别为 23 mL/min,60 mL/min,400 mL/min,检测器温度为 310 ℃,采集频率为 200 谱图/s,溶剂延迟时间为 0 min。

B2. 将步骤 B1 分析后的样品在室温下自然挥发至恒重(等同于氯仿沥青"A"的恒重条件),得到轻烃挥发后的液态产物及其质量 W_1。

B3. 采用二氯甲烷溶解轻烃挥发后的液态产物,然后利用全二维气相色谱-氢火焰离子化检测器分析(该全二维气相色谱-氢火焰离子化检测器的分析条件与步骤 B1 中相同)含有轻烃挥发后的液态产物的二氯甲烷溶液,根据全二维气相色谱图的峰面积计算出饱和烃的质量和芳烃的质量,分别记为 S_2 和 F_2(S_2 和 F_2 的计算公式与步骤 B1 中的 S_1 和 F_1 的计算公式一致);得到的全二维点阵谱图如图 3 所示,全二维 3D 谱图如图 4 所示,图 3 中标记了正十三烷($n-C_{13}$)、甲基萘和标样($C_{24}D_{50}$)的出峰位置,以及饱和烃和芳烃在全二维点阵谱图上的分布,图 4 立体直观地反映了饱和烃和芳烃在全二维谱图上的分布情况。从图 3 可以看出,经自然挥发至恒重,样品中已除去沸点低于正十三烷烃和甲基萘的化合物。

C. 生排烃模拟实验液态产物全组分定量结果的计算。

生排烃模拟实验液态产物总质量 $W=W_1+S_1+F_1-S_2-F_2$,生排烃模拟实验液态产物中含有的非烃类物质的质量 $N=W_1-S_2-F_2$。

其中,S_1,F_1,W_1,S_2,F_2,W 和 N 的数据如表 2 所示。

表 2　样品中共馏峰的定性分析结果

S_1/mg	28.44
F_1/mg	263.04
S_2/mg	1.19
F_2/mg	42.34
W_1/mg	446.67
W/mg	694.61
N/mg	403.14

实施例 2

本实施例提供一种生排烃模拟实验液态产物的定量分析方法。取与实施例 1 相同的样品,采用与实施例 1 相同的方法进行生排烃模拟实验,将生排烃模拟实验的液态烃产物按照与实施例 1 相同的生排烃模拟实验液态产物的定量分析方法进行分析,得到定量结果。采用本实施例的定量结果与实施例 1 的定量结果进行对比,考察本发明生排烃模拟实验液态产物的定量分析方法的重复性。

本实施例的 S_1,F_1,W_1,S_2,F_2,W 和 N 的数据如表 3 所示。

表 3　样品中共馏峰的定性分析结果

S_1/mg	27.64
F_1/mg	331.00
S_2/mg	8.06
F_2/mg	190.45
W_1/mg	563.33
W/mg	723.47
N/mg	364.82

由表 3 的数据可知,本实施例的方法得到的生排烃模拟实验液态产物总质量 W 为 723.47 mg,生排烃模拟实验液态产物中含有的非烃类物质的质量 N 为 364.82 mg。采用本实施例的定量结果与实施例 1 的定量结果进行比较,两次实验得到的液态产物总质量的结果偏差为 2.03%,非烃类物质的质量的结果偏差为 4.99%,均在实验方法允许误差 5% 的范围内。因此,采用本发明的定量分析方法得到的生排烃模拟实验液态产物的定量结果真实、可靠、有可重复性。

说明书附图

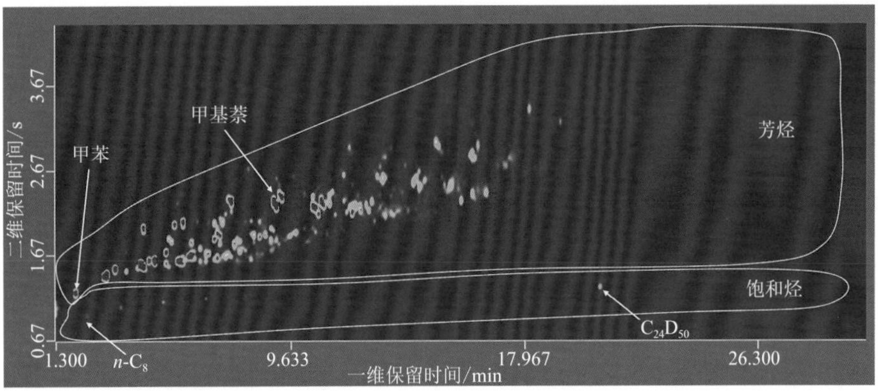

图 1

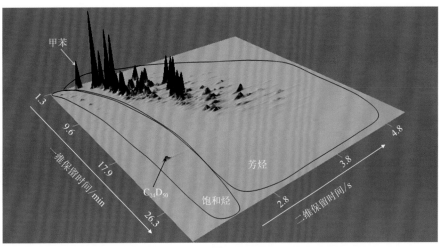

图 2

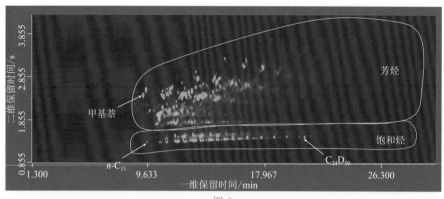

图 3

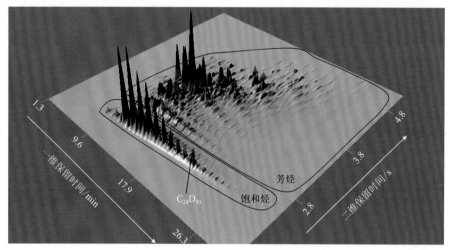

图 4

参考文献

[1] Ahsan A, Karlsen D A, Patience R L. Petroleum biodegradation in the Tertiary reservoirs of the North Sea[J]. Marine and Petroleum Geology, 1997, 14(1):55-64.

[2] Aguiar A, Silva A I, Azevedo D A, et al. Application of comprehensive two-dimensional gas chromatography coupled to time-of-flight mass spectrometry to biomarker characterization in Brazilian oils [J]. Fuel, 2010, 89(10):2 760-2 768.

[3] Armanios C, Alexander R, Kagi R I, et al. Fractionation of sedimentary higher-plant derived pentacyclic triterpanes using molecular sieves[J]. Organic Geochemistry, 1994, 21(5):531-543.

[4] Ávila B M F, Aguiar A, Gomes A O, et al. Characterization of extra heavy gas oil biomarkers using comprehensive two-dimensional gas chromatography coupled to time-of-flight mass spectrometry [J]. Organic Geochemistry, 2010, 41(9):863-866.

[5] Ávila B M F, Pereira R, Gomes A O, et al. Chemical characterization of aromatic compounds in extra heavy gas oil by comprehensive two-dimensional gas chromatography coupled to time-of-flight mass spectrometry[J]. Journal of Chromatography A, 2011, 1218(21):3 208-3 216.

[6] Bastow T P, Aarssen van B G K, Alexander R, et al. Biodegradation of aromatic land-plant biomarkers in some Australian crude oils[J]. Organic Geochemistry, 1999, 30(10):1 229-1 239.

[7] Blomberg J, Schoenmakers P J, Beens J, et al. Comprehensive two-dimensional gas chromatography (GC×GC) and its applicability to the characterization of complex(petrochemical)mixtures[J]. Journal of High Resolution Chromatography, 1997, 20:539-544.

[8] BoothA M, Sutton P A, Lewis C A, et al. Unresolved complex mixtures of aromatic hydrocarbons: thousands of overlooked persistent, bioaccumulative, and toxic contaminants in mussels[J]. Environmental Science & Technology, 2007, 41:457-464.

[9] Booth A M, Scarlett A G, Lewis C A, et al. Unresolved complex mixtures (UCMs) of aromatic hydrocarbons: branched alkylindanes and branched alkyl tetralins are present in UCMs and accumulated by and toxic to, the mussel Mytilus edulis[J]. Environmental Science & Technology, 2008, 42(21): 8 122-8 126.

[10] Bost F D, Frontera-Suau R, McDonald T J, et al. Aerobic biodegradation of hopanes and norhopanes in Venezuelan crude oils[J]. Organic Geochemistry, 2001, 32:105-114.

[11] Dagan S. Comparison of gas chromatography-pulsed flame photometric detection-mass spectrometry, automated mass spectral deconvolution and identification system and gas chromatography-tandem mass spectrometry as tools for trace level detection and identification[J]. Journal of Chromatography A, 2000, 868(2):229-247.

[12] Demir C, Hindmarch P, Brereton R G. Deconvolution of a three-component co-eluting peak cluster in gas chromatography-mass spectrometry[J]. Analyst, 2000, 125:287-292.

[13] Ellis L, Kagi R I, Alexander R. Separation of petroleum hydrocarbons using dealuminated mordenite molecular sieve. Ⅰ. Monoaromatic hydrocarbons[J]. Organic Geochemistry, 1992, 18(5):587-593.

[14] Ellis L, Alexander R, Kagi R I. Separation of petroleum hydrocarbons using dealuminated mordenite molecular-sieve. Ⅱ. Alkylnaphthalenes and alkylphenanthrenes[J]. Organic Geochemistry, 1994, 21(8-9):849-855.

[15] Fazeelat T, Alexander R, Kagi R I. Extended 8,14-secohopanes in some seep oils from Pakistan[J]. Organic Geochemistry,1994,21:257-264.

[16] Frysinger G S, Gaines R B, Xu L, et al. Resolving the unresolved complex mixture in petroleum-contaminated sediments[J]. Environmental Science & Technology,2003,37(8):1 653-1 662.

[17] Fryinger G S, Gaines R B. Separation and identification of petroleum biomarkers by comprehensive two dimensional gas chromatography[J]. Journal of Separation Science,2001,24(2):87-96.

[18] Gough M A, Rowland S J. Characterization of unresolved complex mixtures of hydrocarbons in petroleum[J]. Nature,1990,344(6267):648-650.

[19] Head I M, Jones D M, Larter S R. Biological activity in the deep subsurface and the origin of heavy oil[J]. Nature,2003,426(6964):344-352.

[20] Hussler G, Connan J, Albrecht P. Novel families of tetra- and hexacyclic aromatic hopanoids predominant in carbonate rocks and crude oils[J]. Organic Geochemistry,1984,6:39-49.

[21] Jacquot F, Doumenq P, Guiliano M, et al. Biodegradation of the(aliphatic＋aromatic)fraction of Oural crude oil. Biomarker identification using GC/MS SIM and GC/MS/MS[J]. Talanta,1996,43(3):319-330.

[22] Killops S D, Al-Juboori M A H A. Characterisation of the unresolved complex mixture(UCM)in the gas chromatograms of biodegraded petroleums[J]. Organic Geochemistry,1990,15(2):147-160.

[23] Larter S, Huang H P, Adams J, et al. A practical biodegradation scale for use in reservoir geochemical studies of biodegraded oils[J]. Organic Geochemistry,2012,45:66-76.

[24] Li N X, Huang H P, Jiang W L, et al. Biodegradation of 25-Norhopanes in a Liaohe Basin(NE China)oil reservoir[J]. Organic Geochemistry,2015,78:33-43.

[25] Li M J, Shi S B, Wang T G. Identification and distribution of chrysene,methylchrysenes and their isomers in crude oils and rock extracts[J]. Organic Geochemistry,2012,52:55-66.

[26] Mathur N. Tertiary oils from Upper Assam Basin,India:A geochemical study using terrigenous biomarkers[J]. Organic Geochemistry,2014,76:9-25.

[27] Nazir A, Fazeelat T. Petroleum geochemistry of Lower Indus Basin,Pakistan:Ⅰ. Geochemical interpretation and origin of crude oils[J]. Journal of Petroleum Science and Engineering,2014,122:173-179.

[28] Nytoft H P, Bojesen-Koefoed J A, Christiansen F G. C_{26} and $C_{28}-C_{34}$ 28-norhopanes in sediments and petroleum[J]. Organic Geochemistry,2000,31:25-39.

[29] Nytoft H P, Kildahl-Andersen G, Knudsen T Š, et al. Compound "J" in Late Cretaceous/Tertiary terrigenous oils revisited: Structure elucidation of a rearranged oleanane coeluting on GC with 18β(H)-oleanane[J]. Organic Geochemistry,2014,77:89-95.

[30] Oudot J, Chaillan F. Pyrolysis of asphaltenes and biomarkers for the fingerprinting of the Amoco-Cadiz oil spill after 23 years[J]. Comptes Rendus Chimie,2010,13(5):548-552.

[31] Phillips J B, Xu J Z. Comprehensive multi-dimensional gas chromatography[J]. Journal of Chromatography A,1995,703(1-2):327-334.

[32] Phillips J B, Beens J. Comprehensive two-dimensional gas chromatography:a hyphenated method with strong coupling between the two dimensions[J]. Journal of Chromatography A,1999,856(1-2):331-347.

[33] Schoenmakers P J, Oomen J L M M, Blomberg J, et al. Comparison of comprehensive two-dimensional gas chromatography and gas chromatography-mass spectrometry for the characterization of complex hydrocarbon mixtures[J]. Journal of Chromatography A,2000,892(1-2):29-46.

[34] Pool W G, De Leeuw J W, Van De Graaf B. Automated extraction of pure mass spectra from gas

chromatographic/mass spectrometric data[J]. Journal of Mass Spectrometry,1998,32(4):438-443.

[35] Reddy C M,Nelson R K,Sylva S P,et al. Identification and quantification of alkene-based drilling fluids in crude oils by comprehensive two-dimensional gas chromatography with flame ionization detection[J]. Journal of Chromatography A,2007,1148(1):100-107.

[36] Revill A T,Carr M R,Rowland S J. Use of oxidative degradation followed by capillary gas chromatography-mass spectrometry and multi-dimensional scaling analysis to fingerprint unresolved complex mixtures of hydrocarbons[J]. Journal of Chromatography A,1992,589(1-2):281-286.

[37] Sun H,Shen B X,Lin J C. Oxidative regeneration of deactivated binderless 5A molecular sieves for the adsorption-separation of normal hydrocarbons[J]. Journal of Fuel Chemistry and Technology,2009,37(6):734-739.

[38] Sun Y G,Chen Z Y,Xu S P,et al. Stable carbon and hydrogen isotopic fractionation of individual n-alkanes accompanying biodegradation:evidence from a group of progressively biodegraded oils[J]. Organic Geochemistry,2005,36(2):225-238.

[39] Sutton P A,Lewis C A,Rowland S J. Isolation of individual hydrocarbons from the unresolved complex hydrocarbon mixture of a biodegraded crude oil using preparative capillary gas chromatography[J]. Organic Geochemistry,2005,36(6):963-970.

[40] Theissen K M,Zinniker D A,Moldowan J M,et al. Pronounced occurrence of long-chain alkenones and dinosterol in a 25 000-year lipid molecular fossil record from Lake Titicaca,South America[J]. Geochimica et Cosmochimica Acta,2005,69(3):623-636.

[41] Tran T C,Logan G A,Grosjean E,et al. Use of comprehensive two-dimensional gas chromatography/time-of-flight mass spectrometry for the characterization of biodegradation and unresolved complex mixtures in petroleum[J]. Geochimica et Cosmochimica Acta,2010,74(22):6 468-6 484.

[42] Ventura G T,Kenig F,Reddy C M,et al. Analysis of unresolved complex mixtures of hydrocarbons extracted from Late Archean sediments by comprehensive two-dimensional gas chromatography (GC×GC)[J]. Organic Geochemistry,2008,39(7):846-867.

[43] Wang G L,Wang T G,Simoneit B R T,et al. Investigation of hydrocarbon biodegradation from a downhole profile in Bohai Bay Basin:Implications for the origin of 25-norhopanes[J]. Organic Geochemistry,2013,55:72-84.

[44] Warton B,Alexander R,Kagi R I. Characterisation of the ruthenium tetroxide oxidation products from the aromatic unresolved complex mixture of a biodegraded crude oil[J]. Organic Geochemistry,1999,30(10):1 255-1 272.

[45] Wei Z B,Moldowan J M,Paytan A. Diamondoids and molecular biomarkers generated from modern sediments in the absence and presence of minerals during hydrous pyrolysis[J]. Organic Geochemistry,2006,37(8):891-911.

[46] Wei Z B,Moldowan J M,Zhang S C,et al. Diamondoid hydrocarbons as a molecular proxy for thermal maturity and oil cracking:Geochemical models from hydrous pyrolysis[J]. Organic Geochemistry,2007,38(2):227-249.

[47] Wei Z B,Moldowan J M,Peters K E,et al. The abundance and distribution of diamondoids in biodegraded oils from the San JoaquinValley:Implications for biodegradation of diamondoids in petroleum reservoirs[J]. Organic Geochemistry,2007,38(11):1 910-1 926.

[48] 陈军红,傅家谟,盛国英,等. 金刚烷化合物在石油中的分布特征研究[J]. 自然科学进展,1997,7(3):363-367.

[49] 陈致林,刘旋,金洪蕊,等. 利用双金刚烷指标研究济阳坳陷凝析油的成熟度和类型[J]. 沉积学报,2008,26(4):705-708.

[50] 高儑博,常振阳,代威,等.全二维气相色谱在石油地质样品分析中的应用进展[J].色谱,2014,10:1 058-1 065.

[51] 花瑞香,阮春海,王京华,等.全二维气相色谱法用于不同石油馏分的族组成分布研究[J].化学学报,2002,60(12):2 185-2 191.

[52] 姜乃煌,朱光有,张水昌,等.塔里木盆地塔中 83 井原油中检测出 2-硫代金刚烷及其地质意义[J].科学通报,2007,52(24):2 871-2 875.

[53] 刘虎威.气相色谱方法及应用[M].北京:化学工业出版社,2007.

[54] 刘亚明,谢寅符,马中振,等.Orinoco 重油带重油成藏特征[J].石油与天然气地质,2013,34(3):315-322.

[55] 鹿洪亮,赵明月,刘惠民,等.全二维气相色谱/质谱的原理及应用综述[J].烟草科技/烟草化学,2005(3):22-25.

[56] 鹿洪亮,赵明月,刘惠民,等.全二维气相色谱/飞行时间质谱法测定烟草的中性化学成分[J].色谱,2007,25(1):30-34.

[57] 路鑫,武建芳,吴建华,等.全二维气相色谱/飞行时间质谱用于柴油组成的研究[J].色谱,2004,22(1):5-11.

[58] 陆鑫,蔡君兰,武建芳,等.全二维气相色谱/飞行时间质谱用于卷烟主流烟气中酚类化合物的表征[J].化学学报,2004,62(8):804-810.

[59] 苏焕华,姜乃皇,任冬苓.有机质谱在石油化学中的应用[M].北京:化学工业出版社,2010.

[60] 许国旺.现代实用气相色谱[M].北京:化学工业出版社,化学与应用化学出版中心,2004.

[61] 王春江,傅家谟,盛国英,等.18α(H)-新藿烷及 17α(H)-重排藿烷类化合物的地球化学属性与应用[J].科学通报,2000,45(13):1 366-1 372.

[62] 王光辉,熊少祥.有机质谱解析[M].北京:化学工业出版社,2005.1.

[63] 王汇彤,魏彩云,宋孚庆,等.细硅胶层析柱对饱和烃和芳烃的分离[J].石油实验地质,2009,31(3):312-314.

[64] 王培荣.生物标志物质量色谱图集[M].北京:石油工业出版社,1993.

[65] 王培荣.非烃地球化学和应用[M].北京:石油工业出版社,2002.

[66] 武建芳,陆鑫,唐婉莹,等.全二维气相色谱/飞行时间质谱用于莪术挥发油分离分析特性的研究[J].分析化学研究报告,2004,32(5):582-586.

[67] 赵冰,沈学静.飞行时间质谱分析技术的发展[J].现代科学仪器,2006,4:30-33.

[68] 中华人民共和国石油天然气行业标准 生物标志物谱图 SY 5397—91[M].北京:石油工业出版社,1992-05-01.

[69] 中华人民共和国国家标准 气相色谱-质谱法测定沉积物和原油中生物标志物 GB/T 18606—2017[M].北京:中国标准出版社,2017.

[70] 周友平,史继扬,向明菊,等.沉积有机质中藿烯的成因研究——碳稳定同位素证据[J].沉积学报,1998,16(2):14-19.

[71] 朱书奎,刑钧,吴采樱.全二维气相色谱的原理、方法及应用概述[J].分析科学学报,2005,21(3):332-336.

[72] 朱扬明,钟荣春,蔡勋育,等.川中侏罗系原油重排藿烷类化合物的组成及成因探讨[J].地球化学,2007,36(3):253-260.

[73] 菲利普 R P.化石燃料生物标志物——应用与谱图[M].北京:科学出版社,1987.

[74] Peters K E,Walters C C,Moldowan J M.生物标志化合物指南[M].2 版.张水昌,李振西,等译.北京:石油工业出版社,2011.